Chemical and Biological Synthesis
Enabling Approaches for Understanding Biology

Chemical Biology

Editor-in-chief:
Tom Brown, *University of Oxford, UK*

Series editors:
Kira J. Weissman, *Lorraine University, France*
Sabine Flitsch, *University of Manchester, UK*
Nick J. Westwood, *University of St Andrews, UK*

Titles in the series:

1: High Throughput Screening Methods: Evolution and Refinement
2: Chemical Biology of Glycoproteins
3: Computational Tools for Chemical Biology
4: Mass Spectrometry in Chemical Biology: Evolving Applications
5: Mechanisms of Primary Energy Transduction in Biology
6: Cyclic Peptides: From Bioorganic Synthesis to Applications
7: DNA-targeting Molecules as Therapeutic Agents
8: Protein Crystallography: Challenges and Practical Solutions
9: Oxidative Folding of Proteins: Principles, Biological Regulation and Design
10: Chemical and Biological Synthesis: Enabling Approaches for Understanding Biology

How to obtain future titles on publication:
A standing order plan is available for this series. A standing order will bring delivery of each new volume immediately on publication.

For further information please contact:
Book Sales Department, Royal Society of Chemistry, Thomas Graham House, Science Park, Milton Road, Cambridge, CB4 0WF, UK
Telephone: +44 (0)1223 420066, Fax: +44 (0)1223 420247,
Email: booksales@rsc.org
Visit our website at www.rsc.org/books

Chemical and Biological Synthesis
Enabling Approaches for Understanding Biology

Edited by

Nick J. Westwood
University of St Andrews, UK
Email: njw3@st-andrews.ac.uk

and

Adam Nelson
University of Leeds, UK
Email: A.S.Nelson@leeds.ac.uk

ROYAL SOCIETY
OF CHEMISTRY

Chemical Biology No. 10

Print ISBN: 978-1-78262-948-1
PDF ISBN: 978-1-78801-280-5
EPUB ISBN: 978-1-78801-507-3
Print ISSN: 2055-1975
Electronic ISSN: 2055-1983

A catalogue record for this book is available from the British Library

The Royal Society of Chemistry is a charity, registered in England and Wales, Number 207890, and a company incorporated in England by Royal Charter (Registered No. RC000524), registered office: Burlington House, Piccadilly, London W1J 0BA, UK, Telephone: +44 (0) 207 4378 6556.

For further information see our web site at www.rsc.org

Printed in the United Kingdom by CPI Group (UK) Ltd, Croydon, CR0 4YY, UK

Contents

Chemical Biology No. 10
Chemical and Biological Synthesis: Enabling Approaches for Understanding Biology
Edited by Nick J. Westwood and Adam Nelson

Published by the Royal Society of Chemistry, www.rsc.org

Section 2: Synthetic Approaches to Classes of Modified Biomolecule

Section 1:
Synthetic Approaches to Enable Small-molecule Probe Discovery

CHAPTER 1

Synthetic Tools for the Elucidation of Biological Mechanisms

NICHOLAS WESTWOOD*[a,b] AND ADAM NELSON*[c,d]

[a] School of Chemistry, University of St Andrews, North Haugh, St Andrews KY16 9ST, UK; [b] Biomolecular Sciences Research Complex, University of St Andrews, North Haugh, St Andrews KY16 9ST, UK; [c] School of Chemistry, University of Leeds, Leeds LS2 9JT, UK; [d] Astbury Centre for Structural Molecular Biology, University of Leeds, Leeds LS2 9JT, UK
*Email: njw3@st-andrews.ac.uk; a.s.nelson@leeds.ac.uk

1.1 Context

The elucidation of the molecular basis of biological mechanisms is an enduring theme in biomedical science. Although there are as few as approximately 19 000 protein-coding human genes,[1] whose corresponding proteins have been mapped in many tissue types,[2] biological mechanisms are remarkably complex. This complexity arises, in part, from the functional relevance of macromolecular complexes: the number of pairwise protein–protein interactions in humans has been estimated to be approximately 650 000,[3] the vast majority of which have not been functionally annotated. Moreover, biological macromolecules and processes are highly dynamic,[4] operating across many different timescales, and are often physically separated *via* compartmentalisation. In addition, the regulation of biological mechanisms is often achieved through many different post-translational modifications of the participating proteins.[5]

Chemical Biology No. 10
Chemical and Biological Synthesis: Enabling Approaches for Understanding Biology
Edited by Nick J. Westwood and Adam Nelson

Published by the Royal Society of Chemistry, www.rsc.org

Against this backdrop, increased innovation[6] and productivity[7] have been framed as major challenges for the pharmaceutical sector. Despite spiralling investment, the rate of drug discovery has remained roughly constant over the past 60 years.[8] Future innovation in drug discovery is likely to rely on the identification and validation of new protein targets (and classes of target) that may then be translated into drugs with novel mechanisms of action. However, the biology of proteins has historically been investigated in a highly uneven and unsystematic manner.[9] For example, across three protein classes that are central to disease—protein kinases, ion channels and nuclear hormone receptors—biologists' favourite proteins have remained broadly the same for the last 50 years! Crucially, the use of high-quality chemical probes[10–12] has been shown to be one of the few approaches that can change which proteins are the subject of biomedical research.

This book focuses on synthetic approaches that can facilitate the elucidation of biological mechanisms. The nature of the biomolecules that may be prepared using these approaches is extremely wide-ranging: from small-molecule probes through to large site-specifically modified macromolecular complexes. A wide-ranging toolkit is therefore featured, including synthetic and enzymatic approaches, both in isolation and in combination. Throughout, the power of the synthetic approaches is showcased in the context of the elucidation of specific biological mechanisms.

1.2 Synthetic Approaches to Enable Small-molecule Probe Discovery

The availability of high-quality small-molecule probes can have a disruptive effect on which proteins are the subject of fundamental biomedical research.[10–12] The key characteristics of high-quality probes include high *in vitro* potency and selectivity, a defined mechanism of action and demonstrated utility in cells. In addition, it is important that physical samples of probes are readily available, ideally together with structurally-related controls. The Chemical Probes Portal has been established to assist biomedical scientists to identify appropriate small-molecule probes for specific biological investigations.

The challenge of discovering new small-molecule probes is heightened because chemical space is vast, but has been historically explored very unevenly and unsystematically.[13] However, despite consensus on the vastness of chemical space, estimates of the number of possible drug-like molecules vary extremely widely. In one study,[14] extrapolation from GDB-17 (a database[15] of approximately 10^{11} enumerated molecules with up to 17 heavy atoms) led to an estimation of approximately 10^{33} drug-like small molecules with up to 36 heavy atoms. How, then, can chemical space be explored productively in the quest for novel chemical probes? Section 1 of the book covers a range of strategies that has been devised to address this problem.

Chapter 2 opens by introducing different types of molecular diversity, as well as diversity-oriented synthesis strategies that enable the exploration of diverse regions within chemical space. Alternative approaches that can improve the productivity of exploration of chemical space are then covered. For example, in biology-oriented synthesis (Chapter 3), the focus is placed on scaffolds related to the (necessarily) biologically-relevant frameworks that are found in natural products. In contrast, the searching problem may be reduced by focusing on specific molecular property space (*e.g.* fragment- or lead-like chemical space) relevant to the specific discovery application (Chapter 4). In Chapter 5, the practice of harnessing fragment-based approaches in the discovery of high-quality chemical probes is showcased through a series of case studies. Chapter 6 then describes recently-disclosed discovery approaches in which diverse chemical space is explored in a structure-blind manner, with bioactive ligands emerging (together with associated syntheses) on the basis of function alone. Two complementary technologies that can enable bioactive molecular discovery are then covered: DNA-encoded synthesis, in which the resultant very large libraries of small molecules are attached to a DNA "barcode" (Chapter 7) and automated synthetic approaches, including those that may be integrated with functional evaluation (Chapter 8; the automated synthesis of carbohydrates is covered in Chapter 10). Each of these complementary approaches can provide useful starting points for the discovery of novel chemical probes for interrogating specific biological mechanisms.

1.3 Synthetic Approaches to Classes of Modified Biomolecule

Section 2 of this book introduces complementary approaches for the synthesis of modified analogues of different classes of biomolecule. Here, focus is placed on amino-acid- and sugar-based biomacromolecules; more detailed reports on the chemistry of nucleic acids are reported elsewhere in the RSC Chemical Biology series. The first part of Section 2 focuses on technologies that enable the discovery of biomolecules that reside in beyond-rule-of-five chemical space. Chapter 9 describes platforms for preparing and screening libraries of genetically-encoded cyclic peptides, including split-intein-mediated cyclization of peptides and proteins (SICLOPPS), phage display and mRNA display technologies. The reader is also referred to a related book[16] in this series that is specifically focused on the synthesis and applications of cyclic peptides. In Chapter 10, a comprehensive review of carbohydrate-based compounds is provided, including an up-to-date assessment on the state of automated synthetic approaches. Chapter 11 provides an overview of the modification of biosynthetic pathways to deliver novel compounds that are derivatives of natural products. This chapter is complementary to the synthetic approaches to explore natural-product-related chemical space that are featured in Chapter 3.

Section 2 concludes with an overview of approaches that enable the preparation of site-specifically modified proteins (and their complexes). In Chapter 12, the toolkit of methods that enables the site-specific chemical modification of proteins—both *in vitro* and in a cellular context—is introduced. This toolkit can be exploited in a wide range of applications, for example in the discovery and validation of protein targets using chemical proteomics. In contrast, site-specifically modified analogues of proteins may also be prepared by total synthesis or semi-synthesis (Chapter 13). Such approaches have been shown to have tremendous value in the elucidation of complex biological mechanisms, for example those involving many different post-translationally modified proteins. The methods described in Chapters 12 and 13 are thus highly complementary, and it is hoped that these chapters will help readers to select methods that are most appropriate to the biological mechanism under investigation.

1.4 Summary

There is little doubt that huge scientific challenges lie ahead for the chemical biology and drug discovery research communities. In this book, we have drawn together opinions from a wide range of leading scientists whose work directly addresses many of these challenges. The core theme that runs throughout the book is the underpinning importance of synthesis in all of its forms. We hope that the many examples of how these synthetic approaches can be harnessed to gain new insights into biology will serve as an inspiration to the reader!

References

1. I. Ezkurdia, D. Juan, J. M. Rodriguez, A. Frankish, M. Diekhans, J. Harrow, J. Vazquez, A. Valencia and M. L. Tress, *Hum. Mol. Genet.*, 2014, **23**, 5866–5878.
2. M.-S. Kim *et al.*, *Nature*, 2014, **509**, 575–581.
3. M. P. H. Stumpf, T. Thorne, E. de Silva, R. Stewart, H. J. An, M. Lappe and C. Wiuf, *Proc. Natl. Acad. Sci. U. S. A.*, 2008, **105**, 6959–6964.
4. M. Secrier and R. Schneider, *Briefings Bioinf.*, 2014, **15**, 771–782.
5. C. T. Walsh, S. Garneau-Tsodikova and G. J. Gatto Jr., *Angew. Chem., Int. Ed. Engl.*, 2005, **44**, 7342–7372.
6. B. H. Munos and W. W. Chin, *Sci. Transl. Med.*, 2011, **3**, 89cm16.
7. S. M. Paul, D. S. Mytelka, C. T. Dunwiddie, C. C. Persinger, B. H. Munos, S. R. Lindborg and A. L. Schacht, *Nat. Rev. Drug Discovery*, 2010, **9**, 203–214.
8. B. Munos, *Nat. Rev. Drug Discovery*, 2009, **8**, 959–968.
9. A. M. Edwards, R. Isserlin, G. D. Bader, S. V. Frye, T. M. Willson and F. H. Yu, *Nature*, 2011, **470**, 163–165.
10. S. V. Frye, *Nat. Chem. Biol.*, 2010, **6**, 159–161.
11. J. Blagg and P. Workman, *Cancer Cell*, 2017, **32**, 9–25.

12. C. H. Arrowsmith *et al.*, *Nat. Chem. Biol.*, 2015, **11**, 536–541.
13. A. H. Lipkus, Q. Yuan, K. A. Lucas, S. A. Funk, W. F. Bartelt 3rd, R. J. Schenck, A. J. Trippe and C. A. S. Registry, *J. Org. Chem.*, 2008, **73**, 4443–4451.
14. P. G. Polishchuk, T. I. Madzhidov and A. Varnek, *J. Comput. -Aided Mol. Des.*, 2013, **27**, 675–679.
15. L. Ruddigkeit, R. van Deursen, L. C. Blum and J.-L. Reymond, *J. Chem. Inf. Model.*, 2012, **52**, 2864–2875.
16. *Cyclic Peptides: From Bioorganic Synthesis to Applications*, ed. J. Koenke, J. Naismith and W. A. van der Donk, Royal Society of Chemistry, 2017.

CHAPTER 2

The Application of Diversity-oriented Synthesis in Chemical Biology

NATALIA MATEU, SARAH L. KIDD, THOMAS J. OSBERGER, HANNAH L. STEWART, SEAN BARTLETT, WARREN R. J. D. GALLOWAY, ANDREW J. P. NORTH, HANNAH F. SORE AND DAVID R. SPRING*

University of Cambridge, Department of Chemistry, Lensfield Road, Cambridge, CB2 1EW, UK
*Email: spring@ch.cam.ac.uk

2.1 Introduction

Communicable and non-communicable diseases are prevalent worldwide. Whilst treatments and cures exist for several of these, for many more they are either ineffective or non-existent. Cancer, malaria and antibiotic resistance are some representative examples of conditions that cause huge burdens on worldwide healthcare. However, despite research and development investment being higher than ever, the biomedical community struggles to provide effective solutions for such devastating diseases. This can be attributed to the inability to identify appropriate targets to modulate the disease, the lack of suitable compounds to interact with identified targets, or the failure of compounds to pass through clinical trials. In such a situation, the question remains as to where the problem fundamentally lies and how to address it. This chapter will analyse the importance of screening

Chemical Biology No. 10
Chemical and Biological Synthesis: Enabling Approaches for Understanding Biology
Edited by Nick J. Westwood and Adam Nelson

Published by the Royal Society of Chemistry, www.rsc.org

high-quality small-molecule collections to facilitate the identification of novel biologically-active small molecules and how different diversity-oriented synthesis (DOS) approaches have efficiently addressed this issue.

2.1.1 Small Molecule Screening Collections

Investigations into the function of different biological systems and the role of specific biological targets in a given disease can provide vital information for both chemical genetics and drug discovery. For both of these approaches, the identification of suitable small molecules that can modulate different pathways selectively is highly desirable. Within the field of drug discovery, failures resulting from low observed efficacy of drug candidates in clinical trials are often attributed to the inadequate prediction of the physiological response that may result from modulating a given biological target. Initial investigations into biological targets for their therapeutic potential are often first conducted *via* genetic association studies and knockout models using DNA mutations.[1] However, these predicted effects do not always translate into those observed when a target is modulated using small molecules. In contrast, the use of small molecules to selectively perturb the biological function of macromolecules, such as proteins, has been validated as an effective means to study biological systems and thus may provide more reliable tools for predicting biological responses.[2] Nevertheless, the identification of effective small molecules (*probes* or *'tool compounds'*) capable of modulating a target of interest is significantly challenging. These tool compounds not only provide new understanding of biological targets and pathways related to a given disease (*target identification/validation*), but can also form the first step towards the development of a new medicine (*hit identification*) (Figure 2.1). In cases where the biological target is well defined, rational design of potential modulators is often possible, particularly when the structure of a natural substrate is known. However, this is not possible in the absence of structural information, and hence the discovery of these chemical probes relies on the screening of small-molecule libraries. High-throughput screening (HTS) of large small-molecule libraries is one of the most commonly employed techniques for this purpose.[3] In these assays, the quality of the compound collection, in terms of composition and size, ultimately determines the chances of finding good hits.[4] It is therefore vital to identify the key chemical features that constitute a high-quality screening collection in order to facilitate the discovery of new small-molecule modulators for challenging, and currently underexplored, biological targets.

2.1.1.1 Structural Diversity

An enlightening analysis carried out by GlaxoSmithKline (GSK) after their failure to identify new tractable antibiotic candidates from HTS[5] highlighted an underlying problem within pharmaceutical screening collections. Screening libraries have grown massively in size, but there has not been a

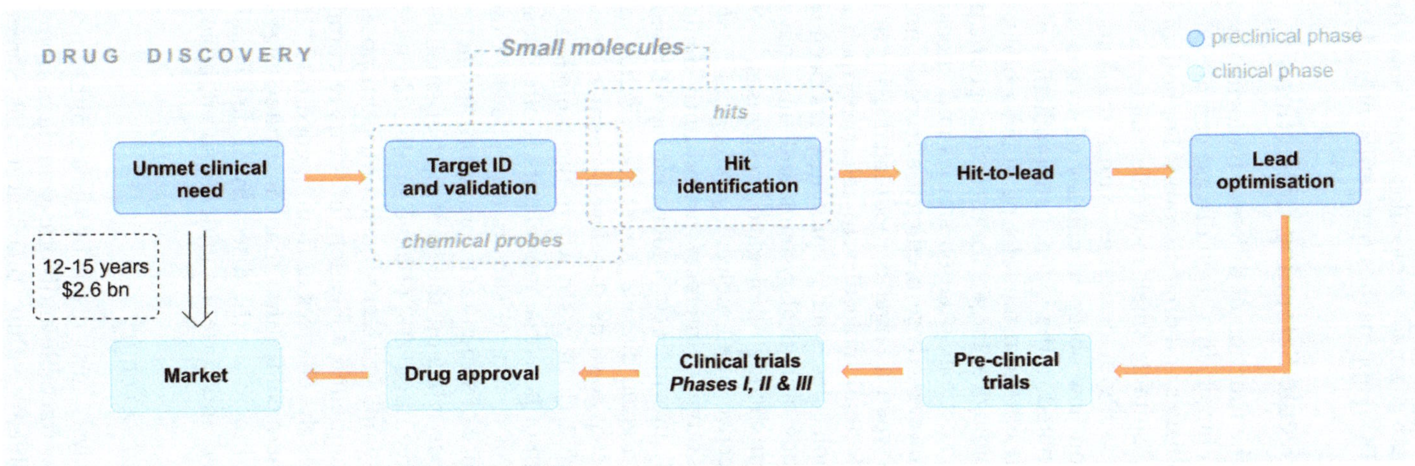

Figure 2.1 The involvement of small molecules in the early stages of drug discovery.

corresponding increase in the number of new structural classes or scaffolds within these collections.[4] Thus, there is a lack of *structural diversity* within these screening collections. It has been proposed that this is a result of the way the libraries are constructed. Combinatorial-type libraries are typically generated by combining a number of building blocks in different ways using the same synthetic methods, resulting in similar scaffolds with varied substituents; hence these libraries possess low levels of structural diversity.[6,7] Analysis supports the suggestion that increasing the library size or number of peripheral substituents does not significantly increase library diversity; instead, the molecular shape distribution of the library as a whole tends to be determined by the nature of the central scaffolds.[8]

An additional factor which has contributed to the structural bias of existing libraries towards similar structures is the fact that medicinal chemistry research over the past few decades has typically focused on a limited set of biological targets.[9] Approximately 29% of all drugs are enzyme modulators and nearly 36% act upon G-protein-coupled receptors.[7,10,11] In addition, even within major target classes, proteins have been studied highly unevenly.[12] As such, most libraries are directed towards identifying modulators of these and other "traditional" targets, including a strong preference for molecules that fulfil Lipinski's 'Rule of five' for oral bioavailability.[13–15] To add to the problem, a significant overlap between synthetic screening libraries across the pharmaceutical industry will be present, since the same chemical methods and vendors have often been used.[5]

2.1.1.2 *Structural Complexity*

In the early 1990s, within drug discovery, drug attrition predominately occurred in early clinical trials as a result of adverse pharmacokinetics and bioavailability.[16] Since the pharmaceutical industry has responded to these issues by improving the physicochemical properties (so called drug-like properties) of the lead candidates, there has been a shift in attrition towards late-stage clinical trials. The primary causes of late-stage attrition currently include efficacy and safety issues, such as toxicology and clinical safety.[16–18]

Studies exploring these causes of late-stage attrition[15–30] revealed that an increase in the *molecular complexity* of a lead is associated with increased potency and decreased toxicity. This was rationalised through the understanding that toxicity is often due to off-target effects.[31] Given that biological macromolecules can be considered as three-dimensional (3D) environments with defined binding regions,[6,7] a given biological macromolecule will only interact with molecules that have a complementary 3D binding surface.[6,7,32,33] Crucially, it is the molecular structure which determines a molecule's 3D shape.[7,33] Thus, by increasing their molecular structural complexity, both drug-candidates or chemical probes should have a better complementarity for a specific target and a decreased chance of binding off-targets, aiding further investigations.[22,34]

Molecular complexity is an intuitive principle that has been quantitatively defined in a number of ways. However, it is most often judged using the fraction of sp^3 carbons (Fsp^3) and the number of chiral centres; increased saturation and stereochemical complexity corresponds to improved selectivity and decreased toxicity in clinical candidates.[19,22,23,34,35] Therefore, going forwards, high-quality screening libraries will need to incorporate these features.

2.1.1.3 *Chemical Space Coverage*

Chemical space is a chemoinformatic concept that refers to all possible molecules that could be, in principle, synthesised as defined by a set of preselected physicochemical and topological descriptors. The chemical space of 'drug-like' molecules (or potential hits) has been estimated to be in the order of 10^{62} achievable compounds.[36] From a biomedicinal point of view, the relationship between biology and chemistry has been described as biologically-relevant chemical space.[37] This defines the regions of chemical space that are populated by biologically active molecules; although some of these regions have been already explored with known bioactive molecules, this space remains largely "unexplored". Accordingly, in the context of hit identification, the "ideal" screening collection would encompass all "drug-like" chemical space and thus access all the biologically relevant regions, but this is physically unattainable. Computational investigations have revealed that the structural diversity and complexity of a screening library indeed correlates to the capability of the collection to achieve broad chemical space coverage and a broad range of distinct biological activities.[19,22,38] In this manner, screening libraries should ideally feature structurally *diverse* and *complex* small molecules in order to maximise the coverage of chemical space and therefore biological space to aid future investigations.

2.1.2 Sources of Complex and Diverse Libraries

To address these issues, there is a need for new screening libraries of *complex* and *diverse* small molecules. However, creating these is no simple task and the question then arises of how to populate a library. There are two main sources of small molecules: natural products and synthetic molecules.

2.1.2.1 *Natural Products*

Natural products are a rich source of complex and diverse bioactive molecules.[39,40] Since natural products result from coevolution with proteins, it has been suggested that their scaffolds are pre-encoded to bind to evolutionarily conserved protein binding sites and feature pre-validated biological activities (see Chapter 3).[41–43] It is not surprising that natural products, characterised by their enhanced molecular complexity, high degree of saturation and chirality content, provide highly specific biological activities and have been widely used as templates for the successful development of drugs that mimic their actions.[40,44] Indeed, these molecules have played a pivotal role in medicine: 33% of small-molecule drugs developed between 1981 and 2014 are derived from natural products.[45]

Despite the fruitful prospects of natural products for the discovery of novel bioactive molecules, difficulties in isolation and purification of these compounds often lead to the screening of mixtures, which can hinder the identification of the active species.[46] Additionally, in order to screen natural products, kilograms of the natural source are often required to isolate only milligrams of the active compound.[47] As such there has been a sustained interest within the synthetic community in designing synthetic routes to construct these complex molecules.[48] Although this has aided our understanding of natural products these synthetic routes are often incredibly challenging and time-consuming, due to the sheer complexity of these molecules. Moreover, analogue generation can be challenging, something that is usually essential during hit optimisation.

2.1.2.2 *Synthetic Small Molecules*

A vital alternative to natural product libraries is the chemical synthesis of small molecules. The over-reliance on combinatorial-type and split-pool syntheses has resulted in vast screening libraries with an inherent lack of structural diversity (see above). These routes usually generate a limited range of molecular scaffolds, predominantly based on achiral or 'flat' architectures.[35] Thus, the development of novel strategies for the generation of screening libraries which feature the two attributes of structural complexity and structural diversity has become a major drive for synthetic chemists. In this manner, the development of novel synthetic strategies that take inspiration from the structural complexity of natural products, but feature improved synthetic tractability and modular aspects has attracted increasing interest. Within the last two decades several different methodologies have emerged, including biology-oriented synthesis (Chapter 3),[41,49–51] DNA-encoded/-templated libraries (Chapter 7)[52–55] and diversity-oriented synthesis (this chapter).[7,10,56]

The subject of structural diversity is a subjective concept. However, four common principles are commonly identified within the literature to guide the development of these libraries:

(1) Skeletal diversity: variation in the molecular architecture/ring structure of a molecule.
(2) Appendage diversity: variation in the structural moieties around a common scaffold.
(3) Stereochemical diversity: variation in the orientation of different elements from the scaffold.
(4) Functional group diversity: variation of the functional groups present which have the potential to interact with a given biological macromolecule.

Accordingly, the ideal library would, therefore, feature molecules that exhibit variation of all four types of diversity. Among the strategies developed, diversity-oriented synthesis has proven to be particularly fruitful for the discovery of novel chemical probes and hit compounds. In the remainder of this section we will discuss the underlying synthetic principles of DOS.

2.1.3 Synthetic Strategies for the Construction of Complex and Diverse Libraries: Diversity-oriented Synthesis

The diversity-oriented synthesis concept, 'DOS', was conceived and developed in the early 2000s.[57,58] DOS describes the deliberate and efficient divergent synthesis of complex small-molecule collections that interrogate large areas of chemical space. This contrasts with traditional retrosynthetic and combinatorial synthetic strategies, which commonly follow a linear process to generate focused libraries around a specific molecular framework occupying a defined area of chemical space. Instead, DOS adopts a forward synthetic analysis approach (Figure 2.2), whereby simple and common starting substrates are converted, through a wide range of complexity-generating transformations, into a collection of different scaffolds that incorporate all four principles of diversity. In particular, the incorporation of skeletal diversity is crucial for ensuring the functional diversity of a DOS library.[7,10]

Principal moment of inertia (PMI) plots provide a useful tool for the visual assessment of the chemical space coverage displayed by compound collection. In this approach, the normalised ratio of principal moments of inertia from the lowest energy conformation of each component of the library is calculated and plotted on a triangular graph. Each vertex of the graph represents one of three representative molecular shapes: rod-like, disk-like, or sphere-like. The broader the distribution within the three corners the compound collection has, the more diverse and complex it is (Figure 2.2). In comparison with combinatorial library synthesis, DOS approaches have been shown to deliver libraries with increased structural diversity.

2.1.3.1 Synthetic Principles in DOS

Broadly speaking, two approaches for the generation of scaffold diversity have been employed in DOS campaigns to date: reagent- and substrate-based DOS (Figure 2.3). In the *reagent-based approach,* branching pathways are carried out on a pluripotent substrate to yield different compounds with distinct molecular scaffolds. In this case, it is the choice of reagents and co-substrates that dictate the stereochemical and skeletal diversity within the final scaffolds. Alternatively, in the *substrate-based approach,* strategically positioned functional groups are reacted together ("paired") so as to "fold" a

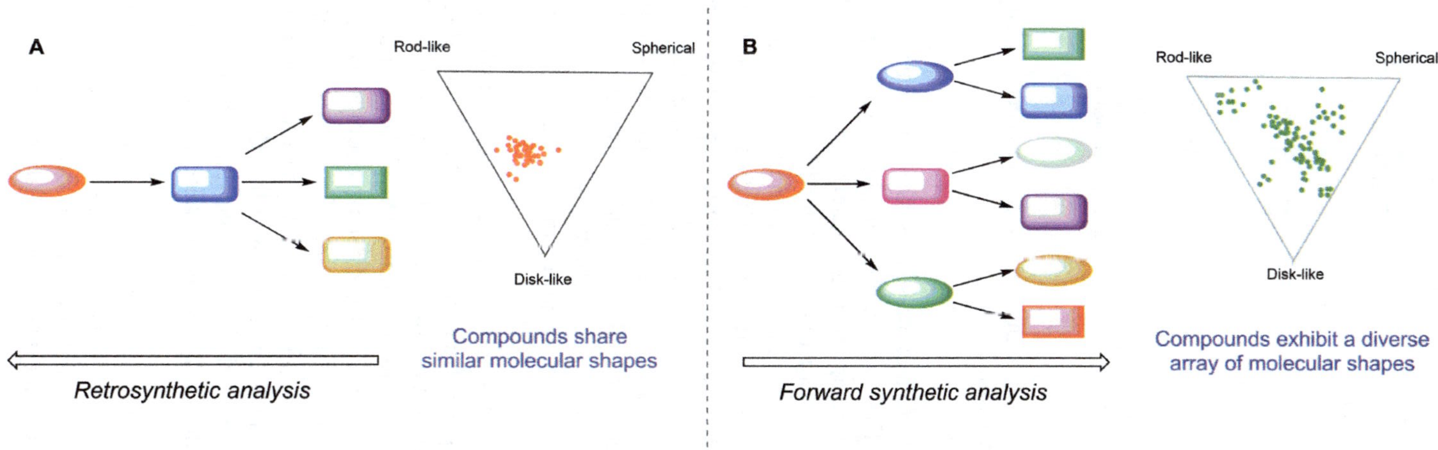

Figure 2.2 Comparison of Combinatorial (A) and DOS (B) strategies for library synthesis and the respective chemical space coverage of each as determined by principal moment of inertia plots.

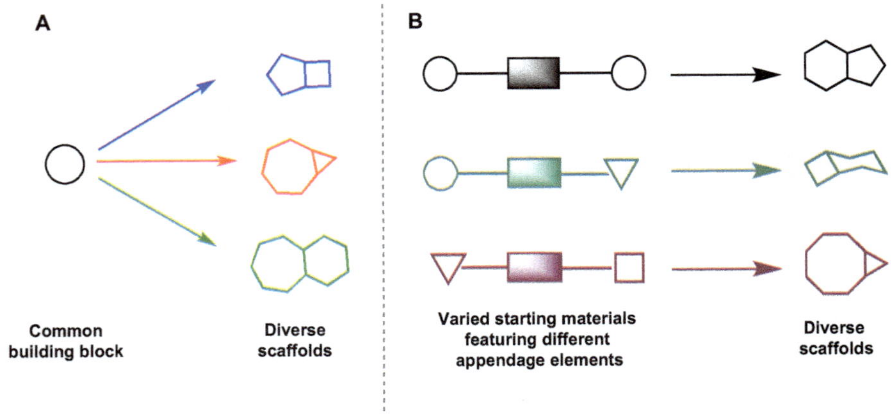

Figure 2.3 DOS strategies to synthesise distinct molecular scaffolds. (A) Reagent-based. (B) Substrate-based.

substrate into distinct molecular scaffolds. In this case, the pre-encoded functionalities within the substrate determine the skeletal composition in one of two ways: (A) starting from a densely-functionalised molecule such that different reaction conditions yield multiple scaffolds from one substrate; or (B) using different starting materials and common reaction conditions such that each starting material will furnish a product based around a different scaffold (Figure 2.3). For both approaches, the challenge is centred on the careful selection of the starting substrates.

These two approaches are not orthogonal and most modern DOS strategies incorporate aspects of both.[6,10,33] Other features of diversity (appendage, functional group and stereochemical) can be introduced into the compound libraries through variation in the starting materials and/or reagents used.[33,59]

Independent of the approach taken, the build–couple–pair (B–C–P) algorithm (Figure 2.4) is a powerful method in DOS.[7,60] This three-phase strategy begins with the *build phase*, involving the synthesis of building blocks containing orthogonal and ideally some chiral functionalities. These building blocks can then be linked together or coupled with other substrates intermolecularly to produce complex and densely functionalised molecules. This step is termed the *couple phase* and provides the basis for the introduction of stereochemical diversity. Finally, the *pair phase* involves the intramolecular coupling of different moieties within the intermediates to produce rigidified scaffolds. This stage results in the skeletal diversity of the library.

In the remainder of this chapter, we will describe several state-of-the-art examples of DOS and how these libraries have been applied to aid the identification of novel biologically active molecules for use as both hits (hit identification) and chemical probes (for chemical genetics or target identification/validation).

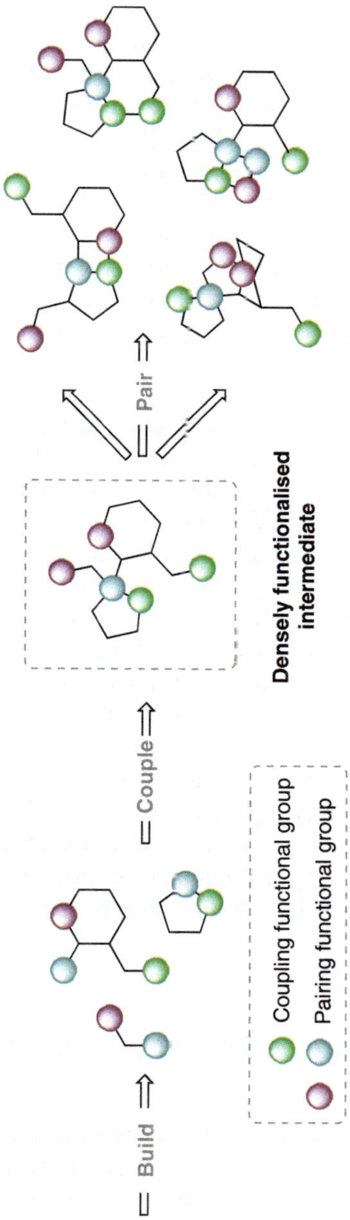

Figure 2.4 Build–Couple–Pair in DOS.

2.2 Application of Diversity-oriented Synthesis for the Identification of Small Molecule Modulators

2.2.1 Structural Diversity and Phenotypic Screening

The identification of small-molecule modulators through the screening of compound collections comprises different *in vitro* or *in vivo* strategies. Among these is phenotypic screening: testing small molecules and observation of phenotypic change in a biological system, which has been successfully used in the pharmaceutical industry for many years.[61–63] Indeed, 56% of "first-in-class" medicines approved by the FDA between 1999 and 2008 have been identified in this manner.[64] In a phenotypic screening approach the molecular mechanism of action and the biological target protein can remain unknown, however, predictive studies can be undertaken by comparing multi-parametric phenotypic profiles or "phenoprints" of the new hits with reference compounds with an established mechanism of action.

Phenotypic screening involves testing small-molecule compound collections in cellular or *in vivo* models directly without any previous target knowledge. In these unbiased processes, where the precise nature of the biological target is unknown, the selection criteria for the "right" small molecule within chemical space are dramatically complicated due to the lack of any structural information.[65] The question then arises, how can the ideal small-molecule collection be selected for these assays without any structural guidelines? As previously discussed, small-molecule collections with a high degree of structural diversity and complexity are predicted to display a higher hit-rate and broader scope of biological activity.[22,66]

There are many examples in the literature where the combination of structurally diverse and complex screening collections, often produced from DOS, with phenotypic screening has resulted in the identification of new biologically active molecules.[67–72] The selected examples discussed later in this chapter (Section 2.2.1.1) are focused on studies that have provided access to new hits related to challenging therapeutic areas, such as antibiotic resistance and cancer.

2.2.1.1 *Efficient Identification of Novel Antibiotics from Diverse Collections*

In the context of DOS, the power of this strategy as a tool to identify new and much-needed antibacterials through phenotypic screening has been validated.[59,73] One of the first case studies was reported by Wyatt *et al.* in 2006 and involved the exploitation of the different reactivity of a two-carbon fluorous-tagged diazoacetate **1** following a reagent-based DOS strategy (Scheme 2.1).[74] In the first set of branching reactions, three key synthetic transformations were applied: (i) three-membered ring formation, (ii) 1,3-dipolar cycloadditions

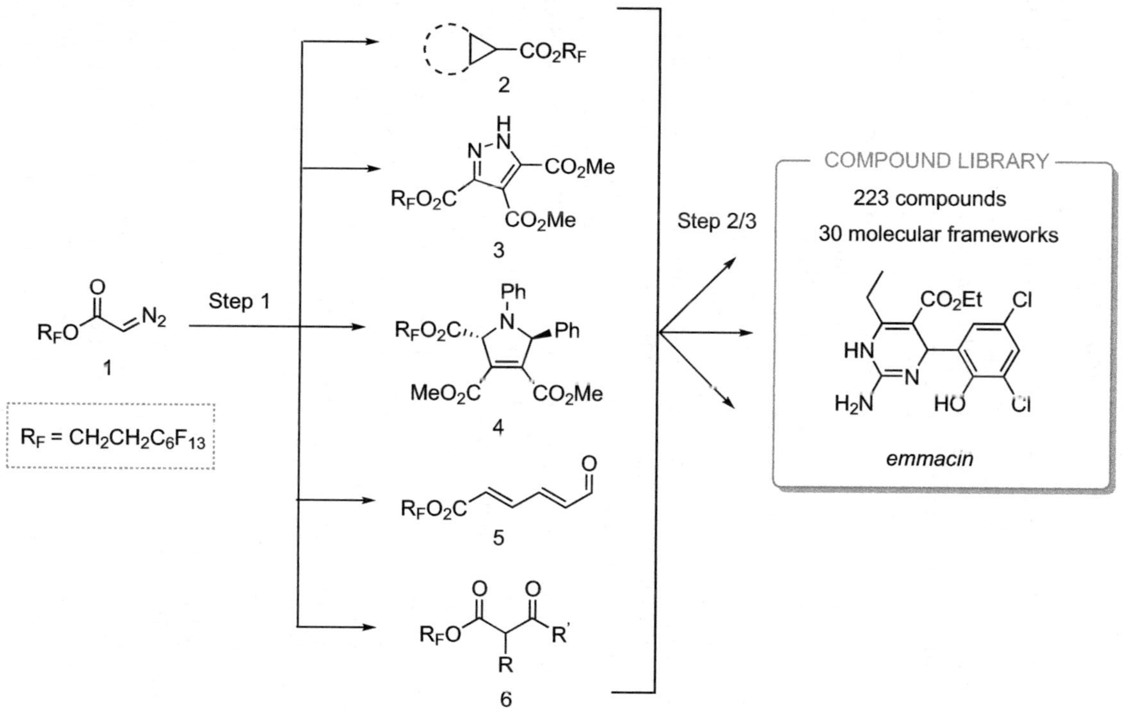

Scheme 2.1 New sub-structural class of bacterial-selective DHFR inhibitor *emmacin* discovered from a DOS library of 223 compounds based on 30 molecular frameworks.

and (iii) α-deprotonation followed by quenching with an electrophile and carbenoid formation. These structurally diverse intermediates **2–6** were subjected to further complexity-generating reactions to diversify the molecular frameworks and thus increase the skeletal diversity of the library (Scheme 2.1, Step 2/3). As a result, a collection of 223 small molecules based around 30 distinct molecular skeletons was efficiently generated in no more than four synthetic steps from the common diazoacetate unit.

Preliminary phenotypic screens revealed that 64 structurally diverse compounds of the DOS library were able to modulate the growth of strains of methicillin-resistant *Staphylococcus aureus* (MRSA).[67] Further investigation revealed that the compound named *emmacin* (shown in Scheme 2.1) was a potent inhibitor of methicillin-susceptible strains of *S. aureus* (MSSA) and two UK epidemic strains of MRSA (EMRSA-15 and EMRSA-16[75,76]). In addition, this compound exhibited no cytotoxic properties in a variety of mammalian-surrogate systems. Target identification revealed that *emmacin* is a prokaryote-selective, uncompetitive and reversible inhibitor of dihydrofolate reductase of EMRSA-16 ($DfrB_{EMRSA16}$).[77,78]

Following a similar strategy, Thomas *et al.*[68] successfully combined the synthesis of a DOS library and further phenotypic screening to identify a new antibiotic with similar potency to those used clinically. The generation of the small-molecule library was achieved starting from a solid-supported phosphonate 7 using a reagent-based DOS strategy (Scheme 2.2). In this case, the reacted phosphonate functionality allowed the *E*-selective formation of highly functionalized α,β-unsaturated acyl imidazolidinones **8** (Step 1) through a Horner–Wadsworth–Emmons reaction with different aldehydes. Following a divergent scheme, **8** was reacted in three catalytic and enantioselective pathways: (i) [2 + 3] cycloaddition; (ii) dihydroxylation, and (iii) [4 + 2] cycloaddition to yield a small set of molecules based on three different molecular frameworks (Step 2). The next steps (Step 3/4) of the synthetic strategy involved a series of branching reactions to further diversify these key branch-point substrates **9–11**. In this manner, a collection of 242 natural-product-like small molecules based on 18 molecular scaffolds, with high levels of skeletal diversity, was synthesized.

Similarly to the previous case study, the resulting DOS collection was screened to study the effect on the growth of three strains of *Staphylococcus aureus* (MSSA, EMRSA 15 and EMRSA 16). Three structurally novel compounds inhibited bacterial growth, but one of them, named *gemmacin*, showed good activity against EMRSA 15 and 16 (Figure 2.5). Stereoselective synthesis and further SAR studies permitted the identification of (−)-*gemmacin*-B, which demonstrated higher levels of antibacterial activity against EMRSA 16.[69] Target identification was achieved after the observation that *gemmacin* generated reactive oxygen species, which indicated that the compound may act as a cell-membrane disruptor. This supposition was confirmed using a membrane disruptor assay where *gemmacin* acted as a selective disruptor of bacterial cell membranes.

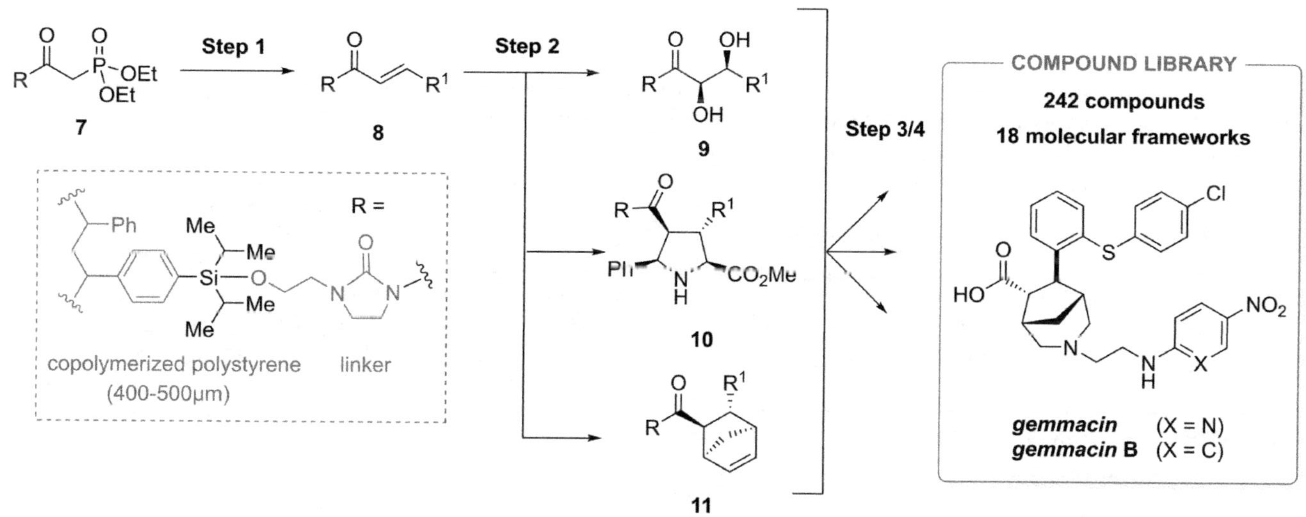

Scheme 2.2 Novel antibacterial *gemmacin* discovered from a DOS library of 242 compounds based on 18 molecular frameworks.

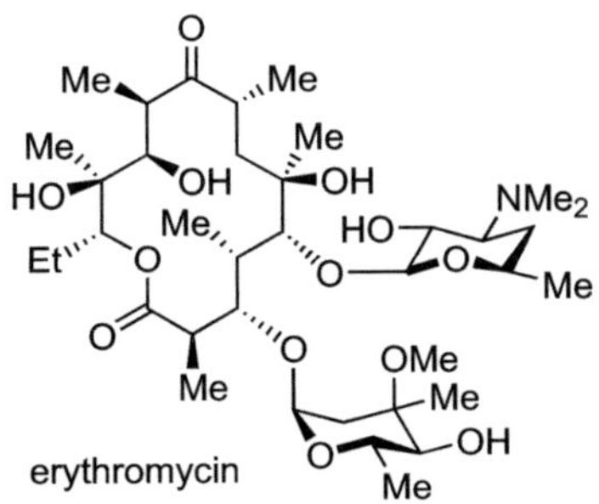

erythromycin

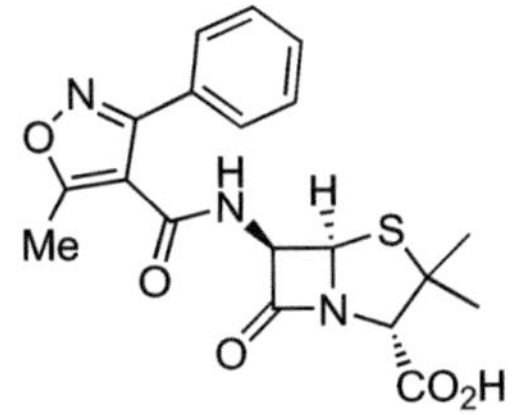

oxacilin

	MIC$_{50}$ (μg ml^{-1})		
	MSSA	EMRSA 15	EMRSA 16
emmacin	**2**	**9**	**9**
(-)-*gemmacin*	**2**	**8**	**16**
(+/-)-*gemmacin* B	**–**	**8**	**8**
erythromycin	0.5	>64	>64
oxacilin	0.5	>32	>32

Figure 2.5 MIC$_{50}$ values for *emmacin, gemmancin* and *gemmacin B* against MSSA, EMRSA 15 and EMRSA 16 are shown. For comparison, the MIC$_{50}$ values for erythromycin and oxacillin are also shown. (MIC$_{50}$ = minimum inhibitory concentration required to inhibit the growth of 50% of organisms).

2.2.1.2 *DOS as a Tool for the Identification of New Anticancer Small Molecules*

Among the many existing techniques developed to understand cancer genetics, the identification of small molecules capable of modifying cancer phenotypes is particularly effective for feeding early drug discovery programs.[79–82] In particular, the screening of structurally diverse compound collections has been demonstrated to be an excellent strategy to identify new modulators of cancer cells.[83] In this context, some examples are discussed below where the combination of DOS collections and phenotypic screening using cancer cell lines has delivered new anticancer hit compounds.

In one of these case studies, rhodium carbenoid chemistry was used as a key step for the reagent-based DOS of a structurally diverse small-molecule collection starting from phenyldiazo ester compounds.[84] Following a divergent scheme, the α-diazo ester **12** was reacted with terminal alkynes, alkenes and allenes *via* rhodium-catalysed cyclopropanation reaction to give rise to a small set of different substituted three-membered rings (Scheme 2.3, **14–15**). Encouraged by the scope of this rhodium(II)-catalysed reaction, the styryl diazo ester derivative **13** was synthesized and treated under similar reaction conditions. Accordingly, after a cyclopropanation-Cope reaction with cyclopentadiene, intermediate **16** was stereoselectively generated. These three restricted and highly functionalized scaffolds **14–16** were considered privileged starting points for further diversification through a wide range of chemical transformations. This strategy delivered a library of 35 three-dimensional and structurally diverse compounds with high sp^3 content and broad coverage of biologically relevant chemical space.

Screening of the resulting compound collection for antimitotic activity in human U2OS osteosarcoma cells[85] was performed, and two compounds

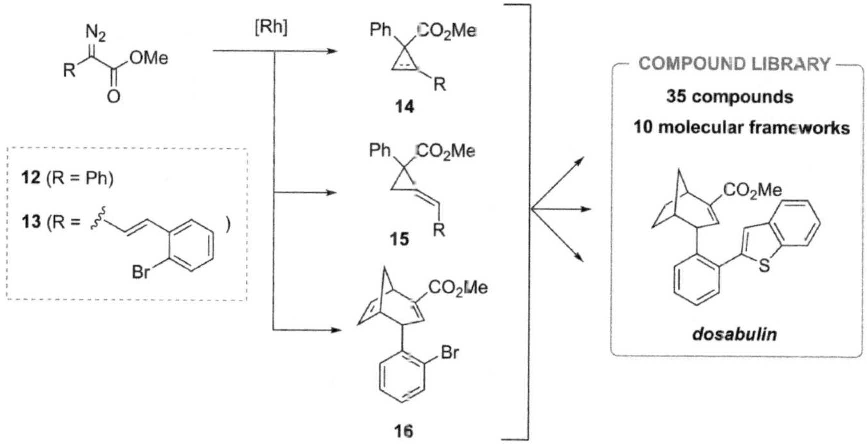

Scheme 2.3 Rhodium carbenoid methodology used for DOS leading to identification of *dosabulin* as an inhibitor of tubulin polymerization.

from the DOS library were observed to cause mitotic arrest. Further chemical modifications carried out on these initial active molecules resulted in the identification of the (*S*)-enantiomer of *dosabulin* as a more potent inhibitor of mitosis [concentration giving 50% effect (EC_{50}) 1.2 μM] causing cell death by apoptosis. Target identification was investigated by competition studies observing the displacement of colchicine or vinblastine from tubulin by (*S*)-*dosabulin* using confocal microscopy. These experiments indicated that *dosabulin* was not binding to the vinblastine site on tubulin, but that the binding site was in the vicinity of, or allosteric to, the colchicine site.

Recently, the use of carbohydrates has attracted interest in the field of DOS[86–89] due to their immense potential for generating stereochemical and structural diversity as well as their biomedical applications.[90–92] In notable recent work in this area, Trabocchi and colleagues[93] described how building blocks derived from D-mannose and glycine could be used to create the structural complexity following a substrate-based build–couple/pair strategy (Scheme 2.4). The protected sugar derivatives **17** and **18**, resulting from D-mannose modifications (*build* phase), were combined in the *couple* step with amino-acetaldehyde dimethyl acetal through three reducing-based reactions. The presence of polyhydroxylated species and the protected carbonyl function coming from the amino acid derivative in intermediates **19–21** allowed intramolecular *trans*-acetalisations as the key pairing strategy. Thus, after the *pair* phase, a collection of 26 polyhydroxylated nitrogen-containing scaffolds based on six molecular frameworks with high levels of structural diversity (demonstrated by computational analysis) was generated.

Further investigations were focused on testing the resulting DOS library in a phenotypic screen using MDA-MB-231 cell lines for the identification

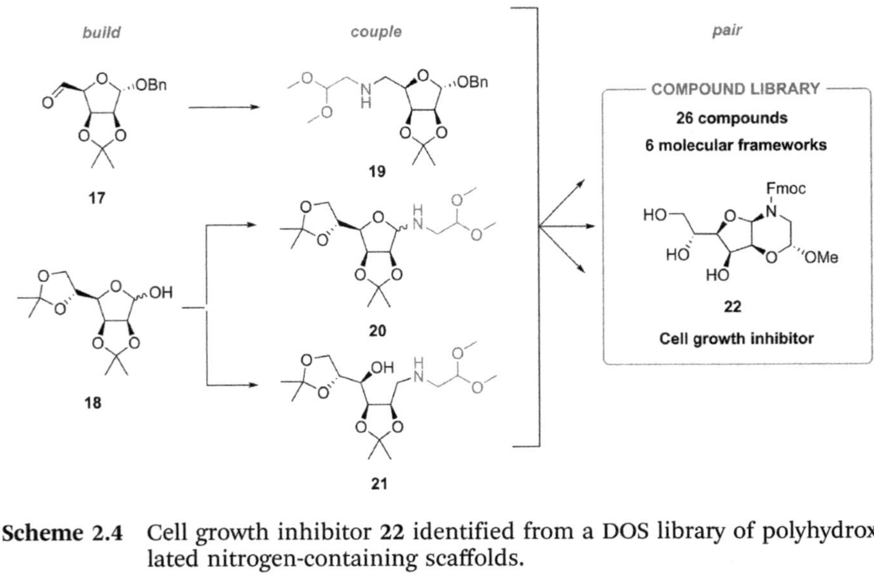

Scheme 2.4 Cell growth inhibitor **22** identified from a DOS library of polyhydroxylated nitrogen-containing scaffolds.

of modulators of the breast carcinoma cell cycle mechanism.[70] These investigations were based on the preliminary evidence regarding the capability of iminosugars to inhibit the growth of breast cancer cells.[94,95] The MDA-MB-231 cell line is a simple model system for the study of triple-negative breast cancer, which shows a major tendency towards early metastasis, not responding to hormonal chemotherapy and accounts for 15% of all breast carcinomas. In this experiment, after 48 hours of incubating members of the DOS library with MDA-MB-231 cells, compound **22** showed the best range of inhibition of cell proliferation (exhibiting 40% inhibition). Despite subsequent synthesis of analogues of **22**, none of the newly synthesised compounds displayed better inhibition values, validating compound **22** as a new anticancer modulator based on a polyhydroxylated scaffold. Further studies towards the identification of the molecular mechanism of action to identify the target related with the phenotypic effect are ongoing.

2.2.2 The Role of DOS in Target Validation through the Discovery of New Chemical Probes

A complementary method for target validation in a drug discovery context is the use of small-molecule chemical probes, which can modulate biological systems in order to predict therapeutic potential.[96] Importantly, as well as small-molecule inhibitors this includes the discovery of probes that can activate signalling pathways, since this can shed light on the workings of complex biological systems.[97] In this manner, the discovery of novel, high-quality small molecules capable of inducing functional pharmacology and proving phenotype perturbation can play a vital role in facilitating target validation. DOS represents an ideal synthetic strategy to deliver novel structurally diverse molecules for use as chemical probes, since the limiting factor in the application of small molecules for this process is often the availability of appropriate chemical modulators. Since DOS aims to populate new areas of chemical space, this provides the potential to widen our 'chemical probe toolbox', enabling us to modulate and interrogate the functional behaviour of challenging, poorly characterized and even novel biological systems to seed new drug-discovery programs.

2.2.2.1 DOS Yields a Novel PPI Modulator

Protein–protein interactions (PPIs) play a critical regulatory role within a range of biological functions within the body and have been implicated in a vast range of disease states, including cancer.[98] Despite the fruitful prospects of PPI inhibition for drug discovery, the challenging nature of inhibiting these interactions using 'traditional' small-molecule screening collections has led to the perception of PPIs as 'undruggable'. Recent work by Kim *et al.*, however, has demonstrated the application of DOS in the identification of a modulator of a key PPI interaction implicated in the amino-acid-dependent activation of the mechanistic target of rapamycin complex 1 (mTORC1), a regulator of cellular growth, proliferation and autophagy with oncogenic implications.[99]

In this work, the authors utilised a privileged-DOS (pDOS) strategy[100] to generate 'biological navigators' using pyrimidodiazepine and pyrimidine cores, considered 'privileged' scaffolds due to their presence within various bioactive molecules and marketed drugs. Additionally, they incorporated a diazepine motif within these fused core scaffolds, to increase the 3D character of the library *via* higher sp^3 content and conformational flexibility. Following a B–C–P algorithm five reactive sites were installed upon the pyrimidodiazepine core, that by utilizing a reagent-based approach were selectively reacted in a pairwise fashion. This produced several tri-cyclic and tetra-cyclic scaffolds in a synthetically efficient manner from **23**. Five different reactive site combinations were then employed (Scheme 2.5)—A–B (**25**), B–C (**27**), C-pair (**24**), C–D (**28**) and D–E (**26**)—to generate 24 compounds representing 16 distinct frameworks. A variety of functional group pairings was employed including intramolecular nucleophilic substitution, ring-closing metathesis (RCM) of pre-installed unsaturated moieties and exploitation of the both the nucleophilic and electrophilic nature of an imine at C. Additionally, the use of a rhodium-catalysed [2 + 2] cycloaddition (to give **24**), a [2,3]-sigmatropic ring expansion (not shown) and RCM (to give (**26**) allowed the formation of challenging ring systems such as a β-lactam, a 6,10-benzoxazecine and a bridge-head [4,3,1] structure, respectively. A comprehensive collection of 3–10 membered carbocycles and hetereocycles was installed upon the key pyrimidine core. The results of chemoinformatic analysis indicated that the resulting library displayed a broad shape distribution, similar to that of 71 bioactive natural products, but an improved distribution compared with 15 pyrimidine-containing FDA approved drugs.[99]

The leucyl-tRNA synthetase (LRS)–Ras-related GTP-binding protein D (RagD) PPI plays an important role in the amino-acid-dependent activation of mTORC1 *via* leucine sensing and signalling to mTORC1.[101] Initial screening for inhibition of this PPI was conducted using ELISA-based HTS using LRS and glutathione-*S*-transferase tagged RagD, leading to the identification of **29a** and **29b** as dose-dependent inhibitors. Subsequent biological experiments focused on target validation of LRS–RagD PPI inhibition through investigations into the effects on mTORC1 activity and cell proliferation. Western blot experiments revealed that **29a** and **29b** suppressed the phosphorylation of p70 ribosomal protein S6 kinase 1 (S6K1)—a known kinase substrate for mTORC1—whilst **29b** proved to selectively down-regulate two further mTORC1 substrates but not mTORC2 and 5′ AMP-activated protein kinase (AMPK) substrates in cancer cell lines. This indicated selectivity for this signalling pathway; however, notable differences in these phosphorylations were observed compared with the known mTORC1 inhibitor rapamycin, which the authors suggested was a result of

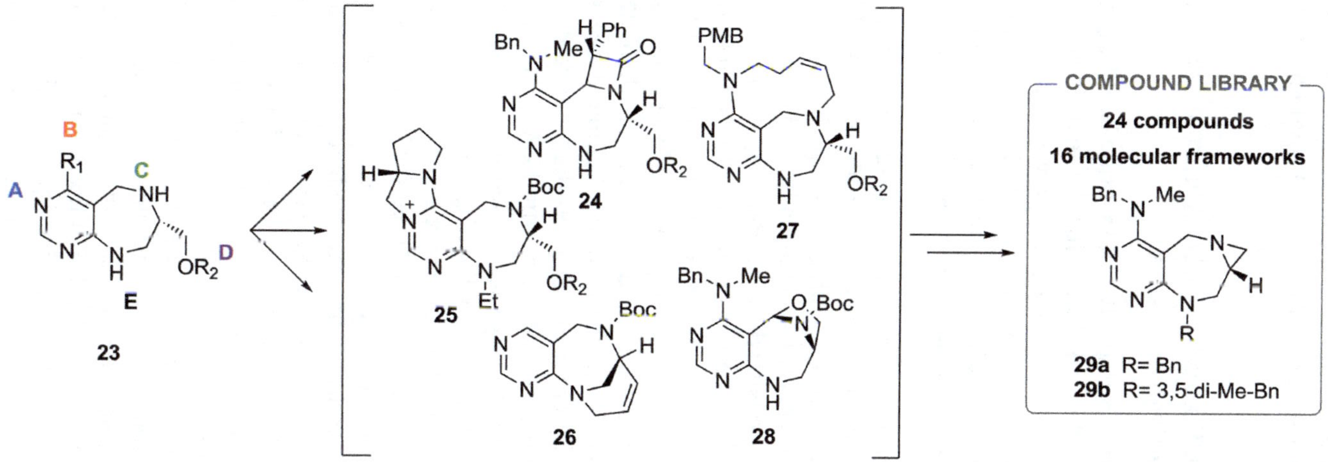

Scheme 2.5 The pDOS strategy utilised by Kim *et al.*[99] using pyrimidodiazepine and pyrimidine core scaffolds. Five reactive sites were added to the cores, allowing functional group pairings to yield 16 distinct compounds (selected examples shown)—two of which proved to inhibit the LRS–RagD PPI.

29a acting *via* an alternative mode of action. Finally western blot, live-cell imaging and cell proliferation assays were used to confirm that **29a** is capable of inducing autophagy and a reduction in proliferation, even in the presence of leucine, as a result of inhibition of the LRS–RagD PPI.[99]

2.2.2.2 A Single DOS Library Yields Multiple Chemical Tools for Multiple Biological Systems

Recent efforts by researchers at the Broad Institute have begun to investigate the application of a single DOS library for the identification of numerous chemical probes. Strikingly, this library has proven to be particularly fruitful for discovery, with several publications detailing the identification of biological hits from this collection in the seven years since the library's publication, including antiparasitic molecules[102] and probes which modulate autophagy.

A seminal report by Marcaurelle *et al.* in 2010[103] first described the methodology used to construct this complex library, describing an elegant strategy for the construction of a collection of medium-to-large sized rings with diverse stereochemical and skeletal features. Following a B–C–P algorithm, the build phase consisted of stereodivergent *syn*- and *anti*-aldol reactions, producing all four stereoisomers of a Boc-protected γ-amino acid (**30**), as well as the separate production of both stereoisomers of a protected alaninol derivative (**31**, Scheme 2.6). Joining of the two blocks *via* amide formation followed by reduction yielded the intermediate **32**. Importantly, the modular nature of the synthetic route facilitated access to all eight diastereomers of **32** and therefore of each resultant scaffold, enabling future stereochemical structure–activity relationship (SSAR) data to be generated.

Finally, a reagent-based strategy was applied in the pair phase, using nucleophilic aromatic substitution (S_NAr), Huisgen triazole formation and RCM chemistries to construct 24 8–14-membered-ring scaffolds stemming from five distinct and complex frameworks. Three further papers[104–106] from the Marcaurelle laboratory described additional extensions of this methodology, using the same stereochemically-rich intermediate and four alternative pairing partners in S_NAr and head to tail (H to T) cyclisations to generate a further five scaffolds using **32**. The innovative incorporation of suitable branching points, such as aromatic halides, aliphatic and aromatic amino functionalities (*via* nitro reduction), across the whole library resulted in the presence of multiple potential exit vectors within each scaffold and allowed facile library expansion. Importantly, the robust nature of the chemistry allowed multi-gram quantities of key intermediates and final scaffolds to be isolated. A combination of these factors enabled significant expansion of the library *via* solid-phase synthesis using combinatorial-type modifications to yield over 44 000 compounds as stereoisomers and analogues of the initial nine frameworks in a matrix-like fashion.

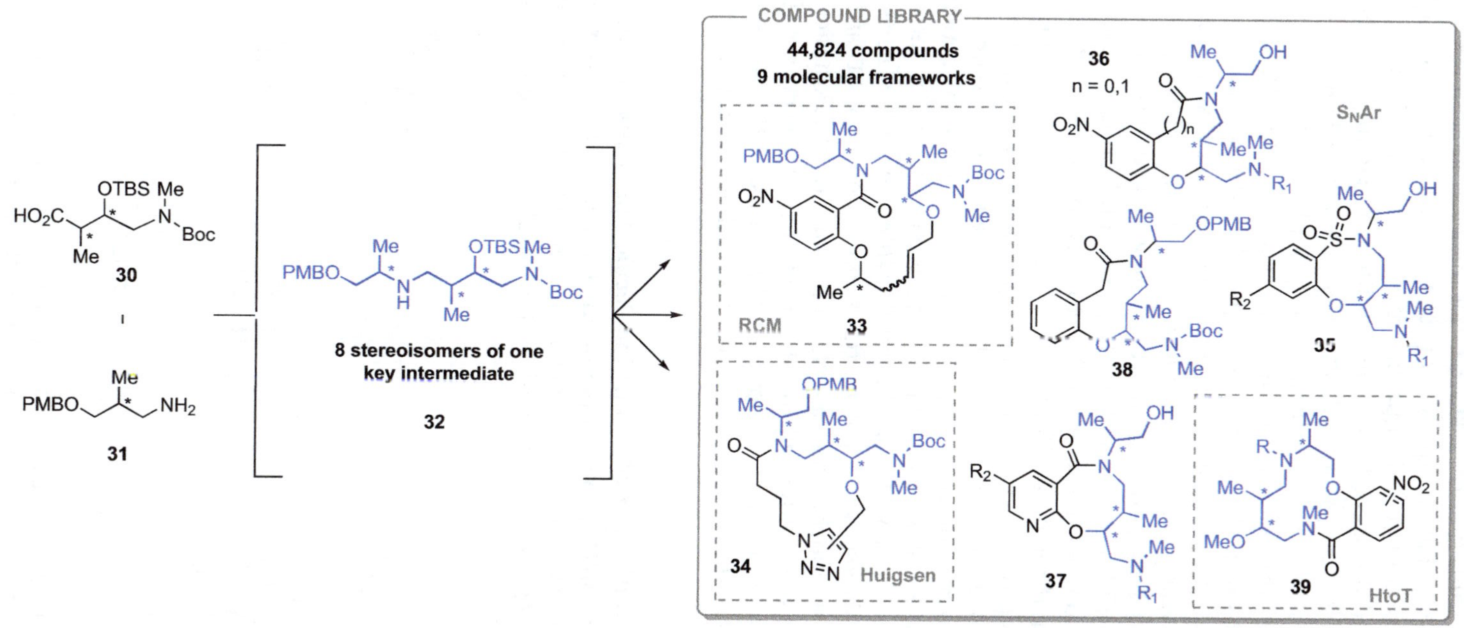

Scheme 2.6 The synthetic strategy employed by Macaurelle *et al.*,[103] Gerard *et al.*,[104] Fitzgerald *et al.*[105] and Chou *et al.*[106] for the formation of nine molecular frameworks, with subsequent library expansion yielding over 44 000 compounds.

2.2.2.2.1 Identification of a Novel Probe for VATPase Function. Aldrich *et al.*[107] investigated the application of the Marcaurelle DOS library for the identification of chemical probes for autophagy. This led to the identification of inhibitors of lysosomal acidification *via* high-content screening followed by target identification studies. BRD1240 (Scheme 2.7) showed low micromolar activity in an initial phenotypic screen detecting modulation of the number of autophagosomes within HeLa cells, with key differences in activity between all possible stereoisomers of the scaffold being noted; only two of the eight proved to be active. Further experiments investigated autophagosomal turnover, where it was found that BRD1240 and Bafilomycin A1 (a known vacuolar-type H^+–ATPase (V-ATPase) inhibitor) inhibited the turnover, indicating a possible mechanism of action. Additionally, BRD1240 was found to modulate lysosomal function through inhibiting lysosomal acidification and protease activity. Structure–activity relationship (SAR) data indicated that the 4-position nitrogen within the pyridine was vital for activity, whilst an electron-rich urea moiety was also required for optimal activity. Using cancer cell line sensitivity profiling and a comparison of BRD1240 and BafA1, the authors concluded that BRD1240 perturbs V-ATPase function. This hypothesis was validated by the suppression of V-ATPase function by BRD1240 in biochemical assays. Surprisingly, the kinetics of these biochemical assays indicated that BRD1240 may act *via* a novel mode of action compared with the known BafA1. Thus, BRD1240 serves as a novel probe for the investigation of lysosomal acidification *via* V-ATPase modulation.

2.2.2.2.2 Identification of a Small-molecule Modulator of Autophagy Independent of mTOR and Lysosomal Function. Compounds in the Marcaurelle library were also shown to yield novel hits that modulated autophagy; however, importantly, the results of studies into their

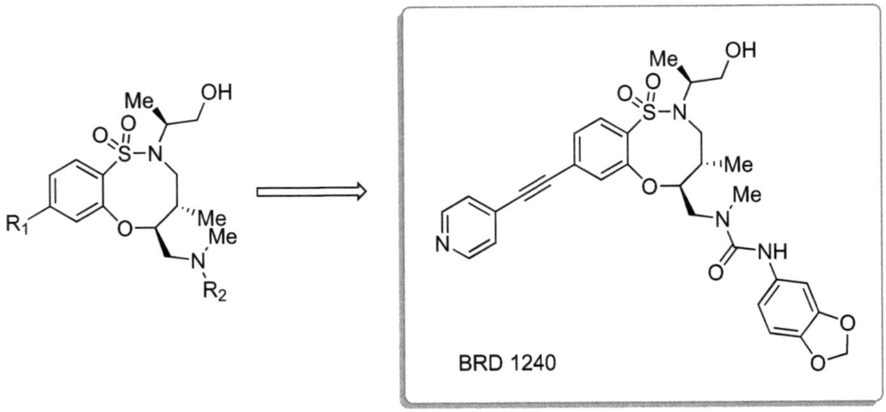

Scheme 2.7 The development of a V-ATPase probe. From initial hits, BRD1240 was chosen as the lead compound for further investigation.

mechanism of action indicated that these compounds in fact promoted autophagy and did not perturb mTOR signalling or lysosomal function. Thus, these small molecules can serve as orthogonal probes for autophagic processes. Kuo *et al.*[108] screened 59 541 DOS compounds for modulation of autophagosome number in HeLa cells. Five hits, which were shown to increase autophagesome number, were selected for further investigations where they were found to promote autophagy, whilst not disrupting mTOR signalling pathways or lysosomal function determined by western blot and protein phosphorylation experiments. Furthermore, the lead hit BRD5631 (Scheme 2.8) was also shown to modulate disease-associated phenotypes, including protein aggregation, cell survival, bacterial replication and inflammatory cytokine production as a result of autophagy activation. Whilst investigations into the precise mode of action and target of BRD5631 are ongoing, this molecule will continue to be useful for illuminating the biological relevance and therapeutic potential of promoting autophagy.

2.2.2.3 Application of DOS for Discovery of Novel Antimalarial Compounds

As part of a pilot investigation into the application of DOS molecules as antimalarials, a phenotypic screen of a DOS-derived library containing a subset of compounds from the Marcaurelle library against multidrug-resistant D2d *Plasmodium falciparum* asexual blood-stage parasites was undertaken.[109] Of the 8000 molecules screened, 560 displayed over 90% D2d growth inhibition. From the initial 560 promising results, 26 molecules exhibited particularly potent inhibition (over 50% at 280 nM), and 20 of these were macrolactam scaffolds synthesized in the Marcaurelle DOS campaign using RCM. Ultimately, compound **40** was identified as the most active in the screen (Scheme 2.9).

Due to the stereochemically comprehensive design and construction of the original DOS library, the stereochemical SAR (SSAR) of all 16 stereoisomers of **40** was easily determined, with the (2*S*,5*R*,6*R*,12*S*) stereoisomer being found to be most potent. Additional SAR of the peripheral substituents identified that switching the hydroxyl to a dimethylamino group increased solubility. These initial efforts led to the designation of ML238 as an antimalarial probe, which inhibits two strains of *P. falciparum* (D2d and 3D7) and exhibits low off-target cytotoxicity, as well as high stability in human plasma.

Taking ML238 as a lead compound, further medicinal chemistry optimization was undertaken to remove unfavourable properties such as potential cardiotoxicity [Human ether-a-go-go-related gene (hERG) product binding], poor stability in microsomes of a mouse model and low phosphate buffered saline (PBS) solubility, factors which could lead to low bioavailability or very high doses in any future therapeutic application. The short,

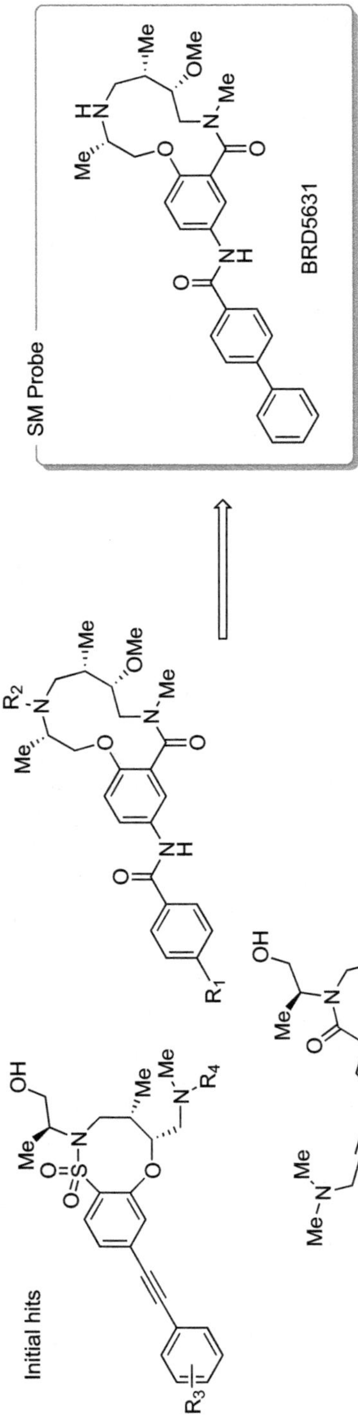

Scheme 2.8 From five prioritised hit compounds, BRD5631 was chosen as the lead compound for further investigation.

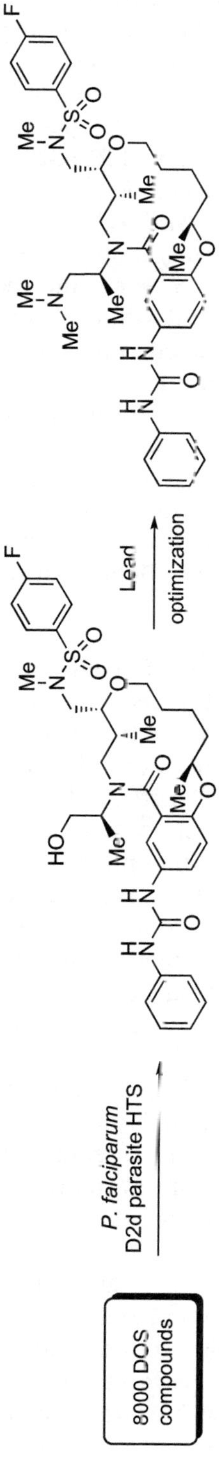

Scheme 2.9 HTS of DOS library yields novel antimalarial probe ML238.

modular synthesis of this class of macrolactams enabled further facile optimization, due to the flexibility of the DOS strategy to modify and evaluate nearly every position of the core ring, including ring size and effects of heteroatom exchange, simply by exchanging building blocks used at the outset of the route.[110] The application of appendage and skeletal SAR ultimately yielded an improved analogue, BRD6323, which was studied *in vivo* and further investigated for its mechanism of action (Scheme 2.10).

In initial experiments ML238 and BRD6323 proved effective against parasite isolates with drug-resistant genotypes, indicating a different mode of action to current treatments. To determine the cellular target of these molecules, resistance selection and whole-genome sequencing were performed on resistant strains, ultimately revealing mutations occurring solely in the *cytb* locus over a number of BRD6323-resistant lines.[111] It was further determined that BRD6323 and ML238 inhibit the ubiquinone reductase site (Q_i) of cytochrome b in *P. falciparum*. When used in concert with known cytochrome b ubiquinol oxidase site (Q_o) inhibitors like atovaquone, a synergistic effect was observed. These observations illuminated a new potential combination therapy strategy due to dual inhibition of an enzyme at two sites, which may prove useful in the fight against malarial infection in the future.

2.2.2.3.1 DOS for the Discovery of Azetidine-based Antimalarials and Identification of Novel Cellular Targets. Encouraged by the results of the pilot-scale HTS of a DOS library yielding new leads with important biological activity and underexplored modes of action, an expanded antimalarial screen of the complete Broad Institute DOS library, comprising nearly 100 000 molecules, was undertaken.[112] Included in this library were a collection of highly substituted azetidine molecules. The azetidine library was constructed based on an ephedrine-like scaffold.[113] All possible stereoisomers of an aryl-containing *N*-allyl amino diol were constructed (Scheme 2.11). Subsequent transformations led to the diastereoselective formation of azetidines, which were elaborated *via* a reagent-based approach to generate a series of bicyclic compounds. This methodology yielded several distinct scaffold classes of bicyclic azetidine molecules, including bridged bicyclic, monoketopiperazine, azocane and azaspirocyclic scaffolds, representing seven molecular scaffolds and over 2000 compounds after a combinatorial-type solid-phase synthesis effort. The features of this library included comprehensive stereochemical information for each scaffold and, as a result, high three-dimensionality of the molecules, as well as high scaffold diversity of the library.

Similar to the pilot-scale study, these and other molecules were screened against *P. falciparum* D2d cells in a phenotypic blood-stage parasite growth-inhibition assay. In this study, counter-screens against drug-resistant clones and parasite isolates were performed to determine if the compounds acted according to a known mechanism of action, in order to prioritize compounds displaying novel modes of action. Additional assays against

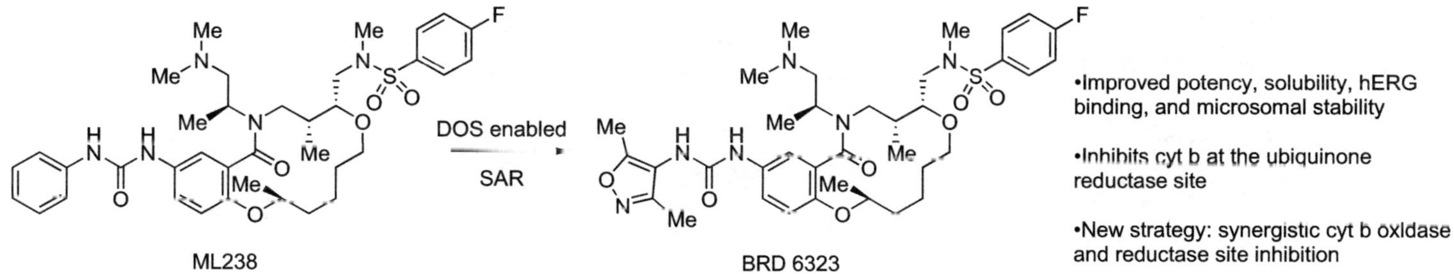

Scheme 2.10 DOS-enabled SAR yields improved antimalarial probe BRD6323.

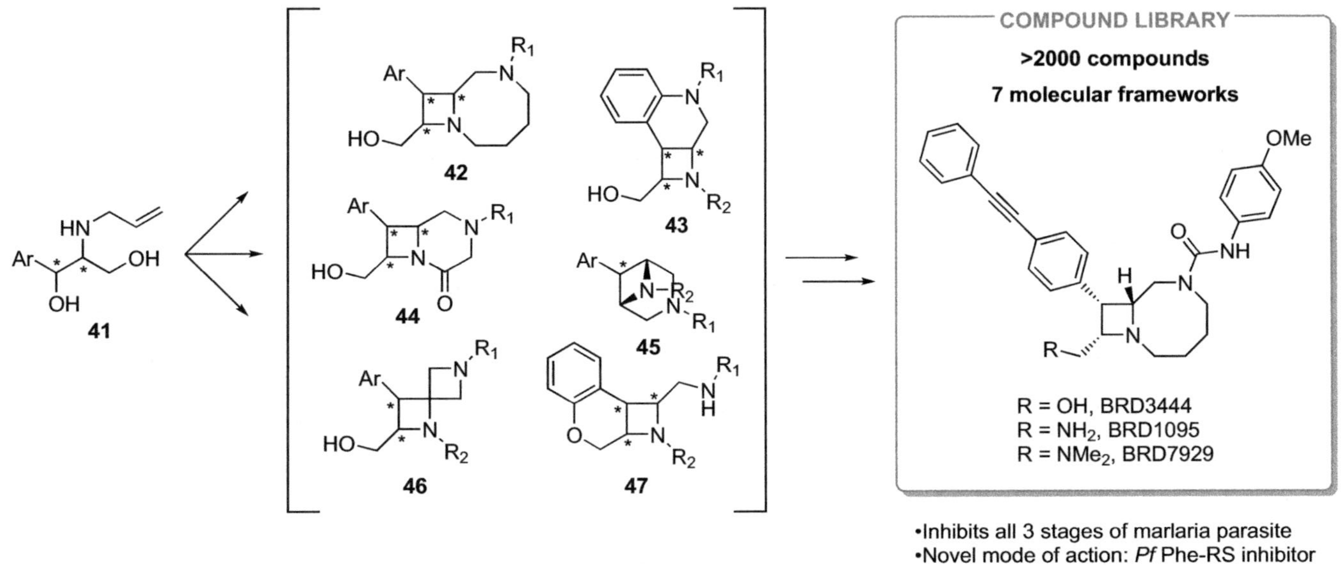

Scheme 2.11 DOS of azetidine compounds leads to discovery of a novel multistage inhibitor of malaria.

liver-stage and transmission-stage parasites identified four compound series that exhibited multistage inhibition. Of these four series, three represented new scaffolds against known targets, and one (BRD3444, Scheme 2.11) was found to inhibit *P. falciparum* through an unknown target. Medicinal chemistry optimization of BRD3444 using the modular synthetic approach developed in the original DOS led to analogues with improved solubility, bioavailability and potency, resulting in molecules BRD7929, with a dimethylamino group replacing a hydroxymethyl group in the original structure, and BRD1095, with a methylamino group replacement.

Using these molecules, resistance selection and whole-genome sequencing of the resistant clones were performed, and analysis predicted the inhibition of cytosolic phenylalanine-tRNA synthetase (*Pf* Phe-RS). This target was confirmed by experiments with purified recombinant protein. Inhibition of this target results in the elimination of asexual blood-, liver- and transmission-stage parasites, a unique effect among antimalarial drugs, with demonstrated *in vivo* efficacy in mouse malaria models. When applied to mouse models, it was demonstrated that treatment with BRD7929 prevents transmission, ensures prophylaxis and provides single-dose cures for malaria infections.

2.3 Conclusions and Outlook

The case studies presented herein demonstrate the ability of DOS-derived molecules to accelerate discovery by generating hits and facilitating derivatisation, to serve as probes to enable new biological insights and to generate lead compounds to address some of the most pressing problems in medicine. The molecules generated using DOS possess complexity and diversity that is reminiscent of natural products whilst maintaining the synthetic tractability of simpler, drug-like compounds. These features have enabled the rapid construction of diverse libraries which, when combined with modern methods for high-throughput screening, present tremendous opportunities for new discoveries in chemistry, biology and medicine. Future efforts towards these goals will be aided by developments in synthetic methodology and strategy, enabling access to new areas of chemical space and increases in synthetic efficiency. As we have shown, DOS libraries access vast, often untapped, potential for finding novel bioactive molecules, often beyond the uses for which the molecules were originally envisioned. Many of the examples discussed herein demonstrate the importance of follow-up studies beyond initial chemical probe discovery, and can lead to novel areas of biological inquiry or reveal new targets or modes of action. Therefore, the continued screening and biological evaluation of existing and new DOS libraries will considerably increase our biological understanding and therapeutic application of these molecules.

Acknowledgements

The Spring lab acknowledges support from AstraZeneca, the European Union, the Engineering and Physical Sciences Research Council (EP/P020291/1) and the Royal Society.

References

1. S. L. Schreiber, J. D. Kotz, M. Li, J. Aubé, C. P. Austin, J. C. Reed, H. Rosen, E. L. White, L. A. Sklar, C. W. Lindsley, B. R. Alexander, J. A. Bittker, P. A. Clemons, A. De Souza, M. A. Foley, M. Palmer, A. F. Shamji, M. J. Wawer, O. McManus, M. Wu, B. Zou, H. Yu, J. E. Golden, F. J. Schoenen, A. Simeonov, A. Jadhav, M. R. Jackson, A. B. Pinkerton, T. D. Y. Chung, P. R. Griffin, B. F. Cravatt, P. S. Hodder, W. R. Roush, E. Roberts, D. H. Chung, C. B. Jonsson, J. W. Noah, W. E. Severson, S. Ananthan, B. Edwards, T. I. Oprea, P. J. Conn, C. R. Hopkins, M. R. Wood, S. R. Stauffer and K. A. Emmitte, *Cell*, 2015, **161**, 1252–1265.
2. C. J. O' Connor, L. Laraia and D. R. Spring, *Chem. Soc. Rev.*, 2011, **40**, 4332.
3. S. Fox, S. Farr-Jones, L. Sopchak, A. Boggs, H. W. Nicely, R. Khoury and M. Biros, *J. Biomol. Screening*, 2006, **11**, 864–869.
4. K. H. Bleicher, H.-J. Böhm, K. Müller and A. I. Alanine, *Nat. Rev. Drug Discovery*, 2003, 2, 369–378.
5. D. J. Payne, M. N. Gwynn, D. J. Holmes and D. L. Pompliano, *Nat. Rev. Drug Discovery*, 2007, **6**, 29–40.
6. K. M. G. O'connell, W. R. J. D. Galloway and D. R. Spring, in *Diversity-Oriented Synthesis:Basics and Applications in Organic Synthesis, Drug Discovery and Chemical Biology*, ed. A. Trabocchi, John Wiley & Sons, Inc, Hoboken, NJ, 2013, pp. 253–287.
7. W. R. J. D. Galloway, J. E. Stokes and D. Spring, in *Small Molecule Medicinal Chemistry: Strategies and Technologies*, ed. W. Czechtizky and P. Hamley, John Wiley & Sons, Inc., 2016, pp. 79–101.
8. W. H. B. Sauer and M. K. Schwarz, *J. Chem. Inf. Comput. Sci.*, 2003, **43**, 987–1003.
9. W. R. J. D. Galloway and D. R. Spring, *Diversity-Oriented Synth.*, 2013, **1**, 21–28.
10. C. J. O' Connor, H. S. G. Beckmann and D. R. Spring, *Chem. Soc. Rev.*, 2012, **41**, 4444–4456.
11. M. Rask-Andersen, M. S. Almén and H. B. Schiöth, *Nat. Rev. Drug Discovery*, 2011, **10**, 579–590.
12. A. M. Edwards, R. Isserlin, G. D. Bader, S. V. Frye, T. M. Willson and F. H. Yu, *Nature*, 2011, **470**, 163–165.
13. C. Lipinski and A. Hopkins, *Nature*, 2004, **432**, 855–861.
14. C. A. Lipinski, F. Lombardo, B. W. Dominy and P. J. Feeney, *Adv. Drug Delivery Rev.*, 2012, **64**, 4–17.

15. P. D. Leeson and B. Springthorpe, *Nat. Rev. Drug Discovery*, 2007, **6**, 881–890.
16. I. Kola and J. Landis, *Nat. Rev. Drug Discovery*, 2004, **3**, 711–716.
17. M. J. Waring, J. Arrowsmith, A. R. Leach, P. D. Leeson, S. Mandrell, R. M. Owen, G. Pairaudeau, W. D. Pennie, S. D. Pickett, J. Wang, O. Wallace and A. Weir, *Nat. Rev. Drug Discovery*, 2015, **14**, 475–486.
18. D. Cook, D. Brown, R. Alexander, R. March, P. Morgan, G. Satterthwaite and M. N. Pangalos, *Nat. Rev. Drug Discovery*, 2014, **13**, 419–431.
19. F. Lovering, J. Bikker and C. Humblet, *J. Med. Chem.*, 2009, **52**, 6752–6756.
20. F. Lovering, *Med. Chem. Commun.*, 2013, **4**, 515–519.
21. W. P. Walters, J. Green, J. R. Weiss and M. A. Murcko, *J. Med. Chem.*, 2011, **54**, 6405–6416.
22. M. M. Hann, A. R. Leach and G. Harper, *J. Chem. Inf. Comput. Sci.*, 2001, **41**, 856–864.
23. A. R. Leach and M. M. Hann, *Curr. Opin. Chem. Biol.*, 2011, **15**, 489–496.
24. F. López-Vallejo, M. A. Giulianotti, R. A. Houghten and J. L. Medina-Franco, *Drug Discovery Today*, 2012, **17**, 718–726.
25. T. J. Ritchie and S. J. F. Macdonald, *Drug Discovery Today*, 2009, **14**, 1011–1020.
26. T. J. Ritchie, S. J. F. MacDonald, R. J. Young and S. D. Pickett, *Drug Discovery Today*, 2011, **16**, 164–171.
27. T. J. Ritchie, S. J. F. Macdonald, S. Peace, S. D. Pickett and C. N. Luscombe, *Med. Chem. Commun.*, 2012, **3**, 1062.
28. T. J. Ritchie, S. J. F. Macdonald, S. Peace, S. D. Pickett and C. N. Luscombe, *MedChemComm*, 2013, **4**, 673.
29. P. D. Leeson and A. M. Davis, *J. Med. Chem.*, 2004, **47**, 6338–6348.
30. A A Shelat and R. K. Guy, *Nat. Chem. Biol.*, 2007, **3**, 442–446.
31. J. D. Hughes, J. Blagg, D. A. Price, S. Bailey, G. A. DeCrescenzo, R. V. Devraj, E. Ellsworth, Y. M. Fobian, M. E. Gibbs, R. W. Gilles, N. Greene, E. Huang, T. Krieger-Burke, J. Loesel, T. Wager, L. Whiteley and Y. Zhang, *Bioorg. Med. Chem. Lett.*, 2008, **18**, 4872–4875.
32. C. J. Gerry and S. L Schreiber, *Nat. Rev. Drug Discovery*, 2018, **17**, 333–352.
33. W. R. J. D. Galloway, A. Isidro-Llobet and D. R. Spring, *Nat. Commun.*, 2010, **1**, 80.
34. O. Méndez-Lucio and J. L. Medina-Franco, *Drug Discovery Today*, 2017, **22**, 120–126.
35. P. A. Clemons, N. E. Bodycombe, H. A. Carrinski, J. A. Wilson, A. F. Shamji, B. K. Wagner, A. N. Koehler and S. L. Schreiber, *Proc. Natl. Acad. Sci. U. S. A.*, 2010, **107**, 18787–18792.
36. R. Bohacek and C. McMartin, *Med. Res.*, 1996, **16**, 3–50.
37. Z. L. Deng, C. X. Du, X. Li, B. Hu, Z. K. Kuang, R. Wang, S. Y. Feng, H. Y. Zhang and D. X. Kong, *J. Chem. Inf. Model.*, 2013, **53**, 2820–2828.

38. A. Schuffenhauer, N. Brown, P. Selzer, P. Ertl and E. Jacoby, *J. Chem. Inf. Model.*, 2006, **46**, 525–535.
39. A. L. Harvey, *Drug Discovery Today*, 2008, **13**, 894–901.
40. T. Rodrigues, D. Reker, P. Schneider and G. Schneider, *Nat. Chem.*, 2016, **8**, 531–541.
41. S. Wetzel, R. S. Bon, K. Kumar and H. Waldmann, *Angew. Chem. - Int. Ed.*, 2011, **50**, 10800–10826.
42. R. S. Bon and H. Waldmann, *Acc. Chem. Res.*, 2010, **43**, 1103–1114.
43. H. Van Hattum and H. Waldmann, *J. Am. Chem. Soc.*, 2014, **136**, 11853–11859.
44. K. Grabowski and G. Schneider, *Curr. Chem. Biol.*, 2007, **1**, 115–127.
45. D. J. Newman and G. M. Cragg, *J. Nat. Prod.*, 2016, **79**, 629–661.
46. R. J. Spandl, A. Bender and D. R. Spring, *Org. Biomol. Chem.*, 2008, **6**, 1149–1158.
47. I. S. Young and M. A. Kerr, *J. Am. Chem. Soc.*, 2007, **129**, 1465–1469.
48. K. C. Nicolaou, C. R. H. Hale, C. Nilewski and H. A. Ioannidou, *Chem. Soc. Rev.*, 2012, **41**, 5185.
49. S. Basu, B. Ellinger, S. Rizzo, C. Deraeve, M. Schürmann, H. Preut, H.-D. Arndt and H. Waldmann, *Proc. Natl. Acad. Sci. U. S. A.*, 2011, **108**, 6805–6810.
50. H. van Hattum and H. Waldmann, *J. Am. Chem. Soc.*, 2014, **136**, 11853–11859.
51. A. Nören-Müller, I. Reis-Corrêa, H. Prinz, C. Rosenbaum, K. Saxena, H. J. Schwalbe, D. Vestweber, G. Cagna, S. Schunk, O. Schwarz, H. Schiewe and H. Waldmann, *Proc. Natl. Acad. Sci. U. S. A.*, 2006, **103**, 10606–10611.
52. A. I. Chan, L. M. McGregor and D. R. Liu, *Curr. Opin. Chem. Biol.*, 2015, **26**, 55–61.
53. R. M. Franzini, D. Neri and J. Scheuermann, *Acc. Chem. Res.*, 2014, **47**, 1247–1255.
54. R. E. Kleiner, C. E. Dumelin, G. C. Tiu, K. Sakurai and D. R. Liu, *J. Am. Chem. Soc.*, 2010, **132**, 11779–11791.
55. Z. J. Gartner, B. N. Tse, R. Grubina, J. B. Doyon, T. M. Snyder and D. R. Liu, *Science*, 2004, **305**, 1601–1605.
56. D. Morton, S. Leach, C. Cordier, S. Warriner and A. Nelson, *Angew. Chem. Ind. Ed.*, 2009, **48**, 104–109.
57. D. Lee, J. K. Sello and S. L. Schreiber, *Org. Lett.*, 2000, **2**, 709–712.
58. S. L. Schreiber, *Science*, 2000, **287**, 1964–1969.
59. W. R. J. D. Galloway, A. Bender, M. Welch and D. R. Spring, *Chem. Commun.*, 2009, 2446–2462.
60. T. E. Nielsen and S. L. Schreiber, *Angew. Chem. - Int. Ed.*, 2008, **47**, 48–56.
61. M. Schirle and J. L. Jenkins, *Drug Discovery Today*, 2016, **21**, 82–89.
62. W. Zheng, N. Thorne and J. C. McKew, *Drug Discovery Today*, 2013, **18**, 1067–1073.
63. J. A. Lee, M. T. Uhlik, C. M. Moxham, D. Tomandl and D. J. Sall, *J. Med. Chem.*, 2012, **55**, 4527–4538.

64. D. C. Swinney and J. Anthony, *Nat. Rev. Drug Discovery*, 2011, **10**, 507–519.
65. D. R. Spring, *Chem. Soc. Rev.*, 2005, **34**, 472–482.
66. M. Hansson, J. Pemberton, O. Engkvist, I. Feierberg, L. Brive, P. Jarvis, L. Zander-Balderud and H. Chen, *J. Biomol. Screening*, 2013, **19**, 727–737.
67. E. Wyatt, W. Galloway, G. Thomas, M. Welch, O. Loiseleur, A. Plowright and D. Spring, *Chem. Commun.*, 2008, **40**, 4962–4964.
68. G. L. Thomas, R. J. Spandl, F. G. Glansdorp, M. Welch, A. Bender, J. Cockfield, J. A. Lindsay, C. Bryant, D. F. J. Brown, O. Loiseleur, H. Rudyk, M. Ladlow and D. R. Spring, *Angew. Chem. - Int. Ed.*, 2008, **47**, 2808–2812.
69. A. Robinson, G. L. Thomas, R. J. Spandl, M. Welch and D. R. Spring, *Org. Biomol. Chem.*, 2008, **6**, 2973–2981.
70. E. Lenci, R. Innocenti, A. Biagioni, G. Menchi, F. Bianchini and A. Trabocchi, *Molecules*, 2016, **21**, 1405.
71. L. Li, Z.-L. Li, F.-L. Wang, Z. Guo, Y.-F. Cheng, N. Wang, X.-W. Dong, C. Fang, J. Liu, C. Hou, B. Tan and X.-Y. Liu, *Nat. Commun.*, 2016, **7**, 13852.
72. S. Oh, H. J. Nam, J. Park, S. H. Beak and S. B. Park, *ChemMedChem*, 2010, **5**, 529–533.
73. A. M. Socha, N. Y. Tan, K. L. Laplante and J. K. Sello, *Bioorg. Med. Chem.*, 2010, **18**, 7193–7202.
74. E. E. Wyatt, S. Fergus, W. R. J. D. Galloway, A. Bender, D. J. Fox, A. T. Plowright, A. S. Jessiman, M. Welch and D. R. Spring, *Chem. Commun.*, 2006, 3296–3298.
75. P. C. L. Moore and J. A. Lindsay, *J. Med. Microbiol.*, 2002, **51**, 516–521.
76. M. R. Brown and A. W. Smith, *J. Antimicrob. Chemother.*, 2001, **48**, 141–142.
77. N. V. Kovalevskaya, E. D. Smurnyi, B. Birdsall, J. Feeney and V. I. Polshakov, *Pharm. Chem. J.*, 2007, **41**, 350–353.
78. K. Bowden, N. V. Harris and C. A. Watson, *J Chemother.*, 1993, **5**, 377–388.
79. S. Hoelder, P. A. Clarke and P. Workman, *Mol. Oncol.*, 2012, **6**, 155–176.
80. V. Lavanya, A. A. Mohamed Adil, N. Ahmed, A. K. Rishi and S. Jamal, *Integr. Cancer Sci. Ther.*, 2014, **1**, 39–46.
81. J. Zhang, P. Yang and N. Gray, *Nat. Rev. Cancer*, 2009, **9**, 28–39.
82. J. L. Adams, J. Smothers, R. Srinivasan and A. Hoos, *Nat. Rev. Drug Discovery*, 2015, **14**, 603–622.
83. I. Collins and A. M. Jones, *Molecules*, 2014, 17221–17255.
84. B. M. Ibbeson, L. Laraia, E. Alza, C. J. O' Connor, Y. S. Tan, H. M. L. Davies, G. McKenzie, A. R. , Venkitaraman and D. R. Spring, *Nat. Commun.*, 2014, **5**, 3155.
85. K. N. Niforou, A. K. Anagnostopoulos, K. Vougas, C. Kittas, V. G. Gorgoulis and G. T. Tsangaris, *Cancer Genomics Proteomics*, 2008, **5**, 63–77.

86. L. D. S. Yadav, V. P. Srivastava, V. K. Rai and R. Patel, *Tetrahedron*, 2008, **64**, 4246–4253.
87. H. Kubota, J. Lim, K. M. Depew and S. L. Schreiber, *Chem. Biol.*, 2002, **9**, 265–276.
88. A. Aravind, P. S. Kumar, M. G. Sankar and S. Baskaran, *Eur. J. Org. Chem.*, 2011, 6980–6988.
89. S. Senthilkumar, S. S. Prasad, P. S. Kumar and S. Baskaran, *Chem. Commun.*, 2014, **50**, 1549–1551.
90. N. Asano, R. J. Nash, R. J. Molyneux and G. W. J. Fleet, *Tetrahedron: Asymmetry*, 2000, **11**, 1645–1680.
91. N. Asano, T. Yamauchi, K. Kagamifuchi, N. Shimizu, S. Takahashi, H. Takatsuka, K. Ikeda, H. Kizu, W. Chuakul, A. Kettawan and T. Okamoto, *J. Nat. Prod.*, 2005, **68**, 1238–1242.
92. Y. Hu, K. Wang and J. B. MacMillan, *Org. Lett.*, 2013, **15**, 390–393.
93. E. Lenci, G. Menchi, A. Guarna and A. Trabocchi, *J. Org. Chem.*, 2015, **80**, 2182–2191.
94. E. M. Sánchez-Fernández, R. Gonçalves-Pereira, R. Rísquez-Cuadro, G. B. Plata, J. M. Padrón, J. M. García Fernández and C. Ortiz Mellet, *Carbohydr. Res.*, 2016, **429**, 113–122.
95. E. M. Sánchez-Fernández, R. Rísquez-Cuadro, M. Chasseraud, A. Ahidouch, C. Ortiz Mellet, H. Ouadid-Ahidouch and J. M. García Fernández, *Chem. Commun.*, 2010, **46**, 5328–5330.
96. C. H. Arrowsmith, J. E. Audia, C. Austin, J. Baell, J. Bennett, J. Blagg, C. Bountra, P. E. Brennan, P. J. Brown, M. E. Bunnage, C. Buser-Doepner, R. M. Campbell, A. J. Carter, P. Cohen, R. A. Copeland, B. Cravatt, J. L. Dahlin, D. Dhanak, A. M. Edwards, S. V. Frye, N. Gray, C. E. Grimshaw, D. Hepworth, T. Howe, K. V. M. Huber, J. Jin, S. Knapp, J. D. Kotz, R. G. Kruger, D. Lowe, M. M. Mader, B. Marsden, A. Mueller-Fahrnow, S. Müller, R. C. O'Hagan, J. P. Overington, D. R. Owen, S. H. Rosenberg, B. Roth, R. Ross, M. Schapira, S. L. Schreiber, B. Shoichet, M. Sundström, G. Superti-Furga, J. Taunton, L. Toledo-Sherman, C. Walpole, M. A. Walters, T. M. Willson, P. Workman, R. N. Young and W. J. Zuercher, *Nat. Chem. Biol.*, 2015, **11**, 536–541.
97. J. A. Zorn and J. A. Wells, *Nat. Chem. Biol.*, 2010, **6**, 179–188.
98. D. E. Scott, A. R. Bayly, C. Abell and J. Skidmore, *Nat. Rev. Drug Discovery*, 2016, **15**, 533–550.
99. J. Kim, J. Jung, J. Koo, W. Cho, W. S. Lee, C. Kim, W. Park and S. B. Park, *Nat. Commun.*, 2016, **7**, 13196.
100. J. Kim, H. Kim and S. B. Park, *J. Am. Chem. Soc.*, 2014, **136**, 14629–14638.
101. J. M. Han, S. J. Jeong, M. C. Park, G. Kim, N. H. Kwon, H. K. Kim, S. H. Ha, S. H. Ryu and S. Kim, *Cell*, 2012, **149**, 410–424.
102. S. Dandapani, A. R. Germain, I. Jewett, S. le Quement, J.-C. Marie, G. Muncipinto, J. R. Duvall, L. C. Carmody, J. R. Perez, J. C. Engel, J. Gut, D. Kellar, J. Lage Siqueira-Neto, J. H. McKerrow, M. Kaiser,

A. Rodriguez, M. A. Palmer, M. Foley, S. L. Schreiber and B. Munoz, *ACS Med. Chem. Lett.*, 2014, **5**, 149–153.

103. L. A. Marcaurelle, E. Comer, S. Dandapani, J. R. Duvall, B. Gerard, S. Kesavan, M. D. Lee, H. Liu, J. T. Lowe, J. C. Marie, C. A. Mulrooney, B. A. Pandya, A. Rowley, T. D. Ryba, B. C. Suh, J. Wei, D. W. Young, L. B. Akella, N. T. Ross, Y. L. Zhang, D. M. Fass, S. A. Reis, W. N. Zhao, S. J. Haggarty, M. Palmer and M. A. Foley, *J. Am. Chem. Soc.*, 2010, **132**, 16962–16976.
104. B. Gerard, J. R. Duvall, J. T. Lowe, T. Murillo, J. Wei, L. B. Akella and L. A. Marcaurelle, *ACS Comb. Sci.*, 2011, **13**, 365–374.
105. M. E. Fitzgerald, C. A. Mulrooney, J. R. Duvall, J. Wei, B. C. Suh, L. B. Akella, A. Vrcic and L. A. Marcaurelle, *ACS Comb. Sci.*, 2012, **14**, 89–96.
106. D. H. C. Chou, J. R. Duvall, B. Gerard, H. Liu, B. A. Pandya, B. C. Suh, E. M. Forbeck, P. Faloon, B. K. Wagner and L. A. Marcaurelle, *ACS Med. Chem. Lett.*, 2011, **2**, 698–702.
107. L. N. Aldrich, S. Y. Kuo, A. B. Castoreno, G. Goel, P. Kuballa, M. G. Rees, B. A. Seashore-Ludlow, J. H. Cheah, I. J. Latorre, S. L. Schreiber, A. F. Shamji and R. J. Xavier, *J. Am. Chem. Soc.*, 2015, **137**, 5563–5568.
108. S.-Y. Kuo, A. B. Castoreno, L. N. Aldrich, K. G. Lassen, G. Goel, V. Dančík, P. Kuballa, I. Latorre, K. L. Conway, S. Sarkar, D. Maetzel, R. Jaenisch, P. A. Clemons, S. L. Schreiber, A. F. Shamji, R. J. Xavier, M. A. Fischbach and E. Latz, *Proc. Natl. Acad. Sci. U. S. A.*, 2015, **112**, E4281–E4287.
109. R. W. Heidebrecht, C. Mulrooney, C. P. Austin, R. H. Barker, J. A. Beaudoin, K. C. Cheng, E. Comer, S. Dandapani, J. Dick, J. R. Duvall, E. H. Ekland, D. A. Fidock, M. E. Fitzgerald, M. Foley, R. Guha, P. Hinkson, M. Kramer, A. K. Lukens, D. Masi, L. A. Marcaurelle, X. Su, C. J. Thomas, M. We, R. C. Wiegand, D. Wirth, M. Xia, J. Yuan, J. Zhao, M. Palmer, B. Munoz and S. Schreiber, *ACS Med. Chem. Lett.*, 2012, **3**, 112–117.
110. E. Comer, J. A. Beaudoin, N. Kato, M. E. Fitzgerald, R. W. Heidebrecht, M. D. Lee, D. Masi, M. Mercier, C. Mulrooney, G. Muncipinto, A. Rowley, K. Crespo-Llado, A. E. Serrano, A. K. Lukens, R. C. Wiegand, D. F. Wirth, M. A. Palmer, M. A. Foley, B. Munoz, C. A. Scherer, J. R. Duvall and S. L. Schreiber, *J. Med. Chem.*, 2014, **57**, 8496–8502.
111. A. K. Lukens, R. W. Heidebrecht, C. Mulrooney, J. A. Beaudoin, E. Comer, J. R. Duvall, M. E. Fitzgerald, D. Masi, K. Galinsky, C. A. Scherer, M. Palmer, B. Munoz, M. Foley, S. L. Schreiber, R. C. Wiegand and D. F. Wirth, *J. Infect. Dis.*, 2015, **211**, 1097–1103.
112. N. Kato, E. Comer, T. Sakata-Kato, A. Sharma, M. Sharma, M. Maetani, J. Bastien, N. M. Brancucci, J. A. Bittker, V. Corey, D. Clarke, E. R. Derbyshire, G. Dornan, S. Duffy, S. Eckley, M. A. Itoe, K. M. J. Koolen, T. A. Lewis, P. S. Lui, A. K. Lukens, E. Lund, S. March, E. Meibalan, B. C. Meier, J. McPhail, B. Mitasev, E. L. Moss, M. Sayes,

Y. VanGessel, M. J. Wawer, T. Yoshinaga, A.-M. Zeeman, V. M. Avery, S. N. Bhatia, J. E. Burke, F. Catteruccia, J. C. Clardy, P. A. Clemons, K. J. Dechering, J. R. Duvall, M. A. Foley, F. Gusovsky, C. H. M. Kocken, M. Marti, M. L. Morningstar, B. Munoz, D. E. Neafsey, A. Sharma, E. A. Winzeler, D. F. Wirth, C. A. Scherer and S. L. Schreiber, *Nature*, 2016, **538**, 344–349.

113. J. T. Lowe, M. D. Lee, L. B. Akella, E. Davoine, E. J. Donckele, L. Durak, J. R. Duvall, B. Gerard, E. B. Holson, A. Joliton, S. Kesavan, B. C. Lemercier, H. Liu, J. C. Marié, C. A. Mulrooney, G. Muncipinto, M. Welzel-O'Shea, L. M. Panko, A. Rowley, B. C. Suh, M. Thomas, F. F. Wagner, J. Wei, M. A. Foley and L. A. Marcaurelle, *J. Org. Chem.*, 2012, **77**, 7187–7211.

CHAPTER 3

Biology-oriented Synthesis

GEORGE KARAGEORGIS AND HERBERT WALDMANN*

Max-Planck-Institute for Molecular Physiology, Department of Chemical Biology, Otto-Hahn str. 11, 44227 Dortmund, Germany
*Email: herbert.waldmann@mpi-dortmund.mpg.de

3.1 Introduction

Small molecules lie at the core of chemical biology research as they are excellent tools for the analysis of complex biological mechanisms. Unlike genetic approaches, small molecules have acute, not chronic, effects which can be conditional and tuned, *e.g.* by dosage, and work reversibly and rapidly.[1] Thus, the development of selective small-molecule probes for proteins which mediate biological processes is of high value for studying and understanding biological phenomena. Current estimates suggest that drug-like chemical space may accommodate *ca.* 10^{60} small molecules,[2] rendering its investigation simply by means of synthesis unfeasible. Thus, there exists the need to selectively pin-point biologically relevant subspaces and identify a map which will allow one to navigate only that fraction of chemical space in the search of bioactive small molecules.

Nature has already addressed this issue by conserving the structural and chemical properties of protein binding sites during their evolution. For a given protein of an average size of 300 amino acids, synthesized from 20 different amino acids, there are more than 10^{390} combinations possible.[2] However, even the complicated human genome has been estimated to encode only 25 000 proteins, containing subdomains which are highly conserved within protein classes. This restriction may be attributed to selection-driven protein formation under physical and chemical evolutionary pressure, which

Chemical Biology No. 10
Chemical and Biological Synthesis: Enabling Approaches for Understanding Biology
Edited by Nick J. Westwood and Adam Nelson

Published by the Royal Society of Chemistry, www.rsc.org

current insights suggest is mostly based on thermodynamic stability rather than function.[3,4] The three-dimensional structure of proteins is also determined by the arrangement of secondary structures, such as α helices and β sheets, which, in turn, result in characteristic fold classes of the individual protein domains together forming the whole protein. Subfolds determine the shape and size of ligand-binding sites and the chemical structure of amino acid residue side chains determines the kind of ligand that can be bound. Furthermore, the ligand-accommodating shapes within single-domain proteins can be represented by about 1300 binding pockets[3] indicating that amino acid diversity ensures structural viability in these conserved pockets, since a given fold type maybe formed by different amino acid sequences and even protein domains with low primary sequence homology can have very similar folds.[5]

This evolutionary pressure has finely tuned the interactions of binding sites with their respective ligands. Thus, in direct analogy, natural products (NPs) have a limited number of core scaffolds which is differentiated by the diversity of groups with which the central core is functionalized. NPs are small-molecule secondary metabolites and have been a major source of inspiration for bioactive molecule discovery.[6] Two important properties which render NPs intrinsically different from compounds resulting from traditional combinatorial synthesis are their increased molecular complexity and the prevalence of stereogenic centres, *i.e.* their increased three-dimensional character. These properties correlate with higher success rates as compounds progress in the pipeline of drug discovery to clinical trials.[7] The biosynthesis of NPs typically proceeds through the sequential reactions of various intermediates that are bound to different enzymes. NPs may also display different biological effects either within the producing organism itself or across different species.[8] These two facts indicate that NPs have evolved to interact with more than one protein. However, diverse substituent patterns in NPs sharing common molecular scaffolds result in their different bioactivity profiles, leading to the conclusion that NPs define evolutionary selected privileged scaffolds[9] which ensure that the members of the collection will be enriched in their ability to interact with multiple proteins whilst selectivity is finely tuned by the functional group substitution patterns for each individual molecule. It is evident that NPs define biologically relevant subspaces and are Nature's solution to the problem of charting and navigating around the vast chemical space.

Taking these arguments into consideration, the principles of evolutionary conservatism in protein and NP structure may be applied to guide the identification and synthesis of novel small molecules which modulate the function and activity of proteins. By analysing the NP core scaffolds, the bioactivity-relevant features may be selected and applied for the synthesis of NP-inspired compound libraries which have simplified structures yet retain biological relevance. Recognizing Nature's conservatism in the evolution of both proteins and their respective naturally occurring small-molecule modulators, the investigation of a possible analogy between the molecular scaffolds of NPs and ligand-binding sites in proteins may reveal these features. Thus, an alternative yet systematic, intellectual and structure-based approach is of high value and will serve as the basis and guide the development of small molecules with novel bioactivities to be used in chemical biology and medicinal chemistry research. This approach is referred to as biology-oriented synthesis (BIOS) and is based on the structural analysis of proteins and their naturally occurring and evolutionarily selected small-molecule ligands, considering Nature's evolutionary conservatism and functional group diversity. As the diversity of ligand-binding sites stems from the amino acid residue variation within a given protein subfold, the development of NP-inspired compound collections with sufficient members ensures that the diversity of the substituents attached to a common core scaffold will match the diversity of amino acid side chains occurring in otherwise structurally similar binding pockets. BIOS is a chemocentric approach which defines prevalidated starting points for the design and synthesis of such compound collections which are enriched in bioactivity and can be used for the development of novel small-molecule modulators for the study of protein function and consequently complex biological phenomena.

3.2 Structural Classification of Natural Products, Protein Structure Similarity Clustering and Scaffold Hunter

Structural classification of natural products (SCONP) is a chemoinformatic tool designed to effectively chart the chemical space covered by natural products and define the structural differences in a tree-like arrangement of the various NP classes.[10] Using the Dictionary of Natural Products (DNP) (Chapman & Hall/CRC Informa, London, Version 14.1, 2005) as a data source, SCONP is employed to reduce the high complexity of NPs to manageable levels by classifying and arranging in a hierarchical manner the core scaffolds rather than the entire molecule. For each scaffold, a branch is generated by the sequential and repeating deconstruction of one ring at a time, guided by a set of rules. This algorithm results in branches wherein each parent scaffold is a substructure of its child and finally are merged together to generate the whole scaffold tree (Figure 3.1). Since numerous synthetic non-natural compounds display bioactivity, the SCONP algorithm was extended to accommodate these structures by developing additional branch-generating rules and principles guided by knowledge and experience of synthetic and medicinal chemistry.[11] The process results in a flexible and intuitive classification which may accommodate any molecule and connect it to others through specific substructure relationships.

Initially, during chemoinformatic tree construction, the chemical space covered by natural products and their substructures was revealed to contain gaps where certain structures had either not been discovered yet or had not been generated through evolution. Such gaps rendered tree construction as

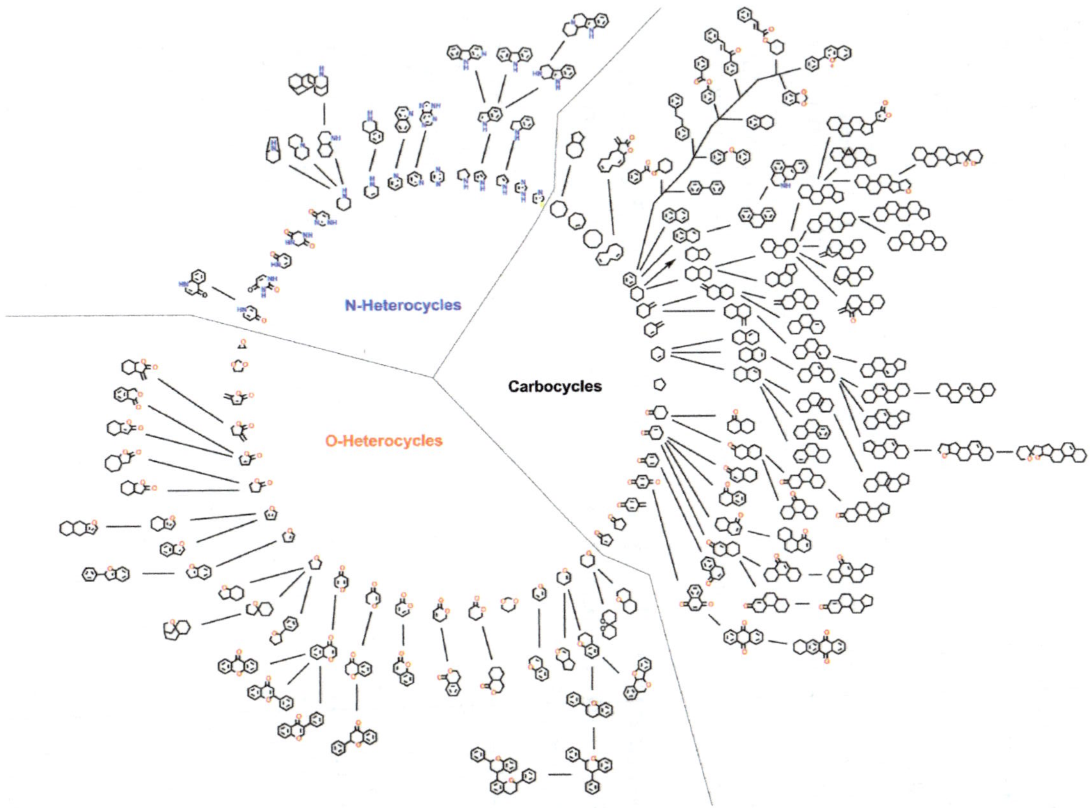

Figure 3.1 The scaffold tree generated by the SCONP algorithm. The application of the algorithm results in branches wherein each parent scaffold is a substructure of its child. SCONP facilitates the reduction in NP scaffold complexity and finally branches are merged together to generate the whole scaffold tree.

well as the overlay of scaffold trees stemming from different sets of molecules virtually impossible. This issue was addressed in the improved version of the scaffold tree by a set of rules which allow for and generate "virtual scaffolds" in order to provide a thoroughly constructed scaffold tree. These virtual scaffolds are not contained in the original data set nor do they represent actual compounds within it but are generated *in silico*, derived from the iterative deconstruction process itself. As these scaffolds had not yet materialized they provided a clear starting point and opportunities for new developments in chemistry and biology.

Moving along the branches of the scaffold tree starting from larger, more complex structures to smaller and structurally more simplified scaffolds is a process termed "brachiation" in analogy to the anthropological term used to describe the movement of gibbons in botanical trees. Brachiation along the branches of a scaffold tree assures that the bioactivity encoded in the derived substructures is retained although it may vary, *e.g.* potency may be lowered. It is important to note that brachiation through NP-derived scaffold trees proceeds with the retention of biological relevance, which distinguishes it from structure simplification based on chemical arguments, such as retrosynthetic limitations or synthetic tractability of smaller substructures. Similarly to fragment-based drug discovery,[12] brachiation is based on the assumption that smaller scaffolds will transfer their biologically relevant aspects to the larger molecules into which they are incorporated. In this manner, brachiation suggested a complementary construction of scaffold trees based on bioactivity instead of strictly on chemical structure.[13] These biology-guided scaffold trees offered a different point of view for chemical space by prioritizing bioactivity as the branch construction principle and consequently led to scaffold sequences where a particular kind of bioactivity was retained, yet graded in various degrees along the tree branches.

In an analogous manner of analysis, the subfolds of the ligand-binding sites of proteins may also be regarded as forming a scaffold, which may thus present similarities between proteins which perform similar functions or accommodate small molecules with similar three dimensional architectures. However, a high degree of sequence similarity between proteins does not necessarily infer a high degree of structural similarity. Such an analysis requires a structure-based method which was termed protein structure similarity clustering (PSSC). The PSSC algorithm pools together proteins according to the degree of similarity of their ligand-binding domains, resulting in clusters of similarly shaped protein subfolds. These clusters may subsequently be used to indicate compound classes with particular scaffolds which may serve as small-molecule modulators for proteins or indicate alternative targets for a given compound class.[14]

The Scaffold Hunter software[15] was developed to serve as the interface of the SCONP algorithm, allowing the visual interaction with the resulting extensively annotated scaffold trees in an interactive manner, facilitating filtering and navigating through scaffold substructures and offering a variety of dynamic options such as zooming, panning, colour-coding *etc.* Scaffold

Hunter allows both chemists and biologists to generate, visualize, analyse and annotate scaffold trees from any set of structures. Chemical-structure-guided scaffold trees may be constructed for any given set of compounds regardless of biological annotation, whilst biological-relevance-guided scaffold trees may incorporate large data sets guiding bioactivity annotation. If the data set at hand allows for it, both approaches should be explored, as they may be used complementarily. The generation of scaffold trees based on entirely virtual scaffolds is also worth exploring as they may provide templates for the design of compound collections with enriched bioactivities. Similarly in biology-guided scaffold trees, virtual scaffolds, representing gaps in existing chemical space lacking biological annotation, may provide starting points for the design of compound collections with particular expected bioactivities.[13]

3.3 Implications and Opportunities for Biology-oriented Synthesis

The underlying principles and concepts which encompass BIOS and SCONP have laid the foundations for the identification of biologically relevant regions in chemical space. Since bioactivity is the primary and key criterion for the selection of scaffolds, the resulting compound collections are expected to be enriched in bioactivity. Thus, the size of the collection does not have to be large, which compensates for the fact that the chemical synthesis of these molecules may be more elaborate compared with collections resulting from combinatorial synthesis.

On the other hand, NP-inspired compounds, selected through BIOS, possess less structural complexity compared with the actual NPs themselves, ensuring reliable and efficient synthetic tractability with retained type of bioactivity. However, BIOS aims to provide the starting points for compound development around a biologically relevant scaffold, for which the already established approaches and strategies of medicinal chemistry will eventually be applied in order to improve potency as well as other physicochemical properties depending on the intended use of the compound as a drug candidate or a chemical biology tool compound.[16] In this regard, the reliable and efficient synthetic tractability of compound collections alleviate problems arising from the difficulties in suppling compounds in sufficient amounts for full biological and pharmacological testing. It is thus evident that at the core of BIOS lies the continuous development of novel and robust chemical methods which are applicable to the parallel and modular syntheses of compounds whilst granting access to structurally complex scaffolds incorporating multiple stereogenic centres with a high degree of selectivity.

Furthermore, the retention of biological relevance does not necessarily ensure that the type of biological activity is also retained. The structural simplification of the original guiding NPs may lead to the resulting compounds targeting multiple proteins with structurally similar ligand-binding pockets. This promiscuity may, however, not be regarded as a negative feature since biological systems are robust and often redundantly interconnected, indicating that modulation of these systems at several signal transduction nodes is required to induce the desired biological effect. NPs have evolved to interact with and modulate such systems in a dynamic manner. These properties are encoded in their structures and therefore into the scaffolds of the resulting NP-inspired compound collections. Target selectivity is an issue which may be addressed by well-known and widely used approaches of medicinal chemistry. As such, BIOS libraries offer unique opportunities for the discovery of potential modulators with special multi-bioactivity profiles, which could be used as tools for systems biology research (Figure 3.2). It is important to note that BIOS may be applied not only to NPs but also to non-natural products as well, since it requires biological relevance of compound classes and not their occurrence in nature as such. Thus, ultimately in BIOS, the analysis of NP structures must be extended to all known bioactive compound classes. Finally, these NP-inspired collections with enriched bioactivities may prove to have added value in target-agnostic and cell-based phenotypic screenings. However, in these cases, effective strategies and technologies for target deconvolution, identification and validation are necessary.[8]

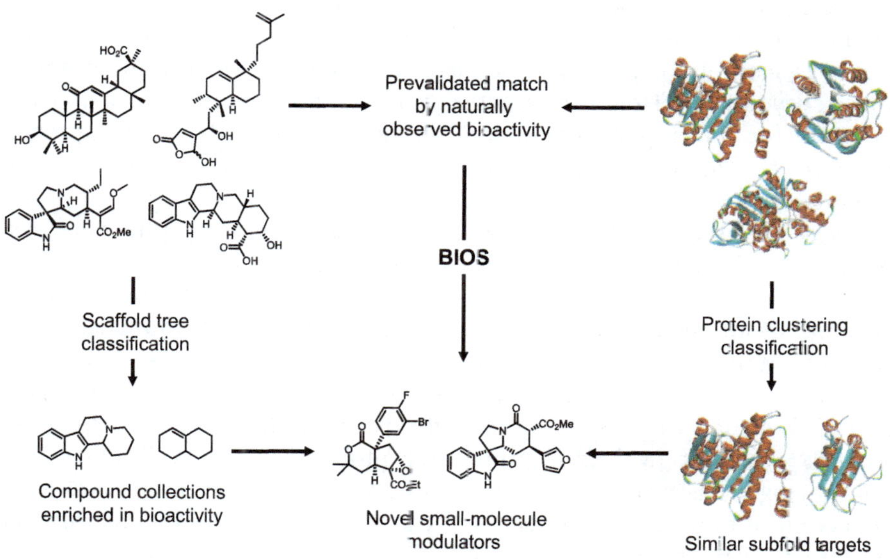

Figure 3.2 Simplified conceptual diagram of the implications of biology-oriented synthesis. Utilizing the prevalidated information provided by the observed bioactivity in nature, either structural classification of chemical structures or ligand-binding domain three-dimensional similarity may be used as hypothesis-driving tools for the design and synthesis of compound libraries, which may lead to the discovery of novel small-molecule modulators.

3.4 Applications of Biology-oriented Synthesis

Several successful applications of biology-oriented synthesis have been reported in the literature and may be categorized by three main guiding principles: chemical-structure- and bioactivity-guided approaches, protein-structure-clustering-guided approaches as well as the most recent natural-product-derived-fragment-based approach. Only a few and selected examples will be outlined for the sake of conciseness. However, the underlying principles, conclusions and implications of these case studies also characterize the other reported BIOS applications. The reader is encouraged to use the references provided for a more complete literature overview.

3.4.1 Chemical-structure- and Bioactivity-guided Approaches

For chemical-structure-guided approaches, the SCONP algorithm and the Natural Product Scaffold Tree were employed to identify NP-derived or inspired scaffolds which were used for the design and synthesis of medium-sized chemical libraries. These libraries were subsequently screened against established biological targets. Extrapolation of structure–activity relationships from these screenings and further derivatization led to the identification of potent and selective inhibitors of a variety of biological targets.

Protein phosphatases are involved in numerous signal transduction cascades, thus acting as key modulators of various biological processes.[17,18] Small molecules which interact with protein phosphatases have proven useful in gaining insights into the mechanisms and biological functions of these enzymes.[19,20] However, target selectivity is challenging. Thus, an early example for the application of BIOS as a working and viable approach involved the identification of novel and selective protein phosphatase modulators.[21]

Seven different protein tyrosine and dual-specificity phosphatases, modulating different biological processes were applied in the screening of a NP-derived and a NP-inspired compound collection, both representing structurally interesting examples of the *N*- and *O*-heterocycle branches of the NP scaffold tree,[10] as well as a small collection of individual NPs. Compound collection 1 (Figure 3.3A), derived from the scaffold of the NP alkaloid cytisine,[22,23] yielded the first reported, selective inhibitor of the phosphatase VE--PTP, which is involved in protein receptor activation,[24] cell adhesion[25] and angiogenesis processes.[26] The hit rate was estimated to be 1.6% for a collection of only 1271 members, and the newly identified low micromolar inhibitors were revealed to have a 40-fold selectivity for VE-PTP over the phosphatases Shp-2 and MptpB, rendering these compounds suitable for the further study of the biological functions of VE-PTP. Compound collection 2 (Figure 3.3B), inspired by the furanodictin class of NPs,[27] yielded potent inhibitors of Shp-2, a potential anti-infective drug target[28] and PTP1B which is involved in insulin signaling.[20,29] A lower hit rate of 0.36% was observed for a library of 1112 members. Inhibition activity was in the low micromolar

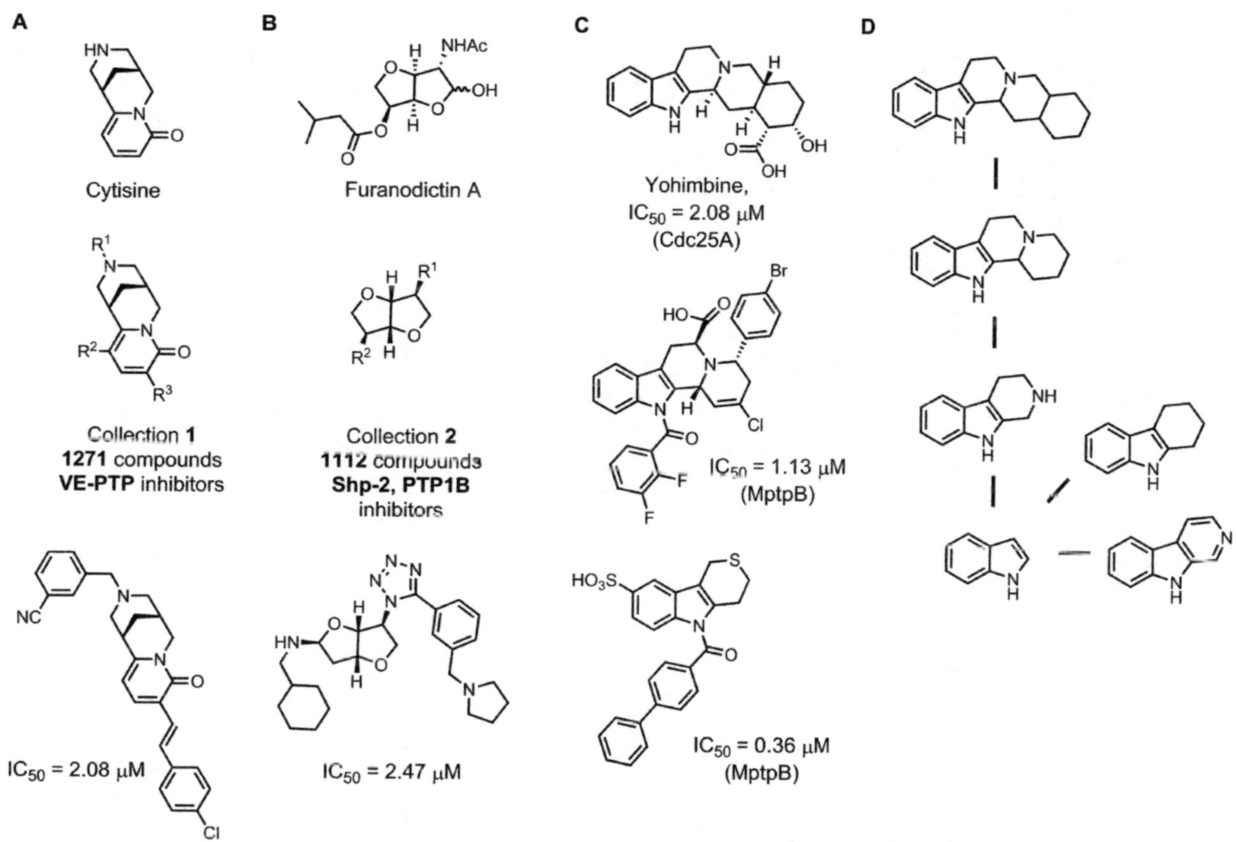

Figure 3.3 Representative chemical structures for the identification of novel and unprecedented protein phosphatases inhibitors. (A) (top to bottom) The alkaloid Cytisine; the Cytisine-derived molecular scaffold; a representative example of the identified inhibitors against VE-PTP. (B) (top to bottom) The natural product Furanodictin A; the Furanodictin-inspired molecular scaffold; a representative example of the identified inhibitors against Shp-2. (C) (top to bottom) Yohimbine was one of the 354 individual NPs included in the screening against seven protein phosphatases; Structures of compounds designed and prepared after brachiation along the yohimbane pentacyclic scaffold branch. (D) Partial schematic representation of the indole branch of the NP tree.

range with at least 20-fold selectivity. Furanodictines had not been identified before as inhibitors of protein tyrosine phosphatases. The small collection of 354 individual NPs, selected on the basis of their structural diversity and complexity, revealed weak inhibitors of the dual-specificity phosphatase Cdc25A, involved in the regulation of the cell cycle and considered to be an anti-cancer target.[19,30] These weak inhibitors belonged to the class of yohimbane alkaloids[31] which had not previously been linked to protein phosphatase inhibition. This discovery provided an opportunity to demonstrate the principle and the value of brachiation. Instead of brachiating along the pentacyclic yohimbane scaffold core, the more synthetically feasible and concise tetracyclic indolo-quinolizidine scaffold core was explored (Figure 3.3C). A solid-phase, six-step synthesis was devised and applied for the preparation of a collection of 450 compounds based on this pre-validated scaffold, yielding two compounds with similar potency against Cdc25A, as the initial NPs. This collection also included selective inhibitors of the MptpB phosphatase which is an important target for the treatment of tuberculosis.[32] Further brachiation towards the tricyclic and bicyclic indole-derived scaffolds suggested the screening of a 188-membered collection for which the synthesis had been earlier described.[33] Two additional compounds with inhibitory activity against Cdc25A at similar level to the initial yohimbane NPs were identified. More importantly, however, seven nanomolar-range inhibitors of MptpB with at least 100-fold selectivity over other tyrosine kinase phosphatases *e.g.* PTP1B or VE-PTP, were also identified. Inhibitors of this enzyme had not been described before and tetracyclic indolo-quinolizidines and tricyclic indole-based compounds had not previously been associated with protein phosphatase inhibition activity.

Chemical cascade reactions which result in the rapid construction of polycyclic three-dimensional core scaffolds are powerful synthetic tools for the investigation of hypotheses suggested by the application of the SCONP algorithm and the scaffold tree. Such reactions facilitate the preparation of a plethora of analogues, enough to construct compound collections which would not be large enough for applications in conventional HTS approaches, yet are sufficiently large for BIOS applications as the bioactivity is, to a certain degree, prevalidated. Furthermore, combinations of cascade reactions able to not only produce polycyclic scaffolds but also to connect or to fuse together such scaffolds would provide additional advantages. Such cascade reactions require less structurally complicated starting materials and fewer purification and chemical handling steps, as well as reduced quantities of the various chemical reagents and equipment. Thus, the development of a 12-step reaction cascade sequence which included nine discrete chemical transformations[34] proved to be remarkably efficient in the preparation of the structurally complex tetracyclic tetrahydroindoloquinolizines (Figure 3.4A), whose scaffold bears significant resemblance to a variety of polycyclic indole alkaloid structures.[35]

This reaction cascade was utilized for the preparation of a focused compound collection of 26 members, which were screened for possible

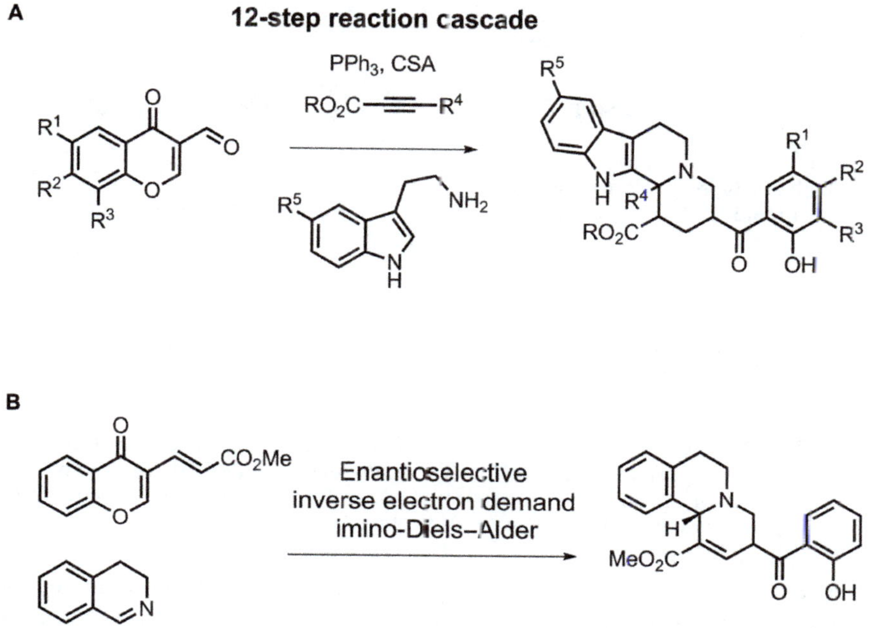

Figure 3.4 Cascade reactions utilised for BIOS applications. (A) Twelve step sequential cascade leading to the formation of novel dual NPM1 and Crm1 inhibitors. (B) Development of an enantioselective synthetic route resulted in the discovery of a fourfold more active molecule compared with the guiding scaffold.

modulation of mitosis, suggested by the susceptibility of the process to small-molecule modulation as well as previously reported indole alkaloid-based molecules.[36,37] Phenotypic screening and confocal microscopy imaging of recombinant cells revealed that at least two compounds resulted in arrested mitosis and multipolar mitotic spindles at low micromolar concentrations. Further chemical proteomics investigation by means of activity-based probe pull down and MS determination or immunoblotting as well as knockdown cell lines lead to the identification of the centrosomal protein NMP1, which is involved in centrosome duplication regulation,[38] and the nuclear export factor Crm1, which is forming a complex with NMP1 during mitosis.[39] Using these novel inhibitors as the guiding molecules an alternative, enantioselective synthetic route (Figure 3.4B) for the preparation of a structurally simplified collection of 41 members was developed.[40] Screening of this collection revealed a novel analogue with fourfold increased potency compared with the initial molecule, validating that the BIOS approach may be also applied to synthetic non-natural yet biologically relevant molecule as well as NPs.

Iridoids comprise an additional class of structurally interesting NPs which have previously been shown to have neuroprotective or neurite-growth-promoting properties.[41,42] As small molecules with such properties are in

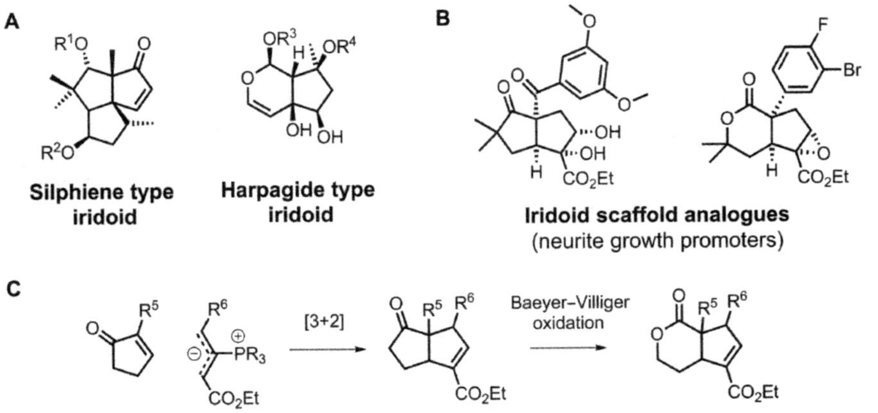

Figure 3.5 Neurite growth promoters inspired by the iridoid class of NPs. (A) Examples of iridoid NPs. (B) Analogues revealed to display neurite-growth-promoting properties. (C) Chemical synthesis sequence to access iridoid scaffold analogues and related molecules.

high demand for the study and treatment of neurodegenerative diseases,[43–45] the BIOS strategy was applied in order to reveal such structurally simplified and synthetically accessible molecules. Employing a [3 + 2] cycloaddition followed by a regioselective Baeyer–Villiger oxidation sequence, two core scaffolds of the general iridoid class of NPs became readily accessible (Figure 3.5).[46] Utilizing an enantioselective version of this sequence as well as individual chemical derivatization steps, a collection of 54 molecules was prepared and screened in two neurite-growth assays: a primary hippocampal neuron outgrowth assay and a mouse embryonic stem cell motor neuron growth assay. The biological testing revealed novel active compounds that promoted neurite growth in different patterns and could be used as tools for the study of these neurodevelopmental processes.

NPs with a different structurally complex and interesting core scaffold, a secoyohimbane scaffold (Figure 3.6A), have also been reported to possess neurite-growth-promoting properties.[47–49] A new organocatalytic, enantioselective synthetic route towards secoyohimbane analogues was developed, granting access to the structurally complex scaffold and a collection of 56 compounds. At least five of these analogues were found to display remarkable neurite-growth-promoting properties when tested with the hippocampal neurite outgrowth and the mouse embryonic stem cell motor neurite growth assays.[50] Although structurally less complex, these seccoyohimbane analogues were shown to maintain the biological relevance and properties compared with the guiding NPs, providing further proof that BIOS is a viable approach in the search for bioactive small molecules.

Apart from linear cascade reactions such as those described, branching reaction cascades are also highly efficient synthetic strategies which may be used to prepare a plethora of diverse molecular scaffolds from only a few common precursors.[51] Such cascades may be applicable to either

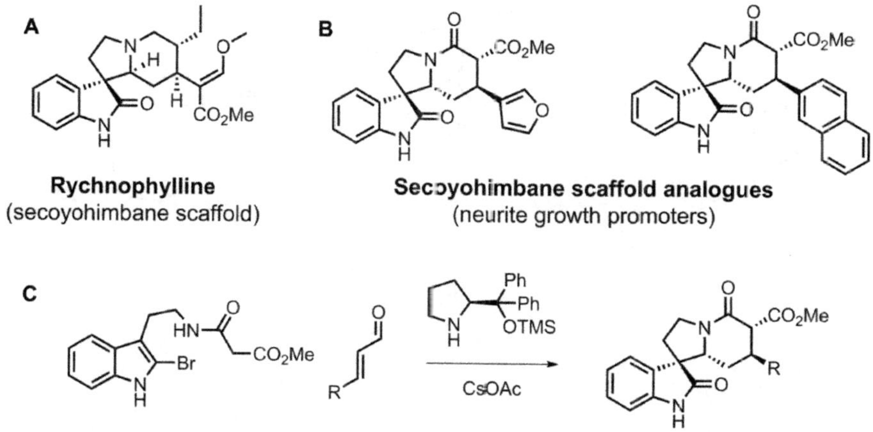

Figure 3.6 Neurite growth promoters inspired by the secoyohimbane scaffold. (A) Rychnophylline is a typical secoyohimbane alkaloid. (B) Secoyohimbane analogues with neurite-growth-promoting properties. (C) Organocatalytic, enantioselective synthesis of the secoyohimbane scaffold.

intermolecular or intramolecular reactions, involving a variety of molecular-complexity-generating chemical transformations in a rapid and concise manner.[52] Thus branching cascades can generate diverse and complex molecular architectures and are therefore suitable for BIOS-guided compound collections. Exploiting the multiple electrophilic positions of a common starting material, a range of bisnucleophiles was introduced yielding 12 diverse and structurally complex fused ring systems (Figure 3.7) including NP-related as well as unprecedented molecular scaffolds.[53]

3.4.2 Protein-structure-clustering-guided Approaches

The application of BIOS guided by the conservation of the scaffolds of a protein ligand binding domain is similar to the concept of chemical-structure-guided approaches, although in an inverted manner. Instead of the chemical structure providing inspiration for the synthesis of structurally similar scaffolds, which can, in turn, be used to target different biological targets, the structure of a protein domain is used to suggest NPs which interact with different and unrelated proteins which display structural similarities in their ligand-binding domains to the guiding protein domain. Such an approach is a useful tool in the cases where potent and selective inhibitors of a certain biological target have not been reported, yet a crystal structural exists that allows comparison with other proteins. Utilising the PSSC algorithm as an investigational tool to identify similarities in subfolds or ligand-binding sites in proteins may lead to the discovery of unprecedented inhibitors, which may subsequently be used as tools to study the biological functions and implications of the protein in question and provide further insights into complex biological phenomena.

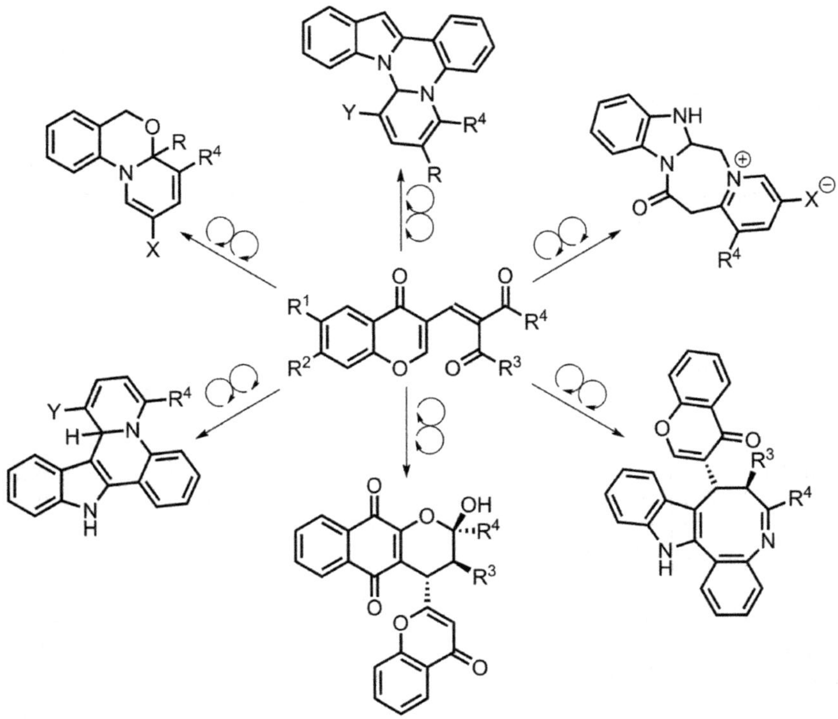

Figure 3.7 Branching cascades leading to the rapid and concise preparation of diverse and complex NP-related and unprecedented molecular scaffolds. Such reaction cascades facilitated the implementation of hypothesis-driven approaches such as BIOS.

The reversible attachment and removal of fatty acids on proteins is a biological process involved in signalling, cellular localization and protein activity modulation.[54–56] With regards to protein palmitoylation, understanding the process of palmitic acid acylation and deacylation would reveal new information about the localization of the Ras oncoprotein, which is linked to unregulated signalling in tumours.[57] Previous attempts at elucidating the cycle of Ras palmitoylation and depalmitoylation indicated that APT1, was involved in the depalmitoylation of Ras;[58,59] however a suitable activity probe or small-molecule modulator of APT1 had not been previously reported. The PSSC algorithm was used to identify other proteins that possessed similar subfolds and structural features in their ligand-binding domains to those displayed in the crystal structure of APT1,[60] indicating dog gastric lipase[61] as a hit with a high degree of structural similarity in spite of their low sequence similarity. This finding indicated that small-molecule lipase inhibitors may be used to modulate the activity of APT1. β-Lactones had been used as lipase inhibitors, even as marketed drugs,[62] thus a collection of β-lactones was synthesized, revealing a novel inhibitor of APT1.[63] Palmitostatin B (Figure 3.8) was found to inhibit APT1 through reversible

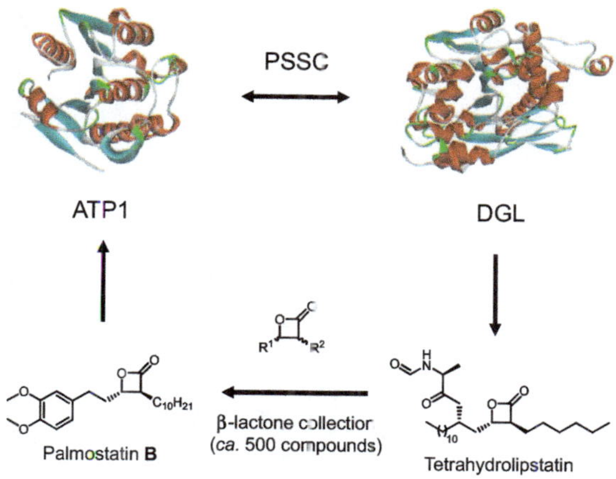

Figure 3.8 Discovery of ATP1 inhibitor Palmostatin B. The PSSC algorithm was used to identify a protein with structurally similar ligand binding site—dog gastric lipase (DGL). The structure of a known inhibitor of DGL inspired the design and synthesis of a library of β-lactones. Screening revealed the potent and selective inhibitor Palmostatin B.

acylation of a nucleophilic serine residue, and was utilized in a series of biological studies revealing the role of APT1 to be as a thioesterase in the acylation cycle of Ras palmitoylation and that it was potentially a new biological target for the treatment of cancer.

11β-Hydroxysteroid dehydrogenase 1 (11βHSD1) is an enzyme involved in the biosynthesis of the glucocorticoid, cortisol.[64] Selective 11βHSD1 inhibitors may be useful in the treatment of type-II diabetes and metabolic syndrome as well as the prevention of lipid layer plaques in the inner wall of arterial blood vessels.[64,65] Applying the PSSC algorithm to the two isoforms 11βHSD1 and 11βHSD2, revealed structural similarities of the ligand-binding site with the dual-specificity phosphatase Cdc25A and acetylcholine esterase, despite their being mechanistically unrelated to 11βHSD1 and their low sequence similarity. Complementing this finding, the application of the SCONP algorithm to the NP dysidiolide, a known Cdc25A inhibitor as well as the 11βHSD1 ligand glycyrhetinic acid (Figure 3.9), suggested two related scaffold cores, a 1,2-dehydrodecalin and a 3,4-dehydrodecalin respectively that might serve as the starting point for the design of 11βHSD1 selective inhibitors. Thus, a collection of 483 decalin analogues was prepared and screened for inhibition of 11βHSD1. Several compounds inhibited 11βHSD1 with concentrations giving 50% inhibition (IC_{50}) values below 10 μM, with four molecules displaying submicromolar activity (Figure 3.9) and selectively inhibiting 11βHSD1 in living cells as well.[10] The hydroxybutenolide part of the Cdc25A inhibitor dysidiolide (Figure 3.9), also served as an inspiration for the synthesis of a smaller collection of 150 members intended to be screened against both acetylcholine esterase and 11βHSD1. Testing revealed three low micromolar inhibitors of acetylcholine esterase as well as seven

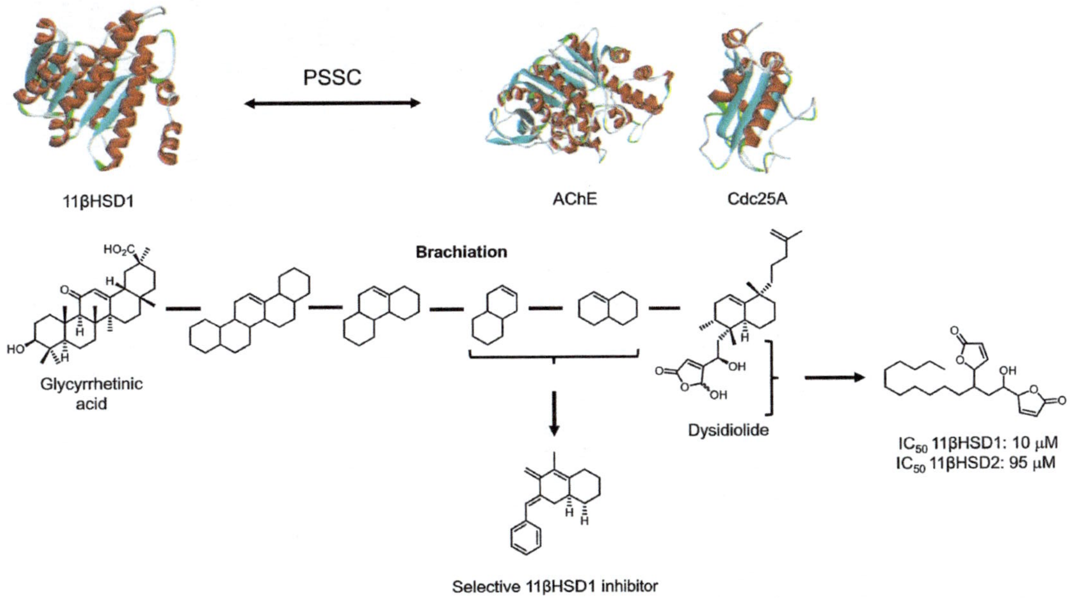

Figure 3.9 Discovery of novel 11βHSD1 inhibitors. The application of the PSSC algorithm revealed AChE and Cdc25A as proteins with similar subfolds to 11βHSD1. Brachiation from both glycyrrhenic acid, a known 11βHSD1 inhibitor, and Dysidiolide, a Cdc25A inhibitor, led to the discovery of a novel, submicromolar and selective 11βHSD1 inhibitor. The hydroxybutenolide segment of Dysidiolide served as a starting point for the design of selective inhibitors of 11βHSD1 over 11βHSD2.

low micromolar hits selective for 11βHSD1 over 11βHSD2,[66] demonstrating the potential of the PSSC in guiding the design of small-molecule modulators of different biological targets.

3.4.3 Natural-product-derived-fragment-based Approaches

Fragments, low molecular weight compounds,[67] facilitate the rapid exploration of large portions of chemical space[68] giving rise to fragment-based drug-discovery.[12] However the majority of fragment libraries are comprised of privileged substructures derived from known drugs thus, covering portions of chemical space which have been previously explored[69,70] and include a low number of complex three-dimensional architectures.[71]

Natural products may be regarded biologically prevalidated by nature, possessing a high degree of complex three dimensional structural characteristics and occupying portions of chemical space that are not typically covered by synthetically accessible compounds.[72] Thus, fragments derived from natural products may be regarded as different compared with synthetically accessible fragments in terms of three-dimensional character and biological relevance, which may provide new opportunities for fragment-based drug discovery. Chemoinformatic analyses of natural product structures according to fragment likeness criteria have confirmed this hypothesis.[73,74] The deconstruction of NPs by the SCONP algorithm facilitates the selection of such NP-derived fragments with the use of an additional filtering algorithm. In an early example of such a BIOS application,[75] the combination of these algorithms led to a virtual library of NP-derived fragments which are characterized by a high degree of structural diversity and complexity, possess a distribution of physical properties similar to that of the NPs themselves and were shown to occupy segments of chemical space that differed significantly from those covered by commercially available fragments. These properties indicated that such NP-derived fragments may give rise to structurally unprecedented modulator classes for established biological targets. In a proof of concept study,[75] bicyclic cytisine–sparteine derivatives were identified as unprecedented inhibitors of p38a MAP kinase and shown to bind to the allosteric site of the protein (Figure 3.10).

The use of NP-derived fragments for fragment-based ligand discovery inspires the design and synthesis of NP-derived compounds, re-iterating the original principles of BIOS. The syntheses of NP-derived fragments are more tractable compared with the multistep procedures required for the preparation of the guiding NPs. Although efficient methodologies for the preparation of the wide variety of these fragments are still being developed, NP-derived fragment libraries may be used to overcome the limitations set by the lack of synthetic accessibility of certain NPs.

The inclusion of certain NP-derived fragments in the core scaffold of a collection of molecules may limit the structural diversity of such a compound collection. A successful strategy to circumvent this potential limitation would be to use a common starting material which already

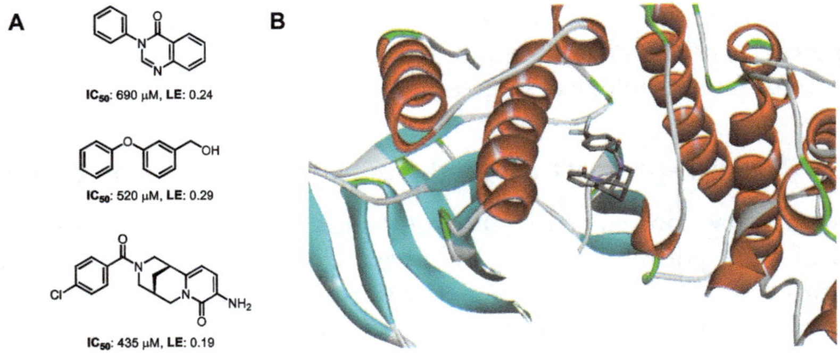

Figure 3.10 Natural-product-derived fragments identified as novel p38a MAP kinase inhibitors by the application of the SCONP and subsequent filtering algorithms. (A) Examples of three different fragment classes. (B) Crystal structure of the p38a MAP kinase with a cytisine–sparteine fragment bound at the allosteric site of the protein (PDB: 4EH9).

incorporates such a fragment as well as certain reactive groups which upon treatment with different reagents or catalyst may undergo different chemical transformations giving rise to structurally diverse molecules. Such a reagent-directed divergent synthesis facilitates access to diverse and structurally enhanced compound collections, which may also benefit from biological relevance enhancement due to the incorporation of NP-derived fragments. In order to implement this strategy, the identification of efficient and versatile chemical transformations which may proceed through more than one possible intermediate and thus give rise to divergent reaction pathways is crucial.

The versatility of gold-catalysed transformations of oxindole-derived 1,6-enynes was successfully harnessed, leading to the formation of three distinct NP-inspired scaffolds (Figure 3.11A). The outcome of gold-mediated rearrangements of enynes may be modulated by the properties of the ligands in the gold complex used in each case[76–78] facilitating the skeletal diversification of the common oxindole-derived 1,6-enyne starting material and the preparation of a compound collection. The oxindole fragment is incorporated in a number of distinct NPs displaying different bioactivities[76,79] and has also been used in a number of medicinal chemistry and drug-discovery-related studies.[80,81] Having established a chemical toolbox which allowed the preparation of each distinct scaffold on demand, a relatively small yet structurally diverse collection of *ca.* 60 compounds was prepared. This collection was subsequently submitted to cellular assays monitoring different biological pathways in order to investigate the translation of the structural diversity to biological activity.

Screening led to the identification of selective and structurally novel modulators of the Hedgehog and Wnt signalling pathways, autophagy and cellular proliferation. The Hedgehog (Hh) pathway is an evolutionary conserved signalling pathway implicated in developmental processes such as cell proliferation and differentiation and tissue repair and regeneration and

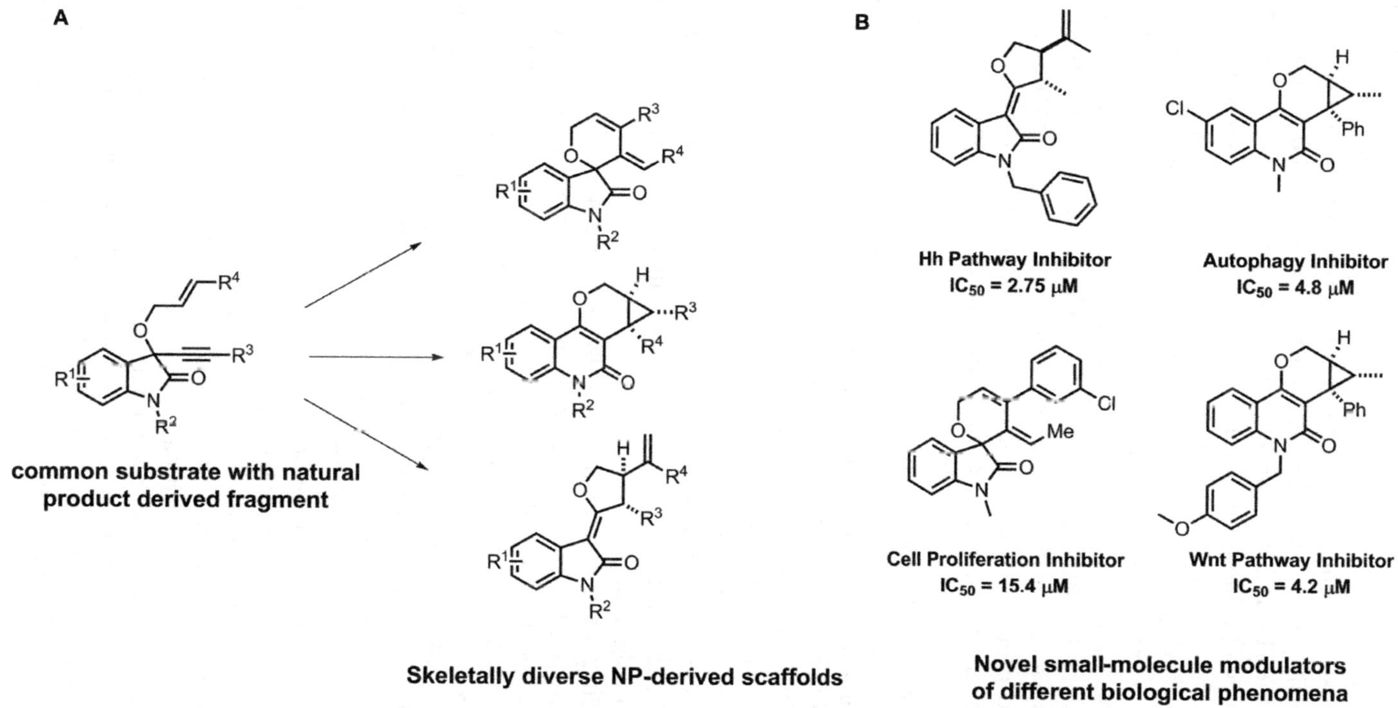

Figure 3.11 NP-derived fragments may be incorporated in diverse scaffolds to enhance biological relevance. (A) Versatile chemical transformations may be utilized to prepare skeletally diverse scaffolds from a common starting material. (B) This strategy led to the identification of novel small-molecule inhibitors of different biological pathways and processes.

is associated with birth defects and cancer.[82] Two molecules were found to display selective, low micromolar inhibitory activity of the Hh pathway without affecting cell viability (Figure 3.11B). Further biological characterization of these molecules revealed that they are reversible binders of the transmembrane protein Smoothened (Smo), an activator of the Hh pathway. Similarly to the Hh pathway, the Wnt signalling pathway is involved in the regulation of cell proliferation and migration as well as tissue regeneration, cell polarity and stem cell renewal, thus playing a major role in cancer cell proliferation.[83] A compound with a different oxindole-derived scaffold was found to be a structurally novel, low micromolar Wnt pathway inhibitor which did not interfere with other signalling pathways, such as the Hh pathway (Figure 3.11B). Autophagy is catabolic process important for cell homeostasis, which involves the self-digestion of cellular components, such as denatured proteins or damaged organelles. Autophagy is regarded as a survival mechanism for cancer cells against metabolic stress caused by the lack of nutrients or oxygen and increased energy demand, and inhibition of autophagy has emerged as a potential cancer-targeting strategy.[84] A structurally different compound was identified as a selective autophagy inhibitor. A novel small-molecule modulator of cancer cell proliferation was also identified from this compound collection using a cell confluency monitoring imaging assay. This compound displayed no activity in the Hh and Wnt pathways as well as autophagy. The incorporation of NP-derived fragments into such scaffold-diversity-oriented syntheses translated to distinct and diverse biological relevance, greatly facilitating the identification of structurally novel and selective small-molecule modulators of different biological processes and signalling pathways.

These findings indicated that biologically relevant small molecules may be identified through the preparation of compounds which share only certain fragments rather than the whole of the core scaffold of the guiding NPs. It may also be suggested that the combinations of these NP-derived fragments may give access to novel non-natural scaffolds which, however, may inherit the biological relevance and physicochemical properties of the NPs themselves. Furthermore, such combinations may lead to the formation of unprecedented small-molecule chemotypes and novel chemical matter which is still biologically relevant and prevalidated, albeit not yet linked to any biological target. Such compounds reach further than the chemical space explored by nature, as their formation may not be achievable by known biosynthetic pathways. Thus, the best evaluation method for such compounds would be target-agnostic cellular assays which monitor phenotypic changes, due to their unbiased nature.

It is important to note that the synthetic methodologies developed for the combination of these fragments also need to address the requirement for the preparation of enantiopure compounds, as the bioactivity of a natural product is often linked to its absolute configuration. In an example of NP-derived scaffold combination a catalytic enantioselective cycloaddition reaction was employed to access fused tropane–pyrrolidine derivatives in an enantiodivergent manner granting, access to both enantiomers of a NP-inspired hybrid scaffold with eight stereocentres.[85] The tropane fragment is abundant in NPs, being part of more than 600 alkaloids which display a plethora of different bioactivities[86–88] and has been the subject of meticulous study in the fields of medicinal chemistry and biological research.[89,90] Pyrrolidine is also a biologically relevant fragment contained in a plethora of alkaloid and other NPs and has been included in multiple drugs and biologically active compounds.[91] In the first step of this approach, the racemic tropane is kinetically resolved with an azomethine ylide in the presence of a chiral copper(I) phosphine catalyst complex to form one enantiomer of the fused hybrid scaffold. The remaining enantiomer is then reacted with a different azomethine ylide in the presence of a racemic copper(I) phosphine catalyst complex in a one-pot procedure, allowing the separation of the two analogues by simple chromatographic separation. The other enantiomer is readily obtained by reversing the addition of the azomethine ylides used (Figure 3.12A). Similar chemistry was utilized to access fused indole and tropane fragments as well.[92] A copper(I) ferrocene complex was used to catalyse the enantioselective [3 + 2] cycloaddition of aryl- and indole-incorporating azomethine ylides with nitro alkenes and to prepare a small yet diverse set of 84 aryl- and indole-tropane fused derivatives, which represented novel and unprecedented small molecule chemotypes (Figure 3.12A). This collection subsequently was screened for bioactivity against signalling cascade assays such as the Wnt and Hedgehog pathways as well as phenotypic change assays such as autophagy, reactive oxygen species induction and lipid droplet formation. At least three indotropanes were found to inhibit the hedgehog signalling pathway with potencies in the low micromolar range (Figure 3.12B) leading to the first tropane derivatives shown to modulate this signalling pathway.

Extending this concept even further, non-natural NP-derived scaffolds may also be prepared by minimal diversification of the guiding NPs. Chemical transformations, such as ring-contraction or expansion, intramolecular cyclisation of functional groups or heteroatom substitution, can lead to new modified scaffolds directly from NPs, which retain their biological relevance and properties[93,94] and may fill potential gaps in chemical space not explored by nature. NPs are suitable starting points because of their structural diversity, even among common classes and they provide excellent substrates for site-selective and stereoselective transformations.[95] Hergenrother and co-workers have demonstrated the applicability of this ring-distortion strategy for the rapid preparation of distinct, diverse and structurally complex small molecules directly from guiding NPs.[96] This strategy allows chemists to capitalize on the structural features, diversity and complexity already bestowed to NPs by Nature.[95] Cinchona alkaloids, such as quinine and quinidine, have been attribute with a plethora of bioactivities[97] and chemical modifications leading to ring distortion and formation of an oxazatwistane scaffold have been reported (Figure 3.13). Applying the ring-distortion strategy and using late-stage C–H functionalization methods and

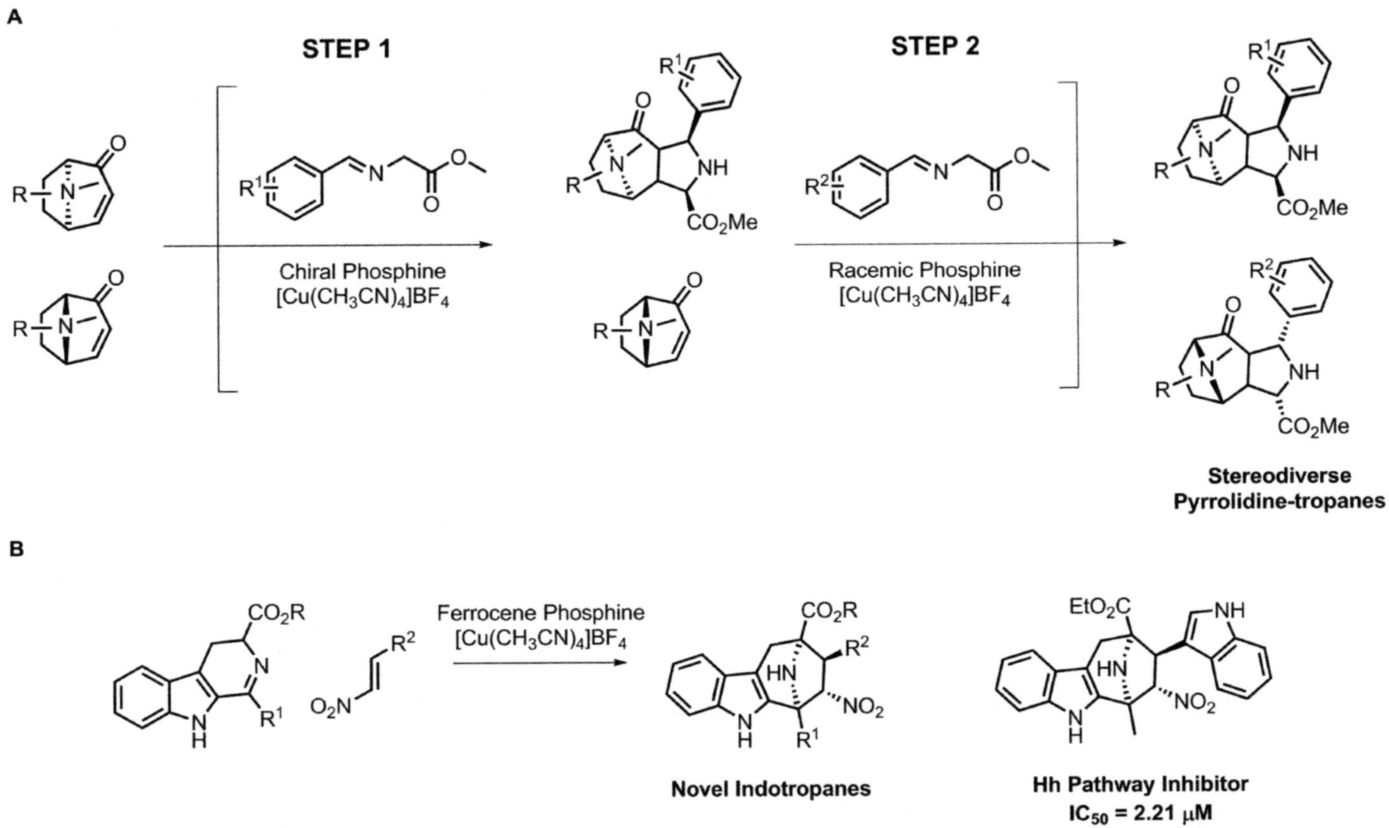

Figure 3.12 (A) One-pot enantiodivergent synthesis of hybrid NP-derived fragment scaffolds enabled by the kinetic resolution of tropanes with azomethine ylides catalysed by a chiral copper(I)–phosphine complex. (B) Enantioselective preparation of indotropane NP-derived fragment scaffold. These molecules displayed diverse bioactivities, leading to the identification of Hedgehog signalling pathway inhibitors.

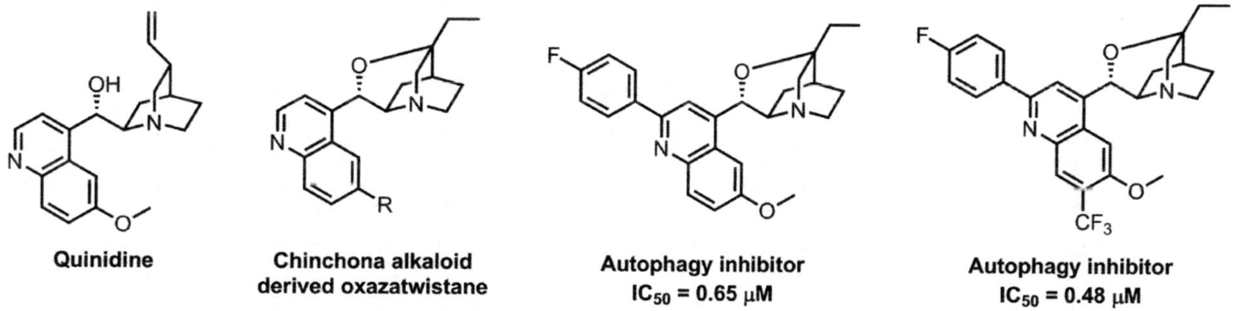

Figure 3.13 Structures of the chinchona alkaloid natural product quinidine and the resulting oxazatwistane scaffold. Substituted oxazatwistanes were found to be submicromolar inhibitors of autophagy.

metal-catalysed cross couplings a collection of about 50 compounds was prepared. Testing for bioactivity in agnostic screenings revealed two novel molecules with submicromolar inhibitory activity against starvation-induced autophagy (Figure 3.13) representing unprecedented autophagy modulators.[98] Notably, neither the unsubstituted oxazatwistane scaffold nor the guiding NPs, quinine and quinidine, displayed any inhibiting activity at this concentration.

3.5 Conclusions and Outlook

Small molecules offer the opportunity to perturb, study and understand complex biological mechanisms and phenomena in a unique and often unprecedented manner due to their rapid, conditional and tunable modes of action. Methods which allow the initial qualitative analysis of chemical space as well as its mapping and eventually its navigation to biologically relevant hot spots are more likely to enable the identification of novel and biologically relevant chemical matter.

This concept underlines the invention of the SCONP algorithm, which analyses the structural scaffolds of NPs that represent biologically pre-validated portions of chemical space. The SCONP algorithm is a hypothesis-generating tool, which was used for the design and synthesis of compound collections shown to be enriched in bioactivity and ultimately yielding potent and selective modulators which may be used as tools for biological research or as starting points for the development of drug discovery projects. In direct analogy, the PSSC algorithm was developed to study the correlation of protein subfold and ligand binding domain similarities with associated small molecules and served as a tool to identify new scaffold classes which modulate these proteins. Examples such as the discovery of Palmostatin have demonstrated the potential of the PSSC algorithm as a hypothesis-generating and small-molecule-discovery tool. The analyses and findings of the applications of both the SCONP and PSSC algorithms indicate that there may by a synergistic effect between conservation and diversity in the evolution of both small molecules and proteins employed by nature. Whilst core scaffolds remain mostly conserved, appendages and substituent functional groups had a much higher level of diversity, indicating that the identification of biologically relevant small molecules is greatly underlined by the matching and compensating between these two elements. Similarly, SCONP and PSSC may be applied in a complementary manner to provide the criteria for the design and synthesis of small-molecule collections based on the concept of BIOS.

BIOS collections are relatively small in size yet display a high degree of three-dimensional architecture as well as sufficient substituent diversity to enable the development of potent and selective small-molecule modulators. The chemical effort needed for the construction of these collections is often higher compared with combinatorial chemistry collections yet the resulting compounds are enhanced in biological relevance, as indicated by the higher hit rates of BIOS collections. Ultimately, at the heart of BIOS lies the application as well as the development of synthetic methods which provide access to such compound collections in a reliable, efficient and flexible manner, rendering BIOS a more chemocentric approach for the discovery of biologically active small molecules.

The established principles of BIOS enabled the utilization of NP-derived fragments towards this goal. Fragments are synthetically more tractable and have been shown to facilitate the exploration of chemical space. Accommodating NP-derived fragments in the SCONP algorithm provided an additional tool for the discovery of biologically active small molecules and indicated the potential which NP-derived fragment combinations may hold, bringing together fragment-based drug discovery (FBDD) and BIOS. This novel concept, coupled with cell-based, unbiased target-agnostic biological screenings has met great success as an idea generating tool leading to the discovery of submicromolar small-molecule modulators of complex biological processes, such as the Hedgehog signalling pathway and autophagy.

In conclusion, BIOS is an approach based on the biological prevalidation of natural products by means of evolutionary conservatism as observed in nature. Natural-product-inspired compound collections prepared using the guiding principles and tools of BIOS, have proven to be enriched in bioactivity and biological relevance and yield novel and unprecedented small-molecule modulators with high hit rates. Thus, BIOS can facilitate overcoming the limitations set by the synthetic tractability of natural products by suggesting NP-inspired compounds.

References

1. E. Zamir and P. I. H. Bastiaens, *Nat. Chem. Biol.*, 2008, **4**, 643–647.
2. C. M. Dobson, *Nature*, 2004, **432**, 824–828.
3. M. Gao and J. Skolnick, *PLoS Comput. Biol.*, 2013, **9**, e1003302.
4. S. C. Patel, L. H. Bradley, S. P. Jinadasa and M. H. Hecht, *Protein Sci.*, 2009, **18**, 1388–1400.
5. R. I. Sadreyev and N. V. Grishin, *BMC Struct. Biol.*, 2006, **6**, 6.
6. D. J. Newman and G. M. Cragg, *J. Nat. Prod.*, 2007, **70**, 461–477.
7. F. Lovering, J. Bikker and C. Humblet, *J. Med. Chem.*, 2009, **52**, 6752–6756.
8. S. Ziegler, V. Pries, C. Hedberg and H. Waldmann, *Angew. Chem. Int. Ed.*, 2013, **52**, 2744–2792.
9. B. E. Evans, K. E. Rittle, M. G. Bock, R. M. DiPardo, R. M. Freidinger, W. L. Whitter, G. F. Lundell, D. F. Veber and P. S. Anderson, *J. Med. Chem.*, 1988, **31**, 2235–2246.
10. M. A. Koch, A. Schuffenhauer, M. Scheck, S. Wetzel, M. Casaulta, A. Odermatt, P. Ertl and H. Waldmann, *Proc. Natl. Acad. Sci. U. S. A.*, 2005, **102**, 17272–17277.
11. A. Schuffenhauer, P. Ertl, S. Roggo, S. Wetzel, M. A. Koch and H. Waldmann, *J. Chem. Inf. Model.*, 2007, **47**, 47–58.

12. C. W. Murray and D. C. Rees, *Nat. Chem.*, 2009, **1**, 187–192.
13. S. Renner, W. A. L. van Otterlo, M. Dominguez Seoane, S. Mocklinghoff, B. Hofmann, S. Wetzel, A. Schuffenhauer, P. Ertl, T. I. Oprea, D. Steinhilber, L. Brunsveld, D. Rauh and H. Waldmann, *Nat. Chem. Biol.*, 2009, **5**, 585–592.
14. M. A. Koch, L.-O. Wittenberg, S. Basu, D. A. Jeyaraj, E. Gourzoulidou, K. Reinecke, A. Odermatt and H. Waldmann, *Proc. Natl. Acad. Sci. U. S. A.*, 2004, **101**, 16721–16726.
15. S. Wetzel, K. Klein, S. Renner, D. Rauh, T. I. Oprea, P. Mutzel and H. Waldmann, *Nat. Chem. Biol.*, 2009, **5**, 581–583.
16. S. V. Frye, *Nat. Chem. Biol.*, 2010, **6**, 159–161.
17. A. Alonso, J. Sasin, N. Bottini, I. Friedberg, I. Friedberg, A. Osterman, A. Godzik, T. Hunter, J. Dixon and T. Mustelin, *Cell*, 2016, **117**, 699–711.
18. T. Hunter, *Cell*, 2016, **100**, 113–127.
19. M. A. Lyon, A. P. Ducruet, P. Wipf and J. S. Lazo, *Nat. Rev. Drug Discovery*, 2002, **1**, 961–976.
20. T. O. Johnson, J. Ermolieff and M. R. Jirousek, *Nat. Rev. Drug Discovery*, 2002, **1**, 696–709.
21. A. Nören-Müller, I. Reis-Corrêa, H. Prinz, C. Rosenbaum, K. Saxena, H. J. Schwalbe, D. Vestweber, G. Cagna, S. Schunk, O. Schwarz, H. Schiewe and H. Waldmann, *Proc. Natl. Acad. Sci. U. S. A.*, 2006, **103**, 10606–10611.
22. C. Reavill, B. Walther, I. P. Stolerman and B. Testa, *Neuropharmacology*, 1990, **29**, 619–624.
23. L. A. Pabreza, S. Dhawan and K. J. Kellar, *Mol. Pharmacol.*, 1991, **39**, 9–12.
24. R. Nawroth, G. Poell, A. Ranft, S. Kloep, U. Samulowitz, G. Fachinger, M. Golding, D. T. Shima, U. Deutsch and D. Vestweber, *EMBO J.*, 2002, **21**, 4885–4895.
25. G. Fachinger, U. Deutsch and W. Risau, *Oncogene*, 1999, **18**, 5948–5953.
26. S. Bäumer, L. Keller, A. Holtmann, R. Funke, B. August, A. Gamp, H. Wolburg, K. Wolburg-Buchholz, U. Deutsch and D. Vestweber, *Blood*, 2006, **107**, 4754–4762.
27. H. Kikuchi, Y. Saito, J. Komiya, Y. Takaya, S. Honma, N. Nakahata, A. Ito and Y. Oshima, *J. Org. Chem.*, 2001, **66**, 6982–6987.
28. M. K. Pathak and T. Yi, *J. Immunol.*, 2001, **167**, 3391–3397.
29. L. Bialy and H. Waldmann, *Angew. Chem. Int. Ed.*, 2005, **44**, 3814–3839.
30. K. Galaktionov, A. K. Lee, J. Eckstein and G. Draetta, *et al.*, *Science*, 1995, **269**, 1575.
31. H. Zinnes and J. Shavel, *J. Org. Chem.*, 1966, **31**, 1765–1771.
32. A. Koul, T. Herget, B. Klebl and A. Ullrich, *Nat. Rev. Microbiol.*, 2004, **2**, 189–202.
33. C. Rosenbaum, P. Baumhof, R. Mazitschek, O. Müller, A. Giannis and H. Waldmann, *Angew. Chem. Int. Ed.*, 2004, **43**, 224–228.
34. H. Dückert, V. Pries, V. Khedkar, S. Menninger, H. Bruss, A. W. Bird, Z. Maliga, A. Brockmeyer, P. Janning, A. Hyman, S. Grimme,

M. Schürmann, H. Preut, K. Hübel, S. Ziegler, K. Kumar and H. Waldmann, *Nat. Chem. Biol.*, 2012, **8**, 179–184.
35. M. Ishikura, K. Yamada and T. Abe, *Nat. Prod. Rep.*, 2010, **27**, 1630–1680.
36. D. T. Hung, T. F. Jamison and S. L. Schreiber, *Chem. Biol.*, 1996, **3**, 623–639.
37. J. R. Peterson and T. J. Mitchison, *Chem. Biol.*, 2002, **9**, 1275–1285.
38. M. J. Lim and X. W. Wang, *Cancer Detect. Prev.*, 2006, **30**, 481–490.
39. S. Hutten and R. H. Kehlenbach, *Trends Cell Biol.*, 2007, **17**, 193–201.
40. V. Eschenbrenner-Lux, P. Küchler, S. Ziegler, K. Kumar and H. Waldmann, *Angew. Chem. Int. Ed.*, 2014, **53**, 2134–2137.
41. M. S. Butler, *Nat. Prod. Rep.*, 2008, **25**, 475–516.
42. T. E. Prisinzano, *J. Nat. Prod.*, 2009, **72**, 581–587.
43. H. J. Jessen, A. Schumacher, T. Shaw, A. Pfaltz and K. Gademann, *Angew. Chem. Int. Ed.*, 2011, **50**, 4222–4226.
44. L. Kuai, X. Wang, J. M. Madison, S. L. Schreiber, E. M. Scolnick and S. J. Haggarty, *ACS Chem.* Neurosci., 2010, **1**, 325–342.
45. A. J. Bauer and B. R. Stockwell, *Chem. Rev.*, 2008, **108**, 1774–1786.
46. P.-Y. Dakas, J. A. Parga, S. Höing, H. R. Schöler, J. Sterneckert, K. Kumar and H. Waldmann, *Angew. Chem. Int. Ed.*, 2013, **52**, 9576–9581.
47. V. Praveen Kumar, R. Gajendra Reddy, D. D. Vo, S. Chakravarty, S. Chandrasekhar and R. Grée, *Bioorg. Med. Chem. Lett.*, 2012, **22**, 1439–1444.
48. X. Cheng, N. Harzdorf, Z. Khaing, D. Kang, A. M. Camelio, T. Shaw, C. E. Schmidt and D. Siegel, *Org. Biomol. Chem.*, 2012, **10**, 383–393.
49. M. Rawat, C. I. Gama, J. B. Matson and L. C. Hsieh-Wilson, *J. Am. Chem. Soc.*, 2008, **130**, 2959–2961.
50. A. P. Antonchick, S. López-Tosco, J. Parga, S. Sievers, M. Schürmann, H. Preut, S. Höing, H. R. Schöler, J. Sterneckert, D. Rauh and H. Waldmann, *Chem. Biol.*, 2013, **20**, 500–509.
51. D. Enders, C. Grondal and M. R. M. Hüttl, *Angew. Chem. Int. Ed.*, 2007, **46**, 1570–1581.
52. T. E. Nielsen and S. L. Schreiber, *Angew. Chem. Int. Ed.*, 2008, **47**, 48–56.
53. W. Liu, V. Khedkar, B. Baskar, M. Schürmann and K. Kumar, *Angew. Chem. Int. Ed.*, 2011, **50**, 6900–6905.
54. R. Misaki, M. Morimatsu, T. Uemura, S. Waguri, E. Miyoshi, N. Taniguchi, M. Matsuda and T. Taguchi, *J. Cell Biol.*, 2010, **191**, 23–29.
55. J. Greaves and L. H. Chamberlain, *J. Cell Biol.*, 2007, **176**, 249–254.
56. M.-J. Bijlmakers and M. Marsh, *Trends Cell Biol.*, 2003, **13**, 32–42.
57. O. Rocks, A. Peyker, M. Kahms, P. J. Verveer, C. Koerner, M. Lumbierres, J. Kuhlmann, H. Waldmann, A. Wittinghofer and P. I. H. Bastiaens, *Science*, 2005, **307**, 1746–1752.
58. D. C. Yeh, J. A. Duncan, S. Yamashita and T. Michel, *J. Biol. Chem.*, 1999, **274**, 33148–33154.
59. J. A. Duncan and A. G. Gilman, *J. Biol. Chem.*, 1998, **273**, 15830–15837.
60. Y. Devedjiev, Z. Dauter, S. R. Kuznetsov, T. L. Z. Jones and Z. S. Derewenda, *Structure*, 2000, **8**, 1137–1146.

61. A. Roussel, N. Miled, L. Berti-Dupuis, M. Rivière, S. Spinelli, P. Berna, V. Gruber, R. Verger and C. Cambillau, *J. Biol. Chem.*, 2002, **277**, 2266–2274.
62. P. Hadváry, W. Sidler, W. Meister, W. Vetter and H. Wolfer, *J. Biol. Chem.*, 1991, **266**, 2021–2027.
63. F. J. Dekker, O. Rocks, N. Vartak, S. Menninger, C. Hedberg, R. Balamurugan, S. Wetzel, S. Renner, M. Gerauer, B. Schölermann, M. Rusch, J. W. Kramer, D. Rauh, G. W. Coates, L. Brunsveld, P. I. H. Bastiaens and H. Waldmann, *Nat. Chem. Biol.*, 2010, **6**, 449–456.
64. Y. Kotelevtsev, M. C. Holmes, A. Burchell, P. M. Houston, D. Schmoll, P. Jamieson, R. Best, R. Brown, C. R. W. Edwards, J. R. Seckl and J. J. Mullins, *Proc. Natl. Acad. Sci. U. S. A*, 1997, **94**, 14924–14929.
65. N. M. Morton, J. M. Paterson, H. Masuzaki, M. C. Holmes, B. Staels, C. Fievet, B. R. Walker, J. S. Flier, J. J. Mullins and J. R. Seckl, *Diabetes*, 2004, **53**, 931–938.
66. M. Kaiser, S. Wetzel, K. Kumar and H. Waldmann, *Cell. Mol. Life Sci.*, 2008, **65**, 1186–1201.
67. G. Chessari and A. J. Woodhead, *Drug Discovery Today*, 2009, **14**, 668–675.
68. S. D. Roughley and R. E. Hubbard, *J. Med. Chem.*, 2011, **54**, 3989–4005.
69. M. G. Siegel and M. Vieth, *Drug Discovery Today*, 2007, **12**, 71–79.
70. K. Babaoglu and B. K. Shoichet, *Nat. Chem. Biol.*, 2006, **2**, 720–723.
71. A. W. Hung, A. Ramek, Y. Wang, T. Kaya, J. A. Wilson, P. A. Clemons and D. W. Young, *Proc. Natl. Acad. Sci. U. S. A.*, 2011, **108**, 6799–6804.
72. K. Grabowski, K.-H. Baringhaus and G. Schneider, *Nat. Prod. Rep.*, 2008, **25**, 892–904.
73. M. Congreve, R. Carr, C. Murray and H. Jhoti, *Drug Discovery Today*, 2003, **8**, 876–877.
74. H. Köster, T. Craan, S. Brass, C. Herhaus, M. Zentgraf, L. Neumann, A. Heine and G. Klebe, *J. Med. Chem.*, 2011, **54**, 7784–7796.
75. B. Over, S. Wetzel, C. Grütter, Y. Nakai, S. Renner, D. Rauh and H. Waldmann, *Nat. Chem.*, 2013, **5**, 21–28.
76. A. Furstner, *Chem. Soc. Rev.*, 2009, **38**, 3208–3221.
77. P. Pérez-Galán, N. J. A. Martin, A. G. Campaña, D. J. Cárdenas and A. M. Echavarren, *Chem. – Asian J.*, 2011, **6**, 482–486.
78. V. López-Carrillo, N. Huguet, Á. Mosquera and A. M. Echavarren, *Chem. – Eur. J.*, 2011, **17**, 10972–10978.
79. J. W. Blunt, B. R. Copp, M. H. G. Munro, P. T. Northcote and M. R. Prinsep, *Nat. Prod. Rep.*, 2005, **22**, 15–61.
80. C. V. Galliford and K. A. Scheidt, *Angew. Chem. Int. Ed.*, 2007, **46**, 8748–8758.
81. K. Ding, Y. Lu, Z. Nikolovska-Coleska, G. Wang, S. Qiu, S. Shangary, W. Gao, D. Qin, J. Stuckey, K. Krajewski, P. P. Roller and S. Wang, *J. Med. Chem.*, 2006, **49**, 3432–3435.
82. J. Briscoe and P. P. Therond, *Nat. Rev. Mol. Cell Biol.*, 2013, **14**, 416–429.
83. J. N. Anastas and R. T. Moon, *Nat. Rev. Cancer*, 2013, **13**, 11–26.

84. K. S. Choi, *Exp. Mol. Med.*, 2012, **44**, 109–120.
85. H. Xu, C. Golz, C. Strohmann, A. P. Antonchick and H. Waldmann, *Angew. Chem.*, 2016, **128**, 7892–7896.
86. G. Grynkiewicz and M. Gadzikowska, *Pharmacol. Rep.*, 2008, **60**, 439–463.
87. S. Cheenpracha, T. Ritthiwigrom and S. Laphookhieo, *J. Nat. Prod.*, 2013, **76**, 723–726.
88. D. O'Hagan, *Nat. Prod. Rep.*, 1997, **14**, 637–651.
89. S. Singh, *Chem. Rev.*, 2000, **100**, 925–1024.
90. F. I. Carroll, S. P. Runyon, P. Abraham, H. Navarro, M. J. Kuhar, G. T. Pollard and J. L. Howard, *J. Med. Chem.*, 2004, **47**, 6401–6409.
91. D. O'Hagan, *Nat. Prod. Rep.*, 2000, **17**, 435–446.
92. R. Narayan, J. O. Bauer, C. Strohmann, A. P. Antonchick and H. Waldmann, *Angew. Chem. Int. Ed.*, 2013, **52**, 12892–12896.
93. R. J. Rafferty, R. W. Hicklin, K. A. Maloof and P. J. Hergenrother, *Angew. Chem. Int. Ed.*, 2014, **53**, 220–224.
94. M. Garcia-Castro, S. Zimmermann, M. G. Sankar and K. Kumar, *Angew. Chem. Int. Ed.*, 2016, **55**, 7586–7605.
95. K. C. Morrison and P. J. Hergenrother, *Nat. Prod. Rep.*, 2014, **31**, 6–14.
96. R. W. Huigens, K. C. Morrison, R. W. Hicklin, T. A. Flood, M. F. Richter and P. J. Hergenrother, *Nat. Chem.*, 2013, **5**, 195–202.
97. V. Amirkia and M. Heinrich, *Phytochem. Lett.*, 2014, **10**, xlviii–liii.
98. L. Laraia, K. Ohsawa, G. Konstantinidis, L. Robke, Y.-W. Wu, K. Kumar and H. Waldmann, *Angew. Chem. Int. Ed.*, 2017, **56**, 2145–2150.

CHAPTER 4

Lead- and Fragment-oriented Synthesis

JAMES D. FIRTH AND PETER O'BRIEN*

Department of Chemistry, University of York, Heslington, York YO10 5DD, UK
*Email: peter.obrien@york.ac.uk

4.1 Introduction

The control of molecular properties is an increasingly important factor in drug discovery, with the properties of clinical candidates being strongly linked to the successful transition to marketed drugs.[1–3] The molecular weight,[1,4] lipophilicity[4–7] and fraction of sp^3 hybridised carbons (Fsp^3)[8,9] have been described as key parameters that should be controlled in the drug discovery process. Indeed, many rules have been proposed to guide medicinal chemists in the control of these key properties.[10–13] For example, Lipiniski's seminal 'rule-of-five' was devised to outline the key properties of orally available drugs,[10] and has since been widely adopted to describe small-molecule drug space. More recently, Hopkins and co-workers have introduced a single metric, the quantitative estimate of drug-likeness (QED), a composite of eight commonly used molecular properties.[14] Whilst the desired molecular properties of drugs[15] are well established, the metrics used are of little use to describe the optimum chemical starting points for the development of such therapeutic entities. In order to describe the starting points for drug discovery, the concepts of lead-like and fragment-like molecules have evolved over the past 20 years to meet the needs of the hit- and lead-discovery communities. The generation of screening libraries

Chemical Biology No. 10
Chemical and Biological Synthesis: Enabling Approaches for Understanding Biology
Edited by Nick J. Westwood and Adam Nelson

Published by the Royal Society of Chemistry, www.rsc.org

containing lead-like or fragment-like compounds is non-trivial and, as a result, the synthetic chemistry community is being encouraged by the pharmaceutical industry to develop methods for the targeted exploration of chemical space.[16–18]

This review focusses on efforts to explore lead-like and fragment-like chemical space for the generation of orally bioavailable clinical candidates.

4.1.1 Introduction to Lead-oriented Synthesis

It has been shown that the hit-to-lead and lead-optimisation processes (in which hits, usually from traditional high-throughput screening (HTS) assays, are optimised to generate a drug candidate) result in an increase in molecular weight and lipophilicity.[19,20] This is an almost inevitable consequence of the increase in structural complexity associated with the medicinal chemist's goals of improving potency, selectivity, metabolism and pharmacokinetics. In order to allow for this drift in molecular properties during lead optimisation, whilst still producing a clinical candidate with drug-like properties, the molecular properties of the lead compounds themselves must be carefully controlled. Consequently, Oprea and co-workers introduced the idea of lead-like compounds, and defined such compounds as having molecular weights (MW) of less than 350 Da and values for the log of the partition coefficient (logP) between 1 and 3.[19] Libraries that comprise such compounds are ideally placed to generate leads with micromolar affinity in high-throughput screens, whilst leaving sufficient room for growth in the lead-optimisation phase. However, over the past two decades, the concept of lead-likeness has been expanded to include molecular parameters other than MW and lipophilicity and the previous value ranges have been modified.[21–23]

In an influential paper from 2012, Churcher and co-workers at GlaxoSmithKline broadly outlined their definition of lead-like space. Firstly, the key physicochemical parameters were defined as: (i) MW in the range of 200–350 Da (14–26 heavy atoms) and (ii) logP values in the range of −1 to +3 (Figure 4.1, red circle).[16] Other physicochemical parameters, such as polar surface area, number of hydrogen bond donors and acceptors and number of rotatable bonds, were not considered. Figure 4.1 shows lead-like chemical space (red circle) in comparison with other commonly used descriptors of chemical space including Lipinski's 'rule-of-five' space for orally bioavailable drugs (blue line),[10] and Astex's original 'rule-of-three'[24] (green line) space for fragments, along with their revised guidelines[17] (yellow oval) (see Section 4.1.2 for discussion of fragment space).

Additionally, as well as MW and lipophilicity, the shape of the lead molecules was also considered to be important. It has been shown that a high degree of three-dimensionality (high Fsp^3 and low aromatic ring count) results in a higher likelihood of molecules successfully progressing through the stages of a drug-discovery project.[6,8] Therefore, Churcher and co-workers[16]

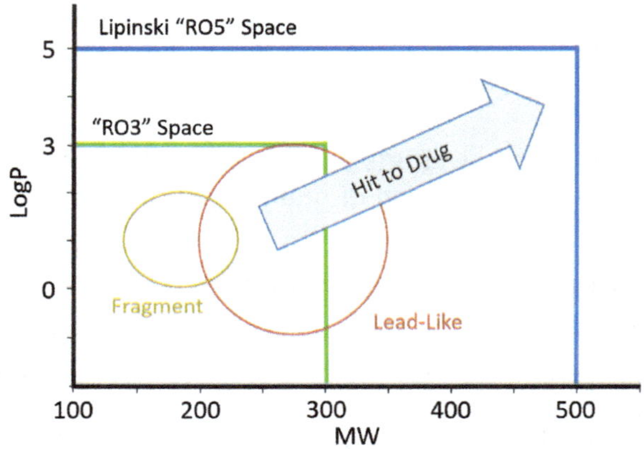

Figure 4.1 Key areas of medicinal chemical space.

have stated that compounds with higher three-dimensional (3D) character and a lower degree of aromatic character should be considered more lead-like than their flatter counterparts. Moreover, they specified that desirable lead-like compounds should not possess problematic functional groups. For example, unstable, electrophilic or redox-active molecules are considered to have a poor lead-likeness.

Churcher and co-workers issued a direct challenge to the academic community to develop synthetic methodology that could deliver diverse lead-like compounds. Such an approach was termed lead-oriented synthesis (LOS) and was introduced in order to assist in the generation of high-quality lead compounds for drug discovery. For a synthetic method to be considered lead-oriented, it should: (i) be able to produce a wide range of lead-like chemical structures; (ii) be efficient, utilising cheap reagents and conditions applicable to array format; (iii) produce molecules without undesirable functional groups; and (iv) be tolerant of a wide range of polar functional groups (*vida infra*).[16]

This call for the development of LOS approaches arises from the fact that many screening libraries and synthetic methods fail to deliver lead-like compounds. One analysis revealed that only 2.6% of almost five million commercially available screening compounds can be considered lead-like.[16] Additionally, analysis of a subset of synthetic methodology papers indicated that the vast majority of compounds synthesised did not fit within lead-like space. Likewise, an analysis of compounds detailed in the *Journal of Medicinal Chemistry* over a 50-year period has shown that the average MW, lipophilicity and 'flatness' of discovery compounds has increased over time.[25] This is partially attributed to the fact that the toolkit of reactions that medicinal chemists routinely use is remarkably narrow[26,27] and biased in terms of capping-group selection.[28,29] Additionally, one analysis revealed that many common methodologies used in array chemistry are more successful with more lipophilic substrates, resulting in successfully-produced compound libraries having higher average logP than the designed compounds, a phenomenon known as 'logP drift'.[16]

These problems for drug discovery are exacerbated by the uneven exploration of chemical space; a 2008 analysis of the Chemical Abstracts Service (CAS) registry showed that around half of all known compounds are based on only 0.25% of the known small-molecule scaffolds,[30] thus leading to a lack of structural diversity in screening libraries.[31] This predominance of the use of unsuitable starting points for drug discovery has resulted in the rise of LOS in an attempt to generate higher quality diverse small molecules for drug discovery. This review will look at the synthetic methods being developed that adhere to the doctrine of lead-oriented synthesis.

4.1.2 Introduction to Fragment-oriented Synthesis

Over the past 20 years, fragment-based drug discovery (FBDD)[32] has progressed from a niche technology to a mainstay of drug discovery.[33] It has developed into a powerful tool for lead generation, with two approved drugs, Vemurafenib[34] and Venetoclax[35] (Figure 4.2) originating from FBDD and an additional 30 compounds having entered clinical trials.[36]

The key to FBDD is the use of low-molecular-weight molecules (typically less than 300 Da) in the screening library, allowing more efficient exploration of chemical space than is possible when using larger molecules.[37] This facilitates the use of small screening libraries (generally 1000–5000 compounds), allows the identification of molecules that bind with high efficiency and potentially helps identify hits for targets that are difficult to drug.

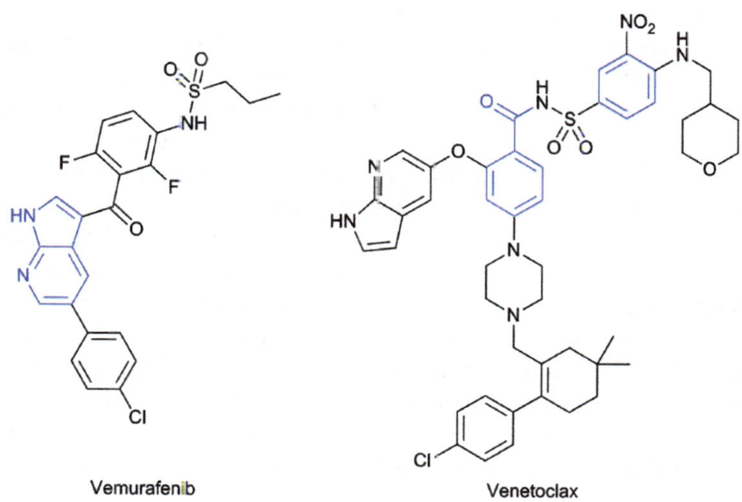

Figure 4.2 Drugs from FBDD projects. The remnants of the original fragments are shown in blue.

The hits from fragment screens are often structurally simple, allowing elaboration into lead compounds that possess scaffolds underrepresented in traditional HTS libraries. Analysis has revealed that fragment screening generates smaller, more polar hits than HTS.[38] However, fragment hits often bind weakly to the protein target (typically with millimolar affinity) and sensitive biophysical screening methods are required. X-ray crystallography is extremely advantageous for fragment elaboration, allowing iterative determination of fragment binding modes and the potential growth vectors. Ultimately, fragment screening adds flexibility to lead-generation programs. However, the careful design of fragment libraries is key to success within FBDD.[39–42]

In 2003, researchers at Astex published their highly influential 'rule-of-three', a set of guidelines (based on analysis of successful fragment hits) describing desirable physicochemical properties for molecules in FBDD screening collections.[24] Key observations were that fragments should have MW of less than 300 Da and group contribution method-based calculated logP (clogP) of less than 3 and three or fewer hydrogen bond donors and acceptors (see Figure 4.1). Importantly, the fragments must be sufficiently soluble for fragment screening, ideally more than 5 mM in 5% DMSO in aqueous media. Astex's guidelines have since been refined to limit MW to 140–230 Da, clogP to between 0 and 2 (see Figure 4.1) and the number of rotatable bonds should be 0–3.[17] It is worth noting that other groups are less stringent with their limits on physicochemical properties of fragments.[43] Additionally, structural considerations, as well as physicochemical properties, now play a significant role in fragment design.[17,44] Each fragment should possess a single pharmacophore (protein binding group), avoid problematic functional groups and have multiple synthetic vectors for fragment growth and a fragment library should contain multiple 3D shapes for each scaffold and pharmacophore.

Whilst there is no strict definition of fragment-oriented synthesis, Astex has recently discussed the need for greater investment in chemistry methodology that can enable FBDD, and outlined the features of FBDD-enabling methodology.[17] This is due to the fact that elaboration of a fragment hit into lead molecules is often difficult and time-consuming. Bespoke chemistry often needs to be developed for the synthesis of the core of the fragment, or to incorporate additional heteroatoms or substituents. A FBDD-enabling method should generate molecules with the preferred physicochemical properties and structural features, namely small, polar molecules with a hydrogen-bonding pharmacophore. Additionally, the methodology should demonstrate (prior to fragment screening) the ability to allow growth of the fragment in multiple directions in the presence of a polar binding group. These desires and limitations make developing synthetic methodology to enable fragment-based drug discovery challenging.

This review will detail achievements in the developing area of fragment-oriented synthesis (see Section 4.3).

4.2 Lead-oriented Synthesis

This section will focus on two classes of synthetic projects. Firstly, those that have attempted to realise lead-oriented synthesis (LOS) by making multiple, diverse and novel scaffolds will be discussed. Secondly, synthetic methodology that generates focussed libraries of lead-like compounds (aligned with the GSK guidelines[16]), which have been proactively analysed will also be detailed.

4.2.1 Diverse and Novel Scaffolds for Lead-oriented Synthesis

In 2015, Nelson, Marsden and co-workers reported an elegant and effective approach to the implementation of lead-oriented synthesis.[45] Their work delivered the unified synthesis of over 50 diverse molecular scaffolds, which, upon decoration with medicinal-chemistry-relevant capping groups, would allow the targeting of broad regions of lead-like space. Notably, computational analysis of the potential products was performed prior to synthesis to ensure that the planned scaffolds would be suitable for the generation of lead-like molecules.

Conceptually, a single key connective reaction initially generates a range of cyclisation precursors armed with functional groups that can undergo further (1 or 2) cyclisation reactions to give a wide range of lead-like scaffolds. In the synthesis phase of the project, 13 cyclisation precursors **3** were prepared from amines **1** (ten alternatives) and allylic carbonates **2** (two alternatives) using a highly stereoselective iridium-catalysed allylic amination, employing phosphoramidate (*R*,*R*,a*R*)-**4** as a ligand (Scheme 4.1A). Importantly, this method had previously been developed to allow the use of highly polar coupling partners, including unprotected amines, thus facilitating the synthesis of highly functionalised cyclisation precursors whilst avoiding 'logP drift'.[46] Next, a toolkit of six cyclisation reactions was employed to transform the 13 cyclisation precursors **3** into 52 lead-like scaffolds (**5–8** are exemplar scaffolds) (Scheme 4.1B). These reactions were Pd-catalysed amino-arylation (Scheme 4.1B, Method A), iodocyclization (Method B), reactions exploiting bifunctional reagents (Methods C and D), (CDI, Method C and chloroacetyl chloride, Method D) ring-closing metathesis (Method E) and lactamisation (Method F). Many of the 13 cyclisation precursors **3** were amenable to several of these methods and, in many cases, two sequential cyclisations (either telescoped or stepwise) could be performed (*e.g.* **3a** → **5**). This led to the formation of a large number of scaffolds from a relatively small number of cyclisation precursors.

The synthesised scaffolds were assessed for their suitability for lead generation. The authors virtually deprotected and decorated the scaffolds either once or twice, choosing from 59 medicinal chemistry capping groups. This created a virtual library of 19 530 probably synthetically tractable compounds. In order to validate this analysis, decoration of scaffold **9** was demonstrated experimentally (Scheme 4.2). Firstly, the Cbz group was

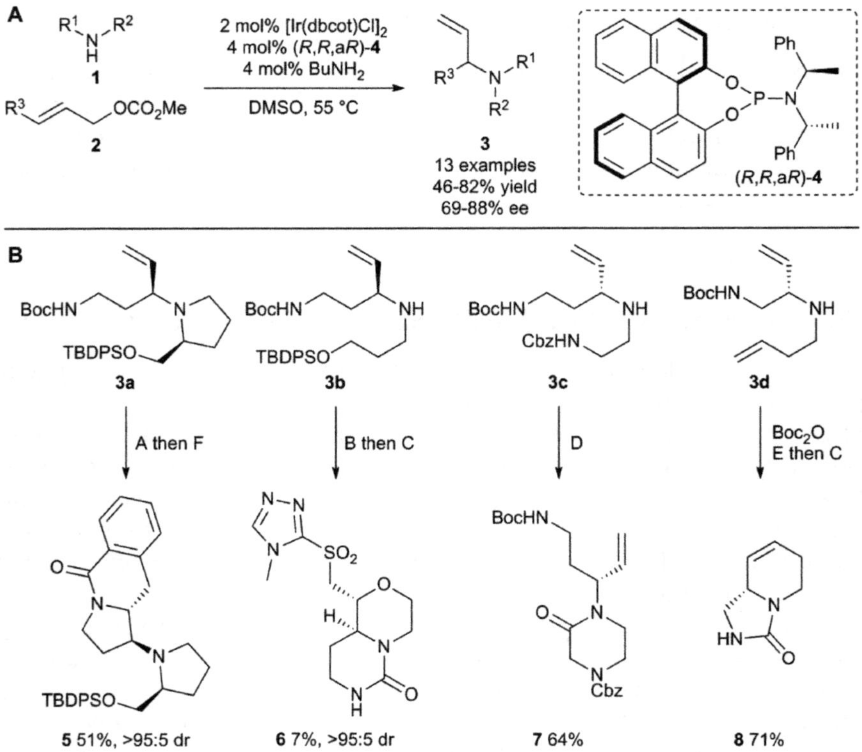

A: Aryl bromide, 5 mol% $Pd(OAc)_2$, 10 mol% DPE-Phos, Cs_2CO_3, 1,4-dioxane, 105 °C; B: (i) NsCl, NEt_3, DMAP, rt, then TBAF, AcOH, THF, rt; (ii) NIS, MeCN, 65 °C; (iii) ArSH, DBU, MeCN, rt; (iv) *m*CPBA, CH_2Cl_2, rt; (v) PhSH, DBU, MeCN, rt; C: CH_2Cl_2, TFA, 0 °C to rt, then CDI, DBU, THF, 50 °C; D: chloroacetyl chloride, NEt_3, CH_2Cl_2, 0 °C to rt, then NaH, NaI, THF, rt; E: 5 mol% Grubbs II, CH_2Cl_2, reflux; F: CH_2Cl_2, TFA, 0 °C to rt, then K_2CO_3, CH_2Cl_2, H_2O, rt;

Scheme 4.1 Lead-oriented synthesis of over 50 scaffolds.

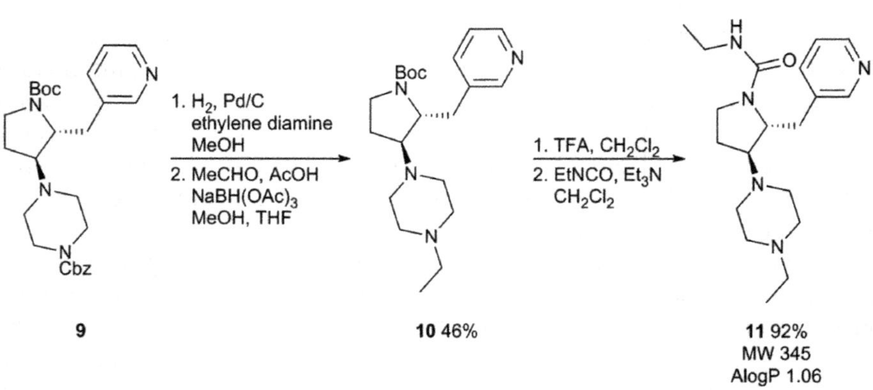

Scheme 4.2 Exemplar decoration of scaffold **9**.

removed and the resulting secondary amine was subjected to reductive amination to give **10** with 46% yield. Then, Boc removal under acidic conditions followed by urea formation gave lead compound **11** with MW 345 Da and atomic logP (AlogP) 1.06.

The lead-likeness of the virtual library was assessed by successively filtering by MW (between 200 and 350 Da), lipophilicity (logP between −1 and +3) and structural features (as outlined by Churcher and colleagues[16]). Over 59% of the virtual compounds were lead-like based upon these criteria; in comparison only 23% of the 9 million compounds in the ZINC database (of commercially available compounds)[47] were considered to be lead-like. Importantly, each of the 52 scaffolds led to significant numbers of lead-like virtual compounds. The shape of the compounds was also assessed and the mean fraction of sp^3 carbons (Fsp^3) for compounds in the virtual library was 0.58, which compares with 0.33 for the ZINC database. Clearly, this LOS approach has the potential to produce compounds that are significantly more three-dimensional in shape than the compounds that are commercially available.

As well as molecular properties, the novelty of the scaffolds was also assessed. In 43 of the 52 cases, the deprotected scaffold was not found as a sub-structure in any compound in the ZINC database. In the nine remaining cases, the compound was not known in the CAS registry. These analyses indicate that the lead-oriented strategy resulted in many novel scaffolds that have the potential to provide large numbers of lead-like compounds for use in drug discovery projects. This piece of work successfully addresses Churcher's four requirements for lead-oriented synthesis: the method developed can produce a wide range of lead-like chemical structures and this was achieved in a way that is efficient, produces molecules without undesirable functional groups and is tolerant of a wide range of polar functional groups.

A second demonstration of lead-oriented synthesis was also reported by Nelson, Marsden and their co-workers.[48] Using small building blocks containing multiple orthogonal functional groups and a toolkit of cyclisation reactions, a wide variety of diverse scaffolds was synthesised rapidly. Conceptually, α-allyl α-amino acid **12** derivatives were appended with functional groups (blue), giving cyclisation precursors **13** (Scheme 4.3A). Subsequent cyclisation on to either ester (green, → **14**) or allyl (red, → **15**) groups generates lead-like scaffolds. Using this approach, scaffold diversity could be generated by using four building blocks **16–19** (Scheme 4.3B).

Boc-protected building blocks **16–19** were used as cyclisation precursors themselves, or deprotected and functionalised at nitrogen to give further substrates for cyclisation. Scheme 4.4 shows five representative examples of such precursors **16** and **20–23**. A small toolkit of reactions was then used to generate 22 lead-like scaffolds in just 49 synthetic operations, ultimately from four simple α-allyl α-amino acids **16–19**. Boc-protected amines (*e.g.* **16**) and ureas (*e.g.* **20**) could be subjected to iodolactonisation (Methods A and B) followed by treatment with NaN_3 to give cyclic carbamates (*e.g.* **24** and **25**).

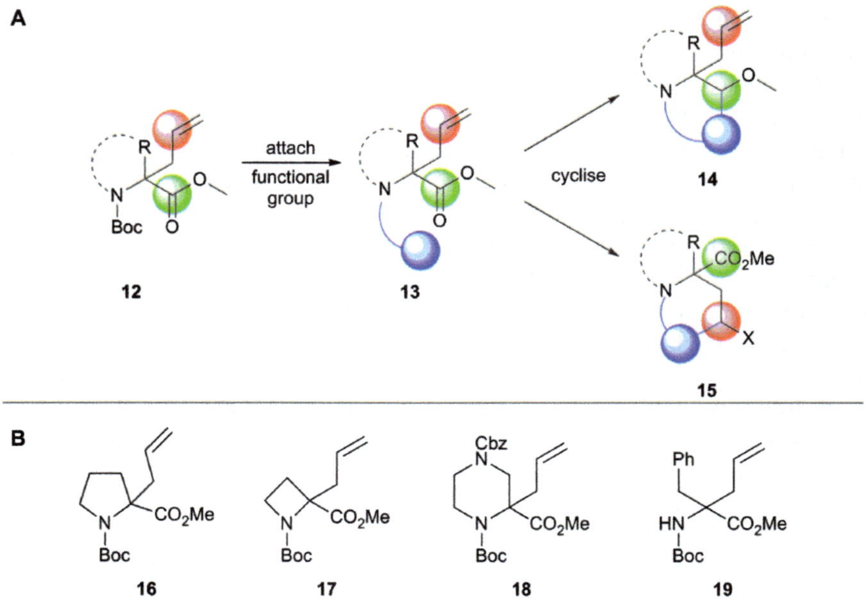

Scheme 4.3 Lead-oriented synthesis using α-allyl α-amino acids. A. Attachment of functional groups (blue) to building block **12**, gives intermediates **13** which may cyclise to give scaffolds **14** and **15**. B. Structures of building blocks **16–19**.

Additionally, ureas were cyclised onto the ester group to generate hydantoins (*e.g.* **20** → **26**, Method C). Likewise, pendant nitrogen nucleophiles could be cyclised on to the esters to give ketopiperazines (*e.g.* **21** → **27**, Method D). Substituents appended to the amine could be cyclised on to the allyl group in two ways. Firstly, allyl-appended substrates could undergo ring-closing metathesis (*e.g.* **22** → **28**, Method E) and secondly, Pd-catalysed Heck cyclisations gave azepanes (*e.g.* **23** → **29**, Method F).

The suitability of the scaffolds for generation of lead-like compounds was assessed. It was found that 66% of 1100 virtual compounds, computationally derived from the 22 scaffolds, occupied lead-like molecular space, by Churcher's definition (*cf.* 23% for the ZINC database). Figure 4.3A shows a comparison of the lead-likeness (MW and AlogP) of both the virtual library (orange) and a random 1% sample of the ZINC database (approximately 91 000 compounds, blue); lead-like space is indicated by the black rectangle.

Additionally, the shape diversity was assessed by analysis of the principal moments of inertia (PMI) (Figure 4.3B).[49] To generate a PMI plot, the normalised principal moments of inertia ratios (I1 and I2) of the lowest energy conformation of the compounds are displayed on a triangular plot, with the three vertices corresponding to rod- (linear-alkyne), disc- (benzene), and spherical- (adamantane) shaped molecules. In general, there is a concentration of commercially available molecules with flat and disc shapes (*i.e.* with low-energy conformations that lie close to the rod–disc axis).[29,50]

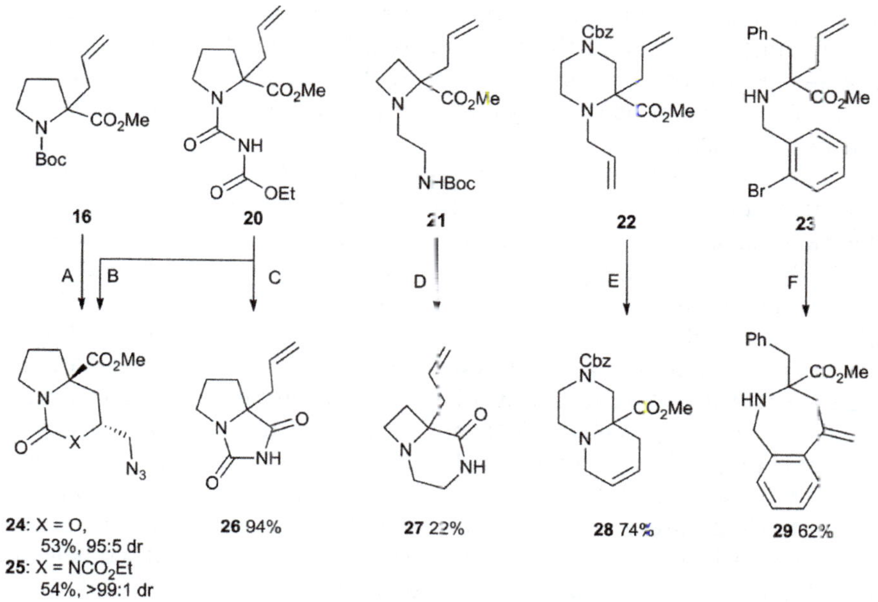

All yields from **16-19**. A: (i) I_2, THF–H_2O; (ii) NaN_3, DMF; B: (i) I_2, Li[Al(O^{t}Bu)$_4$], THF, −5 °C; (ii) NaN_3, DMF; C: NaOMe, 85:15 MeOH–PhMe, 65 °C; D: (i) TFA, CH_2Cl_2; (ii) Cs_2CO_3, DMF, Δ; E: K: 2.5 mol% Grubbs II, TsOH, PhMe, Δ; F: 5 mol% $Pd(PPh_3)_4$, Et_3N, MeCN, μw, 125 °C.

Scheme 4.4 A systematic approach to producing diverse, lead-like scaffolds from α,α-amino acids.

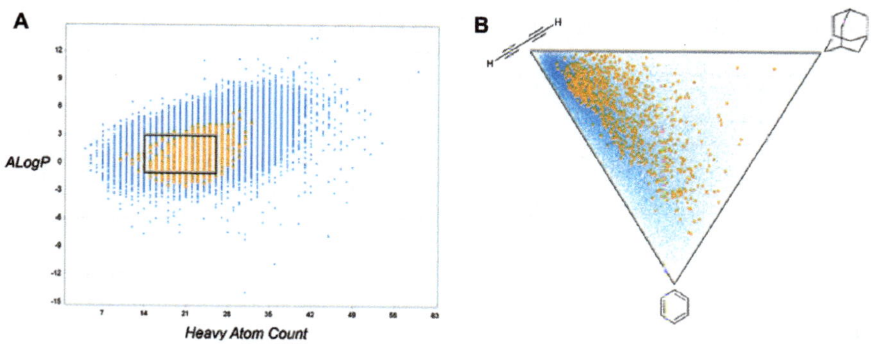

Figure 4.3 Analysis of molecular properties and shape diversity. Based upon the scaffolds prepared using the approach outlined in Scheme 4.4, a virtual library of 1100 compounds (orange) was compared with 90 911 randomly selected compounds from the ZINC database (blue). A. Molecular property distribution (the black rectangle denotes lead-like space). B. Principal moment of inertia (PMI) plot.

Molecules which occupy areas of the PMI plot distant from this well-populated area are of interest as screening compounds. Indeed, the PMI plot of the virtual library (Figure 4.3B, orange) is sufficiently more shape-diverse than that of 1% of the ZINC library (blue), with the ZINC library clustering around the rod–disc axis. Analysis of the fraction of sp^3 carbons (Fsp^3) also shows that the virtual leads were much more three dimensional (mean Fsp^3: 0.57) than compounds from the ZINC database (mean Fsp^3: 0.33).

A novelty assessment against the ZINC and CAS databases revealed that derivatives of only two of the 22 scaffolds were known. Additionally, skeletal diversity was assessed using a hierarchical method[51] that revealed that the scaffolds were structurally diverse and not simply derivatives of each other.

Nelson and co-workers have reported a lead-oriented synthesis of diverse piperazine, 1,4-diazepane and 1,5-diazocane scaffolds.[50] Amino alcohols **30** were deprotonated and added to cyclic sulfamidates **31** (Scheme 4.5A) and after treatment with acid and purification by strong ion exchange chromatography (SCX), open-chain intermediates **32** were isolated. Subsequent exposure to DEAD and PPh_3 resulted in cyclisation to give heterocyclic scaffolds **33** with 8–87% yield. In some cases, purification by SCX was sufficient to yield pure compounds. The use of ion-exchange columns reduced reliance on silica gel column chromatography, expediting scaffold isolation and aligning the methodology with lead-oriented synthesis principles.

A modular synthetic approach was used, combining five different amino alcohols **30** and three cyclic sulfamidates **31**, with 10 final scaffolds being obtained (see Scheme 4.5B for examples **33a–d**). Different combinations of the building blocks resulted in scaffolds with varying ring sizes, carbon substituents and the possibility of fused ring systems.

Analysis of a virtual library derived from the piperazine, 1,4-diazepane and 1,5-diazocane scaffolds **33** showed that 34% of the library was lead-like, whilst shape-diversity analysis by PMI analysis revealed that the library was

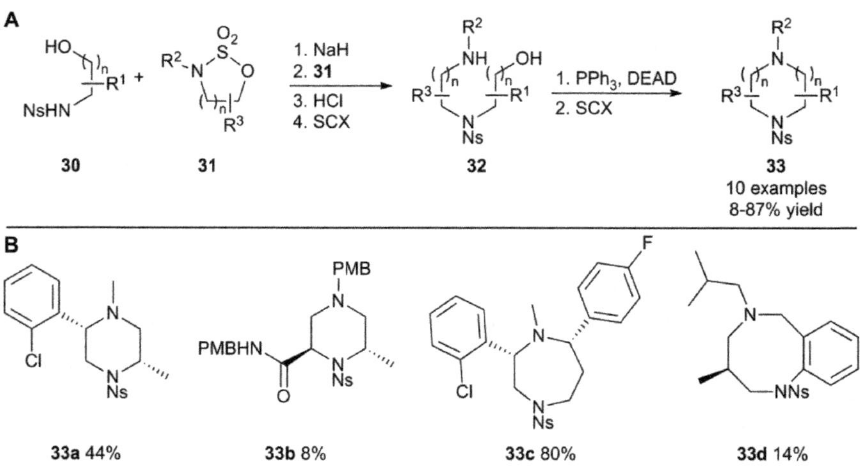

Scheme 4.5 Lead-oriented synthesis of piperazine, 1,4-diazepane and 1,5-diazocane scaffolds. A. General synthetic route to scaffolds **33**. B. Exemplar scaffolds.

significantly more three-dimensional than the ZINC database of commercial compounds.

The assessment of lead-likeness required access to computational tools for the enumeration and analysis of the virtual library, most of which are not available to the majority of academic researchers. To overcome this significant limitation, Nelson, Marsden and co-workers have developed LLAMA (Lead-Likeness And Molecular Analysis; llama.leeds.ac.uk), an open-access tool for the enumeration and analysis of small-molecule scaffolds.[52] The lead-likeness of a library is assessed by assigning a 'lead-likeness penalty' to the virtual compounds, rather than the previously used method of successive filtering based upon strict adherence to molecular property limits. This approach penalises properties and features that are not lead-like (Figure 4.4). For example, a compound with 23 heavy atoms, an AlogP of 3.7 and 3 aromatic rings would incur a lead-likeness penalty of 3.

The analysis of the shape of the virtual library is assessed not only by Fsp[3] but also by PMI[49] and by deviation from the plane of best fit.[53] Novelty is assessed by comparing the corresponding Murcko assemblies[54] with those of, or embedded in, a random 2% of the ZINC database.

The power of LLAMA to assess the lead-likeness of a synthetic approach was demonstrated on a divergent method to convert differentially protected unsaturated diamines into a range of alternative scaffolds (Scheme 4.6). Treatment of orthogonally-protected linear diamine **34a** with iodine, under basic conditions, resulted in cyclisation of the Boc nitrogen onto the intermediate iodonium ion. After treatment with LiHMDS, followed by hydrolysis of the intermediate oxazoline, the pyrrolidine **35** was obtained with 60% yield. In contrast, iodocyclisation of **34b** proceeded through the Boc carbamate. Base-induced rearrangement, followed by displacement of the resulting primary iodide with an aryl thiol and subsequent oxidation gave scaffold **36** at 46% yield over four steps (Scheme 4.6A). Additional scaffolds

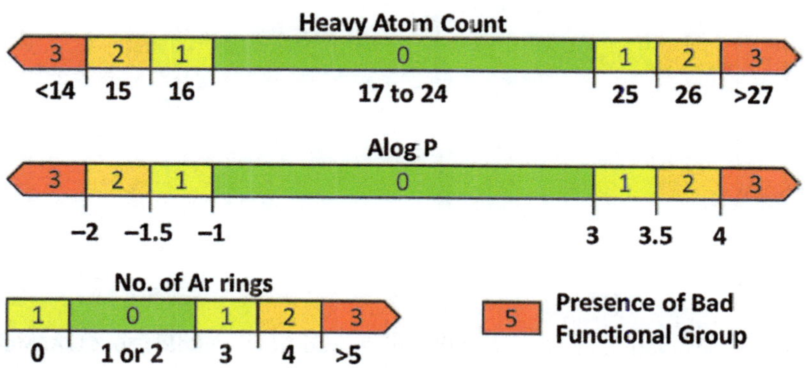

Figure 4.4 Contributions to the lead-likeness penalty used by the LLAMA open-access computational tool.
Adapted from ref. 52 with permission from The Royal Society of Chemistry.

37–39 were derived from pyrrolidine **35** through secondary cyclisation processes, whilst **40** and **41** could be obtained from **34b** using Pd-catalysed amino arylation chemistry (Scheme 4.6B).

LLAMA was used to generate a virtual library of compounds from the scaffolds **36–41**. Analysis of these compounds revealed that the average lead-likeness penalty of the virtual library was 1.57. In comparison, the average score of the ZINC database was 4.17. Figure 4.5A shows a plot of the molecular properties of the virtual library, coloured by lead-likeness penalty (green: 0, orange: 3 Red: 6+) with lead-like space indicated. A significant proportion of the library occupies this space.

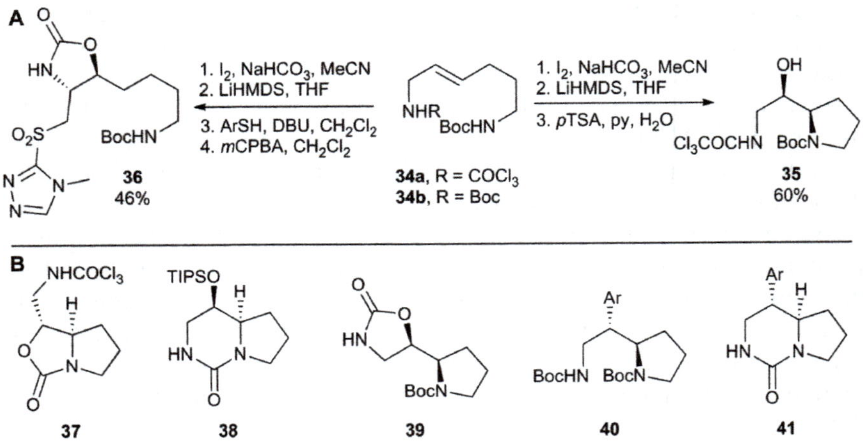

Scheme 4.6 Divergent cyclisations of unsaturated diamines. A. General synthetic route to key intermediate **35** and scaffold **36**. B. Scaffolds **37–41** derived from **35**.

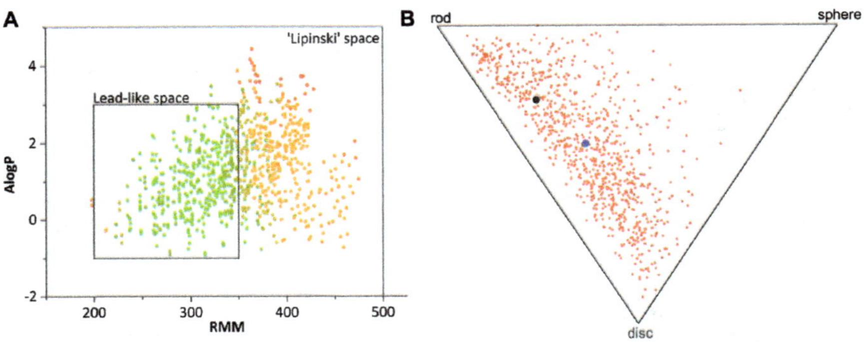

Figure 4.5 Evaluation of the virtual library derived from scaffolds **36–41**. A. Molecular properties (coloured according to lead-likeness penalty: 0, green; 3, orange; 6+ red). B. PMI plot, virtual library (red) and centre of gravities for the virtual library (blue) and 2% of the ZINC database (black). Adapted from ref. 52 with permission from The Royal Society of Chemistry.

Figure 4.5B shows the PMI plot of the virtual library (red dots), whilst the mean of the library is shown with a blue dot. It can be seen that the virtual library is more 3D than the ZINC library (the mean of 2% of the ZINC library is indicated by a black dot).

4.2.2 New Synthetic Methods for Lead-oriented Synthesis

Grygorenko and co-workers have reported simple scaffolds that allow the synthesis of compounds that fall within lead-like space.[55,56] More recently, a lead-oriented one-pot Castagnoli reaction was employed to generate a library of lead-like molecules.[57] A combinatorial approach was used, with condensation of cyclic anhydrides and imines, preformed from primary amines **42** and aldehydes **43**, giving pyrrolidinones and piperidinones **44** (Scheme 4.7A). A 125-member library was synthesised from random combinations of two cyclic anhydrides, 44 amines and 44 aldehydes with 67–88% yields and generally >85:15 dr (see Scheme 4.7B for examples **44a–c**). The molecular properties of the library were assessed, with most of the library falling within lead-like space; the mean molecular weight was 335 Da, the mean clogP was 1.29 and the mean Fsp^3 was 0.51. The simple and rapid synthesis of the library make this one-pot approach suitable for lead-oriented synthesis. Indeed, the approach has recently been adapted to generate seven-membered benzo-azepine analogues.[58]

Our group has employed the directed α-lithiation of *N*-Boc heterocycles for lead-oriented synthesis.[59] Six scaffolds were synthesised based on piperazine, pyrrolidine and piperidine ring systems (Scheme 4.8) and were chosen

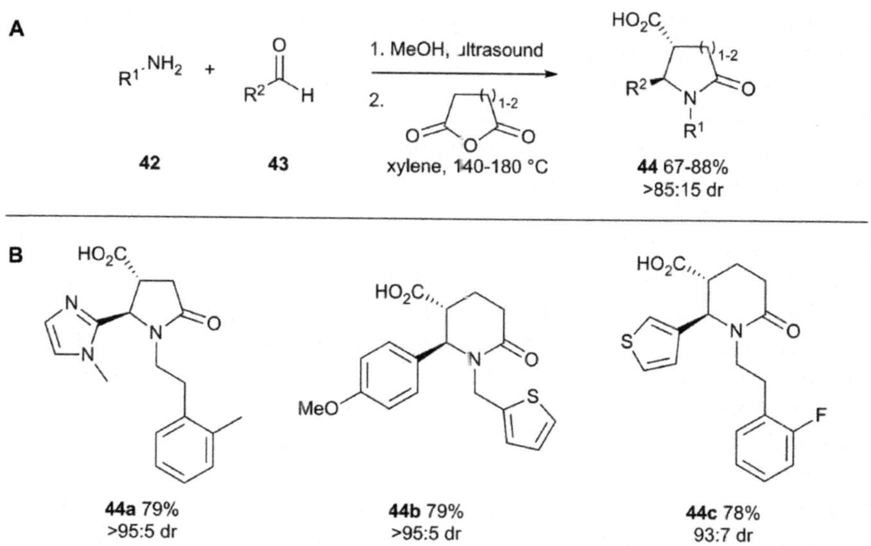

Scheme 4.7 One-pot synthesis of lead-like compounds *via* the Castagnoli reaction. A. General synthetic route to scaffolds **44**. B. Exemplar scaffolds.

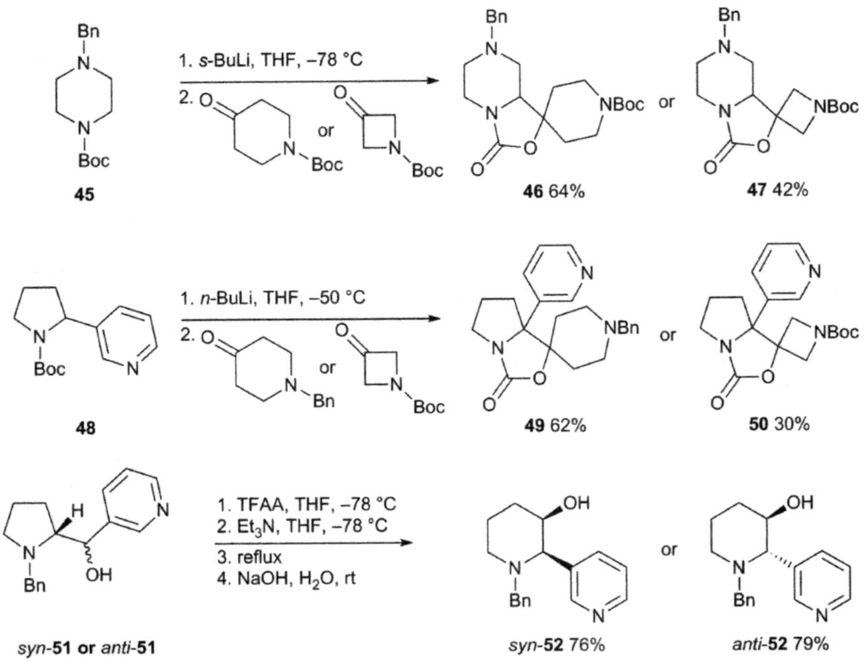

Scheme 4.8 Synthesis of heterocyclic scaffolds *via* α-lithiation.

due to their predicted 3D shape, their ease of synthesis and the fact they had two or three points of diversification. Lithiation of protected *N*-Boc piperazine **45** with *s*-BuLi in THF followed by trapping with heterocyclic ketones gave tricyclic scaffolds **46** and **47** with 64% and 42% yield respectively. Likewise, Boc-directed benzylic lithiation–trapping of pyridinyl *N*-Boc pyrrolidine **48** gave scaffolds **49** and **50.** Piperidines **52** were accessed through a five- to six-ring expansion process from amino alcohols *syn*-**51** and *anti*-**51** (derived from the α-lithiation of *N*-Boc pyrrolidine).

A virtual library of 190 compounds derived from scaffolds **46, 47, 49, 50** and **52**, and close analogues, was assessed for their lead-likeness. Using Churcher's criteria[16] and the extra criterion of total polar surface area (PSA)$\geq$50 Å^2, 92 (48%) of the 190 compounds can be considered lead-like. Shape diversity analysis was performed by assessing the PMI and revealed that the library was shape-diverse, with a significant proportion of virtual compounds occupying underpopulated areas of chemical space. These attributes, along with the compatibility of the lithiation–trapping methodology with the polar lithiation substrates and heterocyclic electrophiles makes organolithium chemistry suitable for lead-oriented synthesis.

Recently, Unsworth used LLAMA to assess medium-sized lactam scaffolds, synthesised through a ring-expansion method.[60] Functionalised β-keto esters **53** underwent *C*-acylation with acid chlorides **54**, and upon removal of the pendant Fmoc group (to give amine **55**), spontaneous ring expansion occurred to give lactams **56** (Scheme 4.9A). Using this method, 30 lactams

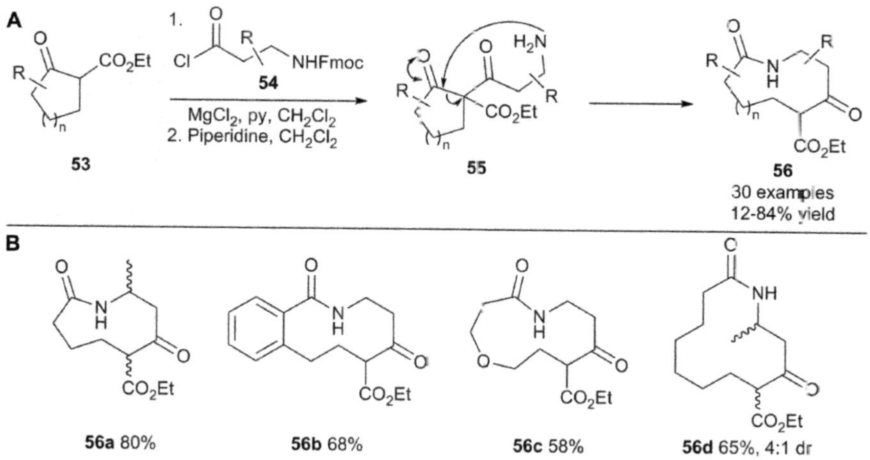

Scheme 4.9 Synthesis of medium ring lactam scaffolds. A. General synthetic route to scaffolds **56**. B. Exemplar scaffold precursors.

were synthesised with a variety of ring sizes (8–12-membered) with 12–84% yield (see Scheme 4.9B for examples **56a–d**). Decarboxylation of the resulting β-keto esters could be achieved under basic conditions giving lead-like scaffolds suitable for library synthesis.

Analysis of the 30 potential scaffolds using LLAMA revealed that of the 402 virtual compounds that were derived from these scaffolds (using a single decoration), 283 (70%) fell within lead-like chemical space (Figure 4.6A), and the library had a mean lead-likeness penalty of 1.30 (*ca.* 4.17 for the ZINC database). A PMI plot of this virtual library (Figure 4.6B) reveals that a wide area of the chemical space is covered and that the average PMI coordinates (blue cross) are well away from the rod–disc axis. Unsurprisingly, given the relatively high MW of scaffolds, a virtual library of compounds derived by decorating the scaffolds twice led to a predominance of compounds that occupy drug-like space rather than lead-like space.

Whilst there are several reports of methodology being analysed for its suitability for lead-oriented synthesis, there are many other synthetic methods that would be suitable but have not been considered in such terms. A complete analysis of existing synthetic methodology is far beyond the scope of this review. However, Nelson and co-workers have performed retrospective analyses of potentially suitable methodology, whilst highlighting the need for synthetic chemists to demonstrate the value of new methodology by synthesising lead-like scaffolds.[29,61] For example, Bode's approach to nitrogen heterocycles using aldehydes or ketones and SnAP reagents[62,63] was assessed.[29] Nelson and co-workers considered the condensation of known amino stannanes (SnAP reagents) **57** with benzaldehyde **58** to give heterocycles (Scheme 4.10A). Using LLAMA, nine possible heterocyclic scaffolds **58a–i** were decorated once and the molecular

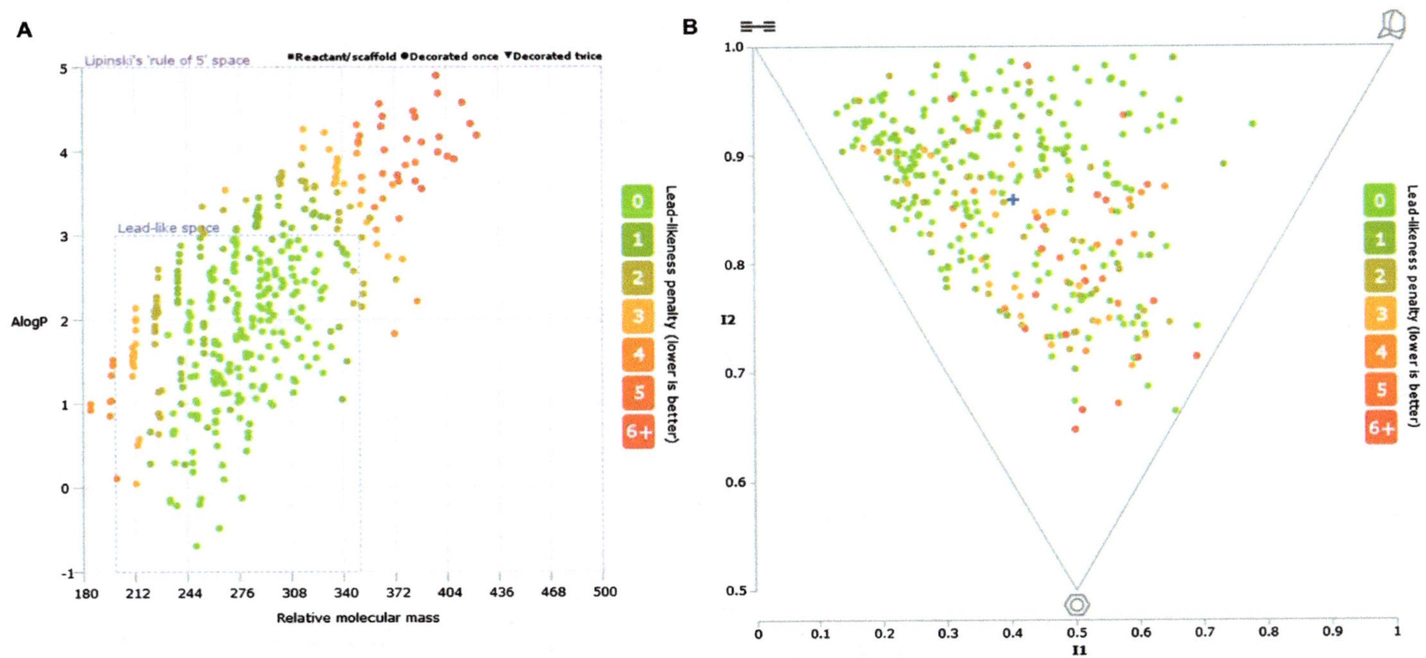

Figure 4.6 Lead-likeness and shape analysis of virtual library of lactams based upon scaffolds prepared using the approach described in Scheme 4.9. A. Molecular properties of the virtual library, coloured according to lead-likeness penalty. B. PMI plot of the virtual library, coloured according to lead-likeness penalty, and centre of gravity for the virtual library (blue cross).
Adapted from L. G. Baud, M. A. Manning, H. L. Arkless, T. C. Stephens and W. P. Unsworth, Ring Expansion Approach to Medium Sized Lactams and Analysis of Their Medicinal Lead Like Properties, *Chem. – Eur. J.*, 2017, **23**, 2225–2230.[60] Copyright Wiley-VCH Verlag GmbH & Co. KGaA. Reproduced with permission.

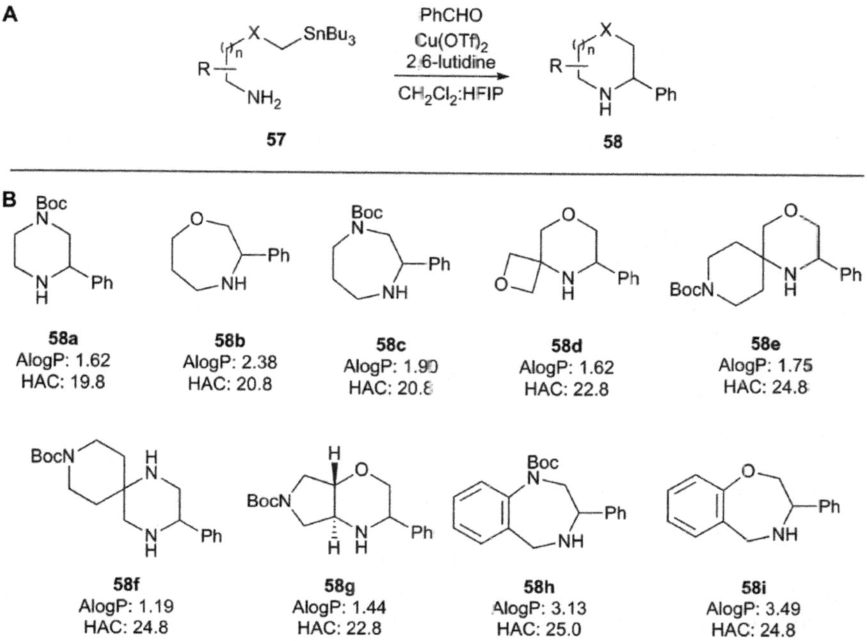

Scheme 4.10 Retrospective analysis of the suitability of SnAP reagents for the synthesis of lead-like scaffolds. A. General synthetic route to scaffolds **58**. B. Exemplar scaffolds with mean AlogP and HAC for compounds derived from each scaffold after deprotection and one decoration.

properties analysed. Scheme 4.10B shows the mean AlogP and heavy atom count (HAC) for the derivatives of scaffolds **58a–i.**

With the exception of benzo-fused scaffolds **58h** and **58i**, all scaffolds resulted in virtual derivatives that fell within lead-like space. Nelson and co-workers also used LLAMA to assess the shape diversity of the virtual library by PMI analysis and showed that the compounds were shape-diverse and very few of the derivatives were close to the rod–disc axis. A similar analysis was performed on a range of azetidine-containing scaffolds devised for CNS drug discovery.[64]

4.2.3 Natural Products as Inspiration for Lead-like Libraries

Natural products have long been a source of drug molecules, with approximately a third of small-molecule drugs approved between 1981 and 2010 being inspired by natural products.[65] They occupy different chemical space to that of commercially available screening libraries,[66–68] having a much higher Fsp^3 and number of chiral centres. They also have lower nitrogen atom count and higher oxygen atom count and often contain ring systems that are not found in synthetic compounds.[69] However, natural products are often structurally complex, making analogue and library generation demanding. Consequently, significant effort has been devoted to the synthesis of natural-product-like libraries for drug discovery.[70–72] In addition, such libraries may be exploited in biology-oriented synthesis (BIOS)[73] to facilitate the discovery of small-molecule tools for interrogating biological mechanisms (Chapter 3).

One approach to developing natural-product-inspired libraries involves retaining the structural features that differentiate natural products from synthetic molecules whilst simplifying the structures sufficiently to enable expedient synthesis. Collins and co-workers applied such an approach to the synthesis of a lead-like library of compounds inspired by the 2,11-cembranoids, a complex family of diterpenoid natural products. They fall into related structural classes of cladiellins, briarellins and asbestinins (Scheme 4.11A) and can be further oxygenated or unsaturated.[74] To develop simplified analogues of the 2,11-cembranoids, Collins chose to exploit the Diels–Alder cyclisation between substituted butadienes **59** and unsaturated lactone **60** to give a simplified 6,5-ring system **61**. These key intermediates were elaborated to either further functionalise the bicyclic core or to install a third medium or large ring **62** (Scheme 4.11B). Ring-closing metathesis, lactonisation and Reformatsky methods were used to install 9–12 membered rings, to give compounds with a range of oxygenation and substitution patterns, for example **62a–c** (Scheme 4.11C). Additionally, several substituted bicyclic analogues were prepared (*e.g.* **62d**).

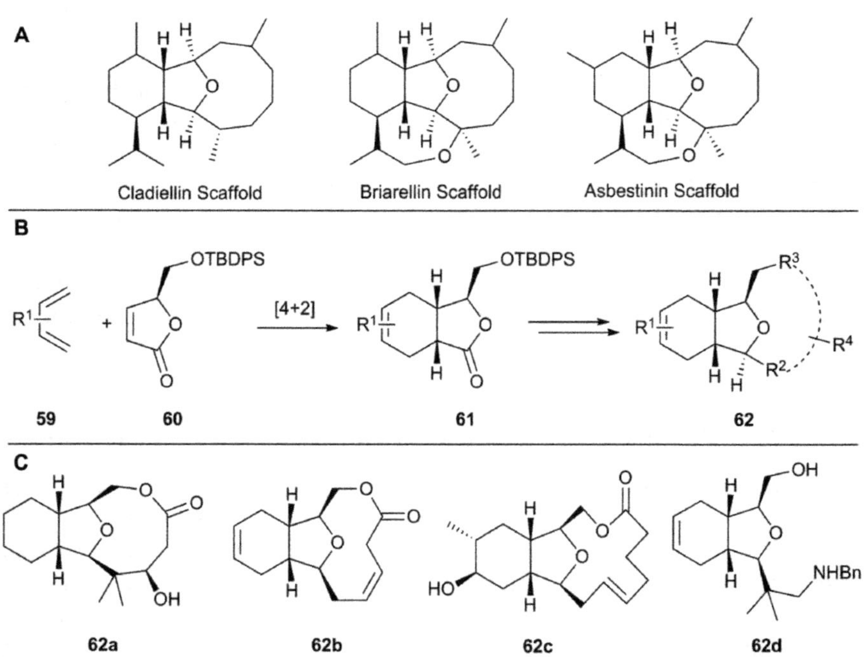

Scheme 4.11 Library of 2,11-cembranoids. A. Structure of cladiellins, briarellins and asbestinins. B. General synthetic route to scaffolds **62**. C. Exemplar scaffolds.

Using these methods, a 44-member library of screening compounds was synthesised. The mean MW of the library was 275 Da, AlogP was 2.6 and the number of stereocentres was 4.9, placing the library firmly within lead-like space. Analysis of the chemical space of the synthetic library by principal component analysis (PCA) showed that the library bridged the chemical space between 2,11-cembranoid natural products and classical screening compounds. One drawback of this synthetic programme is that, whilst being simpler than the corresponding natural products and occupying an interesting area of chemical space, significant synthetic effort was required.

Pascolutti and Quinn have retrospectively analysed 24 natural product derived and inspired libraries for lead-likeness and discussed the strategies required to synthesise libraries with the desired molecular properties.[75] Many of these libraries fall within lead-like chemical space.

It is worth noting that many of the compounds generated in natural-product-like libraries do not adhere to strict lead-like and drug-like criteria. However, these rules were developed with properties and structures of synthetic compounds in mind and there is a school of thought that suggests that these rules should not be applied in a natural-product context.[66]

4.2.4 The European Lead Factory

The European Lead Factory (ELF) is a public–private partnership that provides a unique platform for translation of innovative biology and chemistry into high-quality starting points for drug discovery.[76] As part of this collaboration, the Joint European Compound Library (JECL) has been established, consisting of over 500 000 high-throughput screening compounds.[77] It is envisaged that over 200 000 of these compounds will originate from a collaboration between ten academic groups and six small and medium enterprises (SMEs). The goal is that these compounds will populate biologically-relevant chemical space that is typically not addressed in traditional screening collections. Unlike Churcher's definition of lead-like compounds, with a focus on molecular properties, the compounds synthesised by the ELF consortium are focussed on structural novelty and molecular shape diversity,[78] with the compounds generally designed to fit within Lipiniski chemical space. The scaffolds themselves should generally have two or more points of diversification, an absence of problematic structural features and should be synthetically tractable.

By late 2016, over 120 000 compounds had been synthesised from over 220 scaffolds (and >180 000 compounds based on >300 scaffolds by early 2018).[76] In 2015, an in-depth analysis of the molecular properties of the first 54 831 compounds, synthesised in the first 18 months of the project, was performed.[78] It was found that over 85% of the library compounds had a clogP of less than 4, in comparison with only 62% of a commercially available screening collection. However, 58% of the compounds had MW of over 400 Da. Crucially, the major difference from screening libraries is the shape of the molecules. Analysis showed that 85% of the compounds were chiral, with 62% having two or more chiral centres (*cf.* 3% for a Maybridge screening collection). Additionally, 68% of compounds have Fsp^3 of over 0.4 and analysis using PMI and plane of best fit methods revealed the compounds to be significantly more globular than traditional screening compounds. Novelty assessment indicated that there was no structural overlap with the Maybridge, MLP or ChEMBL libraries. Analysis of the topological and skeletal frameworks indicated that 56% were unique, with the library containing a significant number of unprecedented spiro, bridged and fused ring systems. Even though the majority of compounds fall outside lead-like space, the European Lead Factory has succeeded in generating a significant amount of novel, shape-diverse material for HTS. Importantly, analysis of the first 106 unique scaffolds revealed that they were small (80% with MW of less than 200) and polar (70% with TPSA of 40–80 Å). Whilst the use of two or three diversification steps increases the number and diversity of compounds, it often increases the MW beyond the limits of lead-like space. Given the small size of the scaffolds, it is feasible that performing a single decoration would generate structurally interesting leads that would fit within lead-like space (MW between 200 and 350 Da, logP between 1 and 3).

The European Lead Factory project has resulted in many synthetic papers in which analysis of the molecular properties and/or shape of the compounds has been reported.[79–98] Table 4.1 gives the structure of some selected scaffolds that have been exploited to enhance the JECL.

4.3 Fragment-oriented Synthesis

Astex's call for synthetic methods that can enable FBDD was only made in 2016,[17] and to date only a couple of reports of methodology that allows growth of the fragment in multiple directions have been detailed. Therefore, this section additionally focuses on the published research activities of academia and industry in the generation of fragment molecules that obey the original 'rule-of-three' proposed by Astex.[24] The strategies and approaches to access fragment libraries and the use of natural products are discussed along with single pieces of methodology that make compounds with the appropriate molecular properties.

Interestingly, most of the reports from academic groups have focussed on the synthesis of 3D fragments. These reports often cite the fact that most traditional fragment libraries contain flat structures[99] and that an increased Fsp^3 has been shown to lead to a greater chance of success in a compound's progression through clinical trials.[8] It is widely believed that one potential advantage of 3D fragments is the fact that they may prove more successful at generating hits for demanding targets, such as protein–protein interactions (PPI).[100,101] However, there is evidence that planar fragments generally result in higher hit-rates in X-ray-based fragment screens than more 3D fragments, even against PPI targets.[102]

In 2011, Young and co-workers reported the application of diversity-oriented synthesis (DOS)[103,104] to the generation of a structurally diverse 3D fragment library.[105] DOS seeks to generate skeletal, stereochemical,

Table 4.1 Selected European Lead Factory scaffolds.

Scaffold(s)	Library size	Reference
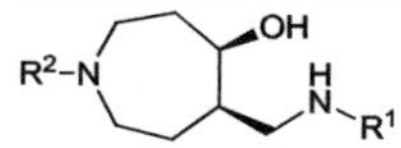	>500	79
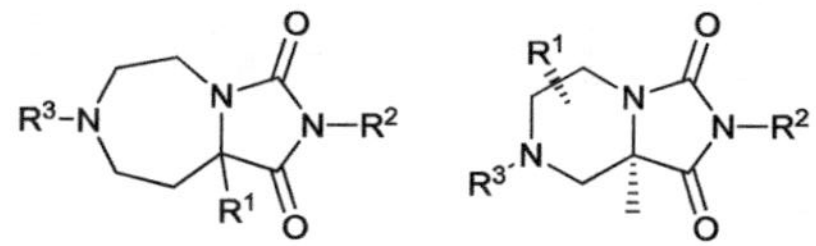	>500	80
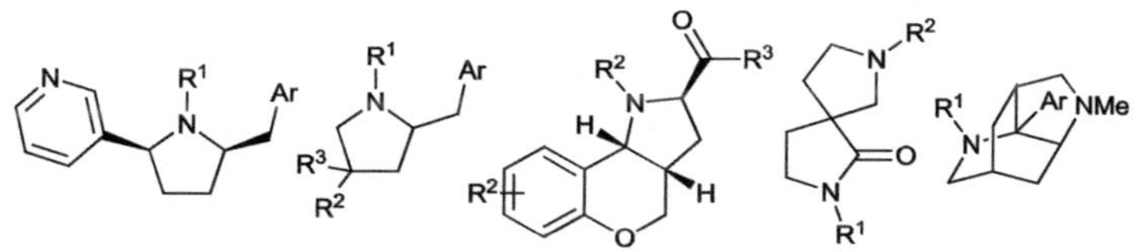	>2500	81
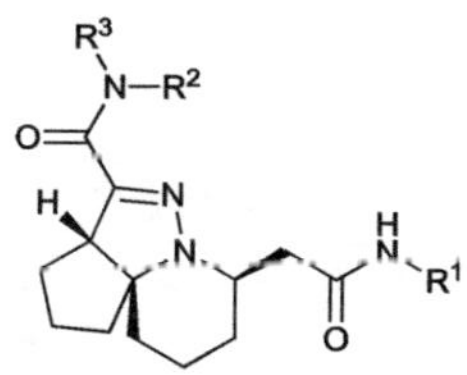	—[a]	98
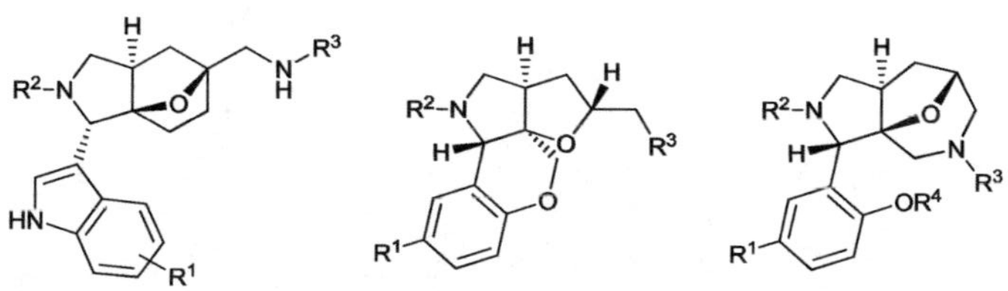	120 + 1617	89, 95
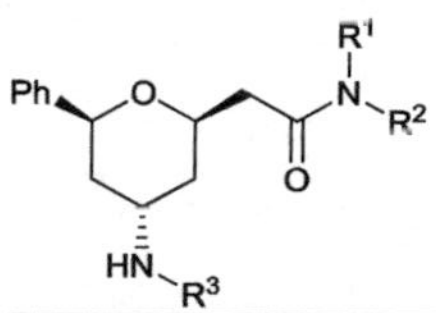	768	92

[a]Not disclosed.

regiochemical and substitution variation using a common synthesis scheme, and to provide broader variation of molecular structure than other approaches (see Chapter 2). Whilst DOS has been used to generate molecules for HTS, the molecules produced are often too large for lead- or fragment-oriented synthesis. However, Young successfully merged DOS and fragment-oriented synthesis, employing the build–couple–pair (B–C–P) strategy[106] to synthesise 35 'rule-of-three' compliant fragments.

Initially, three proline-based building blocks **63–64** were 'built' (Scheme 4.12A). These pyrrolidines were chosen as numerous substituted pyrrolidines can be subsequently obtained for follow-up phases of the

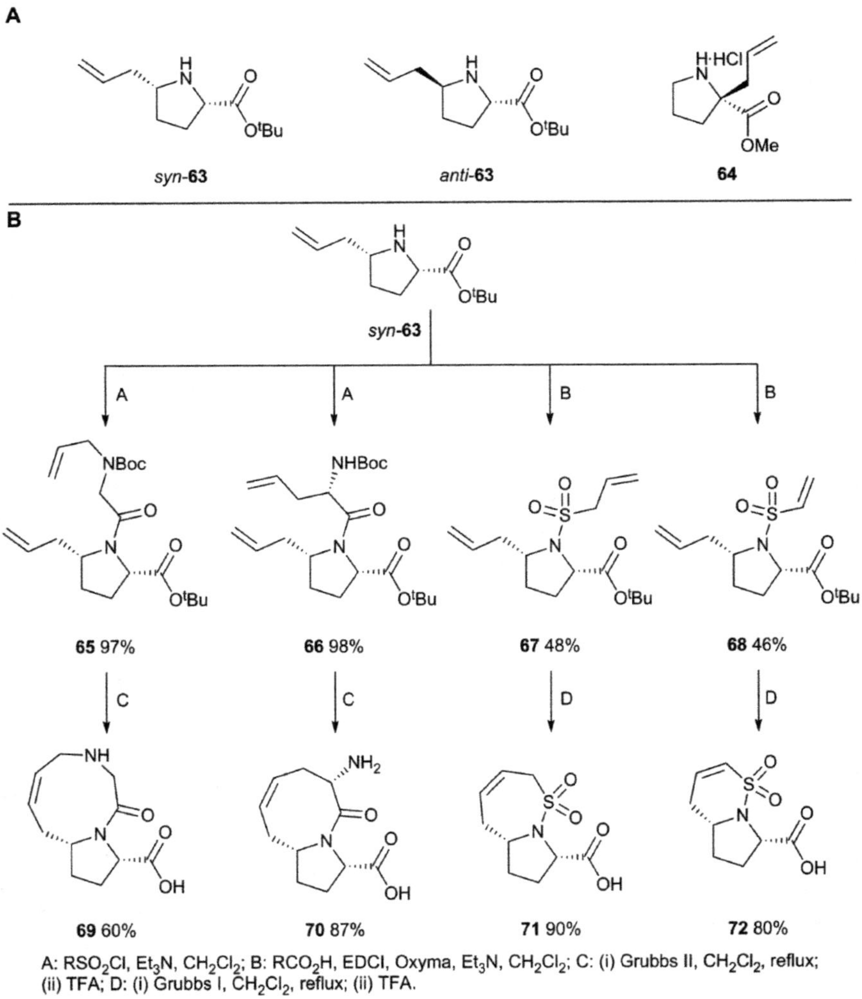

Scheme 4.12 Example of a build–couple–pair method for fragment synthesis. A. Building blocks **63** and **64**. B. Exemplar couple and pair phases, giving fragments **69–72**.

discovery process. In exemplar couple and pair phases, *syn*-**63** was first 'coupled' (through intermolecular reactions) to give multifunctional pyrrolidines **65–68** (Scheme 4.12B). These were then 'paired' (cyclised, in this case using ring-closing metathesis), to generate structurally and functionally diverse bicycles **69–72**. Notably, 5–6, 5–7, 5–8 and 5–9 fused bicyclic ring systems were synthesised.

Similar approaches were applied to the diastereomeric building block *anti*-**63** (giving all possible diastereomers of compounds **69–72**) and to quaternary pyrrolidine **64**, yielding a range of fused and spirocyclic bicyclic fragments. Then, in what was termed a 'postpairing' phase, functional group interconversions were used to increase fragment diversity, namely ester hydrolysis and alkene reduction. Ultimately, the use of this DOS strategy allowed short and efficient routes to 35 fragment molecules with diverse shapes, stereochemistry and physical properties. All fragments had MW of less than 300 Da and AlogP of less than 3 and analysis of their shapes by PMI indicated that they were more 3D than compounds in the ZINC library and covered a wider area of chemical space, a difference that was attributed to the high Fsp^3 of the library. In 2016, the Young group published a second DOS approach to 3D fragments, discovering and optimising hits for glycogen synthase kinase 3β (GSK3β), without the need for structural information.[107]

Fragments derived from natural products can also provide valuable starting points for the discovery of bioactive molecules (see Chapter 3). In one example, workers at Novartis have recently described their efforts to generate 3D natural-product-like fragment libraries.[108] Their strategy was based upon the modification of existing natural products within the Novartis collection and used three approaches. Firstly, *in silico*-guided degradation of larger natural products was investigated. Degradation products were computationally generated by subjecting 17 000 natural products to cleavage reactions including ozonolysis, ester hydrolysis and Baeyer–Villiger oxidation. The resulting compounds were filtered using fragment-like criteria (MW between 150 and 300 Da, clogP less than 3) giving 9000 virtual fragments. Next, PMI analysis and novelty assessment against an in-house collection of fragments was performed to further filter the possible targets. Once synthetic tractability and literature precedent had been considered, fragments of interest were synthesised and unsuitable functional groups (introduced by the cleavage methods) were removed. For example, Sanglifehrin A **73** was subjected to known oxidative cleavage conditions to generate intermediate **74** (Scheme 4.13). Subsequent functional group manipulation gave the natural-product-like fragment **75**.

Secondly, chemical modification of small natural products to generate suitable fragments was performed. For example, modification of Massarigenin C **76** resulted in fragments **77–81** (Scheme 4.14). Reduction *via* hydrogenation gave ketone **77**, whereas reduction with sodium triacetoxyborohydride gave alcohol **78**. Rearrangement of **76** under acidic conditions proceeded by ester hydrolysis, Michael addition then aldol condensation gave fragment **79**. Further modification was then performed to give **80**

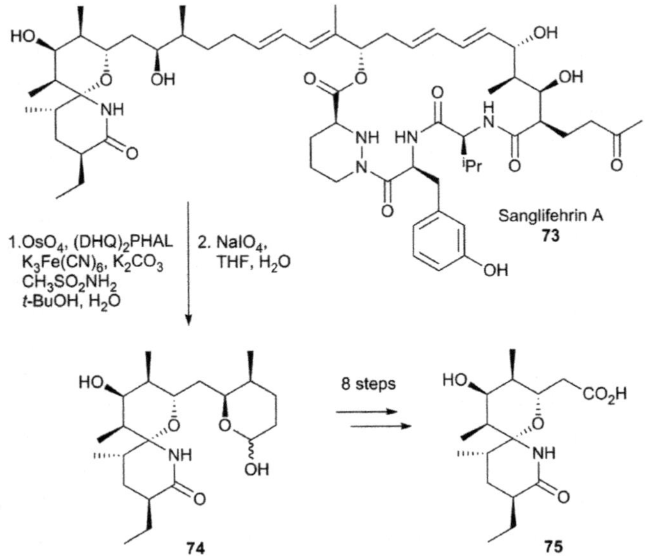

Scheme 4.13 Fragment generation by degradation of a natural product, Sanglifehrin A 73.

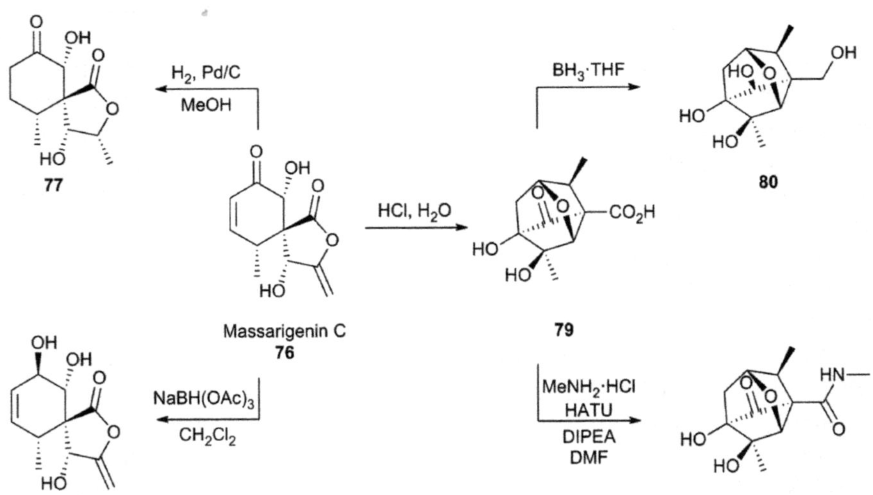

Scheme 4.14 Fragment generation by modification of a low-molecular-weight natural product, Massarigenin C **76**.

(*via* reduction) or **81** (*via* amidation). It is worth noting that several of the fragments identified and synthesised by the Novartis team have molecular weights of over 300, *i.e.* they do not meet the 'rule-of-three' criteria.[108]

Additional candidates for the natural-product-like 3D fragment library were identified by analysis of commercial and in-house compounds using

PMI and a natural-product-likeness scoring system.[67] Using the three described methods a library of over 150 compounds was assembled and shown to cover novel chemical space in comparison with the Novartis in-house fragment library.

Brennan and co-workers developed what they termed a 'poised' fragment collection for hit identification and rapid fragment elaboration.[109] A poised fragment is defined as a fragment synthesised from a robust and general synthetic reaction, such that elaboration of the fragment to generate a library of analogues can be carried out using straightforward and efficient parallel chemistry. Likewise, the idea of a 'poised' bond (or bonds) was used to deconstruct a fragment into two (or more) synthons. Poised reactions must be reliable, tolerate a wide range of substrates and utilise many commercially available building blocks. The products themselves should contain groups for binding to target proteins and avoid problematic functional groups. Brennan chose to use the top ten most commonly used transformations in drug discovery (*e.g.* amide formation, reductive amination and Suzuki coupling) to devise poised scaffolds. The commercially available starting materials for these reactions contain common functionalities. This chemical space was further expanded with 13 common heterocycle-forming reactions. Using these criteria, analysis of 11 677 in-house fragments revealed 2347 compounds that could be considered 'poised'. A subset of 406 poised fragments was selected for screening, ensuring a diversity of chemotypes and the ability to rapidly follow up any hits. Figure 4.7 depicts how the use of a poised-fragment library can be used to rapidly synthesise follow-up analogues.

This poised-fragment library was then assessed by screening against PHIP(2), an atypical bromodomain with no known small-molecule inhibitors, by high-concentration crystal soaking and subsequent X-ray

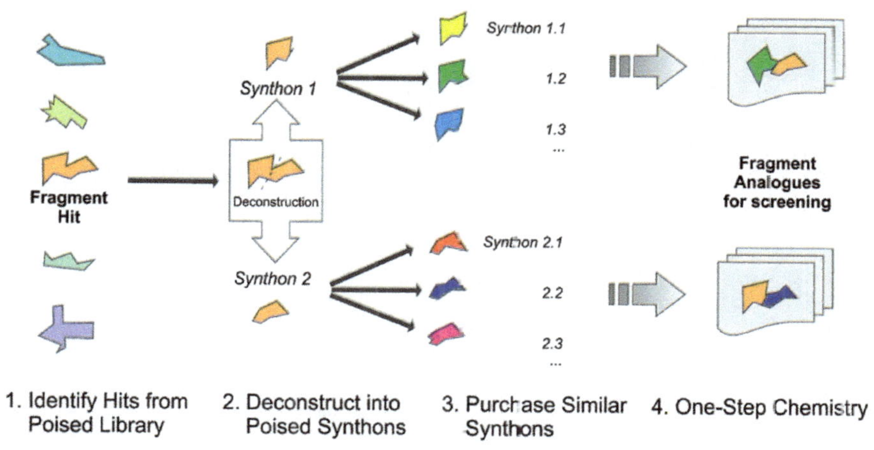

Figure 4.7 The poised fragment concept.
Reproduced from ref. 109 with permission from The Royal Society of Chemistry.

crystallography. This identified four low-affinity binders to the protein, thiourea **82a**, *N*-benzyl amide **83a** and oxazoles **84a** and **84b** (Scheme 4.15A). Due to the poised qualities of these fragments, rapid synthesis of analogues was possible (Scheme 4.15B). Firstly, 29 thioureas **82** were synthesised by sequential addition of amines to thio-CDI, with **82b** having an improved ligand efficiency of 0.45 (*cf.* 0.40 for **82a**). *N*-Benzyl amides **83** were synthesised from reaction of benzyl amines and acyl chlorides; 11 analogues were synthesised, with **83b** having an increased negative log IC_{50} (pIC_{50}) of 3.72 (*cf.* <2.30 for **83a**). Likewise, analogues of the oxazole series **84** were synthesised, from acyl chlorides and amino-malonitrile. The use of poised fragments allowed rapid elaboration of fragment hits weak enough to only be identified by X-ray crystallography. Using this method, the first inhibitors of PHIP(2) were identified, with several compounds showing sub-millimolar

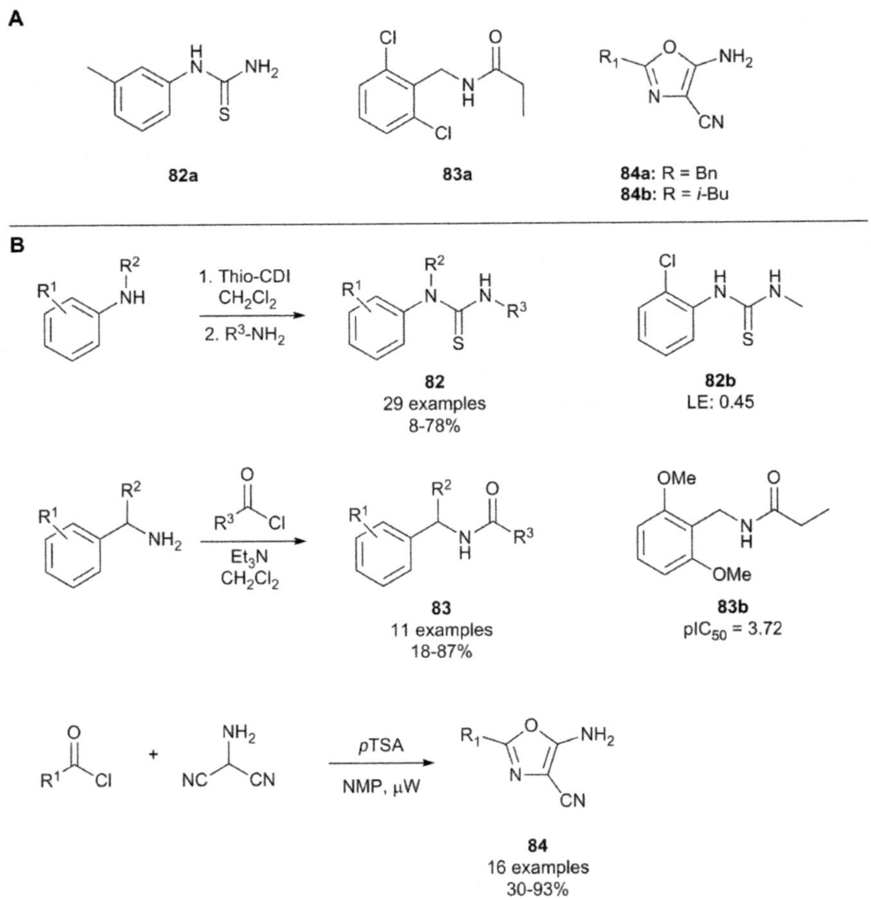

Scheme 4.15 Poised fragment hits against PHIP(2) and subsequent elaboration. A. Low-affinity binders to PHIP(2). B. General synthetic route to fragment analogues **82–84** and improved binders **82b** and **83b**.

activity. The concept of poised fragment libraries can be considered a FBDD-enabling technology.

Astex has recently described methodology for the synthesis of dihydroisoquinolone fragments that allows synthetic elaboration in multiple directions.[110] The attraction of a dihydroisoquinolone template **85** is that it has excellent molecular properties (MW 147 Da, clogP 1.0 and solubility of over 5 mM in aqueous buffer) and elaboration of the core is well precedented and possible in many positions, as indicated in Figure 4.8.

Substituted dihydroisoquinolones **87a** and **87b** can be accessed from hydroxamates **86** using Rh(III) catalysed C–H activation methodology (Scheme 4.16A). Importantly, this chemistry uses simple starting materials, proceeds at room temperature, is scalable and can be run under an air atmosphere. It was envisaged that expansion of the methodology to introduce a polar binding motif (Figure 4.8, red arrows) would be suitable as a demonstration of FBDD-enabling methodology development. Reaction of hydroxamates **86** with monosubstituted alkenes containing (protected) polar groups was therefore investigated, exemplar products are shown in Scheme 4.16B. Use of dimethyl allylamine gave **88a** as a single regioisomer in 59% yield. Use of protected allyl alcohol gave **89a/b** as a 1:1 regioisomeric mixture, with good yield. Benzyl-substituted acrylamide and acrolein diethylacetal gave protected amide **90a** and aldehyde **91a** at 63% and 74% yield respectively. The ability to structurally vary the core of the dihydroisoquinolones (Figure 4.8, blue arrows) was demonstrated using a range of substituted and heteroaromatic hydroxamates. For example, bromide **92a** or ester groups **93a** could be introduced, as could thiophenes **94a/b** and pyridines **95a**. Astex has shown that the Rh-catalysed synthesis of dihydroquinones is highly suitable as methodology for FBDD. Notably, the incorporation of polar side chains to explore binding interactions, the introduction of heteroatoms into the scaffold core and the addition of functional handles for fragment growth have all been demonstrated.

Spring and co-workers have recently reported the synthesis of partially saturated bicyclic heteroaromatics to give an sp^3-enriched fragment collection.[111] Using pyrrole **96** and pyridine **100** building blocks, functionalised either with polar (protected amino) or apolar (chloro) groups, 15

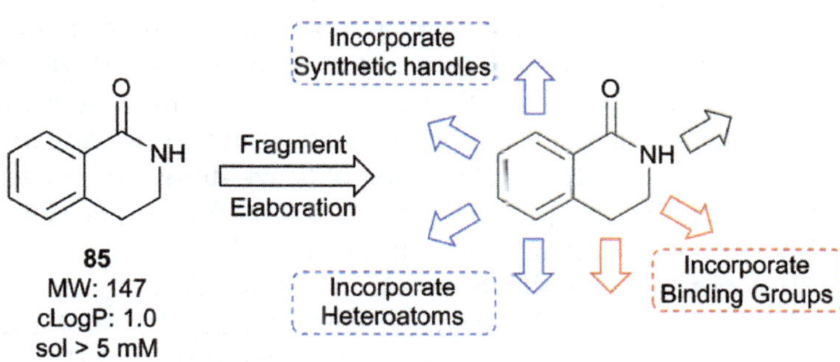

Figure 4.8 Dihydroisoquinolone fragment elaboration strategy.

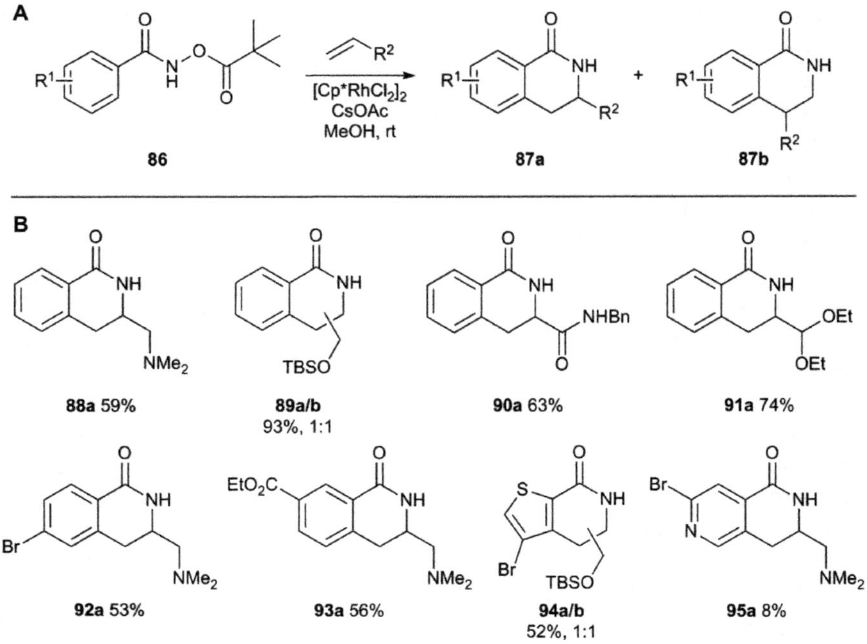

Scheme 4.16 FBDD-enabling Rh(I)-catalysed synthesis of dihydroisoquinolones. A. General synthetic route to fragments **87**. B. Exemplar protected and/or functionalised fragment precursors **88–95**.

fragment precursors **99** and **103** were synthesised (Scheme 4.17A). A modular and divergent approach was used. First, cross-coupling gave the alkenes **97** and **101**. Next, alkylation of either the pyrrole nitrogen or *via* lateral lithiation, gave the ring closing metathesis precursors **98** and **102**. Finally, treatment with Grubbs II or Hoveyda–Grubbs II catalysts gave bicyclic heteroaromatics **99** and **103** containing an alkene for further functionalisation. Simple one-, two-, or three-step modifications were employed to either reduce the olefin or introduce heteroatoms. In all, 40 exemplar fragments were prepared and Scheme 4.17B shows representative examples **104a–e** and **105a–e**.

Analysis of the deprotected fragments showed that almost all of them were compliant with Astex's recent stringent molecular property guidelines.[17] Indeed, the mean molecular weight was 190 Da and mean SlogP was 1.45. The fragments were sp^3-rich, with on average 0.88 chiral centres and a 'fragment aromatic' value of 0.43 (defined as the number of aromatic atoms expressed as a fraction of the total number of heavy atoms). Importantly, the final fragments have a range of 3D growth vectors for further elaboration. Hence, this piece of methodology could be classed as being useful for enabling FBDD, as recently outlined by researchers at Astex.

The Bull group has reported several methods for the synthesis of fragments, including the synthesis of arylsulfonyl oxetanes (Scheme 4.18A).[112,113] Alkylation of ethylene glycol with chloromethyl aryl sulfides **106** gave S,O acetals

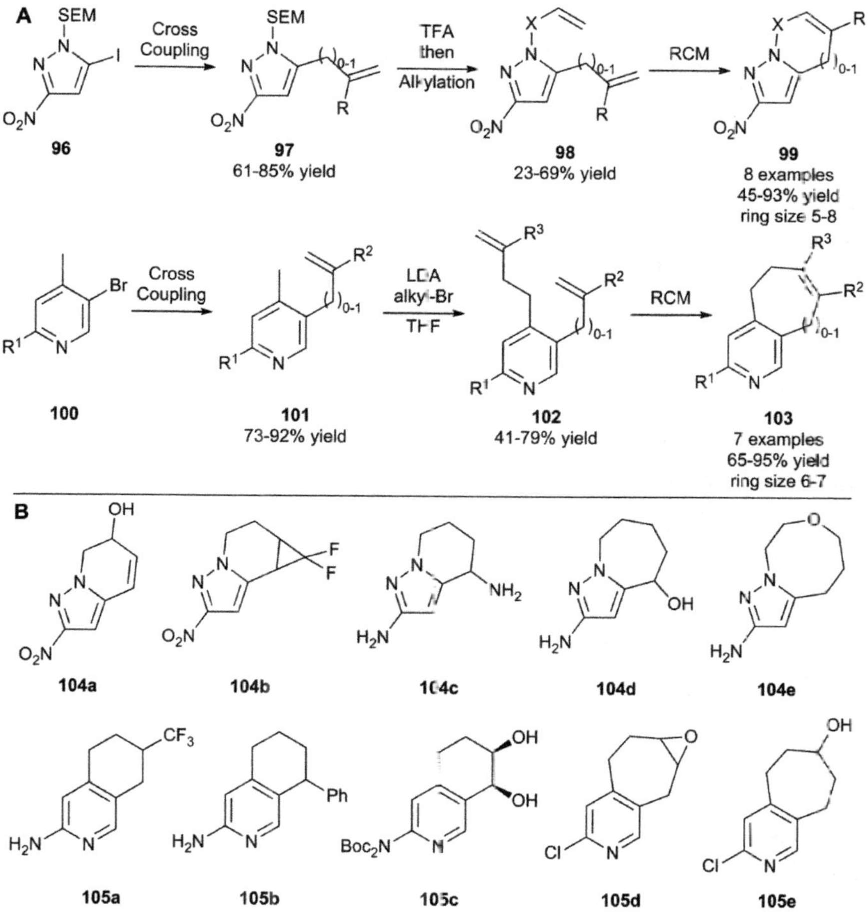

Scheme 4.17 Synthesis of a partially saturated bicyclic heteroaromatic fragment library. A. General synthetic route to fragment precursors **99** and **103**. B. Exemplar fragments.

107 at 58–87% yield. Tosylation and oxidation with *m*CPBA gave sulfones **108** at 82–96% yield. In the key oxetane-forming step, treatment of **108** with LiHMDS resulted in deprotonation adjacent to the anion-stabilizing sulfone and cyclisation occurred to give 11 different 'rule-of-three'-compliant oxetanes **109** at 45–95% yield (see Scheme 4.18B for examples).

A key design feature of the methodology was the ability to grow the fragments through carbon–carbon-bond-forming reactions (Scheme 4.19). Firstly, aryl halides *e.g.* **109c**, could be diversified by cross-coupling, either with Pd-catalysed Suzuki coupling to give biaryls *e.g.* **110**, or by Fe-catalysed coupling with alkyl Grignard reagents giving *e.g.* **111**. Secondly, lithiation of oxetane **109d** followed by trapping with electrophiles gave quaternary oxetanes *e.g.* **112a–c**. Unfortunately, however, quaternary alkylated oxetanes were found to be unstable upon prolonged storage.

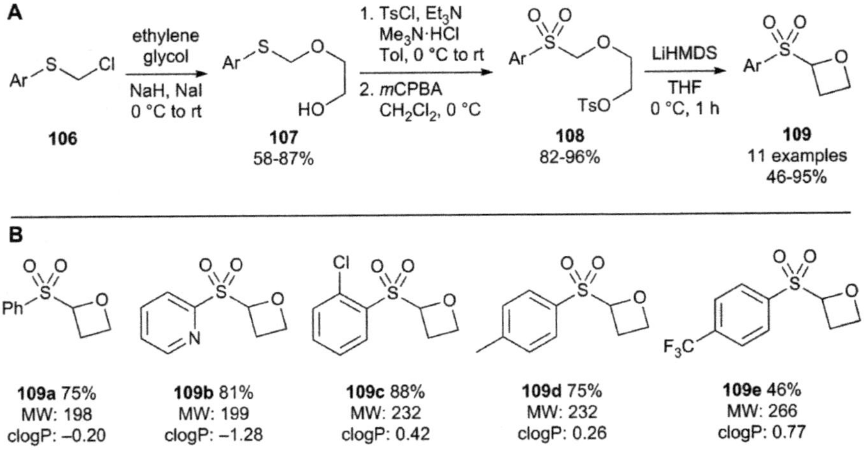

Scheme 4.18 Synthesis of arylsulfonyl oxetanes. A. General synthetic route to fragments **109**. B. Exemplar fragments.

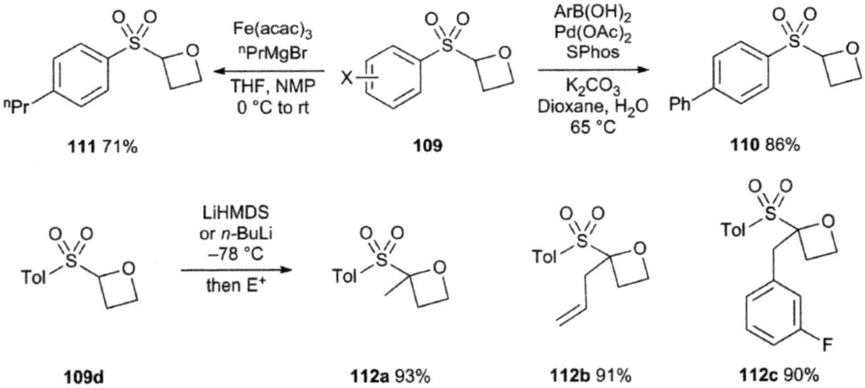

Scheme 4.19 Elaboration of arylsulfonyl oxetane fragments **109**.

Bull and co-workers have also developed a Cu(I)-catalyzed *N*-arylation of imidazolidines **113** to give 2-substituted *N*-arylimidazolines **114** (Scheme 4.20A),[114] which are often non-planar due to puckering of the ring and out-of-plane orientation of the ring substituents. The reaction is tolerant of electron-rich and -poor aromatics, heteroaromatics and various functional groups, giving a range of fragment-like structures (*e.g.* **114a–f**), some with handles (*e.g.* **114b–d** and **116f**) for further fragment growth (Scheme 4.20B).

Willand and co-workers have reported the synthesis of sp^3-rich 3D fragments, many containing spirocyclic motifs.[115,116] They demonstrated the synthesis of 16 novel functionalized 2-isoxazolines **118** using a one-pot method (Scheme 4.21A). Aldoximes **115** were converted into hydroxyiminoyl chloride **116** with NCS *in situ*. Subsequent 1,3-dipolar

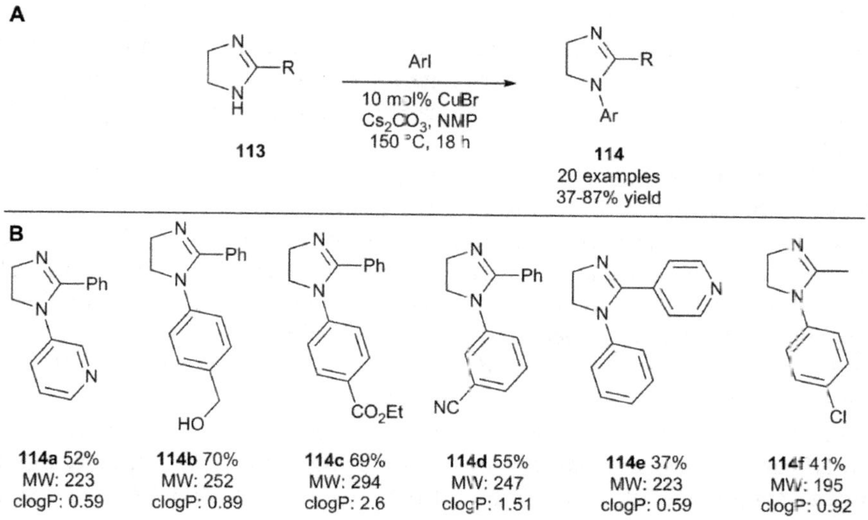

Scheme 4.20 Synthesis of a 2-substituted *N*-aryl imidazoline library *via* Cu(I)-catalysis. A. General synthetic route to fragments **114**. B. Exemplar fragments.

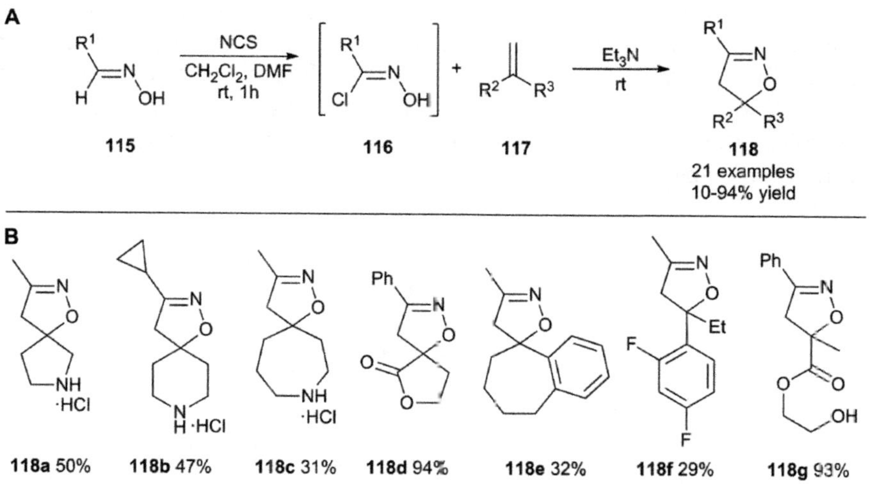

Scheme 4.21 Synthesis of a sp^3-rich 2-isoxazoline fragment library. A. General synthetic route to fragments **118**. B. Exemplar fragments.

cycloaddition with alkenes **117** (derived from the corresponding ketones) gave 21 substituted isoxazolines **118** at 10–94% yield (see Scheme 4.21B for examples). Of particular interest are 5–7-membered amine-containing spirocyclic compounds **118a–c**, obtained as their hydrochloride salts after Boc deprotection. Lactone **118d** could be further functionalized with amines to generate additional structural diversity.

All the compounds had MW of less than 300 Da, high Fsp^3 and, importantly, were generally highly soluble in aqueous buffer. Analysis of their shapes by PMI showed a wide coverage of chemical space and significantly more sphere-like character in comparison with 471 commercially available fragments.

Willand and co-workers then demonstrated the microwave-assisted synthesis of functionalized spirohydantoins **120** as 3D fragments.[116] Treatment of cyclic ketones **119** with ammonium carbonate and potassium cyanide under microwave irradiation gave 27 different spirocyclic hydantoins **120** with moderate to excellent yields (Scheme 4.22A). Notably, 4–7-membered spirocycles **120a–d** were accessible, protected amines could be introduced and the use of α- or β-substituted cyclohexanones led to products **120e** and **120f** as single diastereomers (Scheme 4.22B).

Further elaboration of the fragments was possible, with selective functionalisation of both hydantoin nitrogens being demonstrated. Once again, all the fragments fulfil the 'rule-of-three' and most have MW of less than 220 Da. Most have solubilities greater than 0.8 mM in aqueous buffer and the PMI analysis revealed broad coverage of 3D chemical space, with some compounds occupying underrepresented areas.

Dömling and co-workers have recently reported the synthesis of 40 fragments based upon two isomeric tetrazole libraries.[117] In this simple process, cyano- and isocyanoacetyl methyl esters **121** and **124** were reacted with neat amines to give amides **122** and **125** (Scheme 4.23). Treatment of nitriles **122** with sodium azide and triethylamine hydrochloride resulted in a dipolar cycloaddition, giving 22 5-substituted 1H-tetrazoles **123**. Similarly, isomeric tetrazoles **126** could be obtained by reaction of isonitriles **125** with trimethylsilyl azide.

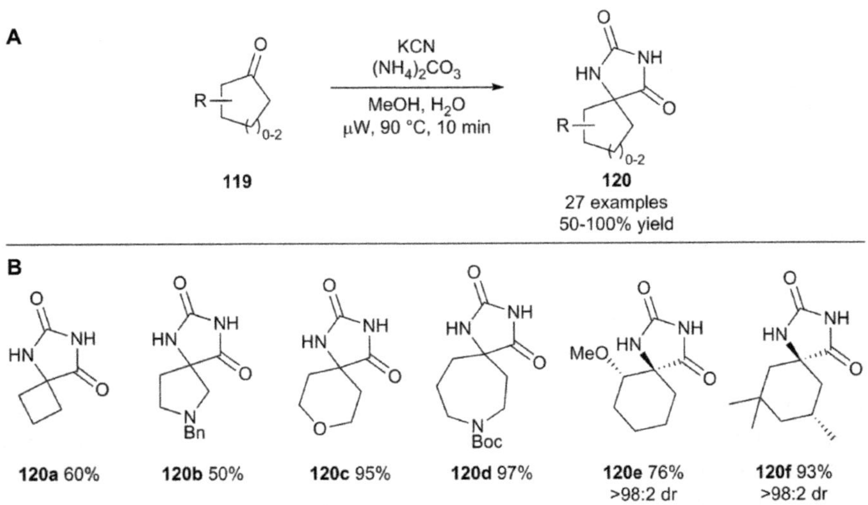

Scheme 4.22 Synthesis of a spirohydantoin library. A. General synthetic route to fragments **120**. B. Exemplar fragments.

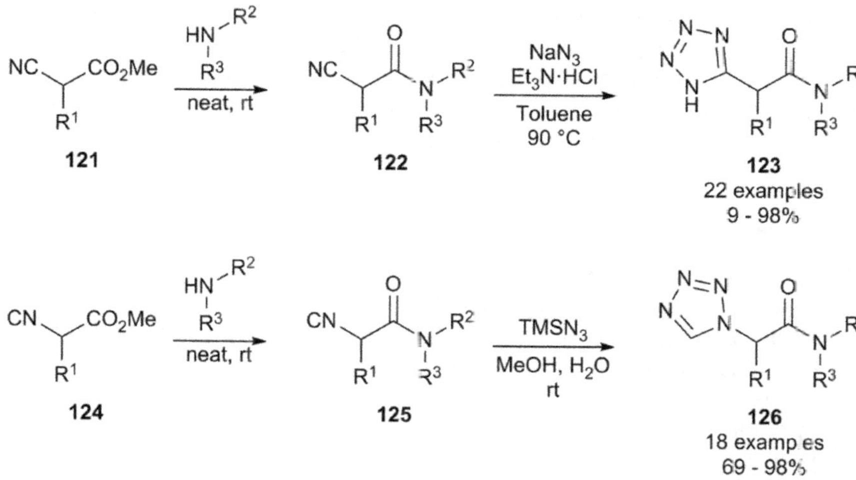

Scheme 4.23 Synthesis of isomeric tetrazole fragment libraries **123** and **126**.

The vast majority of the products were isolated by filtration after aqueous work-up, simplifying purification and allowing the synthesis of large libraries. Nearly all the synthesised fragments had MW of less than 300 Da and clogP of less than 3.

4.4 Conclusions

The calls for lead- and fragment-oriented synthesis were issued to improve the quality of screening libraries for drug-discovery projects. Since the introduction of the concept of lead-oriented synthesis in 2012 there have been multiple reports, by both industry and academia, of the development of methods to meet the requirements of the doctrine. Nelson, Marsden and co-workers have shown that it is possible to design and synthesise multiple novel heterocyclic scaffolds from common simple precursors in an efficient manner whilst successfully targeting lead-like space. They have introduced an open access tool, LLAMA, to allow other researchers to assess the suitability of their chemistry for lead-oriented synthesis, which is being taken up by the community. Additionally, there have been several reports of the development and retooling of specific synthetic methods for the synthesis of compounds that fall within lead-like space, as well as the use of natural products as inspirations for such libraries. Furthermore, the European Lead Factory has generated significant numbers of highly novel, shape-diverse screening compounds for hit identification.

Even though the concept of fragment-oriented synthesis is less well defined, there has been a call for methods that can enable and facilitate fragment-based drug discovery, and several groups have recently risen to this challenge. We expect to see a burgeoning of this area over the coming years. Furthermore, many different approaches to the generation of 'rule-of-three'-compliant libraries have been reported, including diversity-orientated synthesis, the use of natural products and 'poised' chemistries.

Lead- and fragment-oriented methods offer a significant opportunity for the synthetic community to improve the early stages of drug discovery and further research in these areas is encouraged.

References

1. M. C. Wenlock, R. P. Austin, P. Barton, A. M. Davis and P. D. Leeson, *J. Med. Chem.*, 2003, **46**, 1250–1256.
2. M. J. Waring, J. Arrowsmith, A. R. Leach, P. D. Leeson, S. Mandrell, R. M. Owen, G. Pairaudeau, W. D. Pennie, S. D. Pickett, J. Wang, O. Wallace and A. Weir, *Nat. Rev. Drug Discovery*, 2015, **14**, 475–486.
3. N. A. Meanwell, *Chem. Res. Toxicol.*, 2011, **24**, 1420–1456.
4. M. J. Waring, *Bioorg. Med. Chem. Lett.*, 2009, **19**, 2844–2851.
5. P. D. Leeson and B. Springthorpe, *Nat. Rev. Drug Discovery*, 2007, **6**, 881–890.
6. T. J. Ritchie and S. J. F. Macdonald, *Drug Discovery Today*, 2009, **14**, 1011–1020.
7. J. D. Hughes, J. Blagg, D. A. Price, S. Bailey, G. A. DeCrescenzo, R. V. Devraj, E. Ellsworth, Y. M. Fobian, M. E. Gibbs, R. W. Gilles, N. Greene, E. Huang, T. Krieger-Burke, J. Loesel, T. Wager, L. Whiteley and Y. Zhang, *Bioorg. Med. Chem. Lett.*, 2008, **18**, 4872–4875.
8. F. Lovering, J. Bikker and C. Humblet, *J. Med. Chem.*, 2009, **52**, 6752–6756.
9. F. Lovering, *Med. Chem. Commun.*, 2013, **4**, 515–519.
10. C. A. Lipinski, F. Lombardo, B. W. Dominy and P. J. Feeney, *Adv. Drug Delivery Rev.*, 1997, **23**, 3–25.
11. C. A. Lipinski, *Drug Discovery Today: Technol.*, 2004, **1**, 337–341.
12. H. Pajouhesh and G. R. Lenz, *NeuroRx*, 2005, **2**, 541–553.
13. T. T. Wager, R. Y. Chandrasekaran, X. Hou, M. D. Troutman, P. R. Verhoest, A. Villalobos and Y. Will, *ACS Chem. Neurosci.*, 2010, **1**, 420–434.
14. G. R. Bickerton, G. V. Paolini, J. Besnard, S. Muresan and A. L. Hopkins, *Nat. Chem.*, 2012, **4**, 90–98.
15. M. N. Gandy, M. G. Corral, J. S. Mylne and K. A. Stubbs, *Org. Biomol. Chem.*, 2015, **13**, 5586–5590.
16. A. Nadin, C. Hattotuwagama and I. Churcher, *Angew. Chem., Int. Ed.*, 2012, **51**, 1114–1122.
17. C. W. Murray and D. C. Rees, *Angew. Chem., Int. Ed.*, 2016, **55**, 488–492.
18. F. W. Goldberg, J. G. Kettle, T. Kogej, M. W. D. Perry and N. P. Tomkinson, *Drug Discovery Today*, 2015, **20**, 11–17.
19. S. J. Teague, A. M. Davis, P. D. Leeson and T. Oprea, *Angew. Chem., Int. Ed.*, 1999, **38**, 3743–3748.

20. G. M. Keseru and G. M. Makara, *Nat. Rev. Drug Discovery*, 2009, **8**, 203–212.
21. T. I. Oprea, A. M. Davis, S. J. Teague and P. D. Leeson, *J. Chem. Inf. Comput. Sci.*, 2001, **41**, 1308–1315.
22. M. M. Hann and T. I. Oprea, *Curr. Opin. Chem. Biol.*, 2004, **8**, 255–263.
23. R. Brenk, A. Schipani, D. James, A. Krasowski, I. H. Gilbert, J. Frearson and P. G. Wyatt, *ChemMedChem*, 2008, **3**, 435–444.
24. M. Congreve, R. Carr, C. Murray and H. Jhoti, *Drug Discovery Today*, 2003, **8**, 876–877.
25. W. P. Walters, J. Green, J. R. Weiss and M. A. Murcko, *J. Med. Chem.*, 2011, **54**, 6405–6416.
26. T. W. J. Cooper, I. B. Campbell and S. J. F. Macdonald, *Angew. Chem., Int. Ed.*, 2010, **49**, 8082–8091.
27. D. G. Brown and J. Boström, *J. Med. Chem.*, 2016, **59**, 4443–4458.
28. D. G. Brown, M. M. Gagnon and J. Boström, *J. Med. Chem.*, 2015, **58**, 2390–2405.
29. D. J. Foley, A. Nelson and S. P. Marsden, *Angew. Chem., Int. Ed.*, 2016, **55**, 13650–13657.
30. A. H. Lipkus, Q. Yuan, K. A. Lucas, S. A. Funk, W. F. Bartelt, R. J. Schenck and A. J. Trippe, *J. Org. Chem.*, 2008, **73**, 4443–4451.
31. M. Krier, G. Bret and D. Rognan, *J. Chem. Inf. Model.*, 2006, **46**, 512–524.
32. C. W. Murray and D. C. Rees, *Nat. Chem.*, 2009, **1**, 187–192.
33. M. Baker, *Nat. Rev. Drug Discovery*, 2013, **12**, 5–7.
34. G. Bollag, J. Tsai, J. Zhang, C. Zhang, P. Ibrahim, K. Nolop and P. Hirth, *Nat. Rev. Drug Discovery*, 2012, **11**, 873–886.
35. A. J. Souers, J. D. Leverson, E. R. Boghaert, S. L. Ackler, N. D. Catron, J. Chen, B. D. Dayton, H. Ding, S. H. Enschede, W. J. Fairbrother, D. C. S. Huang, S. G. Hymowitz, S. Jin, S. L. Khaw, P. J. Kovar, L. T. Lam, J. Lee, H. L. Maecker, K. C. Marsh, K. D. Mason, M. J. Mitten, P. M. Nimmer, A. Oleksijew, C. H. Park, C.-M. Park, D. C. Phillips, A. W. Roberts, D. Sampath, J. F. Seymour, M. L. Smith, G. M. Sullivan, S. K. Tahir, C. Tse, M. D. Wendt, Y. Xiao, J. C. Xue, H. Zhang, R. A. Humerickhouse, S. H. Rosenberg and S. W. Elmore, *Nat. Med.*, 2013, **19**, 202–208.
36. D. A. Erlanson, S. W. Fesik, R. E. Hubbard, W. Jahnke and H. Jhoti, *Nat. Rev. Drug Discovery*, 2016, **15**, 605–619.
37. L. Ruddigkeit, R. van Deursen, L. C. Blum and J.-L. Reymond, *J. Chem. Inf. Model.*, 2012, **52**, 2864–2875.
38. P. D. Leeson and S. A. St-Gallay, *Nat. Rev. Drug Discovery*, 2011, **10**, 749–765.
39. N. Fuller, L. Spadola, S. Cowen, J. Patel, H. Schönherr, Q. Cao, A. McKenzie, F. Edfeldt, A. Rabow and R. Goodnow, *Drug Discovery Today*, 2016, **21**, 1272–1283.
40. N. Baurin, F. Aboul-Ela, X. Barril, B. Davis, M. Drysdale, B. Dymock, H. Finch, C. Fromont, C. Richardson, H. Simmonite and R. E. Hubbard, *J. Chem. Inf. Comput. Sci.*, 2004, **44**, 2157–2166.

41. N. Blomberg, D. A. Cosgrove, P. W. Kenny and K. Kolmodin, *J. Comput.-Aided. Mol. Des.*, 2009, **23**, 513–525.
42. G. M. Keserű, D. A. Erlanson, G. G. Ferenczy, M. M. Hann, C. W. Murray and S. D. Pickett, *J. Med. Chem.*, 2016, **59**, 8189–8206.
43. H. Köster, T. Craan, S. Brass, C. Herhaus, M. Zentgraf, L. Neumann, A. Heine and G. Klebe, *J. Med. Chem.*, 2011, **54**, 7784–7796.
44. H. Jhoti, G. Williams, D. C. Rees and C. W. Murray, *Nat. Rev. Drug Discovery*, 2013, **12**, 644.
45. R. G. Doveston, P. Tosatti, M. Dow, D. J. Foley, H. Y. Li, A. J. Campbell, D. House, I. Churcher, S. P. Marsden and A. Nelson, *Org. Biomol. Chem.*, 2015, **13**, 859–865.
46. P. Tosatti, J. Horn, A. J. Campbell, D. House, A. Nelson and S. P. Marsden, *Adv. Synth. Catal.*, 2010, **352**, 3153–3157.
47. J. J. Irwin, T. Sterling, M. M. Mysinger, E. S. Bolstad and R. G. Coleman, *J. Chem. Inf. Model.*, 2012, **52**, 1757–1768.
48. D. J. Foley, R. G. Doveston, I. Churcher, A. Nelson and S. P. Marsden, *Chem. Commun.*, 2015, **51**, 11174–11177.
49. W. H. B. Sauer and M. K. Schwarz, *J. Chem. Inf. Comput. Sci.*, 2003, **43**, 987–1003.
50. T. James, P. MacLellan, G. M. Burslem, I. Simpson, J. A. Grant, S. Warriner, V. Sridharan and A. Nelson, *Org. Biomol. Chem.*, 2014, **12**, 2584–2591.
51. A. Schuffenhauer, P. Ertl, S. Roggo, S. Wetzel, M. A. Koch and H. Waldmann, *J. Chem. Inf. Model.*, 2007, **47**, 47–58.
52. I. Colomer, C. J. Empson, P. Craven, Z. Owen, R. G. Doveston, I. Churcher, S. P. Marsden and A. Nelson, *Chem. Commun.*, 2016, **52**, 7209–7212.
53. N. C. Firth, N. Brown and J. Blagg, *J. Chem. Inf. Model.*, 2012, **52**, 2516–2525.
54. G. W. Bemis and M. A. Murcko, *J. Med. Chem.*, 1996, **39**, 2887–2893.
55. A. V. Borisov, V. V. Voloshchuk, M. A. Nechayev and O. O. Grygorenko, *Synthesis*, 2013, **45**, 2413–2416.
56. S. Zhersh, O. V. Karpenko, V. Ripenko, A. A. Tolmachev and O. O. Grygorenko, *Open Chem.*, 2014, **12**, 67–73.
57. S. V. Ryabukhin, D. M. Panov, D. S. Granat, E. N. Ostapchuk, D. V. Kryvoruchko and O. O. Grygorenko, *ACS Comb. Sci.*, 2014, **16**, 146–153.
58. M. I. Adamovskyi, S. V. Ryabukhin, D. A. Sibgatulin, E. Rusanov and O. O. Grygorenko, *Org. Lett.*, 2017, **19**, 130–133.
59. M. Lüthy, M. C. Wheldon, C. Haji-Cheteh, M. Atobe, P. S. Bond, P. O'Brien, R. E. Hubbard and I. J. S. Fairlamb, *Bioorg. Med. Chem.*, 2015, **23**, 2680–2694.
60. L. G. Baud, M. A. Manning, H. L. Arkless, T. C. Stephens and W. P. Unsworth, *Chem. – Eur. J.*, 2017, **23**, 2225–2230.
61. R. Doveston, S. Marsden and A. Nelson, *Drug Discovery Today*, 2014, **19**, 813–819.

62. C.-V. T. Vo, M. U. Luescher and J. W. Bode, *Nat. Chem.*, 2014, **6**, 310–314.
63. M. U. Luescher and J. W. Bode, *Angew. Chem., Int. Ed.*, 2015, **54**, 10884–10888.
64. J. T. Lowe, M. D. Lee, L. B. Akella, E. Davoine, E. J. Donckele, L. Durak, J. R. Duvall, B. Gerard, E. B. Holson, A. Joliton, S. Kesavan, B. C. Lemercier, H. Liu, J.-C. Marié, C. A. Mulrooney, G. Muncipinto, M. Welzel-O'Shea, L. M. Panko, A. Rowley, B.-C. Suh, M. Thomas, F. F. Wagner, J. Wei, M. A. Foley and L. A. Marcaurelle, *J. Org. Chem.*, 2012, **77**, 7187–7211.
65. D. J. Newman and G. M. Cragg, *J. Nat. Prod.*, 2012, **75**, 311–335.
66. T. Rodrigues, D. Reker, P. Schneider and G. Schneider, *Nat. Chem.*, 2016, **8**, 531–541.
67. P. Ertl, S. Roggo and A. Schuffenhauer, *J. Chem. Inf. Model.*, 2008, **48**, 68–74.
68. M. Feher and J. M. Schmidt, *J. Chem. Inf. Comput. Sci.*, 2003, **43**, 218–227.
69. J. Hert, J. J. Irwin, C. Laggner, M. J. Keiser and B. K. Shoichet, *Nat. Chem. Biol.*, 2009, **5**, 479–483.
70. G. L. Thomas and C. W. Johannes, *Curr. Opin. Chem. Biol.*, 2011, **15**, 516–522.
71. A. M. Boldi, *Curr. Opin. Chem. Biol.*, 2004, **8**, 281–286.
72. H. Lachance, S. Wetzel, K. Kumar and H. Waldmann, *J. Med. Chem.*, 2012, **55**, 5989–6001.
73. S. Wetzel, R. S. Bon, K. Kumar and H. Waldmann, *Angew. Chem., Int. Ed.*, 2011, **50**, 10800–10826.
74. A. J. Welford, J. J. Caldwell, M. Liu, M. Richards, N. Brown, C. Lomas, G. J. Tizzard, M. B. Pitak, S. J. Coles, S. A. Eccles, F. I. Raynaud and I. Collins, *Chem. – Eur. J.*, 2016, **22**, 5657–5664.
75. M. Pascolutti and R. J. Quinn, *Drug Discovery Today*, 2014, **19**, 215–221.
76. A. Karawajczyk, K. M. Orrling, J. S. B. de Vlieger, T. Rijnders and D. Tzalis, *Front. Med.*, 2017, **3**, 75.
77. J. Besnard, P. S. Jones, A. L. Hopkins and A. D. Pannifer, *Drug Discovery Today*, 2015, **20**, 181–186.
78. A. Karawajczyk, F. Giordanetto, J. Benningshof, D. Hamza, T. Kalliokoski, K. Pouwer, R. Morgentin, A. Nelson, G. Müller, A. Piechot and D. Tzalis, *Drug Discovery Today*, 2015, **20**, 1310–1316.
79. I. Colomer, O. Adeniji, G. M. Burslem, P. Craven, M. O. Rasmussen, A. Willaume, T. Kalliokoski, R. Foster, S. P. Marsden and A. Nelson, *Bioorg. Med. Chem.*, 2015, **23**, 2736–2740.
80. J. D. Firth, R. Zhang, R. Morgentin, R. Guilleux, T. Kalliokoski, S. Warriner, R. Foster, S. P. Marsden and A. Nelson, *Synthesis*, 2015, **47**, 2391–2406.
81. P. Craven, A. Aimon, M. Dow, N. Fleury-Bregeot, R. Guilleux, R. Morgentin, D. Roche, T. Kalliokoski, R. Foster, S. P. Marsden and A. Nelson, *Bioorg. Med. Chem.*, 2015, **23**, 2629–2635.
82. J. D. Firth, P. G. E. Craven, M. Lilburn, A. Pahl, S. P. Marsden and A. Nelson, *Chem. Commun.*, 2016, **52**, 9837–9840.

83. P. Patil, K. Khoury, E. Herdtweck and A. Dömling, *Bioorg. Med. Chem.*, 2015, **23**, 2699–2715.
84. Y. Wang, P. Patil, K. Kurpiewska, J. Kalinowska-Tluscik and A. Dömling, *ACS Comb. Sci.*, 2017, **19**, 193–198.
85. A. Murali, F. Medda, M. Winkler, F. Giordanetto and K. Kumar, *Bioorg. Med. Chem.*, 2015, **23**, 2656–2665.
86. P. Wu, M. Å. Petersen, R. Petersen, M. O. Rasmussen, K. Bonnet, T. E. Nielsen and M. H. Clausen, *Eur. J. Org. Chem.*, 2015, **2015**, 5633–5639.
87. M. Å. Petersen, M. A. Mortensen, A. E. Cohrt, R. Petersen, P. Wu, N. Fleury-Brégeot, R. Morgentin, C. Lardy, T. E. Nielsen and M. H. Clausen, *Bioorg. Med. Chem.*, 2015, **23**, 2695–2698.
88. P. Wu, M. A. Petersen, A. E. Cohrt, R. Petersen, R. Morgentin, H. Lemoine, C. Roche, A. Willaume, M. H. Clausen and T. E. Nielsen, *Org. Biomol. Chem.*, 2016, **14**, 6947–6950.
89. P. Wu, M. A. Petersen, R. Petersen, T. Flagstad, R. Guilleux, M. Ohsten, R. Morgentin, T. E. Nielsen and M. H. Clausen, *RSC Adv.*, 2016, **6**, 46654–46657.
90. A. Nortcliffe and C. J. Moody, *Bioorg. Med. Chem.*, 2015, **23**, 2730–2735.
91. A. T. Murray, E. Packard, A. Nortcliffe, W. Lewis, D. Hamza, G. Jones and C. J. Moody, *Eur. J. Org. Chem.*, 2017, 138–148.
92. A. Nortcliffe, G. D. S. Milne, D. Hamza and C. J. Moody, *Bioorg. Med. Chem.*, 2017, **25**, 2218–2225.
93. R. Ortega, J. Sanchez-Quesada, C. Lorenz, G. Dolega, A. Karawajczyk, M. Sanz, G. Showell and F. Giordanetto, *Bioorg. Med. Chem.*, 2015, **23**, 2716–2720.
94. S. Stotani, C. Lorenz, M. Winkler, F. Medda, E. Picazo, R. Ortega Martinez, A. Karawajczyk, J. Sanchez-Quesada and F. Giordanetto, *ACS Comb. Sci.*, 2016, **18**, 330–336.
95. T. Flagstad, G. Min, K. Bonnet, R. Morgentin, D. Roche, M. H. Clausen and T. E. Nielsen, *Org. Biomol. Chem.*, 2016, **14**, 4943–4946.
96. M. J. Rawling, T. E. Storr, W. A. Bawazir, S. J. Cully, W. Lewis, M. S. I. T. Makki, I. R. Strutt, G. Jones, D. Hamza and R. A. Stockman, *Chem. Commun.*, 2015, **51**, 12867–12870.
97. S. J. Cully, T. E. Storr, M. J. Rawling, I. R. Abeysena, D. Hamza, G. Jones, C. A. Pearce, A. Quddus, W. Lewis and R. A. Stockman, *Bioorg. Med. Chem.*, 2016, **24**, 5249–5257.
98. T. E. Storr, S. J. Cully, M. J. Rawling, W. Lewis, D. Hamza, G. Jones and R. A. Stockman, *Bioorg. Med. Chem.*, 2015, **23**, 2621–2628.
99. P. J. Hajduk, W. R. J. D. Galloway and D. R. Spring, *Nature*, 2011, **470**, 42–43.
100. A. D. Morley, A. Pugliese, K. Birchall, J. Bower, P. Brennan, N. Brown, T. Chapman, M. Drysdale, I. H. Gilbert, S. Hoelder, A. Jordan, S. V. Ley, A. Merritt, D. Miller, M. E. Swarbrick and P. G. Wyatt, *Drug Discovery Today*, 2013, **18**, 1221–1227.

101. D. C. Fry, C. Wartchow, B. Graves, C. Janson, C. Lukacs, U. Kammlott, C. Belunis, S. Palme, C. Klein and B. Vu, *ACS Med. Chem. Lett.*, 2013, **4**, 660–665.
102. R. J. Hall, P. N. Mortenson and C. W. Murray, *Prog. Biophys. Mol. Biol.*, 2014, **116**, 82–91.
103. S. L. Schreiber, *Science*, 2000, **287**, 1964–1969.
104. M. D. Burke and S. L. Schreiber, *Angew. Chem., Int. Ed.*, 2004, **43**, 46–58.
105. A. W. Hung, A. Ramek, Y. Wang, T. Kaya, J. A. Wilson, P. A. Clemons and D. W. Young, *Proc. Natl. Acad. Sci. U. S. A.*, 2011, **108**, 6799–6804.
106. T. E. Nielsen and S. L. Schreiber, *Angew. Chem., Int. Ed.*, 2008, **47**, 48–56.
107. Y. Wang, J.-Y. Wach, P. Sheehan, C. Zhong, C. Zhan, R. Harris, S. C. Almo, J. Bishop, S. J. Haggarty, A. Ramek, K. N. Berry, C. O'Herin, A. N. Koehler, A. W. Hung and D. W. Young, *ACS Med. Chem. Lett.*, 2016, **7**, 852–856.
108. H. Prescher, G. Koch, T. Schuhmann, P. Ertl, A. Bussenault, M. Glick, I. Dix, F. Petersen and D. E. Lizos, *Bioorg. Med. Chem.*, 2017, **25**, 921–925.
109. O. B. Cox, T. Krojer, P. Collins, O. Monteiro, R. Talon, A. Bradley, O. Fedorov, J. Amin, B. D. Marsden, J. Spencer, F. von Delft and P. E. Brennan, *Chem. Sci.*, 2016, **7**, 2322–2330.
110. N. Palmer, T. M. Peakman, D. Norton and D. C. Rees, *Org. Biomol. Chem.*, 2016, **14**, 1599–1610.
111. D. G. Twigg, N. Kondo, S. L. Mitchell, W. R. J. D. Galloway, H. F. Sore, A. Madin and D. R. Spring, *Angew. Chem., Int. Ed.*, 2016, **55**, 12479–12483.
112. K. F. Morgan, I. A. Hollingsworth and J. A. Bull, *Chem. Commun.*, 2014, **50**, 5203–5205.
113. K. F. Morgan, I. A. Hollingsworth and J. A. Bull, *Org. Biomol. Chem.*, 2015, **13**, 5265–5272.
114. O. A. Davis, M. Hughes and J. A. Bull, *J. Org. Chem.*, 2013, **78**, 3470–3475.
115. N. C. Tran, H. Dhondt, M. Flipo, B. Deprez and N. Willand, *Tetrahedron Lett.*, 2015, **56**, 4119–4123.
116. H. Prevet, M. Flipo, P. Roussel, B. Deprez and N. Willand, *Tetrahedron Lett.*, 2016, **57**, 2888–2894.
117. Y. Wang, P. Patil and A. Dömling, *Synthesis*, 2016, **48**, 3701–3712.

CHAPTER 5

Principles and Applications of Fragment-based Drug Discovery for Chemical Probes

MOSES MOUSTAKIM[a,b] AND PAUL E. BRENNAN*[a,c]

[a] Structural Genomics Consortium (SGC) & Target Discovery Institute (TDI), University of Oxford, NDM Research Building, Roosevelt Drive, OX3 7DQ and OX3 7FZ, Oxford, UK; [b] Department of Chemistry, Chemistry Research Laboratory, University of Oxford, Mansfield Road, Oxford OX1 3TA, UK; [c] ARUK Oxford Drug Discovery Institute, University of Oxford, NDM Research Building, Roosevelt Drive, OX3 7FZ Oxford, UK
*Email: paul.brennan@sgc.ox.ac.uk

5.1 Introduction

5.1.1 What Is a Chemical Probe?

Chemical probes are small molecules that interrogate the function of target proteins. Through the generation of multiple high-quality chemical probes alongside their negative controls, rigorous analysis of the phenotypic outcomes of small-molecule inhibition can be ascertained for a given protein target. Ultimately this allows for early target validation (and invalidation) in drug discovery and is ideally combined with complementary data from technologies such as small interfering RNA[1] and/or clustered regularly interspaced short palindromic repeats–CRISPR associated system 9 (CRISPR–Cas9).[2]

Chemical Biology No. 10
Chemical and Biological Synthesis: Enabling Approaches for Understanding Biology
Edited by Nick J. Westwood and Adam Nelson

Published by the Royal Society of Chemistry, www.rsc.org

There have been a number of reviews written describing in depth the definition,[3,4] uses[5–7] and potential[8] of small-molecule chemical probes. Furthermore there is a growing scientific community in the form of the Chemical Probes Portal[4,9] where a portfolio of chemical probes are highlighted and summarised for end users.

5.1.2 Chemical Probe Characteristics

A number of sources have outlined important characteristics for what a chemical probe is.[3,4] The following is a non-exhaustive list of some of the key characteristics of a chemical probe.

- Potency: engages a target protein with sub-micromolar dissociation constant (K_D) and/or a concentration giving 50% of maximum inhibition (IC_{50})—as determined by a biochemical and/or biophysical assay—preferentially less than 100 nM.
- Selectivity: displays selectivity between closely related targets and broader off-targets of at least 30-fold. Selectivity over targets associated with a strong phenotypic response to modulation is especially important.
- Cellular activity: direct or indirect evidence of target engagement in a cellular context, ideally at low micromolar concentration.
- Control compound: an inactive negative control compound is available and characterised that is structurally similar to the probe, such that parallel use of the active and inactive compounds provide some control over the phenotypic effects of undetermined off-targets.

In order to generate new small-molecule chemical probes for a given target protein, initial chemical matter is required to facilitate the discovery of a probe with the desired characteristics. There are a number of screening methods used to find this initial hit compound including high-throughput screening (HTS),[10] X-ray crystallography,[11] surface plasmon resonance (SPR),[12] differential scanning fluorimetry (DSF),[13] mass spectrometry,[14] *in silico* methods,[15] nuclear magnetic resonance (NMR) screening[16] and others.[17] These techniques are amenable to the use of both elaborated 'drug-like' compound libraries and fragment libraries. Unlike larger HTS compound members, a fragment is smaller and less potent but will ultimately be elaborated into a larger, more potent ligand through iterative cycles of design, synthesis and testing. A formal definition of fragments was proposed by Jhoti *et. al.*, as the "rule of three": fragments should have a molecular weight of less than 300 Da, a log of the partition coefficient calculated using a group contribution method (clogP) of 3 or lower and number of hydrogen bond donors/acceptors of 3 or fewer[18] although this serves as a guideline and should not be overemphasized.[18,19] One of the key advantages of using a fragment library is the much larger proportion of chemical space that the library will cover compared with a corresponding more elaborated typical

HTS compound library.[20] Fragment-based drug discovery (FBDD) has been reviewed extensively[21–24] as have the principles of fragment library design[25,26] and screening methods.[27–32] The following is a short introduction to the library design principles, screening methods and case studies where FBDD has been applied to the discovery of new chemical probes. Some of the key chemical challenges associated with FBDD are described in more detail in Chapter 4.

5.1.3 Fragment Library Design

As previously mentioned one of the main criteria for library design is the extent to which compounds obey the fragment "rule of three".[18] Following this, consideration should be given to the library size, the choice of which is mostly guided by the screening method of choice. Biochemical assays are generally less sensitive than most biophysical assays but benefit from being more compatible with a HTS format. Insensitivity often precludes the use of biochemical assays in screening small weak-affinity fragment molecules, instead favouring more elaborated, and potentially potent, compound libraries. As more sensitive biophysical techniques are generally lower-throughput they are often employed when screening fragment libraries and are complementary to the potentially large and diverse chemical space covered by a small fragment library. Other factors besides collection size that are considered in library design are the inclusion of a diverse array of key pharmacophores,[33] balance of size,[34] sampling of 3D space,[35] sp^3 character,[36] diversity of synthetically tractable chemical starting points[37] and the omission of potentially reactive groups (for non-covalent fragment libraries) including known pan-assay interference compounds (PAINS).[38]

5.1.4 Screening Methods/Fragment Elaboration

Due to the smaller size, lack of chemical complexity and ultimately weaker binding affinity of fragment molecules compared with elaborated lead-like HTS compounds, biophysical techniques are generally employed as a primary assay in a fragment-based chemical probe/drug discovery programme. A recent survey carried out on the blog *Practical Fragments* compared the initial screening methods for fragment finding.[39] It emerged that over half the respondents used SPR or ligand-observed NMR as a primary screen. Once a hit molecule is identified it is then validated through the use of an orthogonal screening method. Often, if the target protein has a tractable crystal system, a co-crystal structure of the hit fragment bound to the protein can allow for the rationalisation of potency and visualisation of notable protein–ligand interactions. Obtaining a co-crystal structure serves as the transition between FBDD and structure-based drug discovery (SBDD); having obtained structural information, new substituents can be designed into the hit fragment and structure–activity relationships (SAR) can be obtained, ultimately guiding increases in potency and selectivity of the chemical series.

5.2 Case Studies

There are many examples of FBDD in chemical probe discovery. The following examples have been chosen to cover a diverse range of biological targets and to highlight key challenges in chemical probe discovery that can be overcome with FBDD.

5.2.1 A-1155463 (BCL-X_L)

BCL-2 (B-cell lymphoma 2) family proteins represent targets for oncology due, in part, to their intricate role in apoptosis and programmed cell death. Overexpression of antiapoptotic BCL-X_L (B-cell lymphoma, extra-large; a member of the BCL-2 protein family) has been correlated with drug resistance and poor prognosis of multiple solid tumours and haematological malignancies.[40] The development of small-molecule inhibitors of BCL-2 proteins represents a valuable goal of the oncology community. One of the first examples of FBDD was the discovery of the inhibitor navitoclax[41,42] which is a dual BCL-2–BCL-X_L inhibitor and showed clinical anti-tumour efficacy against BCL-2-dependent lymphoid malignancies. Dose-limiting thrombocytopenia was observed in trials when navitoclax was administered as a single agent.[43] These issues were offset in later work that led to the discovery of the BCL-2 selective inhibitor, venetoclax, which is now approved for the treatment of lymphocytic leukaemia.[44] Due to the issue of dual BCL-2–BCL-X_L inhibition and observed synergy between chemotherapies and BCL-X_L inhibition[45,46] co-workers at AbbVie and Genentech sought to develop a selective small-molecule inhibitor of the BCL-X_L protein.[47]

Initial efforts were focussed on targeting the P4 binding site of BCL-X_L after the known scaffold **1** had been reported to bind BCL-X_L in an adjacent pocket (P2) (Figure 5.1).[48,49] Using compound **1** as a 'first-site ligand', an NMR screen of 875 fragments was conducted leading to the discovery of heterocycle **2** as a ligand-efficient 'second-site binder'. The results of NMR and docking studies indicated that compound **2** was indeed interacting with the P4 site and so efforts were made to link compounds **1** and **2**. A focussed set of compounds was synthesised in order to link the cores of **1** and **2**; much of the focus was to offset the high human serum (HS) protein binding due to the lipophillic nature of the series. A 10–30-fold attenuation of compound binding to BCL-X_L was observed upon the addition of 1% human serum to the time resolved-Förster resonance energy transfer (TR-FRET) assay employed. Structure-guided SAR were obtained, leading to the discovery of compound **3** (A-1155463, Figure 5.1). Compound **3** displays excellent potency against BCL-X_L [dissociation constant (K_i) <0.01 nM] and over 1000-fold selectivity over BCL-2 and other proteins. Furthermore compound **3** displayed excellent cell killing activity in *mcl-1*$^{-/-}$ MEF cells [concentration giving 50% effect (EC_{50}) 65 nM] in the presence of 10% HS. A co-crystal structure of compound **3** with BCL-X_L (PDB ID 4QVX) confirmed the binding mode and rationalised the SAR obtained. The 2-fluorine substituent of compound **3** makes putative

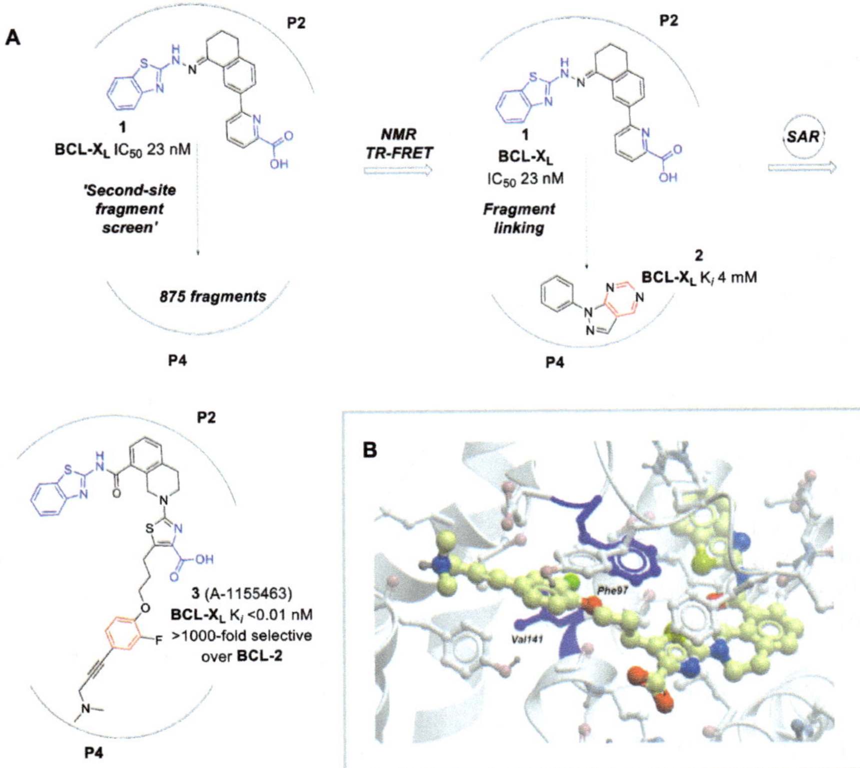

Figure 5.1 (A) Fragment linking of scaffolds **1** and new 'P4' hits leading to the discovery of **3** (A-1155463). (B) Co-crystal structure of BCL-X_L (shown in grey) (PDB ID 4QVX) with chemical probe **3** (yellow sticks). Val141 and Phe97 shown in blue with putative van der Waals interactions with 2-F (green atom) substituent of **3**.

van der Waals contacts with the side chains of Val141 and Phe97: the corresponding 2-H analogue shows tenfold weaker cell-killing activity. Compound **3** displays a good *in vivo* pharmacokinetic profile with good target coverage when SCID-Beige mice were dosed at 5 mg kg^{-1} with a single intraperitoneal (IP) dose. Platelet count was dramatically reduced after 6 h when SCID-Beige mice were treated with compound **3** at 5 mg kg^{-1}. This indirect evidence of BCL-X_L inhibition was concomitant with tumour progression inhibition when mice were inoculated with BCL-X_L-dependent H146 tumour cells. A-1155463 is now available for use as a BCL-X_L-selective chemical probe for investigating associated biology.

5.2.2 CCT244747 (CHK1)

A chemical probe developed using FBDD came out of a drug discovery programme at the Institute of Cancer Research and Sareum Ltd that was aimed at developing selective and orally bioavailable checkpoint kinase 1

(CHK1) inhibitors.[50] CHK1 is a compelling target for oncology in combination therapy or as a single agent[51] due to its role in regulating the cell cycle through phosphorylation of Cdc25C,[52,53] CHK1 inhibition leads to prevention of effective DNA lesion repair and ultimately cell death. At the time this program was initiated, a number of lead inhibitory compounds had suffered from poor selectivity for CHK1 over CHK2 and poor oral bioavailability and dose-limiting toxicity was observed in phase II clinical trials.[54] Selective inhibition over CHK2 was thought to be of benefit compared with dual inhibition.[51] Collins and co-workers sought to develop a selective and orally available CHK1 inhibitor which would allow flexibility for dosing in combination with conventional therapies and be more amenable to frequent single-agent dosing.

Initial work described in 2009 focused on hit finding using *in silico* template screening of over 15 000 compounds from a virtual library.[15] Using a pharmacophore search, the virtual library was filtered to identify features of potential kinase inhibitors matching interactions with five conserved features in multiple ligand-bound crystal structures of CHK1 in its active state. This pharmacophore filter was applied in combination with a filter for potential aggregators as the initial *in vitro* screen was a high-concentration biochemical screen (AlphaScreen™). A total of 361 hit compounds was then selected for high-concentration biochemical screening, leading to the discovery of hit **4** (CHK1 IC_{50} 42 μM). A crystal structure of **4** in complex with CHK1 enabled rationalisation of its binding affinity in terms of interactions with Cys87 and Glu85, and provided structural information on how to explore additional interactions. A variety of analogues, including compound **5**, showed increased potency relative to initial hit **4**, owing to new interactions with the backbone carbonyl Glu85 and the backbone N–H of Cys87 (Figure 2B, PDB ID 2YM3). A subset of the improved compounds was shown to abrogate the G_2–M checkpoint in HT29 cells at concentrations 10–20 fold above their CHK1 IC_{50} values. Some compounds also showed activity in a growth inhibition assay with two-fold to three-fold selectivity for CHK1-mediated growth inhibition over general cytotoxicity.[55]

Compound **5** was later developed through scaffold morphing guided by crystallography to achieve selectivity for CHK1 over CHK2 whilst simultaneously establishing novelty to furnish patentable compounds.[56] Compound **6** (SAR-020106)[57] was shown to be a highly potent and selective inhibitor of CHK1 (CHK1 IC_{50} 13 nM; CHK2 $IC_{50} > 100$ μM). Nanomolar cell activity was also observed in a G_2 checkpoint assay (in HT29 cells; IC_{50} 55 nM). A representative sample of 124 kinases was explored for broader selectivity of compound **6** using a radiometric assay format.[58] Selectivity profiling revealed a moderately clean profile with compound **6** only inhibiting CHK1 and six other kinases over 80% at 1 μM. Good cell permeability was observed in a CaCo-2 assay[59] [apical to basolateral permeability (AB Pe) 1.43×10^{-6} cm s^{-1}] with moderate efflux observed [flux from the basolateral to apical side over the flux from the apical to basolateral side of a cell monolayer (BA/AB) = 4]. Despite a good *in vitro* absorption, distribution, metabolism and excretion (ADME) profile compound **6** suffered from rapid metabolism in both mouse and rat liver microsomes (80% and 92% metabolized at 30 min respectively) which translated into poor oral bioavailability in mice (*F* 5%) in an *in vivo* model. Collins and co-workers then went on to describe pharmacokinetic (PK) optimisation efforts on compound **6** which was hybridised with 2-aminopyridine compound **7** owing to the increased metabolic stability observed for similar compounds. This led to the discovery of an orally active CHK1 inhibitor **8** (CCT244747, Figure 5.2A).[50] Compound **8** features excellent conservation of activity against CHK1 and good selectivity against a representative panel of kinases (>50% inhibition observed for 13 out of 140 enzymes at 1 μM including CHK1). Furthermore, compound **8** displays good cell activity in etoposide-induced G_2 checkpoint arrest in HT29 human colon cancer cells (IC_{50} 29 nM). Cell activity translated into the ability of compound **8** to potentiate the effects of cytotoxic drugs (gemcitabine and irinotecan) when administered in combination *in vivo* using human tumour xenograft models (HT29, SW620 and Calu6). Compound **8** is an excellent research tool to explore CHK1 biology *in vitro* and *in vivo* and as such it features on the Chemical Probes Portal.[60] More recently, a further optimised derivative of compound **8**, CCT245737 has entered two phase I clinical trials for patients with advanced solid tumours and in combination with gemcitabine/cisplatin.[61]

5.2.3 GSK2334470 (PDK1)

Co-workers from GlaxoSmithKline developed a FBDD-derived inhibitor[62] for phosphoinositide-dependent kinase-1 (PDK1) which plays an integral role in the P13K–AKT–mTOR pathway, often misregulated in cancers.[63–66] An initial fragment library (approximately 1000 compounds) was assembled from internal and external sources, focussed on pharmacophores with donor/acceptor motifs capable of targeting the kinase hinge region. A high-concentration biochemical kinase fragment screen was then employed yielding 193 compounds displaying observable binding (>60% at 400 μM). Compounds were then progressed and filtered based on IC_{50} values, synthetic tractability and ligand efficiency (LE > 0.4) to give 36 compounds of which compound **9** (Figure 5.3) showed excellent potency and ligand efficiency. Positive binding was confirmed by NMR saturation transfer difference (STD) experiments and X-ray crystallography. A co-crystal structure of compound **9** and PDK1 (PDB ID 3NUN) explained the observed potency of compound **9**, showing key interactions between the amino-indazole motif and the PDK1 hinge region. Compound **9** was further developed, leading to the discovery of compound **10** (GSK2334470, Figure 5.3), a potent inhibitor of PDK1 which was selective for kinases in the AGC kinase family. Compound **10** showed good selectivity (over 1000-fold) in a panel of 284 protein and lipid kinases with only 24 kinases inhibited over 50% at 10 μM. Cell target engagement was ascertained by an indirect cell assay exploring PDK1-mediated phosphorylation of Thr^{308}–AKT, Ser^{221}–RSK and Ser^{473}–AKT (not phosphorylated

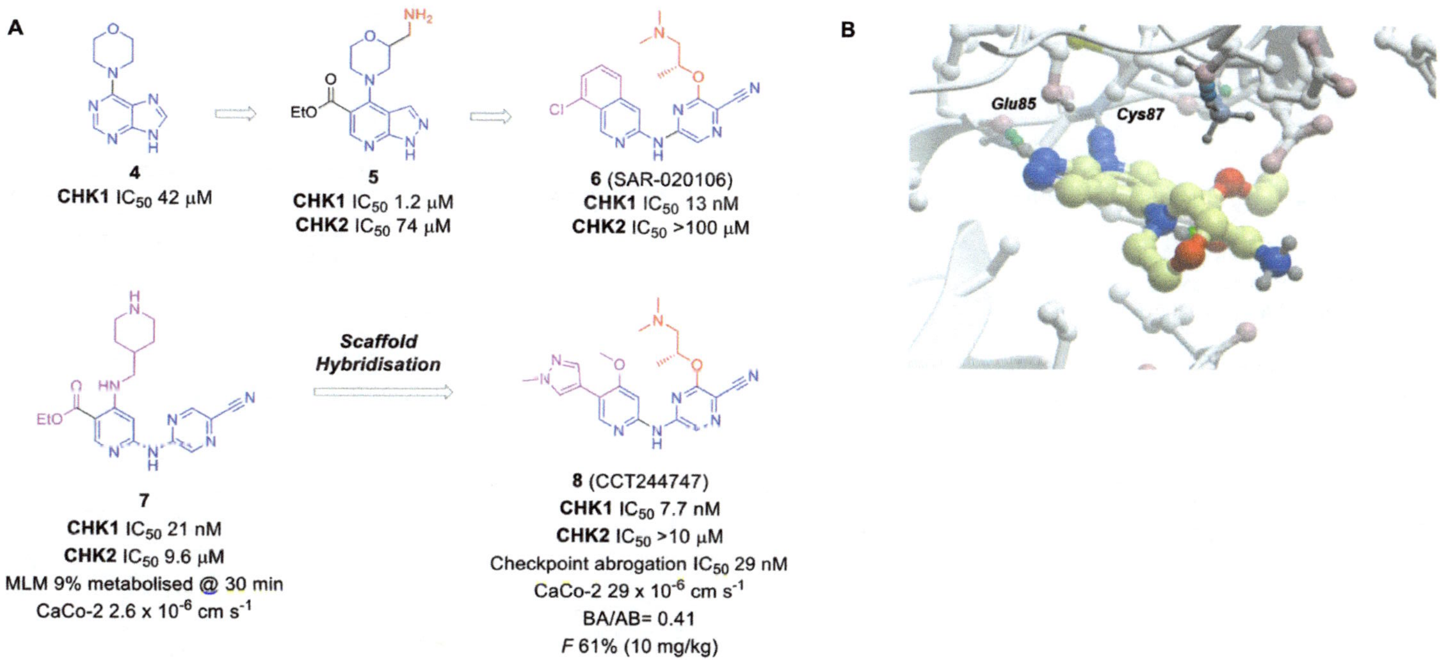

Figure 5.2 (A) Progression of initial hit CHK1 fragment hit **4** to inhibitor **5** and potent lead **6** (structural similarities are highlighted in blue) then hybridised with scaffold **7** to give lead probe **8** (CCT244747). (B) CHK1 co-crystal structure with CHK1 inhibitor **5** (PDB ID 2YM3).

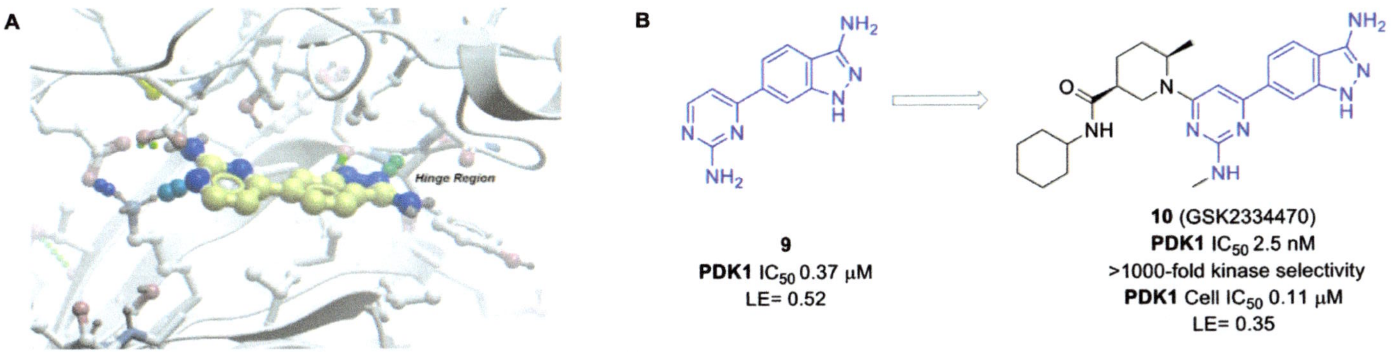

Figure 5.3 (A) Co-crystal structure of fragment hit **9** and PDK1 (PDB ID 3NUN). (B) Initial hit compound **9** was progressed to chemical probe **10** (GSK2334470).

by PDK1). Sub-micromolar inhibitory activity of Thr[308]–AKT/Ser[221]–RSK phosphorylation was observed when PC3 cells were treated with compound **10**, while no activity was observed against phosphorylation of Ser[473]–AKT. Compound **10** also showed moderate activity in *in vivo* tumour growth models (SCID mice harbouring CCI–AML2 tumours). Compound **10** is now available as a chemical probe for use in exploring PDK1-associated biology.

5.2.4 PFI-3 (SMARCA2/4 and PB1)

Bromodomain-containing proteins have received considerable attention in drug discovery as therapeutic targets since seminal work on bromodomain and extra terminal domain protein (BET) bromodomain inhibitors was disclosed by the Structural Genomics Consortium (SGC), Harvard and GlaxoSmithKline (GSK) in 2010.[67,68] Bromodomains are approximately 110-amino-acid domains comprised of four α-helices which fold into an epigenetic 'reading' domain capable of binding to acetylated lysine substrates.[69] Many of the ligands of bromodomain-containing proteins are acetylated histone tails. Histone acetylation is one of the epigenetic post-translational modifications that regulate chromatin remodelling and transcription. Increasingly, the non-BET bromodomains have received attention in chemical probe discovery efforts to enable target validation of these poorly understood proteins.[5,7] Many of these non-BET bromodomains play an integral role in chromatin remodelling and transcription directly or through their functional role as sub-units of larger protein complexes.

SWI/SNF-related, matrix-associated, actin-dependent regulator of chromatin, subfamily A 4 (SMARCA4) [Brahma related gene-1 (BRG1)] and SMARCA2 [Brahma (BRM)] are the key ATPase components of the SWI/SNF complexes (BAF/PBAF in mammals/humans) which play an integral role in cell differentiation and proliferation, and are essential to the core embryonic stem cell (ESC) pluripotency transcriptional network (Figure 5.4).[70] Mutations in components of the SWI/SNF complexes have been linked to oncology and inflammation.[71–73] In addition, the protein polybromo-1 (PB1) features in the polybromo, BRG1-associated factors (PBAF) complex (the human-specific analogue of SWI/SNF) in conjunction with SMARCA2 or SMARCA4. The presence of PB1, SMARCA2 and SMARCA4 in the human SWI/SNF complexes implies the necessity of bromodomains in these essential complexes, but the role of the individual BRD-containing proteins is not clear. This is also potentially the case for other bromodomain-containing proteins, such as the cancer-related BRD9 and BRD7 (Figure 5.4).[74]

Co-workers at Pfizer and SGC sought to uncover the biology and therapeutic potential of the structurally related PB1, SMARCA2 and SMARCA4 Bromodomain-containing proteins through the development of selective chemical probes targeting the associated bromodomains.[75] A fragment screen employing differential scanning fluorimetry (DSF) uncovered compound **11** (salicyclic acid, Figure 5.5) as a weak but ligand-efficient hit

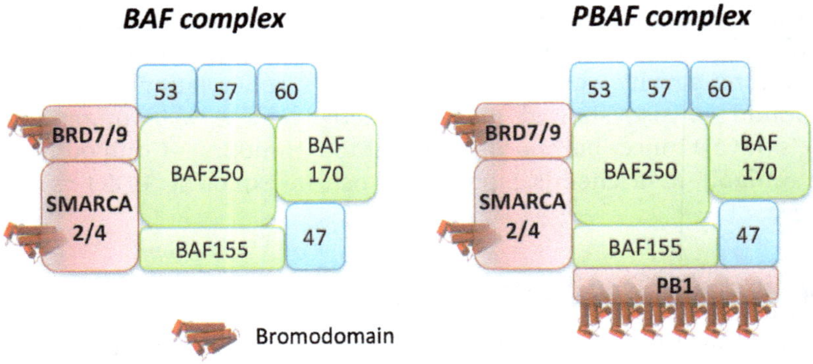

Figure 5.4 Subunits of the BAF/PBAF (SWI/SNF) mammalian ATP-dependent chromatin remodelling complexes including subunits BRD9, BRD7, SMARCA4 (BRG1), SMARCA2 (BRM) and PB1.

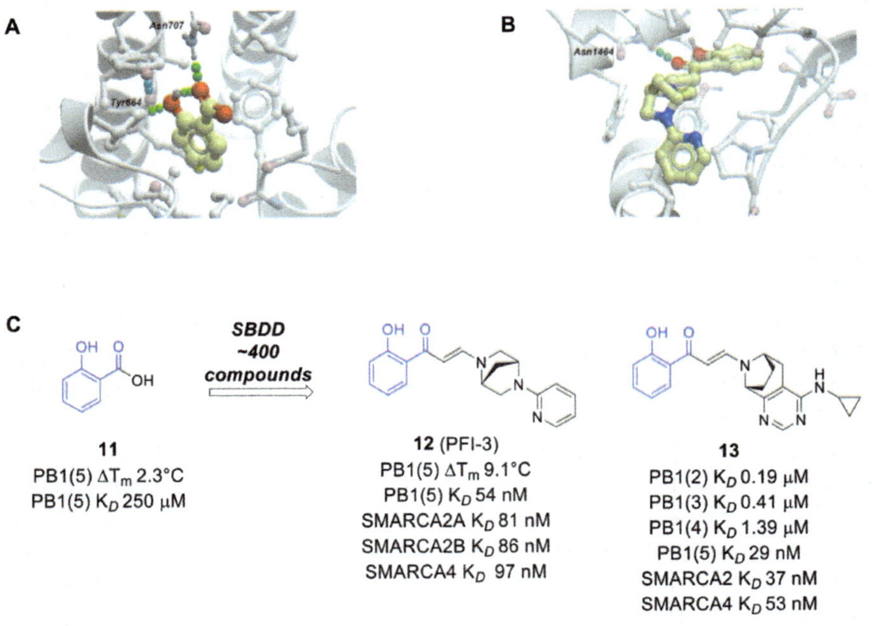

Figure 5.5 (A) Salicyclic acid (**11**) with PB1(5) (Y664 and N707 labeled) PDB ID 4Y03. (B) PFI-3 (**12**) co-crystal structure with SMARCA2 PDB ID 5DKC. (C) Salicyclic acid (**11**) is a ligand-efficient PB1(5) fragment hit which was progressed to chemical probe PFI-3 (**12**) and compound **13**.

molecule for the fifth bromodomain of PB1 [PB1(5)]. Interestingly, such a small molecule also displayed significant selectivity for the Brd subfamily that includes SMARCA2/4/PB1. A crystal structure of fragment **11** with PB1(5) (PDB ID 4Y03) revealed a unique binding mode within the acetyl lysine-binding site, where the phenolic motif displaces the typically conserved water network deep in the binding site, making a series of hydrogen-bond interactions with Y664 and N707 (Figure 5.5A). A combination of screening commercial compounds and analogues from the Pfizer chemical compound library allowed SAR to be built up around compound **11**. Guided by structural information, iterative compound design, synthesis and biophysical characterisation led to the discovery of compound **12** (PFI-3, Figure 5.5C). PFI-3 shows good binding affinity for bromodomains of PB1(5), SMARCA2 and SMARCA4 and shows good selectivity over all other bromodomains, as ascertained by DSF and BROMO*scan* selectivity assays.[76] An additional compound was also designed for broader non-selective inhibition of sub family *VIII*; compound **13** showed good potency against PB1(2) (the second bromodomain of PB1), PB1(5), SMARCA2 and SMARCA4 for use as an additional chemical tool in exploring bromodomain sub family *VIII* biology (Figure 5.5B). The bridged bicyclic ring systems of PFI-3 and compound **13** were incorporated to confer chemical stability on the enamide functionality, which otherwise slowly hydrolyses in aqueous media. This highlights another feature of chemical probes: they must also be chemically stable (to media conditions) in cellular assays. Cell target engagement for PFI-3 was shown in a fluorescence recovery after photobleaching (FRAP) experiment[77] using U2OS cells expressing either GFP–SMARCA2–bromodomain or a N146F inactive mutant. PFI-3 showed accelerated recovery at 1 μM when measuring its ability to displace SMARCA4 from chromatin as compared with an inactive negative control compound.[78] Anti-proliferative effects for PFI-3-mediated SMARCA2/4 and PB1 bromodomain inhibition were investigated in two renal cell carcinoma (ccRCC) cell lines (786-O and Rcc4) in conjunction with RNA interference (RNAi) studies. It was found that despite PFI-3 being a potent, selective SMARCA2/4/PB1 bromdomain inhibitor, treatment of 786-O and Rcc4 cells failed to recapitulate the antiproliferative phenotype shown for target knockdown. In addition to this, the results indicated that the ATPase domain was the relevant therapeutic domain which is capable of delivering a functional effect.[79] PFI-3 is now available as a valuable chemical probe to the scientific community for further exploration of SMARCA2/4/PB1 bromodomain-associated biology.

5.2.5 BI-9564 (BRD9)

Recent work carried out using FBDD to target the additional bromodomain-containing proteins (BRD7 and BRD9) involved in the cancer-related SWI/SNF complexes[80] was reported by co-workers from Boehringer Ingelheim (BI) and SGC.[81] Martin and colleagues sought to develop an *in vivo* chemical probe for the bromodomain of BRD9 following the discovery of the bromodomain's necessity in driving proliferation in acute myeloid leukaemia (AML) cells.[82] Previous chemical probes for BRD9 had been disclosed in the form of LP99,[83] iBRD9[84] and 'compound 28'.[85] Many of these previously described probes had been well characterised, displaying potency, selectivity and cell-target engagement, however, there remained an unmet need for new BRD9 chemical probes with improved pharmacokinetic profiles for *in vivo* use.

Two parallel approaches to the discovery of a new BRD9 chemical probe were employed (Figure 5.6). An initial proprietary BI fragment library (approximately 1700 compounds) was screened using a combination of DSF, SPR and microscale thermophoresis (MST). Initial hit compounds were cross-validated by ^{15}N heteronuclear single quantum coherence–nuclear magnetic resonance (HSQC–NMR) screening to give 77 hit compounds. Further profiling through SPR and X-ray crystallography yielded 12 hit compounds including compound **14** (Figure 5.6) as a weakly active fragment (BRD9 K_D 37.5 μM).

In parallel, a virtual screening campaign was carried out using glide docking[86] and a proprietary high-concentration screening (HiCoS) library of approximately 74 000 compounds yielding 208 compounds (0.3% hit rate). The library was further filtered based on BRD9 pharmacophore mapping and 'fragment profiling' (MW and clogP). Biophysical screening of available compounds by DSF and SPR relative to controls gave a subset of 23 compounds. Of the 23 selected compounds, 11 compounds were validated by SPR (BRD9 K_D less than 100 μM) and were successfully co-crystallised with the BRD9 bromodomain to confirm the mode of binding. Compound **15** (Figure 5.6), a moderately active fragment (K_D 9.1 μM) derived from this virtual screening cascade, was 'fragment merged' with compound **14** aided by structural information gained through crystallography (Figure 5.6A). Both compounds **14** and **15** share a common 1-methylpyridone core and crystallography revealed conserved interactions between both **14** and **15** with Asn100 (Figure 5.6B and C).

Through an iterative process of synthesis and biophysical/biochemical screening, compound **16** (BI-9564, Figure 5.6A) was discovered as a highly potent and selective BRD9 chemical probe. A co-crystal structure of BI-9564 with BRD9 (PDB ID 5F1H) revealed key interactions with Asn100, including a putative C–H hydrogen-bond with the carbonyl of Asn100 (C–N distance 3.14 Å) enabled by the electron-withdrawing effect of the naphthyridinone core. BI-9564 displays excellent potency for BRD9 (K_D 14 nM) with some selectivity over the highly homologous BRD7 (K_D 0.24 μM) and CECR2 (K_D 0.2 μM). Excellent selectivity was observed over other bromodomains (over 1500-fold BET selectivity) in a DiscoverX BROMO*scan* assay alongside a panel of 324 kinases and 53 G-protein-coupled receptors (GPCRs). Cell target engagement was ascertained through a BRD9 FRAP assay (90% inhibition of BRD9 Brd chromatin binding at 1 μM). Compound BI-9564 was screened for efficacy in cancer cell models, showing selective growth inhibition of AML cells (EC_{50} EOL-1 0.8 μM). As BI-9564 had good pharmacokinetics (*F* 88%) its anti-cancer activity was further validated in an *in vivo* disseminated mouse model of AML.

Domain-swapping experiments with the BRD9 bromodomain confirmed the BRD9-mediated anti-proliferative effects of compound BI-9564.[82] BI-9564

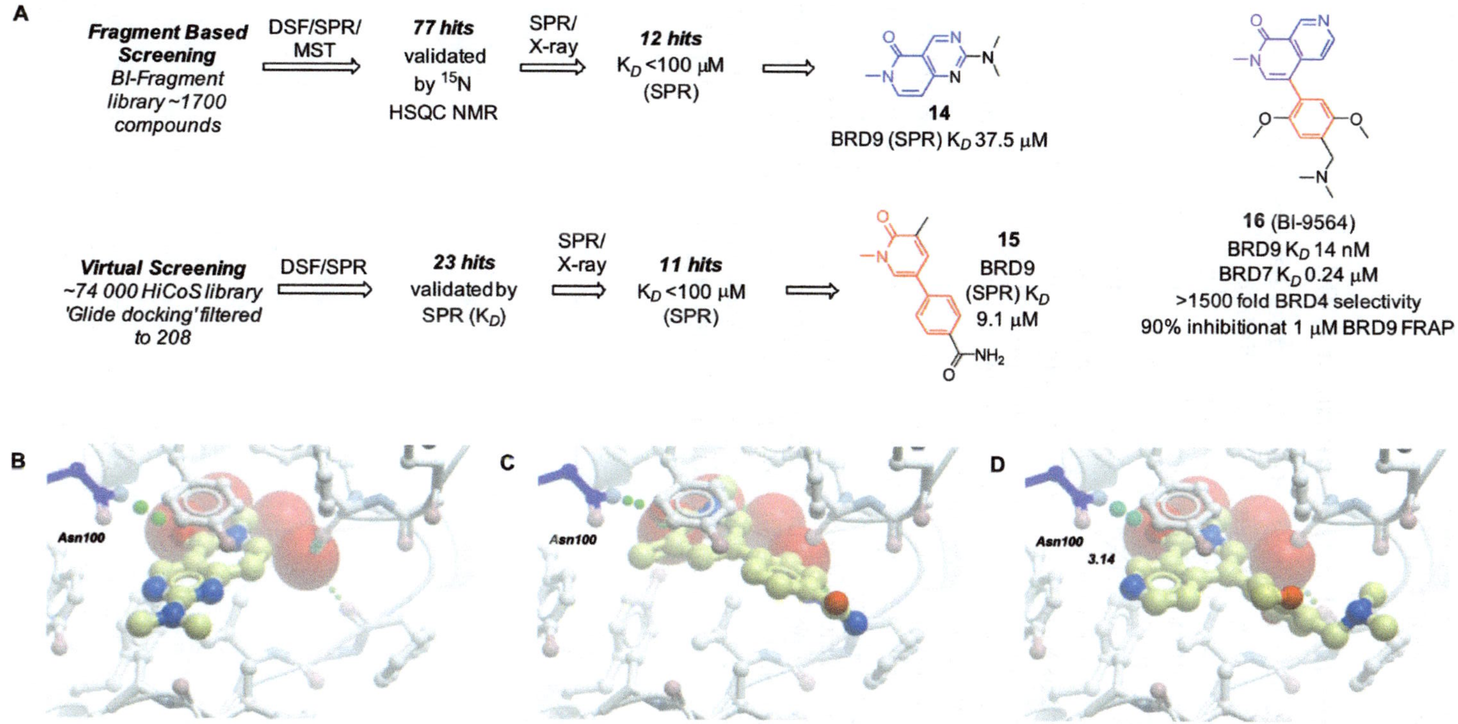

Figure 5.6 (A) Fragment combining of hits **14** and **15** led to the discovery of **16** (BI-9564). (B) BRD9 co-crystal structures of fragment hit **14** (PDB ID 5F2P). (C) BRD9 co-crystal structure of fragment hit **15** (PDB ID 5F25). (D) BRD9 co-crystal structure of chemical probe BI-9564 (**16**) (PDB ID 5F1H). Atom distances for putative C–H bond with Asn100 (blue sticks) shown.

is now available[87] as an *in vitro*/*in vivo* chemical probe for interrogating BRD9 and its associated pathology.

5.2.6 Astex DDR1 Kinase Inhibitor (DDR1/DDR2)

Discoidin domain receptors, DDR1 and DDR2 have been implicated in a number of pathologies, such as fibrotic disorders,[88] atherosclerosis[89] and cancer[90] in part due to the intricate role they play in extracellular remodelling, cell migration, proliferation and adhesion. Genetic interrogation of DDR2 in lung squamous cell cancer has shown that this target is mutated in 4% of cases.[91] Researchers at Astex Pharmaceuticals sought to develop selective DDR1/2 inhibitors to probe the role of these target proteins in diseases such as cancer. An initial set of approximately 15 000 fragments was screened in a high-concentration DDR1 protein thermal shift (DSF) assay. Initial hit compounds were then validated through crystallography with a soakable DDR1 crystal system where approximately 50 ligand-bound structures were obtained, which included some non-selective kinase inhibitors, such as dasatinib **17** (Figure 5.7A). Interestingly, some of the structures obtained revealed fragments, such as urea **18**, bound in the back of the DDR1 pocket—only accessible to the 'DFG-out' inactive form of DDR1 as opposed to binding the hinge region. Fragment **18** made hydrogen-bond interactions with Glu672 and Asp784 through the urea motif and adopted a similar structure and binding mode to other known type II kinase inhibitors. Compound **18** was then implemented into a proprietary computation

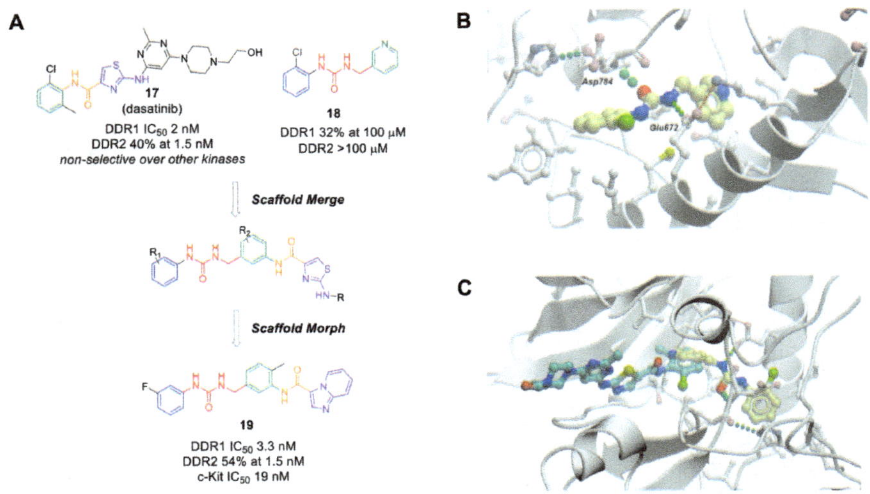

Figure 5.7 (A) Scaffolding 'merging' and 'morphing' strategies leading to the discovery of **19** a potent and selective DDR1/DDR2 chemical probe. (B) Co-crystal structure of DDR1 with fragment **18** (PDB ID 5BVK). (C) Overlay of DDR1/**18** (yellow sticks) co-crystal structure (PDB: 5BVK) with DDR1/**17** (cyan sticks) co-crystal structure (PDB ID 5BVW).

modelling tool, AstexMerge, which is similar to computational design tool BREED.[92] Through AstexMerge, new ligands were designed and scored on the basis of the superposition of atoms and bond distances in similar protein structures. A co-crystal structure of DDR1 with the non-selective kinase inhibitor dasatinib revealed good structural overlay with the DDR1 **18** crystal structure. A subset of new molecules were designed, synthesised and characterised for DDR1/2-binding affinity based on the merging of scaffolds from dasatinib **17** and fragment **18**. Merging of scaffolds led to drastic improvements in activity (over 300-fold DDR2 affinity). Structure-based design of appropriate ligands led to the discovery of compound **19** which shows excellent DDR1/2 potency and good selectivity against a modest set of 24 kinases. Compound **19** also showed acceptable *in vitro* and *in vivo* pharmacokinetics. Indirect cell target engagement was ascertained using a MesoScale Discovery (MSD) assay revealing excellent DDR2 inhibition (EC_{50} 8.7 nM). Compound **19** was then profiled for anti-proliferative effects in DDR2-mutant cell lines, however no effect was observed. It was suggested that the anti-proliferative effects in the same DDR2-mutant cell line induced by treatment with dasatinib **17** or nilotinib are not attributable to DDR2 inhibition, despite them also being potent DDR2 inhibitors, and probably due to the significant polypharmacology of these promiscuous inhibitors. Despite obtained evidence invalidating DDR2 inhibition as a means for obtaining an anti-proliferative phenotype, compound **19** serves as a useful chemical probe for investigating the role of DDR1/DDR2 in other areas.

5.3 Outlook

Chemical probes are increasingly becoming valuable tools for target validation in drug discovery. The ability of a chemical probe programme to validate or invalidate a target early in the drug discovery process will probably improve attrition. FBDD serves as a time- and resource-efficient approach to the development of new chemical probes. As exemplified in the aforementioned case studies, the FBDD approach has been used extensively in many industrial and academic laboratories for chemical probe development. Once hits are found and validated, SBDD can be used to generate potent and selective tools for a given target. Broader selectivity of a chemical probe against a diverse set of drug targets can be readily determined using a number of commercial selectivity panels (*e.g.* Eurofins Safetyscreen™).[93] Target engagement assay technologies are continually being developed, which allow demonstration of compound target interaction in cells, such as the FRAP assay,[77] Promega's nanoBRET assay,[94] phosphoproteomics,[95] cellular thermal shift (CETSA)[96] and "protein pulldown".[97] As chemical probe discovery programmes become increasingly accessible for academic groups, the authors envisage this will catalyse the forging of new collaborative efforts between industry and academia. Ultimately this will create an extensive toolbox of chemical probes to be used in phenotypic assays for target discovery and validation.

Abbreviations

The abbreviated names for cell lines and proteins have not been defined unless the protein is the primary target of the exemplified chemical probes.

Abbreviation	**Definition**
ALPHAScreen	*a*mplified *l*uminescence *p*roximity *h*omogeneous *a*ssay
AML	*a*cute *m*yeloid *l*eukaemia
BA/AB	flux from the *b*asolateral to *a*pical side *over* the flux from the *a*pical to *b*asolateral side of a cell monolayer
BAF	*B*RG1 *a*ssociated *f*actors
BCL-2	*B*-*c*ell *l*ymphoma *2*
BCL-X_L	*B*-*c*ell *l*ymphoma-e*x*tra *l*arge
BET	*b*romodomain and *e*xtra *t*erminal domain protein
BRD7/BRD9	*br*omo*d*omain protein *7* or *9*
BRG-1	*b*rahma *r*elated *g*ene *1*
BRM	*br*ah*m*a gene
BROMOscan	*bromo*domain *scan* (assay)
CETSA	*c*ellular *t*hermal *s*hift *a*ssay
CHK1	*ch*eckpoint *k*inase *1*
CRISPR–Cas9	*c*lustered *r*egularly *i*nterspaced *s*hort *p*alindromic *r*epeats–*C*RIPR *as*sociated *s*ystem *9*
DDR1/DDR2	*d*iscoidin *d*omain *r*eceptor *1* or *2*
DFG-out	aspartate (*D*) phenylalanine (*F*) glycine (*G*) kinase loop in the "*out*" conformation
DSF	*d*ifferential *s*canning *f*luorimetry (assay)
ESC	*e*mbryonic *s*tem *c*ells
F %	*percentage* bioavailability from oral dosing
FBDD	*f*ragment-*b*ased *d*rug *d*iscovery
FRAP	*f*luorescence *r*ecovery *a*fter *p*hoto-bleaching (assay)
G2	2nd *g*rowth phase of the cell cycle
GFP	*g*reen *f*luorescent *p*rotein
GPCRs	*G*-*p*rotein-*c*oupled *r*eceptors
HTS	*h*igh *t*hroughput *s*creen
IC_{50}	compound *c*oncentration that shows *50*% of maximum *i*nhibition
K_D	*d*issociation constant
LE	*l*igand *e*fficiency
MSD	*M*eso*S*cale *D*iscovery (assay)
MST	*m*icro*s*cale *t*hermophoresis (assay)
nanoBRET	*nano*luciferase coupled *b*ioluminescent *r*esonance *e*nergy *t*ransfer (assay)
NMR	*n*uclear *m*agnetic *r*esonance (assay)
PAINS	*pan*-*a*ssay *in*terference (compound*s*)
PB1	*p*oly*b*romodomain protein *1*
PBAF	*PB*1-associated *BAF*

PDK1	*p*hosphoinositide-*d*ependent *k*inase *1*
PK	*p*harmaco*k*inetics
SAR	*s*tructure *a*ctivity *r*elationship
SBDD	*s*tructure-*b*ased *d*rug *d*esign
SMARCA2/ SMARCA4	*SWI/SNF*-related, *m*atrix-associated, *a*ctin-dependent regulator of *c*hromatin, subfamily *A 2* or *4*
SPR	*s*urface *p*lasmon *r*esonance (assay)
STD	*s*aturation *t*ransfer *d*ifference (assay)
SWI/SNF	*swi*tch/*s*ucrose *n*on-*f*ermentable
TR-FRET	*t*ime *r*esolved-*F*örster *r*esonance *e*nergy *t*ransfer

References

1. A. Reynolds, D. Leake, Q. Boese, S. Scaringe, W. S. Marshall and A. Khvorova, *Nat. Biotechnol.*, 2004, **22**, 326–330.
2. J. A. Doudna and E. Charpentier, *Science*, 2014, **346**, 1258096.
3. M. E. Bunnage, E. L. P. Chekler and L. H. Jones, *Nat. Chem. Biol.*, 2013, **9**, 195–199.
4. SGC Chemical Probes, http://www.thesgc.org/chemical-probes, (accessed 29 May 2017).
5. N. H. Theodoulou, N. C. O. Tomkinson, R. K. Prinjha and P. G. Humphreys, *ChemMedChem*, 2016, **11**, 477–487.
6. S. Ackloo, P. J. Brown and S. Muller, *Epigenetics*, 2017, 1–23.
7. M. Moustakim, P. G. K. Clark, D. A. Hay, D. J. Dixon and P. E. Brennan, *Med. Chem. Commun.*, 2016, **7**, 2246–2264.
8. C. H. Arrowsmith, J. E. Audia, C. Austin, J. Baell, J. Bennett, J. Blagg, C. Bountra, P. E. Brennan, P. J. Brown, M. E. Bunnage, C. Buser-Doepner, R. M. Campbell, A. J. Carter, P. Cohen, R. A. Copeland, B. Cravatt, J. L. Dahlin, D. Dhanak, A. M. Edwards, M. Frederiksen, S. V. Frye, N. Gray, C. E. Grimshaw, D. Hepworth, T. Howe, K. V. M. Huber, J. Jin, S. Knapp, J. D. Kotz, R. G. Kruger, D. Lowe, M. M. Mader, B. Marsden, A. Mueller-Fahrnow, S. Muller, R. C. O'Hagan, J. P. Overington, D. R. Owen, S. H. Rosenberg, R. Ross, B. Roth, M. Schapira, S. L. Schreiber, B. Shoichet, M. Sundstrom, G. Superti-Furga, J. Taunton, L. Toledo-Sherman, C. Walpole, M. A. Walters, T. M. Willson, P. Workman, R. N. Young and W. J. Zuercher, *Nat. Chem. Biol.*, 2015, **11**, 536–541.
9. http://www.chemicalprobes.org/, (accessed 29 May 2017).
10. C. Brideau, B. Gunter, B. Pikounis and A. Liaw, *J. Biomol. Screening*, 2003, **8**, 634–647.
11. D. R. Davies, in *Structural Genomics and Drug Discovery: Methods and Protocols*, ed. W. F. Anderson, Springer New York, New York, NY, 2014, pp. 315–323.
12. I. Navratilova and A. L. Hopkins, *ACS Med. Chem. Lett.*, 2010, **1**, 44–48.
13. F. H. Niesen, H. Berglund and M. Vedadi, *Nat. Protoc.*, 2007, **2**, 2212–2221.

14. L. Pedro and R. J. Quinn, *Molecules*, 2016, **21**, 984/981–984/916.
15. S. Ekins, J. Mestres and B. Testa, *Br. J. Pharmacol.*, 2007, **152**, 9–20.
16. C.-O. Ramon, *Curr. Top. Med. Chem.*, 2011, **11**, 43–67.
17. J. P. Hughes, S. Rees, S. B. Kalindjian and K. L. Philpott, *Br. J. Pharmacol.*, 2011, **162**, 1239–1249.
18. M. Congreve, R. Carr, C. Murray and H. Jhoti, *Drug Discovery Today*, 2003, **8**, 876–877.
19. H. Jhoti, G. Williams, D. C. Rees and C. W. Murray, *Nat. Rev. Drug Discovery*, 2013, **12**, 644.
20. R. J. Hall, P. N. Mortenson and C. W. Murray, *Prog. Biophys. Mol. Biol.*, 2014, **116**, 82–91.
21. M. Congreve, G. Chessari, D. Tisi and A. J. Woodhead, *J. Med. Chem.*, 2008, **51**, 3661–3680.
22. C. W. Murray and D. C. Rees, *Nat. Chem.*, 2009, **1**, 187–192.
23. D. A. Erlanson, *Top. Curr. Chem.*, 2012, **317**, 1–32.
24. C. W. Murray, M. L. Verdonk and D. C. Rees, *Trends Pharmacol. Sci.*, 2012, **33**, 224–232.
25. G. M. Keserű, D. A. Erlanson, G. G. Ferenczy, M. M. Hann, C. W. Murray and S. D. Pickett, *J. Med. Chem.*, 2016, **59**, 8189–8206.
26. G. Siegal, E. Ab and J. Schultz, *Drug Discovery Today*, 2007, **12**, 1032–1039.
27. M.-D. Duong-Thi, E. Meiby, M. Bergstroem, T. Fex, R. Isaksson and S. Ohlson, *Anal. Biochem.*, 2011, **414**, 138–146.
28. D. Chen, J. C. Errey, L. H. Heitman, F. H. Marshall, A. P. Ijzerman and G. Siegal, *ACS Chem. Biol.*, 2012, **7**, 2064–2073.
29. J. B. Jordan, L. Poppe, X. Xia, A. C. Cheng, Y. Sun, K. Michelsen, H. Eastwood, P. D. Schnier, T. Nixey and W. Zhong, *J. Med. Chem.*, 2012, **55**, 678–687.
30. H. J. Maple, R. A. Garlish, L. Rigau-Roca, J. Porter, I. Whitcombe, C. E. Prosser, J. Kennedy, A. J. Henry, R. J. Taylor, M. P. Crump and J. Crosby, *J. Med. Chem.*, 2012, **55**, 837–851.
31. F. Sirci, E. P. Istyastono, H. F. Vischer, A. J. Kooistra, S. Nijmeijer, M. Kuijer, M. Wijtmans, R. Mannhold, R. Leurs, I. J. P. de Esch and C. de Graaf, *J. Chem. Inf. Model.*, 2012, **52**, 3308–3324.
32. E. H. Mashalidis, P. Sledz, S. Lang and C. Abell, *Nat. Protoc.*, 2013, **8**, 2309–2324.
33. V. Campagna-Slater, A. G. Arrowsmith, Y. Zhao and M. Schapira, *J. Chem. Inf. Model.*, 2010, **50**, 358–367.
34. S. L. Dixon and R. T. Koehler, *J. Med. Chem.*, 1999, **42**, 2887–2900.
35. A. D. Morley, A. Pugliese, K. Birchall, J. Bower, P. Brennan, N. Brown, T. Chapman, M. Drysdale, I. H. Gilbert, S. Hoelder, A. Jordan, S. V. Ley, A. Merritt, D. Miller, M. E. Swarbrick and P. G. Wyatt, *Drug Discovery Today*, 2013, **18**, 1221–1227.
36. D. G. Twigg, N. Kondo, S. L. Mitchell, W. R. J. D. Galloway, H. F. Sore, A. Madin and D. R. Spring, *Angew. Chem., Int. Ed.*, 2016, **55**, 12479–12483.

37. O. B. Cox, T. Krojer, P. Collins, O. Monteiro, R. Talon, A. Bradley, O. Fedorov, J. Amin, B. D. Marsden, J. Spencer, F. von Delft and P. E. Brennan, *Chem. Sci.*, 2016, 7, 2322–2330.
38. D. A. Erlanson, *J. Med. Chem.*, 2015, **58**, 2088–2090.
39. D. Erlanson, *Practical Fragments*, http://practicalfragments.blogspot.co.uk/2014/01/poll-results-affiliation-fragment.html, (accessed 29 May 2017).
40. S. A. Amundson, T. G. Myers, D. Scudiero, S. Kitada, J. C. Reed and A. J. Fornace, *Cancer Res.*, 2000, **60**, 6101–6110.
41. C. Tse, A. R. Shoemaker, J. Adickes, M. G. Anderson, J. Chen, S. Jin, E. F. Johnson, K. C. Marsh, M. J. Mitten, P. Nimmer, L. Roberts, S. K. Tahir, Y. Xiao, X. Yang, H. Zhang, S. Fesik, S. H. Rosenberg and S. W. Elmore, *Cancer Res.*, 2008, **68**, 3421–3428.
42. L. D. Walensky, *J. Clin. Oncol.*, 2012, **30**, 554–557.
43. W. H. Wilson, O. A. O'Connor, M. S. Czuczman, A. S. LaCasce, J. F. Gerecitano, J. P. Leonard, A. Tulpule, K. Dunleavy, H. Xiong, Y.-L. Chiu, Y. Cui, T. Busman, S. W. Elmore, S. H. Rosenberg, A. P. Krivoshik, S. H. Enschede and R. A. Humerickhouse, *Lancet Oncol.*, 2010, **11**, 1149–1159.
44. A. J. Souers, J. D. Leverson, E. R. Boghaert, S. L. Ackler, N. D. Catron, J. Chen, B. D. Dayton, H. Ding, S. H. Enschede, W. J. Fairbrother, D. C. S. Huang, S. G. Hymowitz, S. Jin, S. L. Khaw, P. J. Kovar, L. T. Lam, J. Lee, H. L. Maecker, K. C. Marsh, K. D. Mason, M. J. Mitten, P. M. Nimmer, A. Oleksijew, C. H. Park, C.-M. Park, D. C. Phillips, A. W. Roberts, D. Sampath, J. F. Seymour, M. L. Smith, G. M. Sullivan, S. K. Tahir, C. Tse, M. D. Wendt, Y. Xiao, J. C. Xue, H. Zhang, R. A. Humerickhouse, S. H. Rosenberg and S. W. Elmore, *Nat. Med.*, 2013, **19**, 202–208.
45. X. Lin, S. Morgan-Lappe, X. Huang, L. Li, D. M. Zakula, L. A. Vernetti, S. W. Fesik and Y. Shen, *Oncogene*, 2006, **26**, 3972–3979.
46. M. Wong, N. Tan, J. Zha, F. V. Peale, P. Yue, W. J. Fairbrother and L. D. Belmont, *Mol. Cancer Ther.*, 2012, **11**, 1026–1035.
47. Z.-F. Tao, L. Hasvold, L. Wang, X. Wang, A. M. Petros, C. H. Park, E. R. Boghaert, N. D. Catron, J. Chen, P. M. Colman, P. E. Czabotar, K. Deshayes, W. J. Fairbrother, J. A. Flygare, S. G. Hymowitz, S. Jin, R. A. Judge, M. F. T. Koehler, P. J. Kovar, G. Lessene, M. J. Mitten, C. O. Ndubaku, P. Nimmer, H. E. Purkey, A. Oleksijew, D. C. Phillips, B. E. Sleebs, B. J. Smith, M. L. Smith, S. K. Tahir, K. G. Watson, Y. Xiao, J. Xue, H. Zhang, K. Zobel, S. H. Rosenberg, C. Tse, J. D. Leverson, S. W. Elmore and A. J. Souers, *ACS Med. Chem. Lett.*, 2014, **5**, 1088–1093.
48. G. Lessene, P. E. Czabotar, B. E. Sleebs, K. Zobel, K. N. Lowes, J. M. Adams, J. B. Baell, P. M. Colman, K. Deshayes, W. J. Fairbrother, J. A. Flygare, P. Gibbons, W. J. A. Kersten, S. Kulasegaram, R. M. Moss, J. P. Parisot, B. J. Smith, I. P. Street, H. Yang, D. C. S. Huang and K. G. Watson, *Nat. Chem. Biol.*, 2013, **9**, 390–397.

49. B. E. Sleebs, W. J. A. Kersten, S. Kulasegaram, G. Nikolakopoulos, E. Hatzis, R. M. Moss, J. P. Parisot, H. Yang, P. E. Czabotar, W. D. Fairlie, E. F. Lee, J. M. Adams, L. Chen, M. F. van Delft, K. N. Lowes, A. Wei, D. C. S. Huang, P. M. Colman, I. P. Street, J. B. Baell, K. Watson and G. Lessene, *J. Med. Chem.*, 2013, **56**, 5514–5540.
50. M. Lainchbury, T. P. Matthews, T. McHardy, K. J. Boxall, M. I. Walton, P. D. Eve, A. Hayes, M. R. Valenti, A. K. de Haven Brandon, G. Box, G. W. Aherne, J. C. Reader, F. I. Raynaud, S. A. Eccles, M. D. Garrett and I. Collins, *J. Med. Chem.*, 2012, **55**, 10229–10240.
51. M. D. Garrett and I. Collins, *Trends Pharmacol. Sci.*, 2011, **32**, 308–316.
52. C.-Y. Peng, P. R. Graves, R. S. Thoma, Z. Wu, A. S. Shaw and H. Piwnica-Worms, *Science*, 1997, **277**, 1501–1505.
53. Y. Sanchez, C. Wong, R. S. Thoma, R. Richman, Z. Wu, H. Piwnica-Worms and S. J. Elledge, *Science*, 1997, **277**, 1497–1501.
54. G. J. Weiss, R. C. Donehower, T. Iyengar, R. K. Ramanathan, K. Lewandowski, E. Westin, K. Hurt, S. M. Hynes, S. P. Anthony and S. McKane, *Invest. New Drugs*, 2013, **31**, 136–144.
55. T. P. Matthews, S. Klair, S. Burns, K. Boxall, M. Cherry, M. Fisher, I. M. Westwood, M. I. Walton, T. McHardy, K.-M. J. Cheung, R. Van Montfort, D. Williams, G. W. Aherne, M. D. Garrett, J. Reader and I. Collins, *J. Med. Chem.*, 2009, **52**, 4810–4819.
56. J. C. Reader, T. P. Matthews, S. Klair, K.-M. J. Cheung, J. Scanlon, N. Proisy, G. Addison, J. Ellard, N. Piton, S. Taylor, M. Cherry, M. Fisher, K. Boxall, S. Burns, M. I. Walton, I. M. Westwood, A. Hayes, P. Eve, M. Valenti, A. de Haven Brandon, G. Box, R. L. M. van Montfort, D. H. Williams, G. W. Aherne, F. I. Raynaud, S. A. Eccles, M. D. Garrett and I. Collins, *J. Med. Chem.*, 2011, **54**, 8328–8342.
57. M. I. Walton, P. D. Eve, A. Hayes, M. Valenti, A. De Haven Brandon, G. Box, K. J. Boxall, G. W. Aherne, S. A. Eccles, F. I. Raynaud, D. H. Williams, J. C. Reader, I. Collins and M. D. Garrett, *Mol. Cancer Ther.*, 2010, **9**, 89–100.
58. MRC Kinase Panel, http://www.kinase-screen.mrc.ac.uk/, (accessed 29 May 2017).
59. Z. Wang, C. E. C. A. Hop, K. H. Leung and J. Pang, *J. Mass Spectrom.*, 2000, **35**, 71–76.
60. CCT244747, http://www.chemicalprobes.org/cct244747, (accessed 29 May 2017).
61. J. D. Osborne, T. P. Matthews, T. McHardy, N. Proisy, K.-M. J. Cheung, M. Lainchbury, N. Brown, M. I. Walton, P. D. Eve, K. J. Boxall, A. Hayes, A. T. Henley, M. R. Valenti, A. K. De Haven Brandon, G. Box, Y. Jamin, S. P. Robinson, I. M. Westwood, R. L. M. van Montfort, P. M. Leonard, M. B. A. C. Lamers, J. C. Reader, G. W. Aherne, F. I. , Raynaud, S. A. Eccles, M. D. Garrett and I. Collins, *J. Med. Chem.*, 2016, **59**, 5221–5237.
62. J. R. Medina, C. W. Blackledge, D. A. Heerding, N. Campobasso, P. Ward, J. Briand, L. Wright and J. M. Axten, *ACS Med. Chem. Lett.*, 2010, **1**, 439–442.

63. A. Mora, D. Komander, D. M. F. van Aalten and D. R. Alessi, *Semin. Cell Dev. Biol.*, 2004, **15**, 161–170.
64. D. R. Alessi, S. R. James, C. P. Downes, A. B. Holmes, P. R. J. Gaffney, C. B. Reese and P. Cohen, *Curr. Biol.*, 1997, 7, 261–269.
65. S. M. Constantine, M. Nicholas and K. Michael, *Curr. Cancer Drug Targets*, 2004, **4**, 235–256.
66. C. Peifer and D. R. Alessi, *ChemMedChem*, 2008, **3**, 1810–1838.
67. P. Filippakopoulos, J. Qi, S. Picaud, Y. Shen, W. B. Smith, O. Fedorov, E. M. Morse, T. Keates, T. T. Hickman, I. Felletar, M. Philpott, S. Munro, M. R. McKeown, Y. Wang, A. L. Christie, N. West, M. J. Cameron, B. Schwartz, T. D. Heightman, N. La Thangue, C. French, O. Wiest, A. L. Kung, S. Knapp and J. E. Bradner, *Nature*, 2010, **468**, 1067–1073.
68. E. Nicodeme, K. L. Jeffrey, U. Schaefer, S. Beinke, S. Dewell, C.-W. Chung, R. Chandwani, I. Marazzi, P. Wilson, H. Coste, J. White, J. Kirilovsky, C. M. Rice, J. M. Lora, R. K. Prinjha, K. Lee and A. Tarakhovsky, *Nature*, 2010, **468**, 1119–1123.
69. P. Filippakopoulos, S. Picaud, M. Mangos, T. Keates, J.-P. Lambert, D. Barsyte-Lovejoy, I. Felletar, R. Volkmer, S. Muller, T. Pawson, A.-C. Gingras, C. H. Arrowsmith and S. Knapp, *Cell*, 2012, **149**, 214–231.
70. L. Ho and G. R. Crabtree, *Nature*, 2010, **463**, 474–484.
71. J. Masliah-Planchon, I. Bieche, J.-M. Guinebretiere, F. Bourdeaut and O. Delattre, *Annu. Rev. Pathol.: Mech. Dis.*, 2015, **10**, 145–171.
72. V. R. Ramirez-Carrozzi, A. A. Nazarian, C. C. Li, S. L. Gore, R. Sridharan, A. N. Imbalzano and S. T. Smale, *Genes Dev.*, 2006, **20**, 282–296.
73. S. J. Cullen, S. Ponnappan and U. Ponnappan, *Mol. Immunol.*, 2009, **47**, 600–605.
74. C. Kadoch and G. R. Crabtree, *Sci. Adv.*, 2015, **1**, e1500447.
75. B. S. Gerstenberger, J. D. Trzupek, C. Tallant, O. Fedorov, P. Filippakopoulos, P. E. Brennan, V. Fedele, S. Martin, S. Picaud, C. Rogers, M. Parikh, A. Taylor, B. Samas, A. O'Mahony, E. Berg, G. Pallares, A. V. Torrey, D. K. Treiber, I. J. Samardjiev, B. T. Nasipak, T. Padilla-Benavides, Q. Wu, A. N. Imbalzano, J. A. Nickerson, M. E. Bunnage, S. Muller, S. Knapp and D. R. Owen, *J. Med. Chem.*, 2016, **59**, 4800–4811.
76. E. Quinn, L. Wodicka, P. Ciceri, G. Pallares, E. Pickle, A. Torrey, M. Floyd, J. Hunt and D. Treiber, *Cancer Res.*, 2013, **73**, 4238.
77. H. Deschout, K. Raemdonck, J. Demeester, S. C. De Smedt and K. Braeckmans, *Pharm. Res.*, 2014, **31**, 255–270.
78. O. Fedorov, J. Castex, C. Tallant, D. R. Owen, S. Martin, M. Aldeghi, O. Monteiro, P. Filippakopoulos, S. Picaud, J. D. Trzupek, B. S. Gerstenberger, C. Bountra, D. Willmann, C. Wells, M. Philpott, C. Rogers, P. C. Biggin, P. E. Brennan, M. E. Bunnage, R. Schule, T. Gunther, S. Knapp and S. Muller, *Sci. Adv.*, 2015, **1**, e1500723/1500721–e1500723/1500711.
79. B. Vangamudi, T. A. Paul, P. K. Shah, M. Kost-Alimova, L. Nottebaum, X. Shi, Y. Zhan, E. Leo, H. S. Mahadeshwar, A. Protopopov, A. Futreal,

T. N. Tieu, M. Peoples, T. P. Heffernan, J. R. Marszalek, C. Toniatti, A. Petrocchi, D. Verhelle, D. R. Owen, G. Draetta, P. Jones, W. S. Palmer, S. Sharma and J. N. Andersen, *Cancer Res.*, 2015, **75**, 3865–3878.
80. J. Masliah-Planchon, I. Bieche, J.-M. Guinebretiere, F. Bourdeaut and O. Delattre, *Annu. Rev. Pathol.: Mech. Dis.*, 2015, **10**, 145–171.
81. L. J. Martin, M. Koegl, G. Bader, X.-L. Cockcroft, O. Fedorov, D. Fiegen, T. Gerstberger, M. H. Hofmann, A. F. Hohmann, D. Kessler, S. Knapp, P. Knesl, S. Kornigg, S. Müller, H. Nar, C. Rogers, K. Rumpel, O. Schaaf, S. Steurer, C. Tallant, C. R. Vakoc, M. Zeeb, A. Zoephel, M. Pearson, G. Boehmelt and D. McConnell, *J. Med. Chem.*, 2016, **59**, 4462–4475.
82. A. F. Hohmann, L. J. Martin, J. L. Minder, J.-S. Roe, J. Shi, S. Steurer, G. Bader, D. McConnell, M. Pearson, T. Gerstberger, T. Gottschamel, D. Thompson, Y. Suzuki, M. Koegl and C. R. Vakoc, *Nat. Chem. Biol.*, 2016, **12**, 672–679.
83. P. G. K. Clark, L. C. C. Vieira, C. Tallant, O. Fedorov, D. C. Singleton, C. M. Rogers, O. P. Monteiro, J. M. Bennett, R. Baronio, S. Müller, D. L. Daniels, J. Méndez, S. Knapp, P. E. Brennan and D. J. Dixon, *Angew. Chem., Int. Ed.*, 2015, **54**, 6217–6221.
84. N. H. Theodoulou, P. Bamborough, A. J. Bannister, I. Becher, R. A. Bit, K. H. Che, C.-W. Chung, A. Dittmann, G. Drewes, D. H. Drewry, L. Gordon, P. Grandi, M. Leveridge, M. Lindon, A.-M. Michon, J. Molnar, S. C. Robson, N. C. O. Tomkinson, T. Kouzarides, R. K. Prinjha and P. G. Humphreys, *J. Med. Chem.*, 2015, **59**, 1425–1439.
85. D. A. Hay, C. M. Rogers, O. Fedorov, C. Tallant, S. Martin, O. P. Monteiro, S. Muller, S. Knapp, C. J. Schofield and P. E. Brennan, *MedChemComm*, 2015, **6**, 1381–1386.
86. R. A. Friesner, J. L. Banks, R. B. Murphy, T. A. Halgren, J. J. Klicic, D. T. Mainz, M. P. Repasky, E. H. Knoll, M. Shelley, J. K. Perry, D. E. Shaw, P. Francis and P. S. Shenkin, *J. Med. Chem.*, 2004, **47**, 1739–1749.
87. BI-9564, http://www.thesgc.org/chemical-probes/BI-9564, (accessed 29 May 2017).
88. W. F. Vogel, R. Abdulhussein and C. E. Ford, *Cell Signaling*, 2006, **18**, 1108–1116.
89. R. R. Valiathan, M. Marco, B. Leitinger, C. G. Kleer and R. Fridman, *Cancer Metastasis Rev.*, 2012, **31**, 295–321.
90. K. Zhang, C. A. Corsa, S. M. Ponik, J. L. Prior, D. Piwnica-Worms, K. W. Eliceiri, P. J. Keely and G. D. Longmore, *Nat. Cell Biol.*, 2013, **15**, 677–687.
91. P. S. Hammerman, M. L. Sos, A. H. Ramos, C. Xu, A. Dutt, W. Zhou, L. E. Brace, B. A. Woods, W. Lin, J. Zhang, X. Deng, S. M. Lim, S. Heynck, M. Peifer, J. R. Simard, M. S. Lawrence, R. C. Onofrio, H. B. Salvesen, D. Seidel, T. Zander, J. M. Heuckmann, A. Soltermann, H. Moch, M. Koker, F. Leenders, F. Gabler, S. Querings, S. Ansén, E. Brambilla, C. Brambilla, P. Lorimier, O. T. Brustugun, Å. Helland, I. Petersen, J. H. , Clement, H. Groen, W. Timens, H. Sietsma, E. Stoelben, J. Wolf,

D. G. Beer, M. S. Tsao, M. Hanna, C. Hatton, M. J. Eck, P. A. Janne, B. E. Johnson, W. Winckler, H. Greulich, A. J. Bass, J. Cho, D. Rauh, N. S. Gray, K.-K. Wong, E. B. Haura, R. K. Thomas and M. Meyerson, *Cancer Discovery*, 2011, **1**, 78–89.

92. A. C. Pierce, G. Rao and G. W. Bemis, *J. Med. Chem.*, 2004, **47**, 2768–2775.
93. Eurofins Safety Screen, https://www.eurofinspanlabs.com/Catalog/Products/ProductDetails.aspx?prodId=WkpXTEM73wg%3D, (accessed 29 May 2017).
94. T. Machleidt, C. C. Woodroofe, M. K. Schwinn, J. Méndez, M. B. Robers, K. Zimmerman, P. Otto, D. L. Daniels, T. A. Kirkland and K. V. Wood, *ACS Chem. Biol.*, 2015, **10**, 1797–1804.
95. S. A. Johnson and T. Hunter, *Nat. Methods*, 2005, **2**, 17–25.
96. R. Jafari, H. Almqvist, H. Axelsson, M. Ignatushchenko, T. Lundbäck, P. Nordlund and D. M. Molina, *Nat. Protoc.*, 2014, **9**, 2100–2122.
97. U. Rix and G. Superti-Furga, *Nat. Chem. Biol.*, 2009, **5**, 616–624.

CHAPTER 6

Function-driven Discovery of Bioactive Small Molecules

SAMUEL LIVER,[a,b] JACOB MASTERS[a,b] AND ADAM NELSON*[a,b]

[a] School of Chemistry, University of Leeds, Leeds LS2 9JT, UK;
[b] Astbury Centre for Structural Molecular Biology, University of Leeds, Leeds LS2 9JT, UK
*Email: a.s.nelson@leeds.ac.uk

6.1 Context

The discovery of biologically-active small molecules is an enduring challenge in both medicinal chemistry and chemical biology. Small molecules continue to dominate Man's ability to treat disease, accounting for nine of the top ten prescribed drugs in the USA in 2014.[1] Moreover, the availability of high-quality chemical probes can help address the historically uneven exploration of the biology of proteins and enable fundamental biomedical science.[2] However, despite escalating investment, the rate of drug discovery has remained roughly constant for over 60 years.[3] Indeed, increased productivity has been identified as the grand challenge for the pharmaceutical industry.[4]

Innovative bioactive small-molecule discovery is enabled by the ability to explore diverse regions within biologically-relevant chemical space. Indeed, the majority of early-stage drug discovery projects in major pharmaceutical companies have been initiated by high-throughput screening.[5] Yet, chemists' historical exploration of chemical space has been highly uneven and unsystematic: half of the small molecules in the Chemical Abstracts Service (CAS) registry are based on around just 0.25% of the known molecular

Chemical Biology No. 10
Chemical and Biological Synthesis: Enabling Approaches for Understanding Biology
Edited by Nick J. Westwood and Adam Nelson

Published by the Royal Society of Chemistry, www.rsc.org

frameworks.[6] Major initiatives can, however, help expand the chemical space accessible for early-stage drug discovery (such as the European Lead Factory consortium, see Chapter 4).[7]

6.1.1 Chemical Approaches That Underpin Bioactive Small-molecule Discovery

The optimisation of the function of small molecules generally involves cycles in which arrays of compounds are designed, prepared, purified and tested. This workflow requires the investment of similar chemistry resources in each compound prepared, irrespective of actual relevance to the targeted activity. Medicinal chemists tend to gravitate towards reactions that generally have predictable outcomes.[8] As a result, the toolkit of reactions for medicinal chemistry has barely changed in the past 30 years,[9] despite an era of unprecedented invention of new synthetic methods. Remarkably, these practices have tended to drive chemists towards more sp^2-rich compounds,[10] despite the recognition that such compounds are linked with higher attrition rates in drug discovery.[11]

6.1.2 Evolution of Biosynthetic Pathways to Natural Products

Natural products provide a source of biologically-relevant chemical diversity, and have inspired, for example, around a third of the small-molecule drugs approved since 1980.[12] Their structural features are typically distinct from those found in synthetic compounds including, generally, a higher fraction of sp^3-hybridised carbons.[13] In biology-oriented synthesis (Chapter 3), the biological relevance of natural product frameworks is exploited in the design of productive small-molecule screening libraries.[14,15]

The emergence of natural products contrasts starkly with the discovery of synthetic bioactive compounds. The evolution of biosynthetic pathways[16,17] is function-driven and structure-blind. Natural products arise on the basis of their benefit to the host organism (Figure 6.1). Crucially, evolutionary

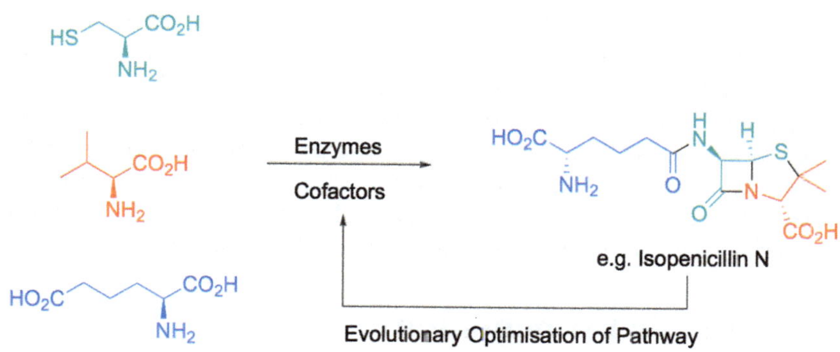

Figure 6.1 The emergence of natural products is structure-blind and function-driven.

pressure provides a feedback loop that serves to optimise the range of natural products produced.

6.1.3 Scope of This Chapter

This chapter focuses on function-driven approaches for the discovery of bioactive small molecules. Although only recently developed for small-molecule discovery, such approaches are well established for the discovery of functional proteins, peptides and oligonucleotides. Directed evolution has, for example, been widely exploited in the discovery of proteins with novel catalytic or binding functions. Moreover, platforms have been established that can enable the discovery of cyclic peptides[18] (see Chapter 9) and oligonucleotides[19] that bind to specific target molecules. In each of these cases, large libraries of potentially functional molecules are constructed, and then screened (or selected) for a required function. These combinatorial approaches are, however, generally limited to the discovery of functional oligomers constructed from building block classes that are found in nature and whose structures may be genetically-encoded.

Dynamic combinatorial chemistry (DCC) has also emerged for the synthesis of functional small molecules from building blocks based on their affinity for a protein. Building blocks, armed with functional groups that allow reversible bond formation, are incubated with a target protein which acts as a template for the formation of functional small molecules.[20] While DCC allows the focus to be placed firmly on small molecules that bind the target protein, it has significant limitations. First, stoichiometric amounts of the target protein are required. In addition, it is necessary to exploit reversible chemical reactions that operate under conditions under which the target protein is stable, which limits the diversity of the products, and tends to leave a functional group (*e.g.* hyrazone, disulfide) as an obvious vestige of the discovery approach.

This chapter focuses on two recently-described function-directed approaches for the discovery of bioactive small molecules: synthetic fermentation (Section 6.2) and activity-directed synthesis (Section 6.3). In each case, the structures of the molecules prepared are encoded by the specific components used in the synthesis. The mixtures of the products are not purified (or elucidated) at this stage: instead, the mixtures are directly assayed, allowing the reactions that yield the most active products to be identified. In both cases, subsequent optimisation of the function-directed synthesis is possible. Finally, the products of the most promising reactions are purified, if necessary after scale-up, to enable structural elucidation and characterisation of the functional small-molecule products.

6.2 Synthetic Fermentation

Huang and Bode have developed an approach—synthetic fermentation—that has been exploited in the discovery of functional β-peptides.[21] Although the products are reminiscent of those produced by non-ribosomal peptide synthases (NRPSs), the approach does not require the use of enzymes or organisms. Instead, the products are prepared by reacting simple building blocks in arrays of plate-based reactions which they term "cultures".[21]

Synthetic fermentation to yield β-peptides exploited aqueous-tolerant amide-forming ligations. To start with, α-keto acid initiators (I) react chemoselectively with isoxazolidine elongation monomers (M) to induce oligomerisation to yield α-keto acid intermediates. Oligomers of varying length and sequence are produced until the elongation process is completed by addition and reaction of a terminator (T) building block. At the end of this process, the reactions in each well contain mixtures of products, which are screened directly without any purification. Subsequent investigation can determine the specific combination of monomers that is needed for the synthesis of each bioactive product.[21]

6.2.1 Discovery of a Protease Inhibitor

The value of synthetic fermentation has been exemplified through the discovery of hepatitis C virus (HCV) NS3/4A protease inhibitors. Initially, an array of reactions was performed to identify combinations of monomers that yielded bioactive products. An initiator (chosen from I^1–I^6) and three elongation monomers (chosen from M^1–M^8) were added to each well and, after 2 h, a terminator building block (chosen from T^1–T^9) was added. The crude reaction mixtures were directly screened for protease inhibition. Selected examples of reactions investigated in one preliminary array are shown in Figure 6.2. In particular, the combinations of monomers that yielded bioactive products are highlighted in green. Further arrays of similar reactions were then performed, and the products screened, including at lower concentration, to optimise the combination of building blocks. Overall, this process enabled the optimal combination of an initiator, three elongation monomers and a terminator building block to be identified (I^3, M^2, M^4, M^5 and T^4).[21]

At this stage, however, it was not clear whether all three of the elongation monomers (M^2, M^4 and M^5) were required for biological activity. Thus, more focused cultures were performed in which I^3 and T^4 were combined with just one or two of the selected elongation monomers. It was found that I^3, M^5 (in excess) and T^4 was the optimal combination. The reaction was then repeated on a larger scale: the β-peptide **1** was isolated with 24% yield (Scheme 6.1), and was shown to be an inhibitor of HCV NS3/4A protease [concentration giving 50% of maximum inhibition (IC_{50}) = 1.0 μM].

6.2.2 Conclusion

In a proof-of-concept study, a total of just 23 building blocks was exploited: six initiators (I^1–I^{46}), eight elongation monomers (M^1–M^8) and nine terminator building blocks (T^1–T^9). By performing five arrays of reactions, it was

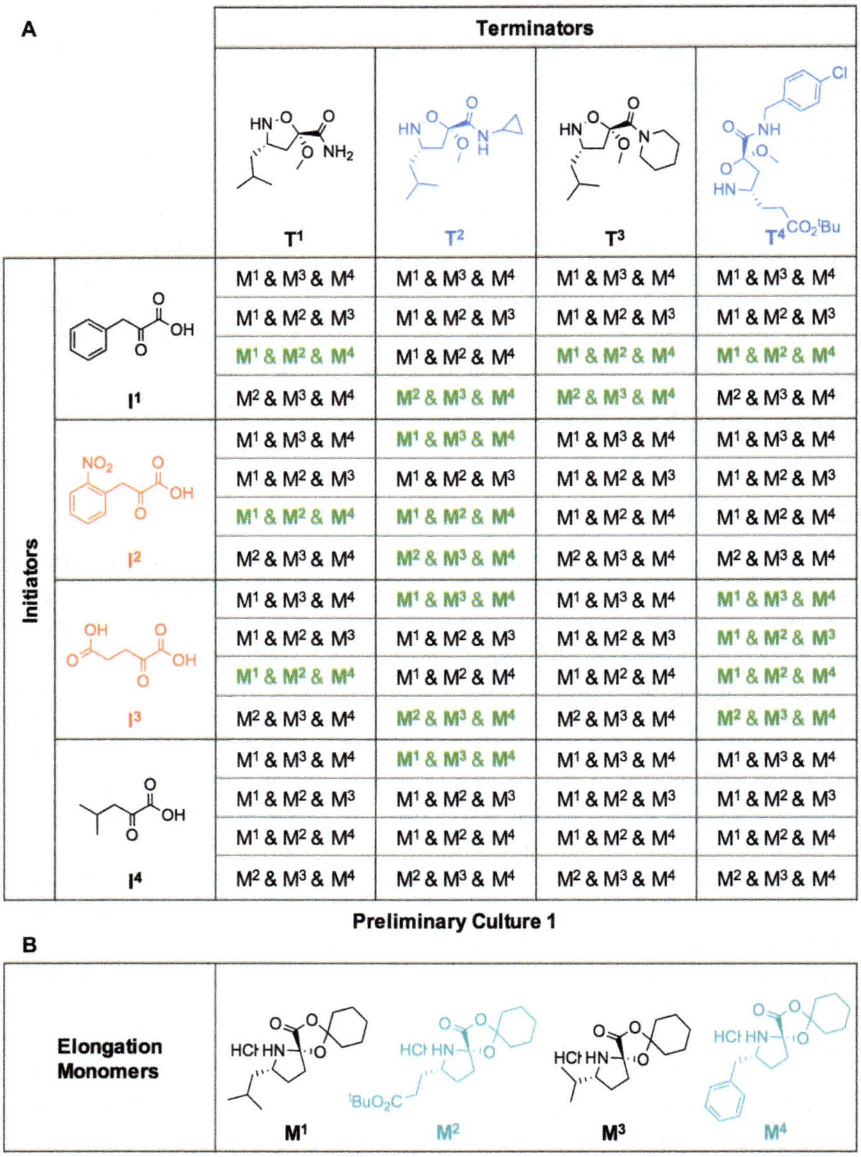

Figure 6.2 Preliminary Reaction Array 1. Each well in the reaction array had a unique combination of one initiator (from I^1–I^4, 1.0 eq.), three elongation monomers (from M^1–M^4, total 2.0 eq.) and, after two hours, one terminator (from T^1–T^4, 1.0 eq.). The crude reaction mixtures were screened directly against HCV protease. Active combinations are highlighted in green, from which individual building blocks—initiators I^2 and I^3 (orange), elongation monomers M^2 and M^4 (cyan) and terminators T^2 and T^4 (blue)—were selected for further study. From a second preliminary reaction array I^2, I^3, I^5, M^5, T^5, T^8 and T^9 were also taken forward for further study.

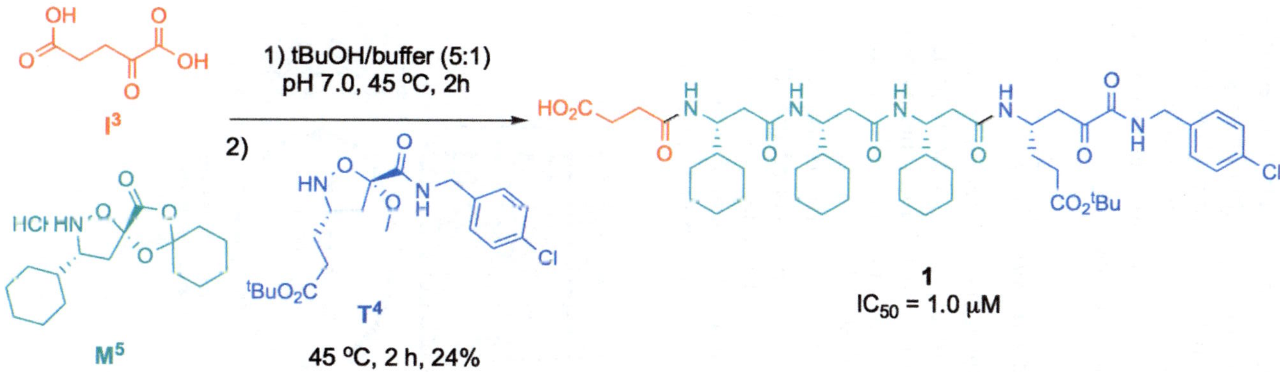

Scheme 6.1 Optimised reaction leading to the formation of a micromolar inhibitor of HCV NS3/4A protease.

possible to identify an optimised combination of building blocks and, hence, to discover a novel micromolar inhibitor of hepatitis C virus (HCV) NS3/4A protease.[21]

In the future, the potential of synthetic fermentation might be extended by expanding the range of underpinning chemistry that may be exploited. The authors have already identified a much broader range of building blocks and an additional reaction type that may also be used. Such additions to the underpinning chemistry will enable broader swathes of chemical space to be explored in bioactive molecule discovery.

6.3 Activity-directed Synthesis

Activity-directed synthesis (ADS)[22,23] harnesses reactions with inherently unpredictable outcomes to explore diverse regions of chemical space. In stark contrast to conventional medicinal chemistry workflows, it is preferable to exploit reactions with many possible outcomes. Initially, arrays of reactions are performed in which the inputs (substrates, catalysts, solvents *etc.*) are widely varied (Figure 6.3). The crude reaction mixtures are scavenging and evaporated and then screened directly for biological function. Subsequent reaction arrays are then designed by varying the inputs of those reactions that yield functional products. At this stage, the structures of the bioactive products are unknown although their synthesis has already been

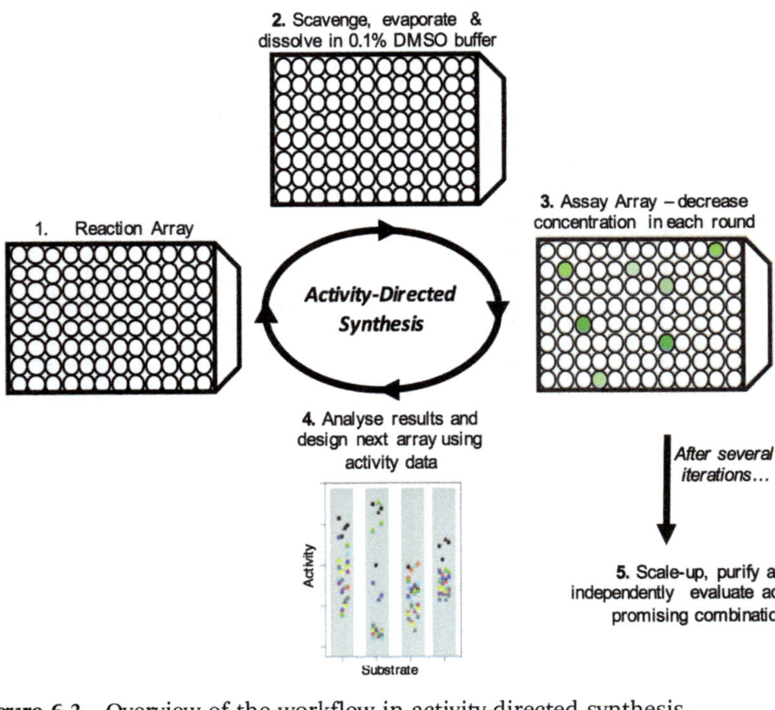

Figure 6.3 Overview of the workflow in activity-directed synthesis.

optimised! Finally, the most promising reactions are scaled up, the products are purified and structurally elucidated and their biological function characterised.

ADS thus borrows many concepts from the evolution of biosynthetic pathways to natural products. The approach is iterative, and enables the structure-blind and function-driven optimisation of the syntheses of bioactive products. Crucially, the approach enables the discovery of biologically-active small molecules in parallel with associated synthetic routes.

6.3.1 Discovery of Androgen Receptor Agonists Using Intramolecular Reactions

The value of ADS was initially demonstrated through the discovery of novel androgen-receptor agonists.[22] Metal-catalysed carbenoid chemistry was chosen because its many alternative patterns of reactivity were expected to enable evaluation of a wide range of different chemotypes. In this proof-of-concept study, the diazo substrates were each armed with a known (4-cyano 3-trifluoromethyl phenyl) binding group.[22] Crucially, the substrates were designed such that many different cyclisation modes were, in principle, possible.

In the first array, the 36 reactions exploited combinations of diazo substrate (100 mM; 12 variants: **2–7**) and catalyst [1 mol%; $Rh_2(OAc)_4$, Rh_2-(*S*-DOSP)$_4$ or Rh_2(5*S*-MEPY)$_4$] using dichloromethane as solvent (Figure 6.4). After 48 h, the reactions were scavenged to remove metal catalysts, evaporated and screened for agonism of the androgen receptor (total product concentration: 10 μM). Four of the substrates (**2b**, **3a**, **4b** and **5b**) yielded some product mixtures with significant biological activity.

In the second array, the four promising substrates (**2b**, **3a**, **4b** and **5b**) were taken forward with, as a control, two substrates (**7a** and **7b**) that had not yielded active products. The six substrates were explored in combination with an expanded range of eight catalysts and four solvents. Following reaction, scavenging and evaporation, the product mixtures were assayed at tenfold lower concentrations (total product concentration: 1 μM). Two of the substrates (**2b** and **4b**) yielded product mixtures with higher biological activity, but only in combination with rhodium carboxylate catalysts and in three specific solvents. This observation indicated that the specific combination of catalyst and solvent had a critical role in steering these substrates towards bioactive products.

The third array built on promising combinations of catalyst and solvent that had been identified in the previous round. The reaction array exploited the two most promising substrates (**2b** and **4b**) along with four new structurally-related analogues (**8–11**); six rhodium carboxylate catalysts; and three different solvents. The product mixtures were screened at tenfold lower concentration (total product concentration: 100 nM). Reactions of three of the substrates yielded bioactive products, though only under specific reaction conditions.

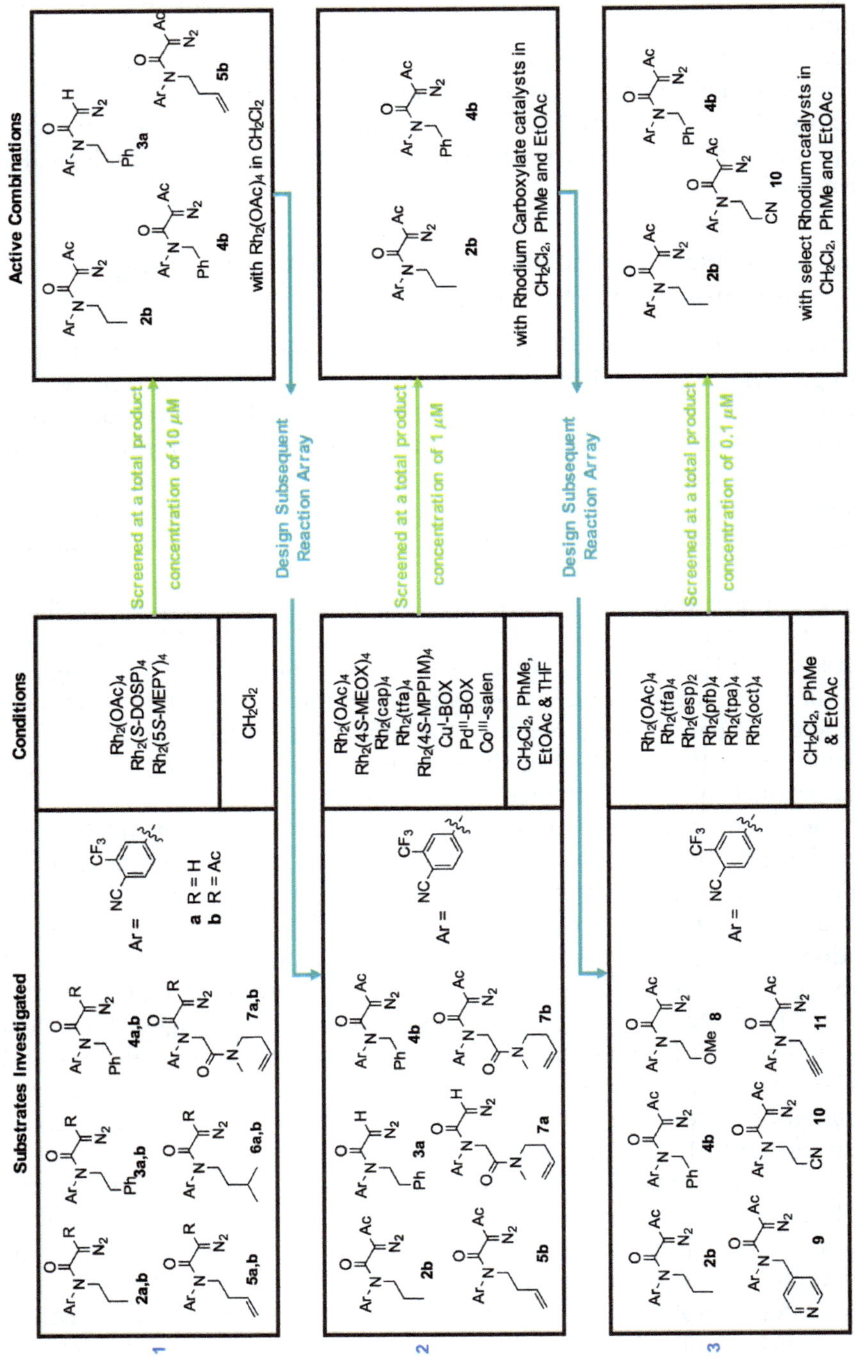

Figure 6.4 Discovery of androgen receptor agonists and associated synthetic routes by intramolecular activity-directed synthesis.

On the basis of the observed biological activity in the third array, ten reactions involving three of the substrates were selected for scale-up. In all ten reactions, a major product was produced with good yield, and its structure elucidated (in total, three distinct products were discovered). The diazo substrates **4b** and **10** underwent intramolecular C–H insertion to yield the β-lactams **12** and **13** (Scheme 6.2). The diazo substrate **2b** also underwent intramolecular C–H insertion, but yielded the γ-lactam **14.** All three products were sub-micromolar modulators of the androgen receptor: **12** and **13** were agonists [with concentrations giving 50% of maximum effect (EC_{50}) of 350 nM and 440 nM, respectively], whilst **14** was a partial agonist (EC_{50}: 470 nM). Remarkably, none of the chemotypes that were identified had previously annotated activity against the androgen receptor. Moreover, the ADS approach was remarkably efficient: although a total of 336 reactions were performed, only three products, all of which had sub-micromolar activity, needed to be purified. Isolation of these products was facile, in part because the ADS approach tends to optimise the yield of bioactive products.

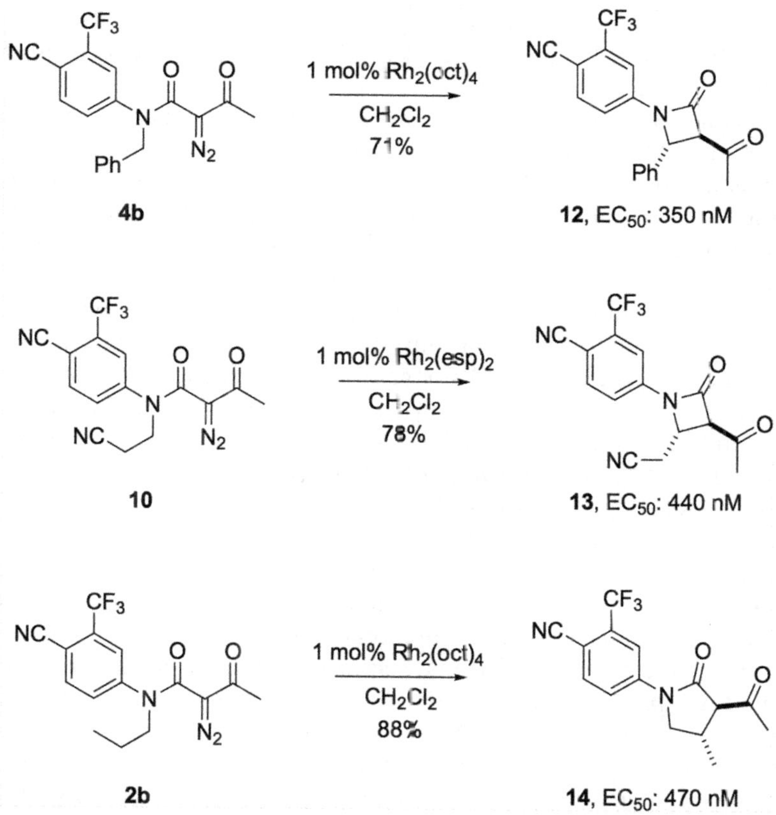

Scheme 6.2 Activity-directed synthesis of androgen receptor agonists using intramolecular carbenoid chemistry.

6.3.2 Discovery of Androgen Receptor Agonists Using Intermolecular Reactions

Following the initial proof-of-concept, the possibility of exploiting intermolecular reactions was explored. It was envisaged that significant areas of relevant chemical space could be explored using combinations of a few armed diazo substrates and commercially-available co-substrates.[23]

Four diazo substrates, each armed with a 4-cyano 3-trifluoromethyl phenyl group as before (Section 6.2), were designed to suppress the possibility of intramolecular reaction. In the first reaction array, 192 reactions (randomly chosen from 480 possible combinations) were carried out using combinations of diazo substrates (100 mM, four variants: **15a–b** and **16a–b**), commercially-available co-substrates (1.0 M, 10 variants: **17a–i** or no co-substrate), catalysts [1 mol%; $Rh_2(tfa)_4$, $Rh_2(oct)_4$, Rh_2(*S*-DOSP)$_4$, Rh_2(5*R*-MEPY)$_4$, Rh_2(4*S*-MEOX)$_4$ or $Rh_2(cap)_4$] and solvents (dichloromethane or toluene) (Figure 6.5). After 48 h, the reactions were scavenged to remove metal catalysts, evaporated and the crude product mixtures screened for agonism of the androgen receptor (total product concentration: 10 μM). Only two of the 192 reactions resulted in product mixtures that showed significant biological activity: both reactions exploited the same diazo substrate (**16a**), catalyst [Rh_2(*S*-DOSP)$_4$] and solvent (CH_2Cl_2) with two different co-substrates (cyclohexene **17a** or indole **17g**).

Consequently, a second array was designed on the basis of the combinations that yielded active product mixtures in the first round. A total of 86 reactions was performed (from 360 possibilities) using combinations of two diazo substrates (**15a** and **16a**), 18 co-substrates related to either cyclohexene or indole (**17a, 17g, 17j–y**), five catalysts [Rh_2(*S*-DOSP)$_4$, $Rh_2(OAc)_4$, $Rh_2(pfb)_4$, $Rh_2(tpa)_4$, $Rh_2(esp)_2$] and two solvents (dichloromethane and toluene). After 48 h, scavenging and evaporation, the crude product mixtures were screened for activity at twofold lower concentrations (total product concentration: 5 μM) to increase the threshold for activity. Five reactions yielded product mixtures that were significantly more active than those identified in the first round: the reactions involved the diazo substrate **16a** with dihydronapthalene **17p**, dihydropyran **17r** or indene **17x**; and the diazo substrate **15a** with indole **17g** or 7-azaindole **17q**.

The third array of 48 intermolecular reactions employed the diazo substrate **16a** with every possible combination of 12 co-substrates (**17a, 17g, 17p, 17q, 17r, 17x, 17z, 17a′–17f′**) and four catalysts [Rh_2(*S*-DOSP)$_4$, Rh_2(*R*-DOSP)$_4$, $Rh_2(OAc)_4$, $Rh_2(esp)_2$] in dichloromethane. After scavenging and evaporation, the crude product mixtures were screened at fivefold lower concentration (total product concentration: 1 μM). Interestingly, four reactions involving the diazo substrate and **17a** gave product mixtures that displayed different levels of activity: this result indicates that ADS may result in optimisation of the yield of the most bioactive product produced. In addition, the reaction involving **16a** and the dihydropyran **17e′** resulted in an active product with Rh_2(*R*-DOSP)$_4$; but same substrates with the enantiomeric

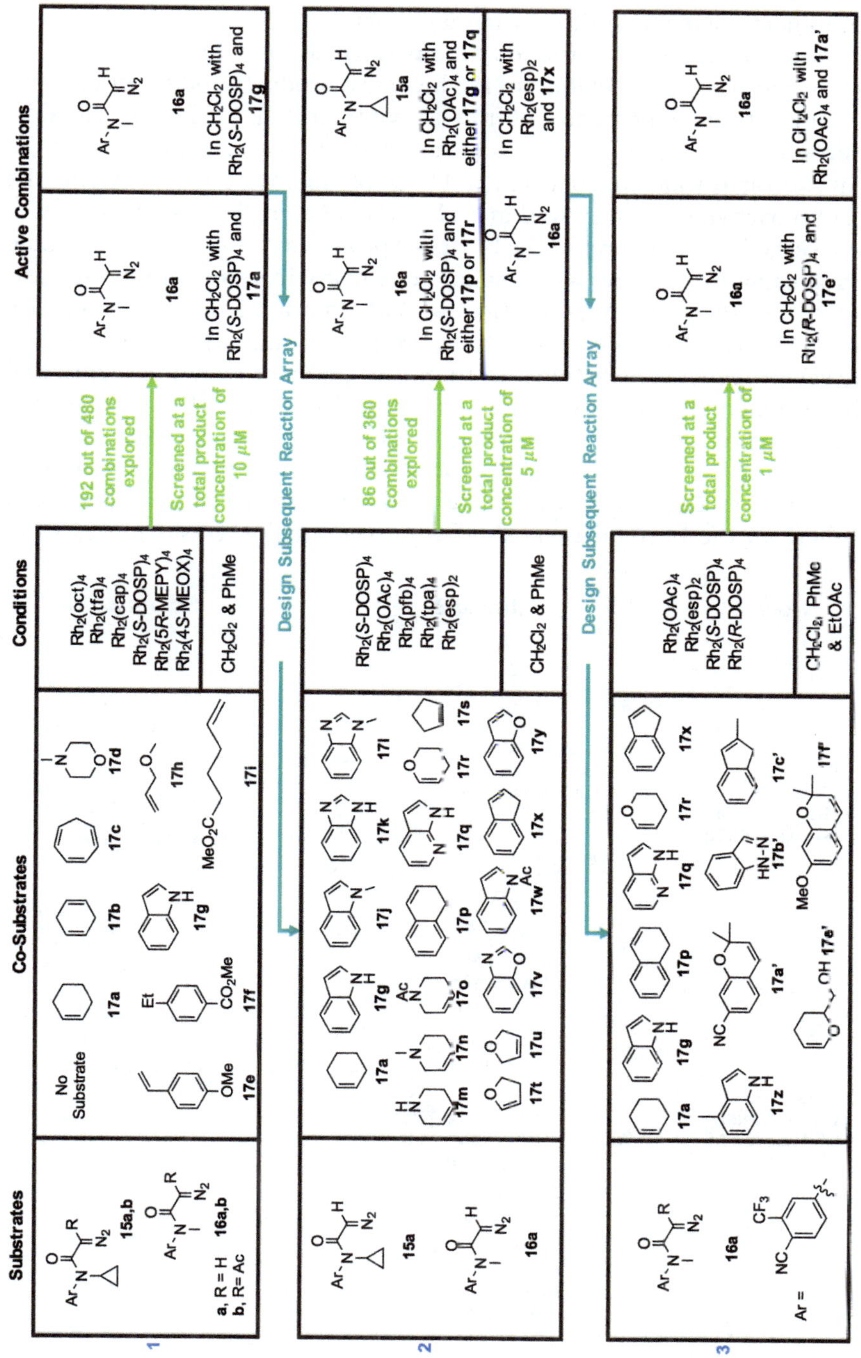

Figure 6.5 Discovery of androgen receptor agonists and associated intermolecular syntheses by activity-directed synthesis.

catalyst did not. This observation indicates that the reaction was enantioselective and the bioactive product was chiral, with one of the enantiomers having significantly higher activity.

Reactions from all three rounds were performed on a 50-fold larger scale to enable structural elucidation of the bioactive products and to understand the basis of the emergence of the activity-directed syntheses. Notably, the products had been prepared with good yields (71–82%), except for the α,β-unsaturated γ-lactam **20** which had been obtained at only 18% yield. The level of bioactivity was improved in each round of activity-directed synthesis.

Analysis of the structures of the bioactive products demonstrated the value of harnessing intermolecular carbenoid chemistry in which alternative patterns of reactivity are possible (Figure 6.6). In round one, the bioactive amides **18** and **19** resulted respectively from intermolecular C–H insertion and cyclopropanation reactions. In round two, cyclopropanation continued to be productive, and variation of the co-substrate enabled significantly more active products (**21**, **22** and **23**) to be discovered. The outcome of round three particularly highlighted the value of harnessing inherently promiscuous chemistry. The dihydropyran **17e′** and the benzopyran **17a′** had been exploited on the basis of their similarity to promising co-substrates from

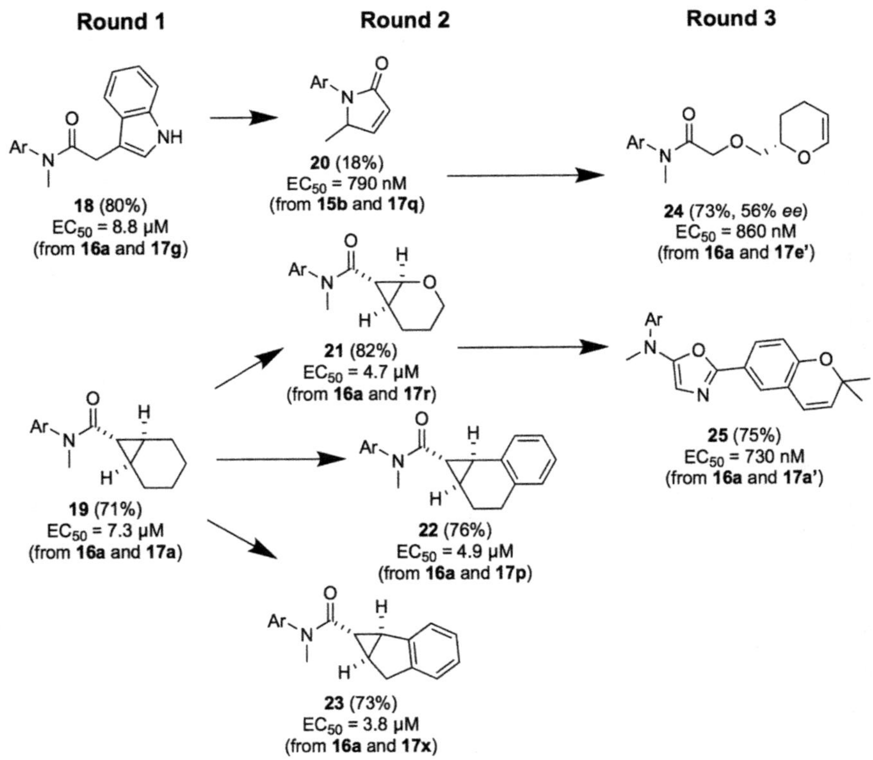

Figure 6.6 Emergence of activity-directed syntheses of androgen receptor agonists prepared using intermolecular carbenoid chemistry.

round two. However, the bioactive molecules discovered stemmed from different modes of reactivity: the amide **24** was the product of an enantioselective O–H insertion reaction, while the oxazole **25** was formed by reaction of the nitrile group in the co-substrate. Remarkably, Rh-catalysed enantioselective O–H insertion was not previously known: this is believed to be the first asymmetric reaction to be discovered on the basis of the bioactivity of the product alone! Over three rounds of ADS, 326 reactions had been performed, yet only the eight bioactive products needed to be purified.

6.3.3 Conclusion

ADS offers a valuable and effective platform for the discovery of bioactive small molecules. Inherently unpredictable chemistry with many possible outcomes is harnessed to explore diverse chemical space. The diverse reaction products are not purified at this stage but, instead, are assessed directly for biological activity using a high-throughput assay. ADS thus enables the efficient exploration of chemical space to identify chemotypes that have the targeted biological function. It is likely that the scope of the discovery approach could be increased by expansion of the range of underpinning chemistries.

6.4 Outlook

The discovery of bioactive small molecules is currently dominated by rounds in which small molecules are designed, synthesised, purified and tested. Function-driven approaches have potential to streamline molecular discovery because they allow resources to be focused on the most bioactive small molecules. Furthermore, the ability to exploit a wider toolkit of underpinning chemistry broadens the range of chemotypes that can be discovered. Synthetic fermentation and activity-directed synthesis are in their infancy. Indeed, the realisation of the full value of these approaches will involve expansion of the underpinning chemistry toolkits, and application to a much broader range of protein targets. Nonetheless, both approaches do have the potential to improve the efficiency of molecular discovery, and to increase the range of functional chemotypes that may be discovered.

References

1. M. Aitken, *Medicines Use and Spending Shifts: A Review of the Use of Medicines in the U.S. in 2014*, The IMS Institute for Healthcare Informatics, New Jersey, 2015.
2. A. M. Edwards, R. Isserlin, G. D. Bader, S. V. Frye, T. M. Willson and F. H. Yu, *Nature*, 2011, **470**, 163.
3. B. Munos, *Nat. Rev. Drug Discovery*, 2009, **8**, 959–968.

4. S. M. Paul, D. S. Mytelka, C. T. Dunwiddie, C. C. Persinger, B. H. Munos, S. R. Lindborg and A. L. Schacht, *Nat. Rev. Drug Discovery*, 2010, **9**, 203–214.
5. R. Macarron, M. N. Banks, D. Bojanic, D. J. Burns, D. A. Cirovic, T. Garyantes, D. V. Green, R. P. Hertzberg, W. P. Janzen, J. W. Paslay, U. Schopfer and G. S. Sittampalam, *Nat. Rev. Drug Discovery*, 2011, **10**, 188–195.
6. A. H. Lipkus, Q. Yuan, K. A. Lucas, S. A. Funk, W. F. Bartelt 3rd, R. J. Schenck, A. J. Trippe and C. A. S. Registry, *J. Org. Chem.*, 2008, **73**, 4443–4451.
7. A. Karawajczyk, F. Giordanetto, J. Benningshof, D. Hamza, T. Kalliokoski, K. Pouwer, R. Morgentin, A. Nelson, G. Muller, A. Piechot and D. Tzalis, *Drug Discovery Today*, 2015, **20**, 1310–1316.
8. T. W. Cooper, I. B. Campbell and S. J. Macdonald, *Angew. Chem., Int. Ed.*, 2010, **49**, 8082–8091.
9. D. G. Brown and J. Boström, *J. Med. Chem.*, 2016, **59**, 4443–4458.
10. W. P. Walters, J. Green, J. R. Weiss and M. A. Murcko, *J. Med. Chem.*, 2011, **54**, 6405–6416.
11. F. Lovering, J. Bikker and C. Humblet, *J. Med. Chem.*, 2009, **52**, 6752–6756.
12. D. J. Newman and G. M. Cragg, *J. Nat. Prod.*, 2006, **70**, 461–477.
13. T. Rodrigues, D. Reker, P. Schneider and G. Schneider, *Nat. Chem.*, 2016, **8**, 531–541.
14. S. Wetzel, R. S. Bon, K. Kumar and H. Waldmann, *Angew. Chem., Int. Ed.*, 2011, **50**, 10800–10826.
15. H. van Hattum and H. Waldmann, *J. Am. Chem. Soc.*, 2014, **136**, 11853–11859.
16. R. A. Maplestone, M. J. Stone and D. H. Williams, *Gene*, 1992, **115**, 151–157.
17. R. D. Firn and C. G. Jones, *Nat. Prod. Rep.*, 2003, **20**, 382.
18. A. D. Foster, J. D. Ingram, E. K. Leitch, K. R. Lennard, E. L. Osher and A. Tavassoli, *J. Biomol. Screening*, 2015, **20**, 563–576.
19. J. Zhou and J. Rossi, *Nat. Rev. Drug Discovery*, 2017, **16**, 181–202.
20. M. Mondal and A. K. Hirsch, *Chem. Soc. Rev.*, 2015, **44**, 2455–2488.
21. L. Huang and J. W. Bode, *Nat. Chem.*, 2014, **6**, 877–884.
22. G. Karageorgis, S. Warriner and A. Nelson, *Nat. Chem.*, 2014, **6**, 872–876.
23. G. Karageorgis, M. Dow, A. Aimon, S. Warriner and A. Nelson, *Angew. Chem., Int. Ed.*, 2015, **54**, 13538–13544.

CHAPTER 7

DNA-encoded Library Technology (ELT)

GANG YAO,* XIAOJIE LU, CHRIS PHELPS AND GHOTAS EVINDAR

GlaxoSmithKline, Platform of Technology and Science, 200 Cambridge Park Drive, Cambridge, MA 02140, USA
*Email: gang.x.yao@gsk.com

7.1 Introduction

The field of chemical biology entails application of chemical techniques and tool compounds produced through synthetic chemistry to study and/or modulate biological systems. One of the focuses in chemical biology is on generating small-molecule tools (chemical biology probes) that can help elucidate the roles of the targeted protein in healthy and diseased cells and tissues.[1,2] High-quality chemical biology probes have served both as powerful research tools and as leads in development of new medicines.[3]

Many techniques have been used to discover chemical biology probes including high-throughput screening (HTS) campaigns, fragment-based screening (see Chapters 4 and 5), computer-assisted "virtual screening" and diversity-oriented synthesis (see Chapter 2).[4–7] Among these, HTS is the most widely used method, which can involve millions of chemical compounds assayed individually in a biochemical assay in a plate-based format. However, the costs of assembling, maintaining and screening a library collection by HTS places a practical upper limit on the number of compounds and therefore the molecular diversity that can be screened. Recently DNA-encoded library technology (ELT) has emerged as a powerful new

Chemical Biology No. 10
Chemical and Biological Synthesis: Enabling Approaches for Understanding Biology
Edited by Nick J. Westwood and Adam Nelson

Published by the Royal Society of Chemistry, www.rsc.org

platform to generate large DNA-encoded libraries (DELs) and to discover small-molecule probes for a target protein through affinity selection.[8–12] The concept of encoding chemically synthesized compound collection with DNA was first proposed by Brenner and Lerner in 1992.[13] DELs are composed of chimeric compounds that consist of a small molecule covalently linked to a single-stranded (ss) or double-stranded (ds) DNA. In comparison to HTS, ELT offers two main advantages. First, the combination of split-and-pool chemistry and stepwise DNA-tagging enables the construction of DELs that are several orders of magnitude larger than any HTS collections, resulting in libraries with sizes ranging from 10^5 to 10^{12} novel encoded small molecules. Second, due to the PCR amplifiable DNA-tag, DELs can be screened as a pool by affinity selection, obviating the need for costly HTS infrastructure. The decoding of the binders is typically achieved by PCR amplification and sequencing of the associated DNA-tags which has been greatly facilitated by advances in access to next-generation sequencing.[14]

One unique feature of ELT-derived hits is the ready availability of a position on the small molecule compound for further modification and attachment of linkers. Crystallography studies for many ELT hits have revealed that the DNA tag-attaching sites on the small molecule always pointed towards the protein surface and were solvent-exposed.[14–18] Because of this feature of ELT molecules, hits from ELT can serve as ideal precursors for making chemical probes by replacing the DNA tag with other functional groups. Attachment of a fluorophore dye would facilitate biochemical assay development and allow for visualization of the target protein in its cellular environment.[19–22] Immobilizing an ELT small molecule through biotin or a chemically reactive group will enable a pull down probe for proteomic studies.[23] Additional advantages of an ELT molecule including the proteolysis-targeting chimeras (PROTACS) approach are illustrated in Figure 7.1.

Screening of DELs has yielded a large number of bioactive molecules with drug-like properties that have been developed into high-quality chemical biology probes and even clinical candidates. An example of this is **GSK2256294**, a potent inhibitor of soluble epoxide hydrolase (sEH) and the first small molecule discovered by ELT to enter human clinical studies as a potential candidate for cardiovascular and respiratory conditions.[16,28,29] Likewise, a family of inhibitors of receptor-interacting protein 1 (RIP1) kinase (*e.g.* **GSK2880182, GSK2982772** *etc.*) derived from a benzoxazepinone DEL, has shown favourable potency, selectivity and pharmacokinetic properties and have progressed into clinical trials for the treatment of inflammatory diseases.[26,30] Here, we will describe the general practice of this emerging ELT platform and how it was applied in the discovery of chemical biology probes.

7.2 Overview of the ELT Process

Four key steps are involved in the ELT process (Figure 7.2).[28] The first step is the design and production of DELs. DELs are composed of organic

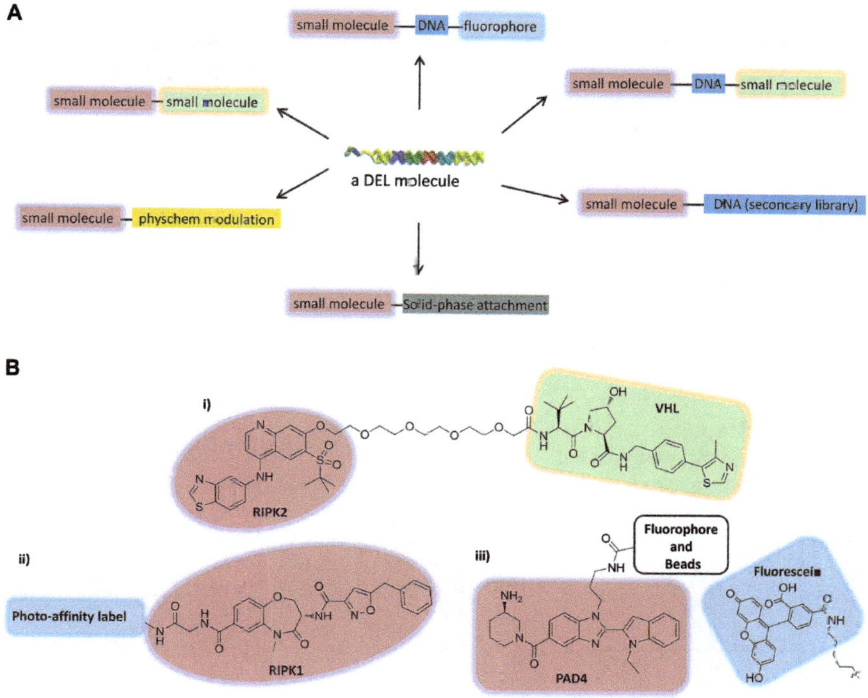

Figure 7.1 A. Potential utilities of an ELT small molecule. Molecules prepared off-DNA can be directly conjugated to probes through their original DNA attachment site. Alternatively, compounds can be prepared on-DNA and ligation can be used to attach DNA to be used for quantitative molecular biology [*i.e.*, quantitative PCR (qPCR) or sequencing] or probes prepared on DNA.[24] B. Selected examples of ELT molecules and the use of a DNA attachment site in the literature.[25–27]

molecules, covalently linked to a unique DNA tag serving as an amplifiable identification bar code. The DNA sequence contains terminal PCR-primer regions and internal coding regions unambiguously identifying the small molecule. The most often used approach for the synthesis of DELs is the DNA-recorded combinatorial split-and-pool method.[8] This approach allows the facile generation of libraries with sizes of between 10^5 and 10^{12} encoded small molecules. The second step involves the screening of DNA-encoded libraries which is generally performed by affinity selection. The selection assay requires the immobilization of an affinity-tagged, purified target protein in complex with bound ligands on the surface of a resin and the washing away of unbound ligands. Usually, a pooled DEL is incubated at micromolar to nanomolar concentrations of total library members with the target protein (see Section 7.4). Elution of the binders is followed by PCR amplification. The DNA barcode tags of the amplicon are then read by DNA sequencing as illustrated by step 3 in Figure 7.2. On-DNA hits are identified by counting the sequences relative to controls, *e.g.* matrix-only selection experiments. Finally

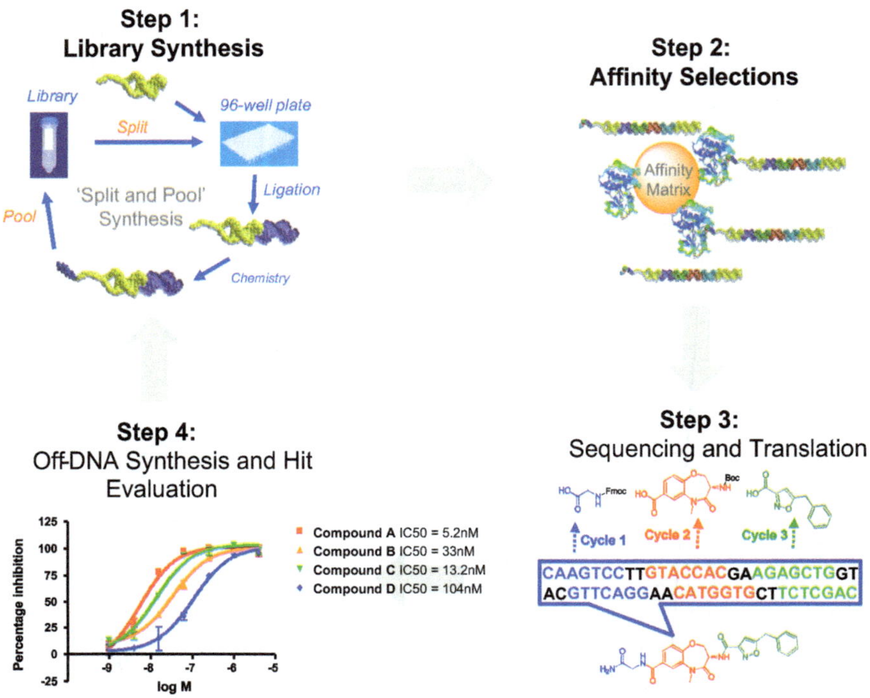

Figure 7.2 ELT process. Step 1: Library synthesis; Step 2: Affinity selections; Step 3: Sequencing and translation; Step 4: Off-DNA synthesis and hit evaluation.

(step 4), the identified on-DNA hits are re-synthesized without the DNA tag at sufficient quantity and purity for evaluation in conventional biochemical and/or cellular assays.

7.3 DEL Design and on-DNA Reaction Development

7.3.1 On-DNA Reaction Development

The DNA encoded library technology platform starts with the design and synthesis of a DNA encoded library (DEL).[31–33] So far, there are several reported strategies to prepare DELs: DNA recorded chemical synthesis (GSK,[8,34] X-Chem[35,36] and NuEvolution[37]); DNA-templated library synthesis by Liu and colleagues;[10,38–40] DNA-routed synthesis by Halpin, Harbury and colleagues;[41–44] encoded self-assembling chemical (ESAC) libraries by Neri and colleagues[9,45] at ETH and YoctoReactor technology by Vipergen.[46] The DNA recorded synthesis method follows a traditional split-and-pool combinatorial chemistry strategy. Due to the space limits herein we focus on describing DNA recorded library synthesis, which is currently the most widely practiced synthesis strategy.[47,48]

As illustrated in Figure 7.3, GSK's starting material (referred as Head Piece, HP-NH_2) was specifically designed for the DEL synthesis.[8,34] It could simply be viewed as stabilized ds DNA with two base pairs overhang on one end positioned for enzymatic ligation of encoding DNA tags employing T4 ligase, with an aliphatic primary amine on the other end to react with chemical reagents to construct the small molecular warhead. After several cycles of enzymatic coding reactions with DNA tags and chemical reactions by adding synthetic building blocks, DELs with millions to billions of DNA-encoded small molecules can be efficiently synthesized by using a split-and-pool strategy as described in Figure 7.4. For example, after three cycles of both ligation and chemistry reactions using one 96-well plate of building blocks per round, a DEL of almost 1 million members can be easily produced.

Although it is possible to build even bigger libraries with more library cycles, the corresponding small molecules' properties tend to become less attractive as the warheads become bigger and more lipophilic (see Chapter 4). To balance the library size and physiochemical properties, it is better to limit the libraries to two or three cycles of chemistry, which requires robust on-DNA chemistry reactions for a very large number of synthetic building blocks in each library cycle. For example, at least 1000 building blocks are required to yield a 1 million warhead library in two cycles of chemistry. Therefore, on-DNA chemistry reactions that are compatible with many synthetic building blocks are needed to achieve this design. In GSK we routinely use large numbers of synthetic building blocks for the DEL synthesis. For example, amines (approximately 9000), carboxylic acids (approximately 9000) and aldehydes (approximately 2000) are common synthetic building blocks applied for DEL production.[33]

Unlike traditional organic reactions, the on-DNA substrate is so dilute that large equivalents of reagents are needed to obtain a useful reaction rate. The typical reaction concentration is less than 1 mM for the DNA tag. In order to keep DNA in solution, a certain amount of water is needed for the reaction. Up to 70–80% organic co-solvent is tolerated for the on-DNA reactions and CH_3CN, *N,N*-dimethylformamide (DMF) and *N,N*-dimethylacetamide (DMA) are the most commonly used. In order to minimize damage to the DNA, the on-DNA reaction conditions are usually mild, and certain harsh conditions, such as strong oxidants and low pH, are not tolerated. Even given these constraints, many on-DNA reaction conditions have been reported in the literature. Table 7.1 summarizes the on-DNA reaction toolbox which has been developed by GSK over the last ten years.[28] These include some of the most commonly used reactions in medicinal chemistry,[49] such as Suzuki coupling[50] and reductive amination.[51]

Pd-promoted C–C cross coupling is widely used in the pharmaceutical industry, especially Pd-catalysed Suzuki couplings. Recently Ding *et al.* reported new on-DNA Suzuki cross coupling reactions for tough heterocyclic chloride substrates.[52] Employing POPd catalyst with Ligand **1**, the on-DNA Suzuki reaction worked efficiently with the general substrate scope of

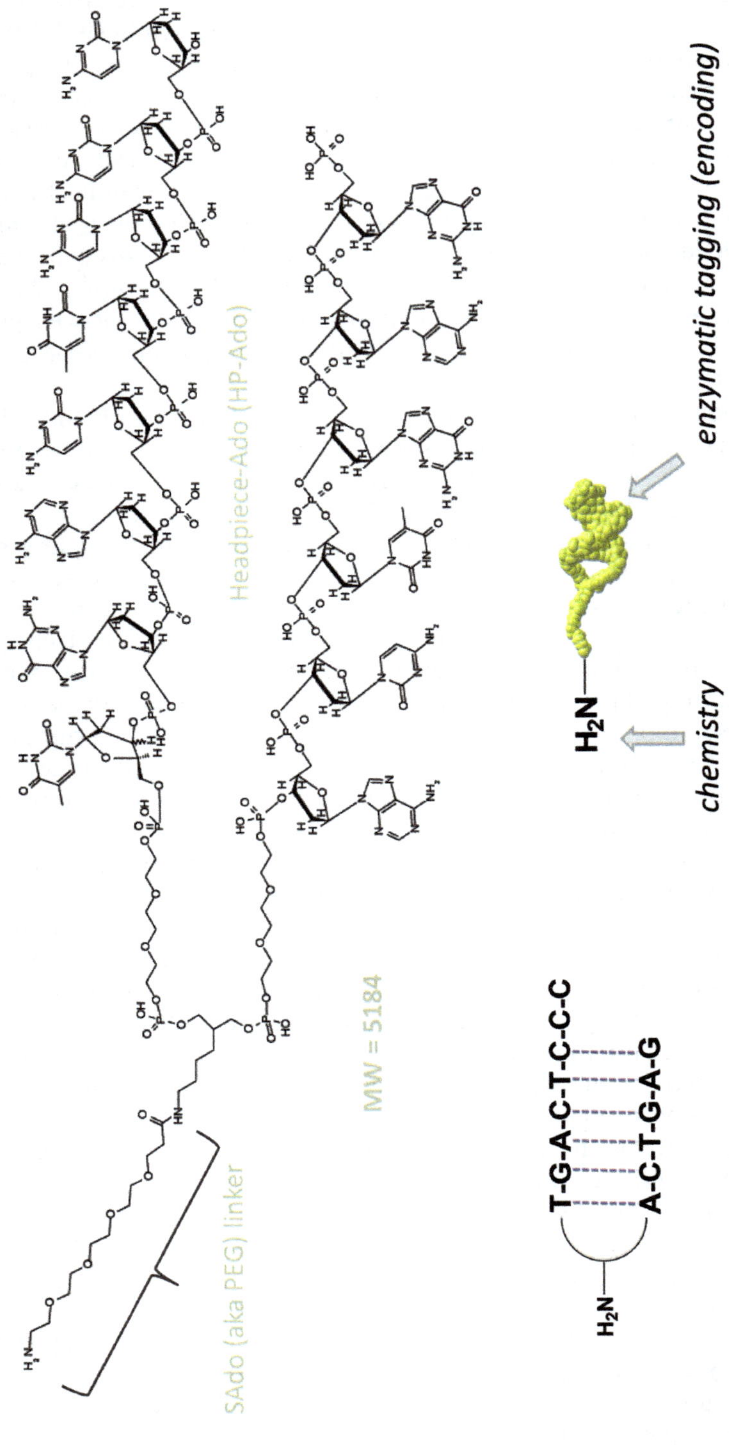

Figure 7.3 Starting material ('Headpiece'/HP) for library synthesis, the two base overhang is designed for enzymatic ligation of coding tag.

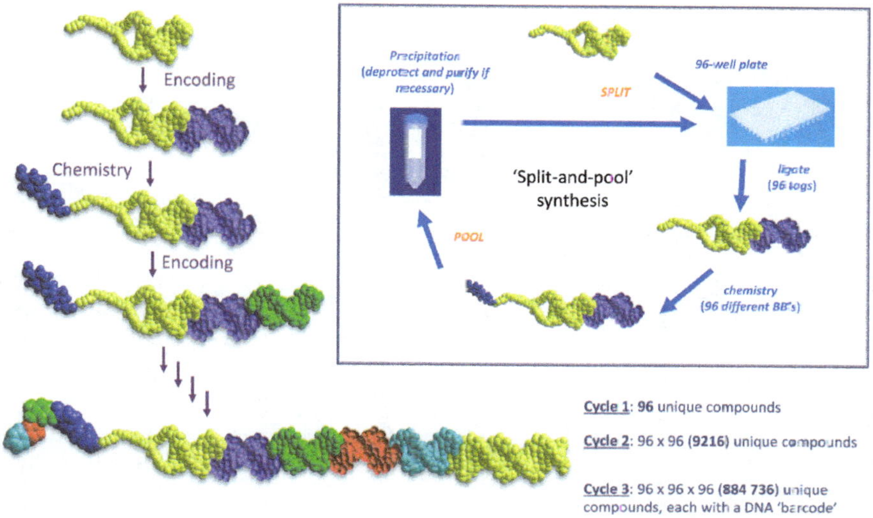

Figure 7.4 Split-and-pool synthesis of DELs.

aromatic boronic acid and esters. On-DNA Sonogashira coupling was also successfully developed by applying the Pd catalyst.[28,53] Unlike boronic acids and esters, there are a limited number of alkynes available for the Sonogashira coupling. A "premix" strategy to generate the alkynes from aldehydes *in situ* has been implemented and the crude alkyne was successfully applied to the on-DNA Sonogashira coupling with good substrate scope. Around 1900 alkynes, the majority from the premix strategy, have been used for on-DNA Sonogashira coupling.[28] Pd-promoted C–N cross coupling has become an efficient strategy for building aromatic amine compounds, which are an important scaffold in many drugs and drug candidates. Thus, it is highly desirable to have on-DNA C–N cross coupling reactions for different kinds of DELs. Recently, GSK disclosed the first on-DNA cross coupling reactions taking advantage of, now widely used, Buchwald precatalysts.[28] General amine substrate scope has been established. This newly developed C–N cross coupling reaction has been applied to a resynthesis of DEL62. Recently, Thomas *et al.* also reported the first enzymatic glycosylation reactions and oxidations.[62] They not only demonstrated that biocatalysts could be successfully applied to the on-DNA reactions but also illustrated the great potential for the design and synthesis of a DNA-encoded carbohydrate library based on the new enzymatic transformations. In addition, Fan and Davie also reported Lewis-acid-promoted epoxide-opening reactions with general amine substrate scope.[63] All of these new developments show the potential for further broadening the scope of DNA-compatible chemistry.

Employing these reactions, many DELs have been reported, which are summarized in Figure 7.5. There are three main types of libraries: linear, scaffolded and macrocyclic.[15,17,22,26,54,56] Each represents a unique chemical

Table 7.1 On-DNA reactions developed by GSK.

on-DNA Reaction Types	Reacent reaction Examples and References
C-C bond formation: Suzuki–Miyaura, Sonogashira, Heck Wittig, Aldol condensation	Suzuki–Miyaura (~850 boronates) 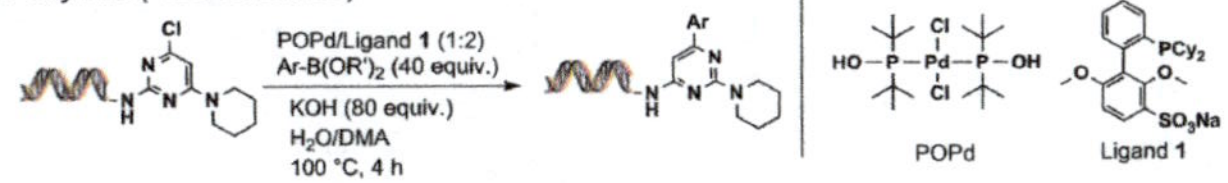
C–N bond formation: Acylation, Reductive alkylation Urea/thiourea formation, Buchwald–Hartwig amination Ullmann coupling S_NAr, S_N2 Epoxide opening by amine Michael addition, cyanamide formation Guanidine formation	Premix Sonogashira coupling (1900 aldehydes mainly from insitu generation of alkye from aldehydes) 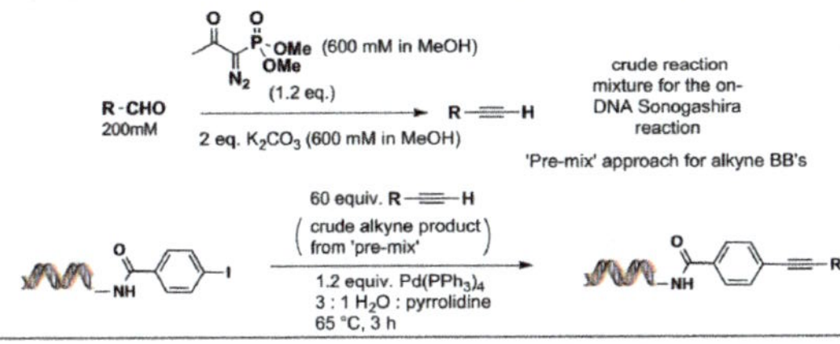
C–O and C–S bond formation: S_NAr, Sulfonylation Ullmann coupling Enzymatic glycosylation	Buchwald–Hartwig amination 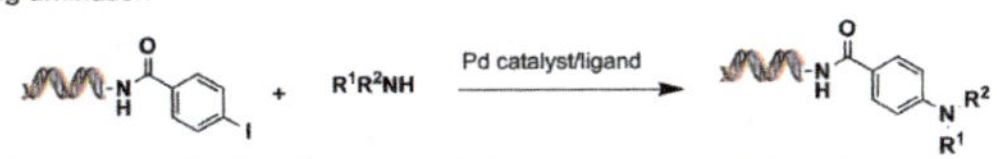
Cyclizations and annulations Diels–Alder, aza-Diels–Alder, Huisgen cycloaddition, Cushman Formation of: benzimidazoles, (aza)indoles, thiazoles, pyrazoles, pyrazolines, oxadiazoles, tertrazoles, imidazolidinones, pyridones, (imino)hydantoins, imidazoles, imidazolidinones, purines, benzothiazepinones Annulation amino alcohols and diamines Ring-closing metathesis (RCM)	Epoxide opening by amine 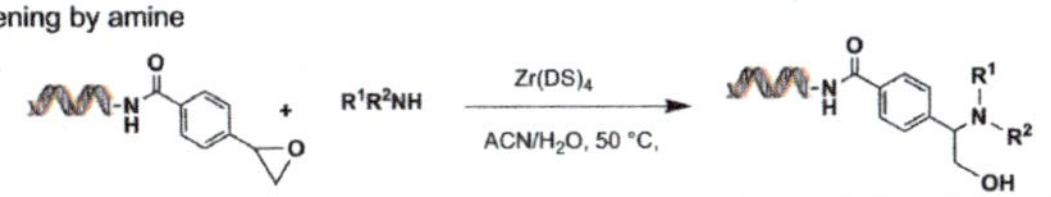
Oxidation and Reductions **Oxidations**: thioether to sulfoxide or sulfone; enzymatic oxidation of alcohol **Reduction**: NO_2, azide, disulfide, Olefin dihydroxylation and cleavage	Enzymatic glycosylation 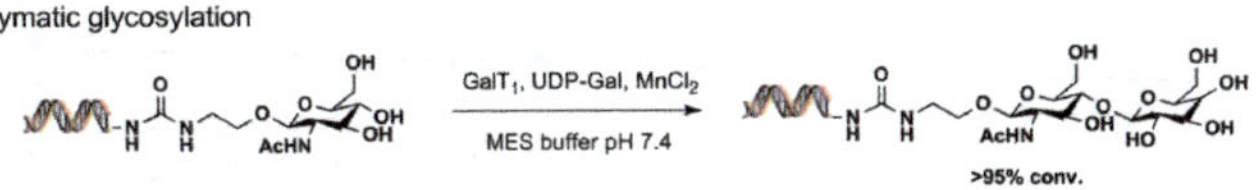
Protection Groups **Amines**: Boc, Fmoc, Alloc, voc, trifluoroacetamide **Esters**: Me, Et, *t*-Bu	Enzymatic oxidation of alcohol 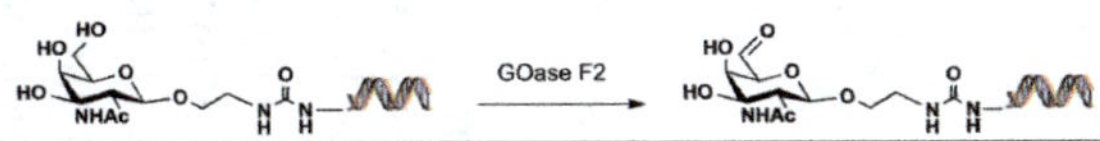

space based on its topological and structural features. A survey of literature indicated that the DELs cover quite unique chemical space as compared with other known compound collections.[12,64] However, it is clear that additional reaction development is needed to access the chemical space occupied by natural products. To achieve this goal, more on-DNA compatible reactions are needed, such as ring closing metathesis to access related types of natural product DELs.

7.3.2 DEL Design and Production

The whole process of DEL construction could be divided into three stages: library design, development and production. In the library design stage, there are several factors for consideration including library size, library cycles, split size of each library cycle, available on-DNA chemistry reactions and coding strategies. Here we use DEL55 (Figure 7.5), a carboxylic acid iodide amine (CIA) library, as an example to demonstrate the DEL synthesis process.[58] DEL55 is a three-cycle scaffold-based library. Cycle 1 involves attaching the trifunctional scaffolds through a free acid to the HP-NH_2 by acylation, cycle 2 is Suzuki coupling with aromatic iodides and cycle 3 is 9-fluorenylmethoxycarbonyl (FMOC) deprotection followed by amine capping with different functional capping groups, such as acids and aldehydes. During the library development stage, each cycle 1 trifunctional scaffold needed to be validated first for the acylation chemistry, then the cycle 2 Suzuki coupling chemistry and, lastly, the cycle 3 amine capping chemistry. For the cycle 2 Suzuki coupling reaction validation, the general practice of using a single standard boronic acid to validate all the aryl iodide scaffolds that had passed the first acylation step was employed. Liquid chromatography–mass spectrometry (LC–MS) is the standard method used for monitoring and validating on-DNA reactions. Scaffolds that passed both the acylation and the Suzuki coupling steps were then validated with a simple acid building block, such as benzoic acid, to confirm the last step amine capping chemistry.

After achieving this proof of concept of the synthetic sequence, building blocks for the Suzuki coupling and amine capping steps in library synthesis were selected. In general, the building block sets need to be validated experimentally against a standard model reaction, while extra filters, such as molecular weights and number of aromatic rings, can be applied to further filter the list of potential building blocks (BBs) *in silico*. Once the BB sets are chosen, the exact library synthetic scheme is established before the library production initiates. Normally, enzymatic ligation is carried out before the chemistry step in each cycle. However, in some cases, the chemistry step will need to be carried out first due to the synthetic challenge of chemistry reactions after the enzymatic ligation. After determining the synthetic sequence of each step including chemistry and ligation, the detailed library production schemes can be finalized (Figure 7.6).

Linear Library:

DEL39

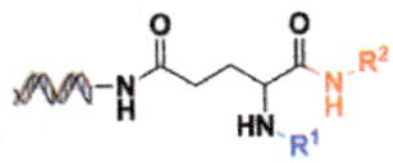

Cycle 1: 937 amine-capping BBs (encoded deletion)
Cycle 2: 597 amines (encoded deletion)

Library size: 0.6 million

DEL50

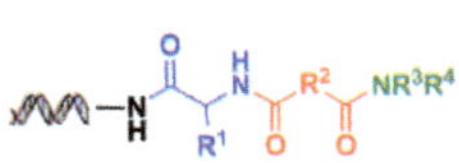

Cycle 1: 112 Fmoc-AA (encoded deletion)
Cycle 2: 18 diacids
Cycle 3: 638 amines

Library size: 1.29 million

DEL53

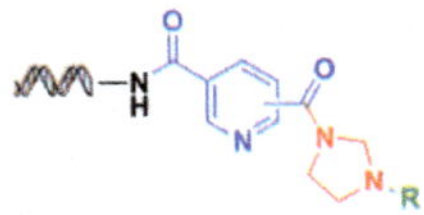

Cycle 1: 16 diacids
Cycle 2: 134 diamines
Cycle 3: 570 hetAryls

Library size: 1.22 million

Scaffolds Library:

DEL33

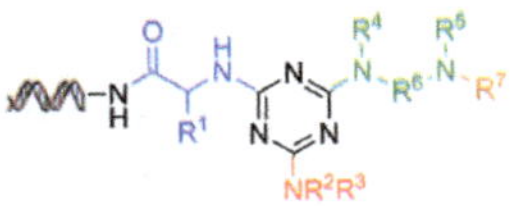

Cycle 1: 192 Fmoc-AAs
Cycle 2: 479 amines
Cycle 3: 96 diamines
Cycle 4: 459 amine-capping BBs (carboxylic acids, aldehydes, sulfonyl chlorides, isocyanates) + 4 blanks

Library size: 4.1 billion

DEL35

Cycle 1: 64 amino acids
Cycle 2: 854 amines
Cycle 3: 758 amines (encoded deletion)

Library size: 41.4 million

DEL38

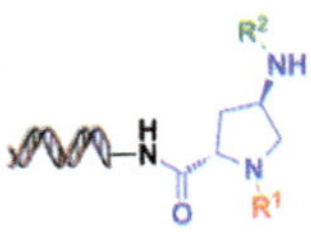

Cycle 1: 22 diamino acids (1 shown)

Cycle 2: 855 amine-capping BBs (carboxylic acids, aldehydes, sulfonyl chlorides, isocyanates) + 3 blanks

Cycle 3: 857 amine-capping BBs (carboxylic acids, aldehydes, sulfonyl chlorides, isocyanates) + 3 blanks

Library size: 16.2 million

DEL55

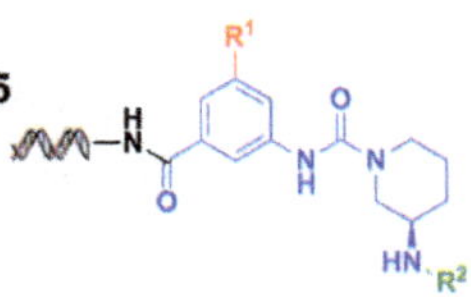

Cycle 1: 44 ClA scaffolds
Cycle 2: 257 boronates, alkynes (1 blank)
Cycle 3: 2993 amine-cappings (1 blank)

Library size: 16.5 million

DEL62

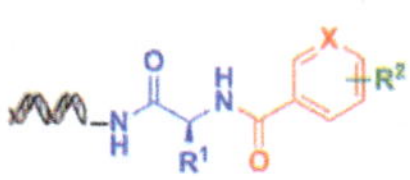

Cycle 1: 694 amines & AAs (3 blanks)
Cycle 2: 95 carboxylic acid/Ar-I scaffolds (1 blank)
Cycle 3: 245 boronic acids/esters (+1 blank)

Expanded library of Del 62

Library size: 16.5 million
Cycle 3: 840 boronic acids/esters (+1 blank)
1849 amines, Pd promoted (+1 blank)
Library size: 180 million

Figure 7.5 DELs reported by GSK including Del33,[54] 35,[55] 38,[17] 39,[55] 43,[56] 47,[55] 50,[55] 53,[15] 54,[57] 55,[58] 61,[59] 62,[28] 66,[56] 68,[60] 92,[26] 99[56] and macrocycle library.[61]

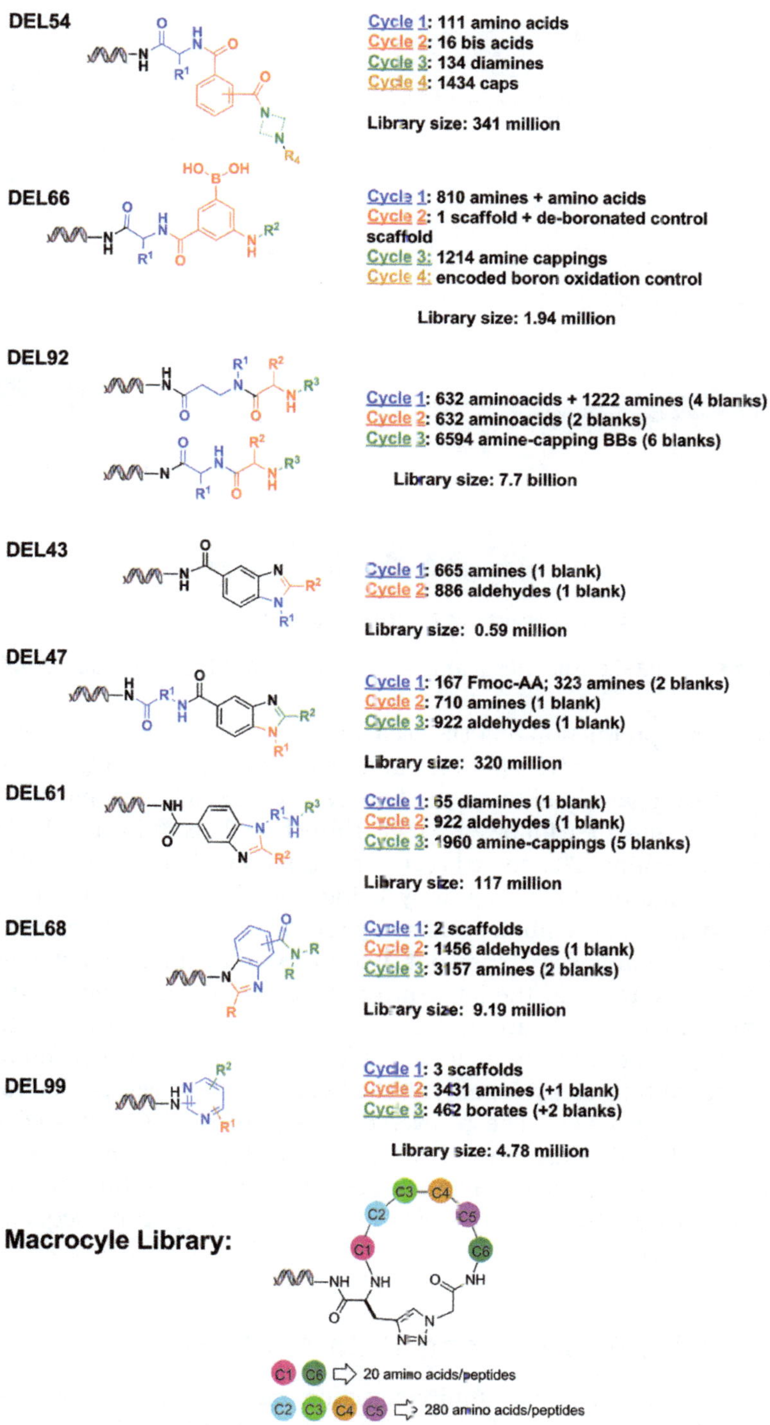

Figure 7.5 Continued.

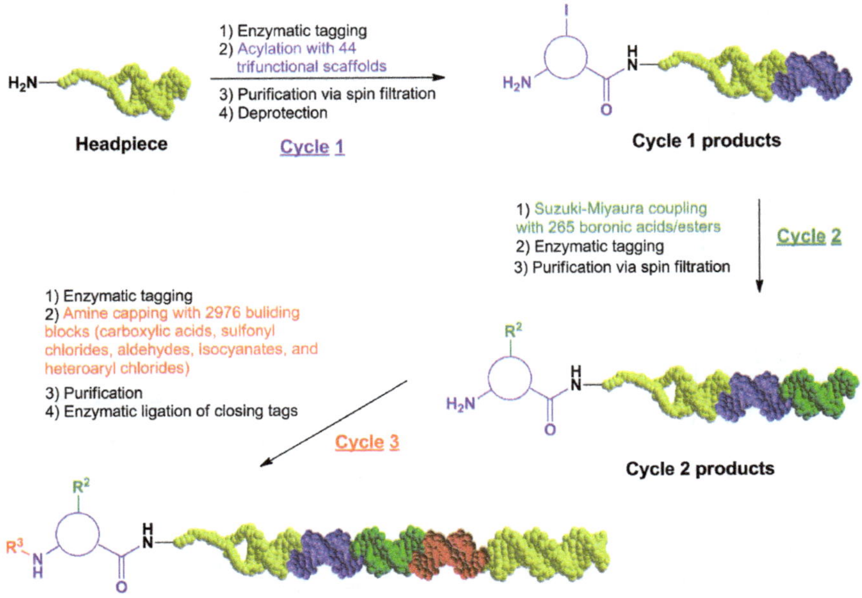

Figure 7.6 Del 55: Carboxylic acid iodide amine (CIA) library production scheme.

The library production of a DEL is a technically demanding process which requires comprehensive planning and stringent quality control to yield a library of sufficient quality. In the first cycle of DEL55, enzymatic ligations were carried out first followed by 44 trifunctional scaffold installations by acylation reactions. The pooled cycle 1 material was then cleaned up by size-exclusion column chromatography using spin-filtration. After the amine deprotection step, the library was ready for the cycle 2 chemistry. In cycle 2, the Suzuki reaction was carried out first, followed by the enzymatic ligation. This was done to avoid the low conversion encountered when ligation was performed first. After ligation, the cycle 2 material was pooled for clean up again by spin-filtration. In cycle 3, ligation was carried out first followed by the amine capping steps. Finally, all the library material was pooled together and purified by HPLC. The recovered library material was ligated with a closing tag for identification of the library so that it could be pooled for downstream affinity selection and sequencing activities. A final QC step was performed to confirm the integrity of the DNA tagging by sequencing a sample of diluted library.

7.4 ELT Selections and Data Analysis

Like most other encoded affinity selection methods [*i.e.*, phage display[65] and SELEX (systematic evolution of ligands by exponential enrichment[66])], the key steps in the selection of DELs are exposure of the target of interest

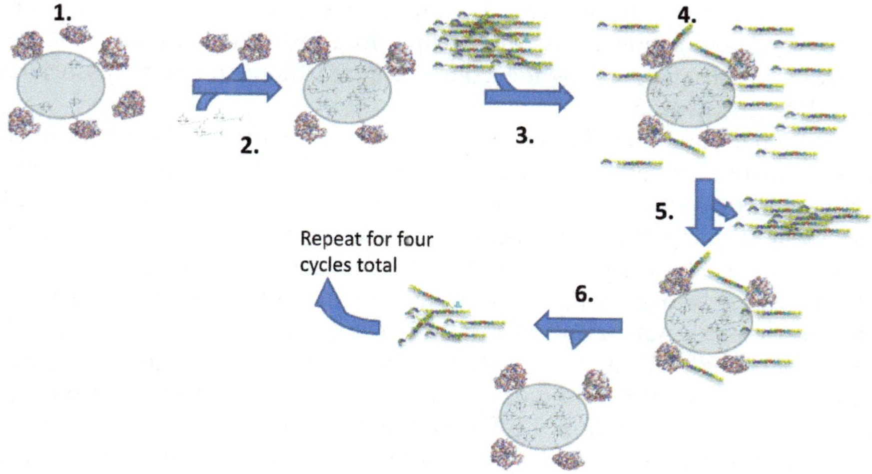

Figure 7.7 Selection scheme. For each round of selection: 1. Biotinylated-BCATm is captured on streptavidin beads. 2. Un-biotinylated BCAtm is washed away and unoccupied stretavidin is blocked with free biotin. 3. Pooled DEL libraries are added to immobilized BCATm beads and 4. Allowed to reach equilibrium. 5. Unbound DEL molecules are washed away. 6. Finally, bound DEL molecules are eluted by thermal denaturation of BCATm. The DEL output of each round is used as the input for subsequent rounds of selection.

to the encoded libraries followed by purification of the target–ligand complexes away from unbound library members, and the identification of the bound ligands through their encoding (Figure 7.7). The purification of the target–ligand complexes is typically achieved using an affinity tag on the target of interest and the appropriate affinity resin. ELT selections require very little of the target protein per round of selection (typically 5–10 μg), and microscale columns (for example Phynexus tips) or magnetic beads are used to immobilize the target–ligand complexes for removal of unbound library and subsequent washing.[8,67] Commonly used affinity tags/matrices include, but are not limited to, poly-His/IMAC (immobilized metal affinity chromatography),[8,18,68,69] flag–anti-flag,[23,70] streptavidin binding peptide–streptavidin[15] and GST–glutathione.[26,27] In cases where an affinity tag is not present, chemical biotinylation[17,22,54,58,59] for selection on streptavidin matrices and covalent attachment directly to beads[21,71,72] of the target protein have also successfully been applied to selections.

Multiple rounds of selection can be employed to further purify the binders from the library,[18,58,59] although binders can be identified in a single round of selection.[72] During iterative rounds of the selection the yield of DEL recovered is frequently tracked using quantitative PCR (qPCR).[73] Selections are considered complete once the recovered yield can be reasonably covered by DNA sequencing (1×10^7–1×10^8 recovered molecules). The samples are then amplified by PCR to install the necessary DNA sequences for sequencing, and sequenced using compatible primers. The advent of next-generation sequencing and the improvements driven by the drive for a low-cost human genome[74] is what allows libraries of the diversity achieved in DELs to be screened, while continuing to lower the costs associated with data acquisition.

To further illustrate the selection process we will use the selection of mitochondrial branched chain aminotransferases (BCATm) inhibitors from the CIA library described above.[58] In this case, BACTm was chemically biotinylated using *N*-hydroxysuccinimidobiotin (NHS-biotin) for immobilization on 5 μl of agarose streptavidin matrix in microscale columns (Phynexus). As a first step, 100 μl of 0.05 μg ml^{-1} biotinylated BCATm was washed over the matrix for 20 minutes. A matrix-only control was run in parallel through all steps of selection to enable the flagging of matrix binders in the final target data set. Following the target capture the tips were washed in selection buffer supplemented with 1 mg ml^{-1} heat-inactivated sheared salmon sperm DNA and 1 mM biotin. These wash steps serve to remove any unbiotinylated target protein as well as to block potential sites of non-specific DNA binding and the free streptavidin binding sites. In our experience, DELs often contain potent streptavidin binders, and the matrix should be blocked prior to exposure to the libraries to prevent the streptavidin binders from dominating the final selection output (unpublished data).

Once washed and blocked the target and matrix-only control columns were incubated with 5 nmol of DELs in 100 μl (50 μM final concentration). As long as individual libraries are encoded with distinct sequences, multiple libraries can be pooled together and selected simultaneously on the same column to facilitate a higher throughput.[55] After incubation, the columns were thoroughly washed ten times with 20 times the column volume. Once washed, the DEL molecules bound to the target or matrix were eluted by thermal denaturation. For the columns this was achieved by incubating 100 μl of selection buffer at 80 °C and washing the heated buffer over the column for 10 minutes to complete the first round of selection. The eluent was then incubated with fresh immobilized protein to start the next round of affinity selection. Three rounds of selection were run in total, and the progress tracked by qPCR following each round. After the third round the yield was approximately 1×10^9 total molecules for the target condition. The target and the matrix-only control samples were amplified by PCR with primers complementary to sequences in the DEL molecule which also contained sequences compatible with cluster generation on a standard Illumina flow cell. Approximately 2×10^7 sequences were collected for each sample and used for analysis.

The collected sequencing data for BCATm and the matrix-only control were decoded from the DNA barcodes to the associated encoded molecules. On the basis of the total diversity of the library and the number of sequences obtained, the noise level was calculated for each selection. Signal was calculated as a value relative to the noise level (*i.e.* a signal value of 10 represents 10-fold greater copies than the noise level of copies). All data points with signal greater than 2 were included in subsequent data analysis steps. This output was then filtered to remove chemotypes that had signal in both the target and the matrix-only control. The remaining enriched, target-specific chemotypes can then be further analysed to select discrete compounds to synthesize off DNA. While the selections run for BCATm were a straightforward comparison of the target to the matrix-only control, other variations can be applied to the basic selection method to help generate more information to drive decision making when selecting compounds for off-DNA synthesis.

The ability to run multiple similar selection conditions in parallel on pooled libraries allows for the rapid evaluation of the effects of different selection conditions on potential chemotypes. For example, in order to identify potent, active-site-binders against Bruton's tyrosine kinase (BTK), Cuozzo *et al.* employed a strategy where in addition to the target protein, conditions were run where the active site was saturated with either ATP or the kinase inhibitor dasatinib.[69] These competition conditions enabled the identification of chemotypes competitive with known active-site binders. An additional condition at low protein concentration increased the stringency and indicated which chemotypes were likely to bind with higher affinity. The combination of these selection data allowed for the rapid identification of compounds that spanned multiple binding modes across a range of potencies. A similar strategy was also used to identify potent active site soluble epoxide hydrolase (sEH) inhibitors.[24]

One recently-described variation on the affinity selection methods is to use intact cells expressing the target of interest themselves as the matrix. Targets expressed on the plasma membrane of cells can be exposed to DEL libraries and washed similarly to a target immobilized on an agarose bead. Wu *et al.* reported this method and used it to identify novel NK3 tachykinin receptor antagonists.[55]

Rather than running multiple conditions on a single target, multiple targets can be run in parallel to assess their relative tractability and to identify tools for chemical target validation. GSK has applied this technique to screens against genetically validated targets for pathogenic organisms. These "buckets" of targets allow for the identification of the most chemically tractable targets in the bucket as well as tool compounds for target validation linking target inhibition to minimum inhibitory concentration (MIC).[56]

7.5 Post Selection Chemistry (PSC)

Post-selection chemistry involves decoding the selection output (*i.e.* enriched warheads) into small-molecule hits, which are derived from the enriched warheads, lack the DNA tag and thus can be evaluated in relevant biochemical and/or cellular assays. High-throughput DNA sequence and subsequent data analysis routinely generates large sets of enriched warheads, but only a small number of the more enriched and chemically interesting small-molecule hits are synthesized in order to reduce the time and cost of interrogating the entire selection output.

The most common practice for picking the more enriched and chemically interesting small-molecule hits is through identifying enriched chemotypes and the selection-structure–activity relationships (SARs) around each chemotype. Once the target-specific chemotypes are identified by statistical methods as described previously (Section 7.4),[85,86] further analysis can be done visualizing the data in a cubic scatter plot where each building block at each cycle is a point on an axis, and points are sized by the copies measured by sequencing. This procedure enables a convenient assessment of selection-SAR. The Spotfire program (Perkin Elmer) is a commonly used visualization tool for this purpose.

In their effort to identify novel antagonists of the NK3 tachykinin receptor through cell-based selection of GSK's proprietary DEL collection, Wu *et al.* visualized the selection outcome from DEL35, a three-cycle library based on a triazine scaffold (Figure 7.8), as a cubic scatter plot.[55] Here, encoded small-molecule compounds enriched at the level of two independent DNA sequences or greater are shown with the cube axes representing building blocks used in cycle 1, cycle 2, and cycle 3 chemistry, respectively. This visualization allows families of structurally related molecules to appear as lines (two building blocks in common) or planes (one building block in common). The lines (coloured red, green and purple in Figure 7.8) indicated that only a small number of cycle-1 and cycle-3 building blocks were selected with NK3-positive cells, while variability in cycle-2 was more widely tolerated. The plane (coloured blue in Figure 7.8) defined by cycle-3 building block **5** suggests this building block is important for NK3 affinity. This conclusion is further augmented by the fact that the structurally related des-methyl building block **6** is much less enriched (green line) indicating clear selection-SAR for block **5** over block **6**. The encoded structure **2** of the most prominent line predicts that the combination of the highly preferred cycle-1 building block 7 and the cycle-3 building block **5** may be especially important for binding. The cycle 1 selection-SAR is also identifiable by the observation that structure **3** from the other line on the same plane (coloured purple in Figure 7.8 highlight) differs from structure **2** only by a single methyl group (compare 7 with **8**). In contrast to cycles 1 and 3, almost all 854 cycle-2 amines were found in the line with encoded structure **2**, indicating that cycle-2 building blocks are not essential for binding to NK3. A small set of compounds was synthesized without the DNA tags and subjected to biochemical and functional assays. These compounds have biological activity and the assay results support the selection-SAR observation.

Figure 7.8 summarizes the potency data for a representative set of compounds tested in the NK1 functional FLIPR assays. Compound **9**, which corresponded to the on-DNA compound **2** and had the highest copy count in the NK3 selection, demonstrated low nanomolar potency in NK3 functional cell-based assays. Truncations of the non-selected cycle-2 provided lower-molecular-weight compounds that remained potent NK3 antagonists (compound **10**). An additional 10-fold potency loss was seen with compound **11**, which corresponded to the cycle-2 truncated on-DNA compound **4** (a simple

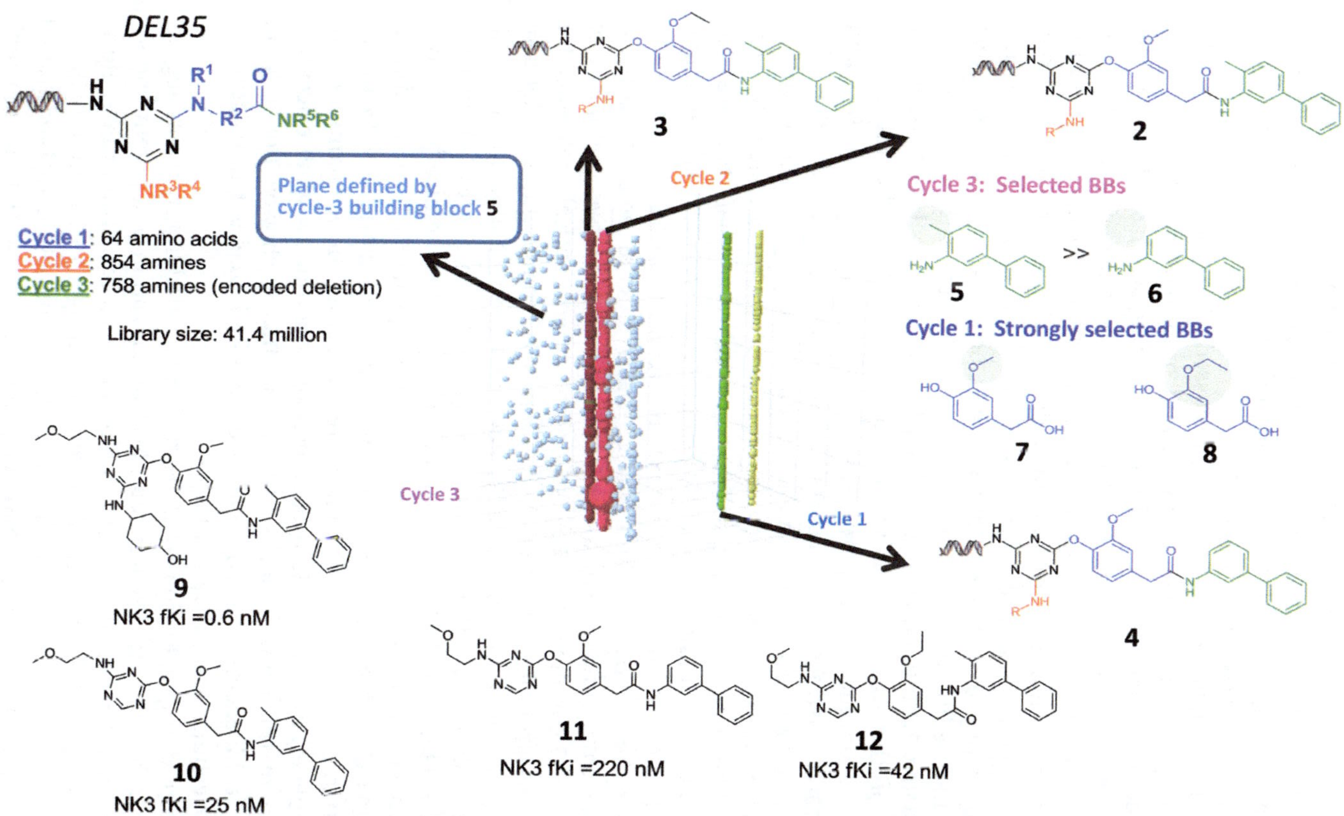

Figure 7.8 Three-cycle triazine library (DEL35) and chemotypes from NK3 cell-based selection. Adapted with permission from Z. Wu, T. L. Graybill, X. Zeng, M. Platchek, J. Zhang, V. Q. Bodmer, D. D. Wisnoski, J. Deng, F. T. Coppo, G. Yao, A. Tamburino, G. Scavello, G. J. Franklin, S. Mataruse, K. L. Bedard, Y. Ding, J. Chai, J. Summerfield, P. A. Centrella, J. A. Messer, A. J. Pope and D. I. Israel, ACS Comb. Sci., 2015, 17, 722–731. Copyright 2015 American Chemical Society.[55]

combination of strongly selected building blocks **7** and moderately selected building blocks **6**). The SAR trends predicted by the selection cube are largely consistent with the rank-order potency of the "off-DNA" compounds, that is, Compound **10** *versus* compound **11** corresponding to cycle-3 building block **5** *versus* **6** and compound **10** *versus* compound **12** corresponding to cycle-1 building block **7** *versus* **8**, respectively.

In another paper, Harris *et al.*[26] discussed the discovery of benzo-[*b*][1,4]oxazepin-4-ones as highly potent and monoselective receptor interacting protein 1 kinase inhibitors through screening against GSK's propriety collection of DNA-encoded small-molecule libraries. Further optimization of this hit series led to a RIP1 clinical candidate, which is being evaluated for the treatment of multiple inflammatory diseases.[30] The affinity selections was done against the RIP1 kinase domain (1–375) and a structurally intriguing family of enriched warheads was observed from a three-cycle amino acid core library (DEL92) which contained a total warhead diversity of approximately 7.7 billion. This family may be visualized as three lines within a plane of the DEL92 cubic scatter plot (Figure 7.9). The three lines contained a unique enantiopure benzo[*b*][1,4]-oxazepin-4-one as BB2 and one of three distinct but structurally related amine caps as BB3. The specificity of the selection experiment was quite remarkable, as only one out of 632 amino acids as BB2 and three out of 6594 amine caps as BB3s were selected. The appearance of the lines also indicated that there was no preference for BB1, indicating that BB1 contributes very little to the binding. Taking into account all of the information inferred from the data cube and truncating the carboxyl linker for BB2, the GSK team prepared benzoxazepinone compounds **13-16** as preliminary representative exemplars of the BB2 + BB3 combinations. To confirm the stereochemical preference for RIP1, benzoxazepinone **16**, the *R* enantiomer of compound **15**, was also prepared. The benzoxazepinone analogues **13-15** were evaluated in a fluorescence polarization (FP) binding assay. The three off-DNA benzoxazepinones (**13**, **14** and **15**) were found to be potent biochemically, with two analogues (**14** and **15**) having concentrations giving 50% of maximum inhibition (IC_{50}) of less than 100 nM. Comparison of enantiomers **15** and **16** demonstrated a clear stereochemical preference, as the *R* enantiomer **16** was completely inactive. *N*-Methylated benzoxazepinone **17**, a more accurate reflection of the combination of **BB2-1** and **BB3-1**, showed not only an increase in biochemical activity but also exhibited excellent translation in the U937 cellular assay ($IC_{50} = 10$ nM).

In a third example, Encinas *et al.* used DNA-encoded library technology as a source of hits for the discovery and lead optimization of a potent and selective class of bactericidal direct inhibitors of mycobacterium tuberculosis InhA.[17] A putative hit was identified from screening a 16.1 million-member encoded aminoproline scaffold library against an immobilizing biotinylated InhA protein in the presence of NADH cofactor. A cubic scatter plot in which each axis represents a cycle of diversity in the library was used again to analyse and visualize the enriched warheads (Figure 7.10). Individual points,

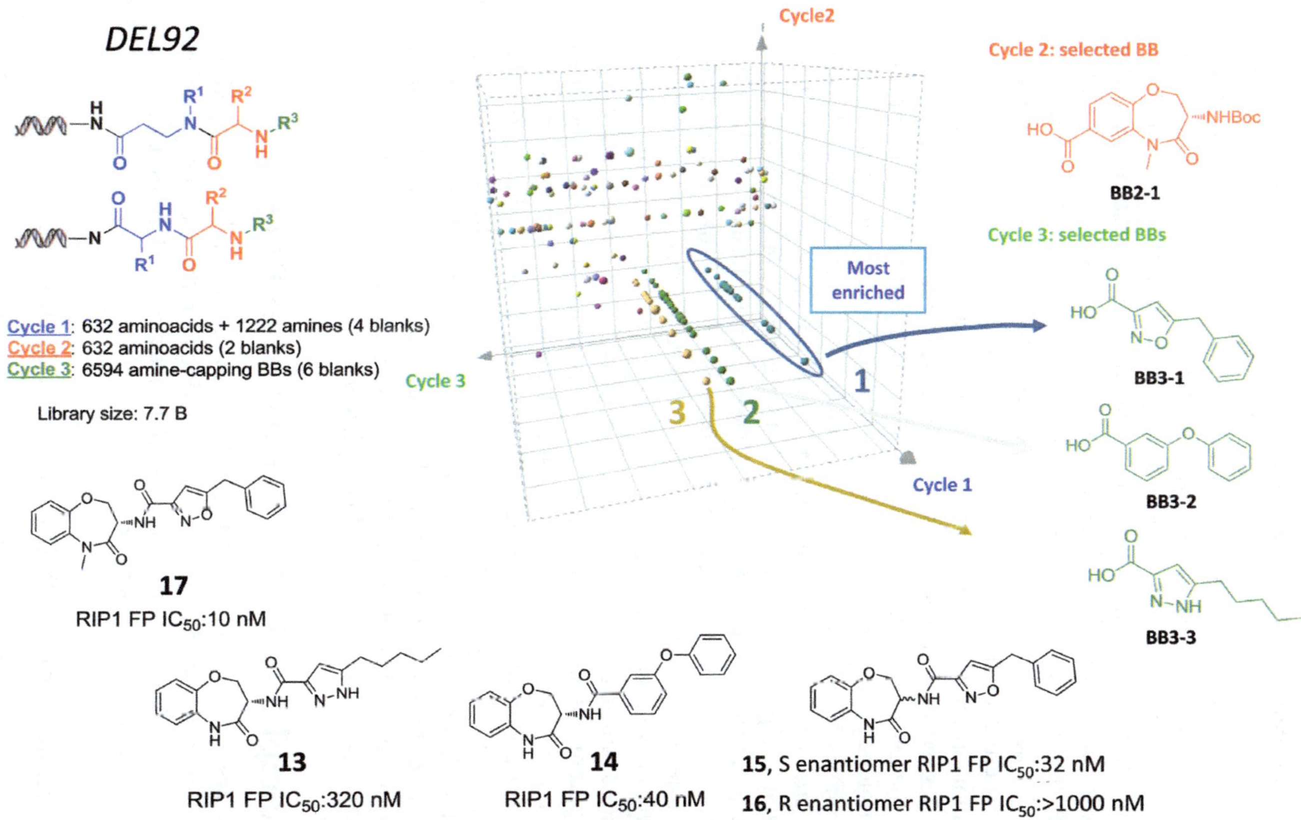

Figure 7.9 Three-cycle amino acid Library (DEL92) and (right) cube view of the RIP1 selection output. Adapted with permission from P. A. Harris, B. W. King, D. Bandyopadhyay, S. B. Berger, N. Campobasso, C. A. Capriotti, J. A. Cox, L. Dare, X. Dong, J. N. Finger, L. C. Grady, S. J. Hoffman, J. U. Jeong, J. Kang, V. Kasparcova, A. S. Lakdawala, R. Lehr, D. E. McNulty, R. Nagilla, M. T. Ouellette, C. S. Pao, A. R. Rendina, M. C. Schaeffer, J. D. Summerfield, B. A. Swift, R. D. Totoritis, P. Ward, A. Zhang, D. Zhang, R. W. Marquis, J. Bertin and P. J. Gough, *J. Med. Chem.*, 2016, **59**, 2163–2178. Copyright 2016 American Chemical Society.[26]

corresponding to discrete enriched small-molecule warheads in the library, are shown in pink and sized according to the number of unique instances recorded by DNA sequencing. The display is set to a minimum of two unique copies per warhead, which should indicate significantly enriched binders under these selection conditions. In this analysis, the enriched chemotype of interest is represented by a plane defined by the cycle 1 building block (2*S*,4*R*)-4-aminopyrrolidine-2-carboxylic acid (**BB1-1**), which was attached to DNA through its carboxylic acid functional group. Within the selected plane, there are multiple lines due to preferred disynthon combinations of **BB1-1** with specific cycle 3 building blocks. The most prominent line is defined by cycle 3 building blocks 3-ethyl-1-methyl-1H-pyrazole-5-carboxylic acid (**BB3-1**). This specific combination of **BB1-1** and **BB3-1** (disynthon) provided the core structure of the InhA chemotype selected from this library as a single stereoisomer. There are a number of BB2s selected within the line, each represented by a single point; therefore, the BB2 is a variable component of the selected scaffold that provides additional SAR in the form of possible moieties tolerated by the site of interaction on the protein. This selection-SAR is a valuable tool in compound prioritization for off-DNA synthesis and can provide exquisite guidance in early chemotype lead optimization. In preparation for the off-DNA activity confirmation, three exemplars (compounds **19**, **20** and **21**) were chosen based on the total copy number identified for that library molecule in the **BB1-1** and **BB3-1** disynthon line. The prepared compounds were then assayed for their activity against *Mycobacterium tuberculosis* InhA. The three off-DNA compounds demonstrated sub-nanomolar inhibitory activity in the InhA assay with the most potent compound **19** giving an IC_{50} potency of 34 nM. The DNA attachment point was also explored, and the methyl ester analogue **22** showed reduced inhibitory activity of 215 nM whereas the carboxylic analogue **23** lost potency even more with single-digit μM inhibitory activity. These results indicate the amide moiety of the DNA attachment point is contributing to the binding affinity to the InhA protein. This observation is in stark contrast to that of the RIP1 case where the amide moiety in the enriched warhead can be truncated completely.

7.6 Examples of Probes Derived from DELs

Over the years, many hits from DELs have been successively developed as chemical probes to provide insight on functions of their target proteins in phenotypes. A few recent examples of probes derived from DELs are described in the following sections.

7.6.1 Protein Arginine Deiminase 4-Inhibitor

Protein arginine deiminases (PADs) catalyse the hydrolysis of peptidyl-arginine to peptidyl-citrulline, they are composed of five, calcium-dependent isozymes (PADs 1–4 and 6). PADs play an important role in the

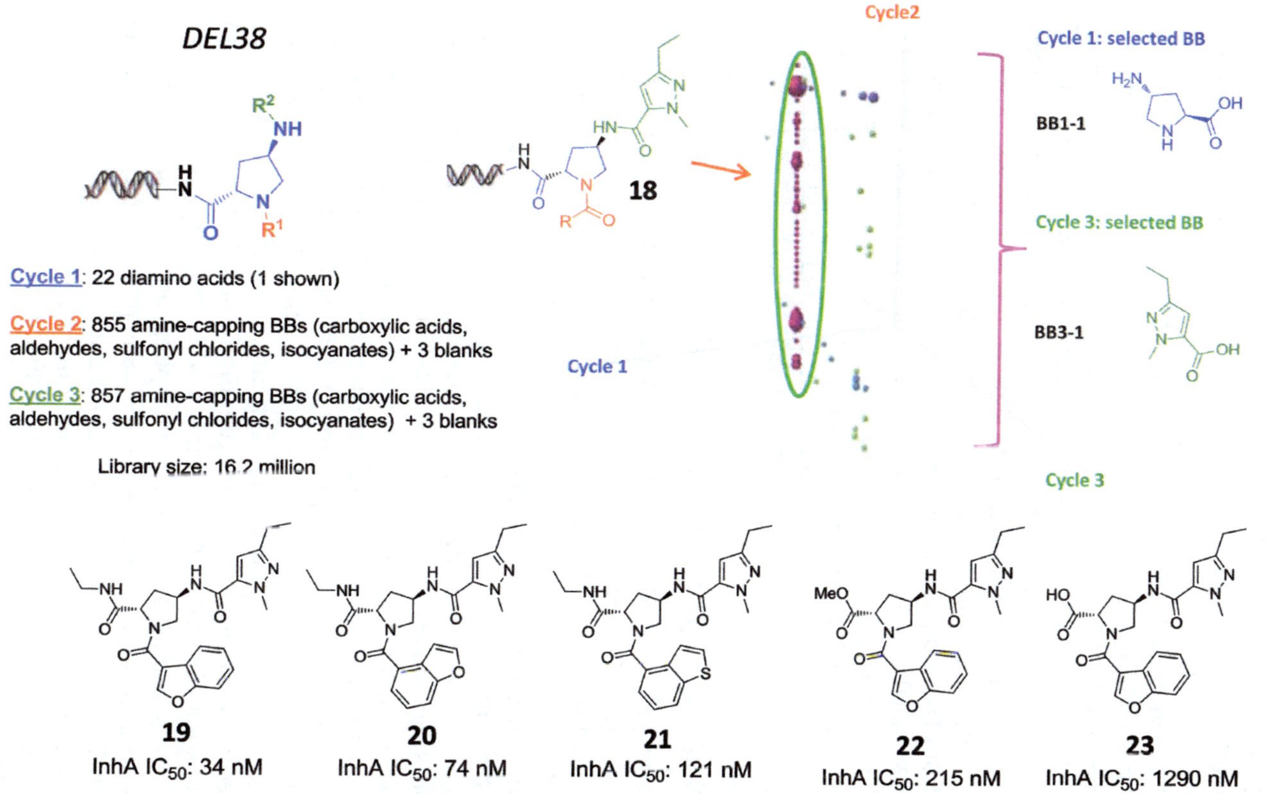

Figure 7.10 Three-cycle aminoproline scaffold library and cube view of the InhA selection output. Adapted with permission from L. Encinas, H. O'Keefe, M. Neu, M. J. Remuinan, A. M. Patel, A. Guardia, C. P. Davie, N. Perez-Macias, H. Yang, M. A. Convery, J. A. Messer, E. Perez-Herran, P. A. Centrella, D. Alvarez-Gomez, M. A. Clark, S. Huss, G. K. O'Donovan, F. Ortega-Muro, W. McDowell, P. Castaneda, C. C. Arico-Muendel, S. Pajk, J. Rullas, I. Angulo-Barturen, E. Alvarez-Ruiz, A. Mendoza-Losana, L. Ballell Pages, J. Castro-Pichel and G. Evindar, *J. Med. Chem.*, 2014, 57, 1276–1288. Copyright (2014) American Chemical Society.[17]

post-translational modifications of histones. PAD4, predominantly expressed in granulocytes, is strongly linked to diverse diseases, including rheumatoid arthritis (RA), Alzheimer's disease (AD), multiple sclerosis (MS), lupus, Parkinson's disease and cancer. Screening of a DEL against the calcium-bound and calcium-free conformation of PAD4 in a head-to-head manner and subsequent compound optimization identified the indolyl-substituted benzimidazole inhibitor **24** (Figure 7.11), a selective, reversible PAD4 inhibitor with nanomolar potency against PAD4.[27] Crystallization of the inhibitor-bound protein showed an induced-fit binding mode probably unique to PAD4. This binding mode explained the isoform selectivity of compound **24**. The ease with which **24** could be fluorescently labelled (compound **25**) and also converted into a functional pull-down probe for chemoproteomics again illustrated one advantageous feature of probes identified from DELs: a position for compound labelling is known *a priori*. Compound **24** was then successfully employed for *in vivo* studies. It validated the crucial role of PAD4 catalytic activity in the formation of neutrophil extracellular traps (NETs), demonstrated that PAD4 is a druggable target for small-molecule inhibitors, and established that the compound qualifies as a useful chemical biology probe for PAD4 in disease models.

7.6.2 Allosteric Wip1 Phosphatase Inhibitor

The wild-type p53-induced phosphatase (Wip1, encoded by PPM1D) is an oncogenic type 2C serine/threonine phosphatase common to multiple cancers. It negatively regulates key proteins in the DNA damage–response pathway including p53, p38 MAPK, ATM, Chk1, Chk2, Mdm2 and histone H2AX1. Wip1 overexpression is believed to promote tumorigenesis by inactivating the tumour-suppressor functions of multiple substrates. Thus Wip1 is an attractive therapeutic target in oncology. However, development of selective small-molecule phosphatase inhibitors has been very challenging due to the high sequence homology and polarity of the catalytic centre across many members of this enzyme class. An affinity selection was run using the full-length Wip1 protein against a pool of DELs. The ELT screen and subsequent structure optimization for *in vivo* studies yielded compound **26** (Figure 7.11).[23] Inhibition by **26** was found to be non-competitive with respect to fluorescein diphosphate (FDP), which is an artificial substrate for WIP1. This contrasts with the reported competitive behaviour of a substrate-derived cyclic phosphopeptide inhibitor of Wip1. These observations indicate that compound **26** is probably binding outside of the catalytic active site. Photolabelling experiments indicated a binding site distal from the catalytic centre in a subdomain called flap near the Wip1 catalytic site that renders Wip1 structurally divergent from other members of the protein phosphatase 2C (PP2C) family and that thereby confers selectivity for Wip1 over other phosphatases. Treatment of tumour cells with the inhibitor **26** increases phosphorylation of Wip1 substrates and causes growth inhibition in both hematopoietic tumour cell lines and

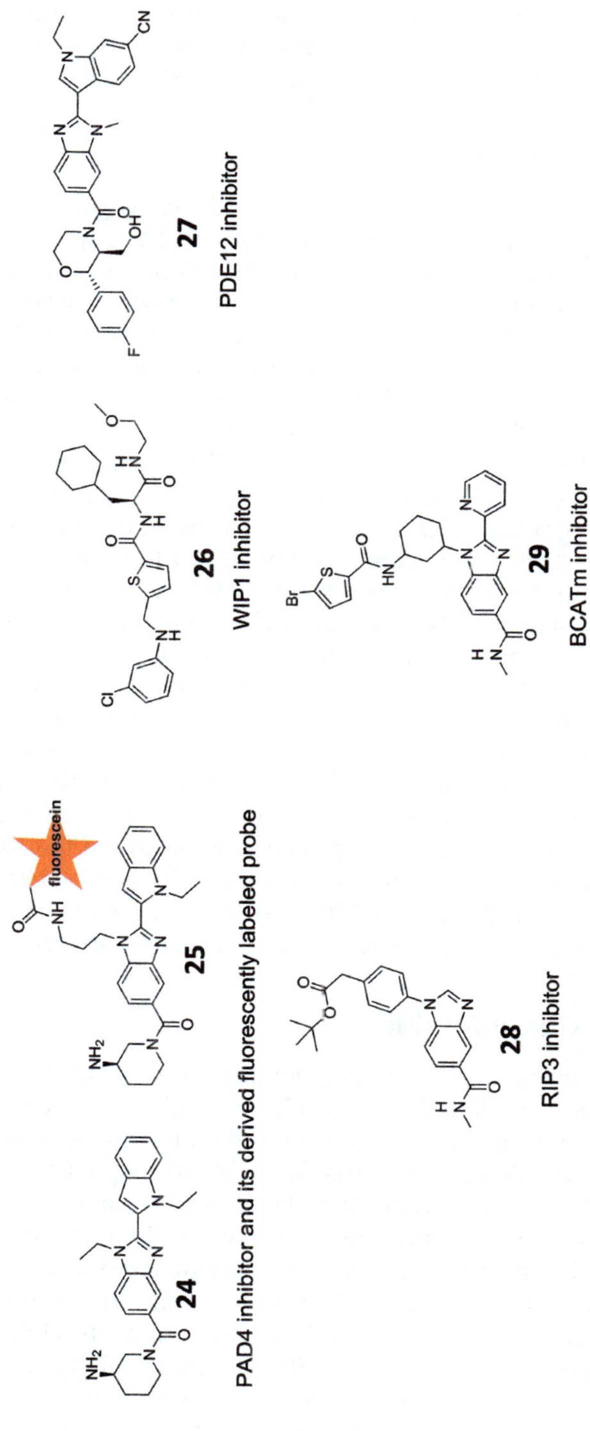

Figure 7.11 Examples of chemical biology probes from DELs.

Wip1-amplified breast tumour cells harbouring wild-type TP53. Oral administration of Wip1 inhibitors in mice results in expected pharmacodynamic effects and causes inhibition of lymphoma xenograft growth. Thus, **26** is a valuable tool compound to investigate Wip1 in disease models.

7.6.3 PDE12 Inhibitor

PDE12 degrades 2′,5′-oligoadenylate, a second messenger involved in the antiviral action of interferon. Inhibition of PDE12 may up-regulate the OAS/RNase-L pathway in response to viral infection, resulting in increased resistance to a variety of viral pathogens. 2′,5′-Oligoadenylate synthetase (OAS) enzymes and RNase-L play a major role in interferon (IFN)-mediated antiviral defence mechanisms. OAS produces a unique oligonucleotide second messenger, 2′,5′-oligoadenylate (2–5A), which binds to an ankyrin repeat regulatory domain of Rnase-L that triggers the dimerization and activation of the RNase. This pathway is down-regulated by virus and host-encoded enzymes that degrade 2–5A. Phosphodiesterase 12 (PDE12) was the first cellular 2–5A-degrading enzyme to be purified and characterized at a molecular level. To further validate the role of PDE12 inhibition in antiviral defence mechanism and the druggability of PDE12, the PDE12 protein was screened against a pool of DELs, which yielded compound **27** (Figure 7.11).[60] Compound **27** inhibited PDE12 with nanomolar potency *in vitro* in a 2–5A-competitive fashion, and had high selectivity *versus* CNOT6, a closely related enzyme that cleaves 3′-5′-oligoadenylates. Treatment of host cells with these inhibitors mimics gene inactivation. The first crystal structure of PDE12 bound to an inhibitor has been reported. The inhibitor occupies nearly the equivalent volume and position in PDE12 as the substrate occupies in CNOT6. Compound **27** produced a maximum fourfold increase in 2–5A levels, which is similar to what was seen with gene inactivation. This result indicates that PDE12 inhibitors exert an antiviral effect through modulation of the IFN-induced OAS/RNase-L effector pathway.

7.6.4 RIP3 Kinase Inhibitor

Necroptosis is an alternative form of programmed cell death that is triggered when apoptosis is inhibited. The results of studies indicated that RIP3-dependent necrotic death is crucial for inflammation. A benzoimidazole RIP3 inhibitor **28** (Figure 7.11) was identified through DEL screening.[70] Compound **28** ($IC_{50} = 0.9$ nM) binds to the RIP3 kinase domain potently and exhibits high selectivity against most of the 300 human protein kinases tested, and blocked necroptosis in primary human neutrophils isolated from whole blood. The benzoimidazole **28** was active in human but inactive in mouse cells, indicating that species differences dictated the ability of RIP3 to bind this compound. The ability of RIP3 kinase inhibitors to block death in a wide variety of human and murine cell types reinforces the prospect of employing such inhibitors for therapeutic intervention in inflammatory diseases. The results of studies using these tool compounds also indicate that RIP3 kinase inhibitors prevent death from a broader range of stimuli than RIP1 inhibitors. These compounds have been shown to interact with RIP3 to activate caspase 8 (Casp8) *via* RHIM-driven recruitment of RIP1 (RIPK1) to assemble a Casp8–FADD–cFLIP complex completely independent of pronecrotic kinase activities and mixed lineage kinase domain like pseudokinase (MLKL). The human-specific RIP inhibitor compound **28** was use to demonstrate that MLKL is dispensable for RIP3i-induced apoptosis.

7.6.5 BACTm Inhibitor

Mitochondrial branched chain aminotransferases (BCATms) catalyse transamination of branched-chain amino acids (BCAAs) leucine, isoleucine and valine to their respective α-keto acids, which are further converted to succinyl-CoA and acetyl-CoA and participate in the tricarboxylic acid cycle and glycolysis. It has been indicated by the results of genomic study that targeting inhibition of BACTm can protect people from obesity. To further validate BCATm as a therapeutic target for the intervention of metabolic disorders, a suitable small-molecular inhibitor with potency, selectivity and *in vivo* activity is highly desirable. A chemically biotinylated human BCATm protein construct was screened against a pool of more than 14 billion DNA-encoded compounds. Subsequent compound optimization identified the pyridyl substituted benzimidazole inhibitor compound **29** (Figure 7.11),[59] a selective BACTm inhibitor [100-fold selective for BCATm over cytosolic BCAT (BCATc,)] that showed both *in vitro* ($IC_{50} = 50$ nM) and *in vivo* potency. After oral administration, compound **29** raised mouse blood levels of all three branched-chain amino acids as a consequence of BCATm inhibition. Compound **29** is thus a valuable pharmacological tool for investigating the biological roles of BCATm in metabolic diseases.

The few examples shown here only represent a small number of cases where ELT has been successfully used for identifying chemical biology probes. As highlighted by the increasing number of recent publications,[37,53,64,75–84] ELT is coming of age as a powerful screening tool 25 years after its initial conception.[13] Screening of DELs has yielded a large number of bioactive molecules with drug-like properties that have been progressed developed into high-quality chemical biology probes and even clinical candidates.[12,24,30] As the chemical space covered by DELs continues to expand, selection and sequencing technology continue to improve and adoption continues to increase, ELT will undoubtedly play an even bigger role in the discovery of chemical biology probes as well as lead molecules for drug discovery in the coming years.

References

1. S. L. Schreiber, J. D. Kotz, M. Li, J. Aube, C. P. Austin, J. C. Reed, H. Rosen, E. L. White, L. A. Sklar, C. W. Lindsley, B. R. Alexander,

J. A. Bittker, P. A. Clemons, A. de Souza, M. A. Foley, M. Palmer, A. F. Shamji, M. J. Wawer, O. McManus, M. Wu, B. Zou, H. Yu, J. E. Golden, F. J. Schoenen, A. Simeonov, A. Jadhav, M. R. Jackson, A. B. Pinkerton, T. D. Chung, P. R. Griffin, B. F. Cravatt, P. S. Hodder, W. R. Roush, E. Roberts, D. H. Chung, C. B. Jonsson, J. W. Noah, W. E. Severson, S. Ananthan, B. Edwards, T. I. Oprea, P. J. Conn, C. R. Hopkins, M. R. Wood, S. R. Stauffer and K. A. Emmitte, *Cell*, 2015, **161**, 1252–1265.

2. M. Schenone, V. Dancik, B. K. Wagner and P. A. Clemons, *Nat. Chem. Biol.*, 2013, **9**, 232–240.
3. C. H. Arrowsmith, J. E. Audia, C. Austin, J. Baell, J. Bennett, J. Blagg, C. Bountra, P. E. Brennan, P. J. Brown, M. E. Bunnage, C. Buser-Doepner, R. M. Campbell, A. J. Carter, P. Cohen, R. A. Copeland, B. Cravatt, J. L. Dahlin, D. Dhanak, A. M. Edwards, M. Frederiksen, S. V. Frye, N. Gray, C. E. Grimshaw, D. Hepworth, T. Howe, K. V. Huber, J. Jin, S. Knapp, J. D. Kotz, R. G. Kruger, D. Lowe, M. M. Mader, B. Marsden, A. Mueller-Fahrnow, S. Muller, R. C. O'Hagan, J. P. Overington, D. R. Owen, S. H. Rosenberg, B. Roth, R. Ross, M. Schapira, S. L. Schreiber, B. Shoichet, M. Sundstrom, G. Superti-Furga, J. Taunton, L. Toledo-Sherman, C. Walpole, M. A. Walters, T. M. Willson, P. Workman, R. N. Young and W. J. Zuercher, *Nat. Chem. Biol.*, 2015, **11**, 536–541.
4. R. Macarron, M. N. Banks, D. Bojanic, D. J. Burns, D. A. Cirovic, T. Garyantes, D. V. Green, R. P. Hertzberg, W. P. Janzen, J. W. Paslay, U. Schopfer and G. S. Sittampalam, *Nat. Rev. Drug Discovery*, 2011, **10**, 188–195.
5. F. N. Edfeldt, R. H. Folmer and A. L. Breeze, *Drug Discovery Today*, 2011, **16**, 284–287.
6. J. Bajorath, *Nat. Rev. Drug Discovery*, 2002, **1**, 882–894.
7. H. K. Binz and A. Pluckthun, *Curr. Opin. Biotechnol.*, 2005, **16**, 459–469.
8. M. A. Clark, R. A. Acharya, C. C. Arico-Muendel, S. L. Belyanskaya, D. R. Benjamin, N. R. Carlson, P. A. Centrella, C. H. Chiu, S. P. Creaser, J. W. Cuozzo, C. P. Davie, Y. Ding, G. J. Franklin, K. D. Franzen, M. L. Gefter, S. P. Hale, N. J. Hansen, D. I. Israel, J. Jiang, M. J. Kavarana, M. S. Kelley, C. S. Kollmann, F. Li, K. Lind, S. Mataruse, P. F. Medeiros, J. A. Messer, P. Myers, H. O'Keefe, M. C. Oliff, C. E. Rise, A. L. Satz, S. R. Skinner, J. L. Svendsen, L. Tang, K. van Vloten, R. W. Wagner, G. Yao, B. Zhao and B. A. Morgan, *Nat. Chem. Biol.*, 2009, **5**, 647–654.
9. R. M. Franzini, D. Neri and J. Scheuermann, *Acc. Chem. Res.*, 2014, **47**, 1247–1255.
10. R. E. Kleiner, C. E. Dumelin and D. R. Liu, *Chem. Soc. Rev.*, 2011, **40**, 5707–5717.
11. R. A. Goodnow, *A Handbook for DNA-Encoded Chemistry: Theory and Applications for Exploring Chemical Space and Drug Discovery*, John Wiley & Sons, Inc., Hoboken, NJ, 1st edn, 2014.

12. H. Salamon, M. Klika Skopic, K. Jung, O. Bugain and A. Brunschweiger, *ACS Chem. Biol.*, 2016, **11**, 296–307.
13. S. Brenner and R. A. Lerner, *Proc. Natl. Acad. Sci. U. S. A.*, 1992, **89**, 5381–5383.
14. G. Gentile, G. Merlo, A. Pozzan, G. Bernasconi, B. Bax, P. Bamborough, A. Bridges, P. Carter, M. Neu, G. Yao, C. Brough, G. Cutler, A. Coffin and S. Belyanskaya, *Bioorg. Med. Chem. Lett.*, 2012, **22**, 1989–1994.
15. J. S. Disch, G. Evindar, C. H. Chiu, C. A. Blum, H. Dai, L. Jin, E. Schuman, K. E. Lind, S. L. Belyanskaya, J. Deng, F. Coppo, L. Aquilani, T. L. Graybill, J. W. Cuozzo, S. Lavu, C. Mao, G. P. Vlasuk and R. B. Perni, *J. Med. Chem.*, 2013, **56**, 3666–3679.
16. R. K. Thalji, J. J. McAtee, S. Belyanskaya, M. Brandt, G. D. Brown, M. H. Costell, Y. Ding, J. W. Dodson, S. H. Eisennagel, R. E. Fries, J. W. Gross, M. R. Harpel, D. A. Holt, D. I. Israel, L. J. Jolivette, D. Krosky, H. Li, Q. Lu, T. Mandichak, T. Roethke, C. G. Schnackenberg, B. Schwartz, L. M. Shewchuk, W. Xie, D. J. Behm, S. A. Douglas, A. L. Shaw and J. P. Marino, Jr., *Bioorg. Med. Chem. Lett.*, 2013, **23**, 3584–3588.
17. L. Encinas, H. O'Keefe, M. Neu, M. J. Remuinan, A. M. Patel, A. Guardia, C. P. Davie, N. Perez-Macias, H. Yang, M. A. Convery, J. A. Messer, E. Perez-Herran, P. A. Centrella, D. Alvarez-Gomez, M. A. Clark, S. Huss, G. K. O'Donovan, F. Ortega-Muro, W. McDowell, P. Castaneda, C. C. Arico-Muendel, S. Pajk, J. Rullas, I. Angulo-Barturen, E. Alvarez-Ruiz, A. Mendoza-Losana, L. Ballell Pages, J. Castro-Pichel and G. Evindar, *J. Med. Chem.*, 2014, **57**, 1276–1288.
18. H. Yang, P. F. Medeiros, K. Raha, P. Elkins, K. E. Lind, R. Lehr, N. D. Adams, J. L. Burgess, S. J. Schmidt, S. D. Knight, K. R. Auger, M. D. Schaber, G. J. Franklin, Y. Ding, J. L. DeLorey, P. A. Centrella, S. Mataruse, S. R. Skinner, M. A. Clark, J. W. Cuozzo and G. Evindar, *ACS Med. Chem. Lett.*, 2015, **6**, 531–536.
19. R. M. Franzini, T. Ekblad, N. Zhong, M. Wichert, W. Decurtins, A. Nauer, M. Zimmermann, F. Samain, J. Scheuermann, P. J. Brown, J. Hall, S. Graslund, H. Schuler and D. Neri, *Angew. Chem., Int. Ed. Engl.*, 2015, **54**, 3927–3931.
20. F. Buller, Y. Zhang, J. Scheuermann, J. Schafer, P. Buhlmann and D. Neri, *Chem. Biol.*, 2009, **16**, 1075–1086.
21. S. Melkko, L. Mannocci, C. E. Dumelin, A. Villa, R. Sommavilla, Y. Zhang, M. G. Grutter, N. Keller, L. Jermutus, R. H. Jackson, J. Scheuermann and D. Neri, *ChemMedChem*, 2010, **5**, 584–590.
22. C. S. Kollmann, X. Bai, C. H. Tsai, H. Yang, K. E. Lind, S. R. Skinner, Z. Zhu, D. I. Israel, J. W. Cuozzo, B. A. Morgan, K. Yuki, C. Xie, T. A. Springer, M. Shimaoka and G. Evindar, *Bioorg. Med. Chem.*, 2014, **22**, 2353–2365.
23. A. G. Gilmartin, T. H. Faitg, M. Richter, A. Groy, M. A. Seefeld, M. G. Darcy, X. Peng, K. Federowicz, J. Yang, S. Y. Zhang, E. Minthorn, J. P. Jaworski, M. Schaber, S. Martens, D. E. McNulty, R. H. Sinnamon,

H. Zhang, R. B. Kirkpatrick, N. Nevins, G. Cui, B. Pietrak, E. Diaz, A. Jones, M. Brandt, B. Schwartz, D. A. Heerding and R. Kumar, *Nat. Chem. Biol.*, 2014, **10**, 181–187.
24. S. L. Belyanskaya, Y. Ding, J. F. Callahan, A. L. Lazaar and D. I. Israel, *ChemBioChem*, 2017, **18**, 837–842.
25. D. P. Bondeson, A. Mares, I. E. Smith, E. Ko, S. Campos, A. H. Miah, K. E. Mulholland, N. Routly, D. L. Buckley, J. L. Gustafson, N. Zinn, P. Grandi, S. Shimamura, G. Bergamini, M. Faelth-Savitski, M. Bantscheff, C. Cox, D. A. Gordon, R. R. Willard, J. J. Flanagan, L. N. Casillas, B. J. Votta, W. den Besten, K. Famm, L. Kruidenier, P. S. Carter, J. D. Harling, I. Churcher and C. M. Crews, *Nat. Chem. Biol.*, 2015, **11**, 611–617.
26. P. A. Harris, B. W. King, D. Bandyopadhyay, S. B. Berger, N. Campobasso, C. A. Capriotti, J. A. Cox, L. Dare, X. Dong, J. N. Finger, L. C. Grady, S. J. Hoffman, J. U. Jeong, J. Kang, V. Kasparcova, A. S. Lakdawala, R. Lehr, D. E. McNulty, R. Nagilla, M. T. Ouellette, C. S. Pao, A. R. Rendina, M. C. Schaeffer, J. D. Summerfield, B. A. Swift, R. D. Totoritis, P. Ward, A. Zhang, D. Zhang, R. W. Marquis, J. Bertin and P. J. Gough, *J. Med. Chem.*, 2016, **59**, 2163–2178.
27. H. D. Lewis, J. Liddle, J. E. Coote, S. J. Atkinson, M. D. Barker, B. D. Bax, K. L. Bicker, R. P. Bingham, M. Campbell, Y. H. Chen, C. W. Chung, P. D. Craggs, R. P. Davis, D. Eberhard, G. Joberty, K. E. Lind, K. Locke, C. Maller, K. Martinod, C. Patten, O. Polyakova, C. E. Rise, M. Rudiger, R. J. Sheppard, D. J. Slade, P. Thomas, J. Thorpe, G. Yao, G. Drewes, D. D. Wagner, P. R. Thompson, R. K. Prinjha and D. M. Wilson, *Nat. Chem. Biol.*, 2015, **11**, 189–191.
28. C. C. Arico-Muendel, *MedChemComm*, 2016, **7**, 1898–1909.
29. A. L. Lazaar, L. Yang, R. L. Boardley, N. S. Goyal, J. Robertson, S. J. Baldwin, D. E. Newby, I. B. Wilkinson, R. Tal-Singer, R. J. Mayer and J. Cheriyan, *Br. J. Clin. Pharmacol.*, 2016, **81**, 971–979.
30. P. A. Harris, S. B. Berger, J. U. Jeong, R. Nagilla, D. Bandyopadhyay, N. Campobasso, C. A. Capriotti, J. A. Cox, L. Dare, X. Dong, P. M. Eidam, J. N. Finger, S. J. Hoffman, J. Kang, V. Kasparcova, B. W. King, R. Lehr, Y. Lan, L. K. Leister, J. D. Lich, T. T. MacDonald, N. A. Miller, M. T. Ouellette, C. S. Pao, A. Rahman, M. A. Reilly, A. R. Rendina, E. J. Rivera, M. C. Schaeffer, C. A. Sehon, R. R. Singhaus, H. H. Sun, B. A. Swift, R. D. Totoritis, A. Vossenkamper, P. Ward, D. D. Wisnoski, D. Zhang, R. W. Marquis, P. J. Gough and J. Bertin, *J. Med. Chem.*, 2017, **60**, 1247–1261.
31. G. Zimmermann, Y. Li, U. Rieder, M. Mattarella, D. Neri and J. Scheuermann, *ChemBioChem*, 2017, **18**, 853–857.
32. R. A. Goodnow, Jr., C. E. Dumelin and A. D. Keefe, *Nat. Rev. Drug Discovery*, 2017, **16**, 131–147.
33. L. H. Yuen and R. M. Franzini, *ChemBioChem*, 2017, **18**, 829–836.
34. M. A. Clark, *Curr. Opin. Chem. Biol.*, 2010, **14**, 396–403.
35. A. Litovchick, M. A. Clark and A. D. Keefe, *Artif. DNA PNA XNA*, 2014, **5**, e27896.

36. A. Litovchick, C. E. Dumelin, S. Habeshian, D. Gikunju, M. A. Guie, P. Centrella, Y. Zhang, E. A. Sigel, J. W. Cuozzo, A. D. Keefe and M. A. Clark, *Sci. Rep.*, 2015, **5**, 10916.
37. S. Ahn, A. W. Kahsai, B. Pani, Q. T. Wang, S. Zhao, A. L. Wall, R. T. Strachan, D. P. Staus, L. M. Wingler, L. D. Sun, J. Sinnaeve, M. Choi, T. Cho, T. T. Xu, G. M. Hansen, M. B. Burnett, J. E. Lamerdin, D. L. Bassoni, B. J. Gavino, G. Husemoen, E. K. Olsen, T. Franch, S. Costanzi, X. Chen and R. J. Lefkowitz, *Proc. Natl. Acad. Sci. U. S. A.*, 2017, **114**, 1708–1713.
38. Z. J. Gartner and D. R. Liu, *J. Am. Chem. Soc.*, 2001, **123**, 6961–6963.
39. Z. J. Gartner, B. N. Tse, R. Grubina, J. B. Doyon, T. M. Snyder and D. R. Liu, *Science*, 2004, **305**, 1601–1605.
40. A. I. Chan, L. M. McGregor and D. R. Liu, *Curr. Opin. Chem. Biol.*, 2015, **26**, 55–61.
41. D. R. Halpin and P. B. Harbury, *PLoS Biol.*, 2004, **2**, E173.
42. D. R. Halpin and P. B. Harbury, *PLoS Biol.*, 2004, **2**, E174.
43. D. R. Halpin, J. A. Lee, S. J. Wrenn and P. B. Harbury, *PLoS Biol.*, 2004, **2**, E175.
44. S. J. Wrenn, R. M. Weisinger, D. R. Halpin and P. B. Harbury, *J. Am. Chem. Soc.*, 2007, **129**, 13137–13143.
45. J. Scheuermann and D. Neri, *Curr. Opin. Chem. Biol.*, 2015, **26**, 99–103.
46. M. H. Hansen, P. Blakskjaer, L. K. Petersen, T. H. Hansen, J. W. Hojfeldt, K. V. Gothelf and N. J. Hansen, *J. Am. Chem. Soc.*, 2009, **131**, 1322–1327.
47. T. Gura, *Science*, 2015, **350**, 1139–1140.
48. A. Mullard, *Nature*, 2016, **530**, 367–369.
49. S. D. Roughley and A. M. Jordan, *J. Med. Chem.*, 2011, **54**, 3451–3479.
50. Y. Ding and M. A. Clark, *ACS Comb. Sci.*, 2015, **17**, 1–4.
51. Y. Ding, G. J. Franklin, J. L. DeLorey, P. A. Centrella, S. Mataruse, M. A. Clark, S. R. Skinner and S. Belyanskaya, *ACS Comb. Sci.*, 2016, **18**, 625–629.
52. Y. Ding, J. L. DeLorey and M. A. Clark, *Bioconjugate Chem.*, 2016, **27**, 2597–2600.
53. A. L. Satz, J. Cai, Y. Chen, R. Goodnow, F. Gruber, A. Kowalczyk, A. Petersen, G. Naderi-Oboodi, L. Orzechowski and Q. Strebel, *Bioconjugate Chem.*, 2015, **26**, 1623–1632.
54. H. Deng, H. O'Keefe, C. P. Davie, K. E. Lind, R. A. Acharya, G. J. Franklin, J. Larkin, R. Matico, M. Neeb, M. M. Thompson, T. Lohr, J. W. Gross, P. A. Centrella, G. K. O'Donovan, K. L. Bedard, K. van Vloten, S. Mataruse, S. R. Skinner, S. L. Belyanskaya, T. Y. Carpenter, T. W. Shearer, M. A. Clark, J. W. Cuozzo, C. C. Arico-Muendel and B. A. Morgan, *J. Med. Chem.*, 2012, **55**, 7061–7079.
55. Z. Wu, T. L. Graybill, X. Zeng, M. Platchek, J. Zhang, V. Q. Bodmer, D. D. Wisnoski, J. Deng, F. T. Coppo, G. Yao, A. Tamburino, G. Scavello, G. J. Franklin, S. Mataruse, K. L. Bedard, Y. Ding, J. Chai, J. Summerfield, P. A. Centrella, J. A. Messer, A. J. Pope and D. I. Israel, *ACS Comb. Sci.*, 2015, **17**, 722–731.

56. C. A. Machutta, C. S. Kollmann, K. E. Lind, X. Bai, P. F. Chan, J. Huang, L. Ballell, S. Belyanskaya, G. S. Besra, D. Barros-Aguirre, R. H. Bates, P. A. Centrella, S. S. Chang, J. Chai, A. E. Choudhry, A. Coffin, C. P. Davie, H. Deng, J. Deng, Y. Ding, J. W. Dodson, D. T. Fosbenner, E. N. Gao, T. L. Graham, T. L. Graybill, K. Ingraham, W. P. Johnson, B. W. King, C. R. Kwiatkowski, J. Lelièvre, Y. Li, X. Liu, Q. Lu, R. Lehr, A. Mendoza-Losana, J. Martin, L. McCloskey, P. McCormick, H. P. O'Keefe, T. O'Keeffe, C. Pao, C. B. Phelps, H. Qi, K. Rafferty, G. S. Scavello, M. S. Steiginga, F. S. Sundersingh, S. M. Sweitzer, L. M. Szewczuk, A. Taylor, M. F. Toh, J. Wang, M. Wang, D. J. Wilkins, B. Xia, G. Yao, J. Zhang, J. Zhou, C. P. Donahue, J. A. Messer, D. Holmes, C. C. Arico-Muendel, A. J. Pope, J. W. Gross and G. Evindar, *Nat. Commun.*, 2017, **8**, 16081.
57. C. Arico-Muendel, Z. Zhu, H. Dickson, D. Parks, J. Keicher, J. Deng, L. Aquilani, F. Coppo, T. Graybill, K. Lind, A. Peat and M. Thomson, *Antimicrob. Agents Chemother.*, 2015, **59**, 3450–3459.
58. H. Deng, J. Zhou, F. S. Sundersingh, J. Summerfield, D. Somers, J. A. Messer, A. L. Satz, N. Ancellin, C. C. Arico-Muendel, K. L. Sargent Bedard, A. Beljean, S. L. Belyanskaya, R. Bingham, S. E. Smith, E. Boursier, P. Carter, P. A. Centrella, M. A. Clark, C. W. Chung, C. P. Davie, J. L. Delorey, Y. Ding, G. J. Franklin, L. C. Grady, K. Herry, C. Hobbs, C. S. Kollmann, B. A. Morgan, L. J. Pothier Kaushansky and Q. Zhou, *ACS Med. Chem. Lett.*, 2015, **6**, 919–924.
59. H. Deng, J. Zhou, F. Sundersingh, J. A. Messer, D. O. Somers, M. Ajakane, C. C. Arico-Muendel, A. Beljean, S. L. Belyanskaya, R. Bingham, E. Blazensky, A. B. Boullay, E. Boursier, J. Chai, P. Carter, C. W. Chung, A. Daugan, Y. Ding, K. Herry, C. Hobbs, E. Humphries, C. Kollmann, V. L. Nguyen, E. Nicodeme, S. E. Smith, N. Dodic and N. Ancellin, *ACS Med. Chem. Lett.*, 2016, **7**, 379–384.
60. E. R. Wood, R. Bledsoe, J. Chai, P. Daka, H. Deng, Y. Ding, S. Harris-Gurley, L. H. Kryn, E. Nartey, J. Nichols, R. T. Nolte, N. Prabhu, C. Rise, T. Sheahan, J. B. Shotwell, D. Smith, V. Tai, J. D. Taylor, G. Tomberlin, L. Wang, B. Wisely, S. You, B. Xia and H. Dickson, *J. Biol. Chem.*, 2015, **290**, 19681–19696.
61. Z. Zhu, A. Shaginian, L. C. Grady, T. O'Keeffe, X. E. Shi, C. P. Davie, G. L. Simpson, J. A. Messer, G. Evindar, R. N. Bream, P. P. Thansandote, N. R. Prentice, A. M. Mason and S. Pal, *ACS Chemical Biology*, 2018, **13**, 53–59.
62. B. Thomas, X. Lu, W. R. Birmingham, K. Huang, P. Both, J. E. Reyes Martinez, R. J. Young, C. P. Davie and S. L. Flitsch, *ChemBioChem*, 2017, **18**, 858–863.
63. L. Fan and C. P. Davie, *ChemBioChem*, 2017, **18**, 843–847.
64. R. M. Franzini and C. Randolph, *J. Med. Chem.*, 2016, **59**, 6629–6644.
65. M. Hamzeh-Mivehroud, A. A. Alizadeh, M. B. Morris, W. B. Church and S. Dastmalchi, *Drug Discovery Today*, 2013, **18**, 1144–1157.
66. R. Stoltenburg, C. Reinemann and B. Strehlitz, *Biomol. Eng.*, 2007, **24**, 381–403.

67. W. Decurtins, M. Wichert, R. M. Franzini, F. Buller, M. A. Stravs, Y. Zhang, D. Neri and J. Scheuermann, *Nat. Protoc.*, 2016, **11**, 764–780.
68. Y. Ding, H. O'Keefe, J. L. DeLorey, D. I. Israel, J. A. Messer, C. H. Chiu, S. R. Skinner, R. E. Matico, M. F. Murray-Thompson, F. Li, M. A. Clark, J. W. Cuozzo, C. Arico-Muendel and B. A. Morgan, *ACS Med. Chem. Lett.*, 2015, **6**, 888–893.
69. J. W. Cuozzo, P. A. Centrella, D. Gikunju, S. Habeshian, C. D. Hupp, A. D. Keefe, E. A. Sigel, H. H. Soutter, H. A. Thomson, Y. Zhang and M. A. Clark, *ChemBioChem*, 2017, **18**, 864–871.
70. P. Mandal, S. B. Berger, S. Pillay, K. Moriwaki, C. Huang, H. Guo, J. D. Lich, J. Finger, V. Kasparcova, B. Votta, M. Ouellette, B. W. King, D. Wisnoski, A. S. Lakdawala, M. P. DeMartino, L. N. Casillas, P. A. Haile, C. A. Sehon, R. W. Marquis, J. Upton, L. P. Daley-Bauer, L. Roback, N. Ramia, C. M. Dovey, J. E. Carette, F. K. Chan, J. Bertin, P. J. Gough, E. S. Mocarski and W. J. Kaiser, *Mol. Cell*, 2014, **56**, 481–495.
71. R. M. Franzini, A. Nauer, J. Scheuermann and D. Neri, *Chem. Commun.*, 2015, **51**, 8014–8016.
72. F. Buller, M. Steiner, K. Frey, D. Mircsof, J. Scheuermann, M. Kalisch, P. Buhlmann, C. T. Supuran and D. Neri, *ACS Chem. Biol.*, 2011, **6**, 336–344.
73. Y. Li, G. Zimmermann, J. Scheuermann and D. Neri, *ChemBioChem*, 2017, **18**, 848–852.
74. M. L. Metzker, *Nat. Rev. Genet.*, 2010, **11**, 31–46.
75. J. Ottl, in *A Handbook for DNA-Encoded Chemistry*, John Wiley & Sons, Inc., 2014, DOI: 10.1002/9781118832738.ch14, pp. 319–347.
76. W. H. Connors, S. P. Hale and N. K. Terrett, *Curr. Opin. Chem. Biol.*, 2015, **26**, 42–47.
77. J. A. Feinberg and Z. Peng, in *A Handbook for DNA-Encoded Chemistry*, John Wiley & Sons, Inc., 2014, DOI: 10.1002/9781118832738.ch9, pp. 201–212.
78. X. Lu, L. Fan, C. B. Phelps, C. P. Davie and C. P. Donahue, *Bioconjugate Chem.*, 2017, **28**, 1625–1629.
79. G. R. Markus, *BIOspektrum*, 2017, **23**, 142–145.
80. A. L. Satz, R. Hochstrasser and A. C. Petersen, *ACS Comb. Sci.*, 2017, **19**, 234–238.
81. Y. Ding, G. J. Franklin, J. L. DeLorey, P. A. Centrella, S. Mataruse, M. A. Clark, S. R. Skinner and S. Belyanskaya, *ACS Comb. Sci.*, 2016, **18**, 625–629.
82. X. Tian, G. S. Basarab, N. Selmi, T. Kogej, Y. Zhang, M. Clark and R. A. Goodnow Jr, *MedChemComm*, 2016, **7**, 1316–1322.
83. M. L. Malone and B. M. Paegel, *ACS Comb. Sci.*, 2016, **18**, 182–187.
84. A. D. Keefe, M. A. Clark, C. D. Hupp, A. Litovchick and Y. Zhang, *Curr. Opin. Chem. Biol.*, 2015, **26**, 80–88.
85. L. Kuai, T. O'Keeffe and C. Arico-Muendel, *SLAS Discovery*, 2018, **23**, 405–416.
86. A. L. Satz, *ACS Chem. Biol.*, 2015, **10**, 2237–2245.

CHAPTER 8

Engineering Chemistry to Enable Bioactive Small Molecule Discovery

STEVEN V. LEY,* DANIEL FITZPATRICK AND CLAUDIO BATTILOCCHIO

Department of Chemistry, University of Cambridge, Lensfield Road, Cambridge CB2 1EW, UK
*Email: svl1000@cam.ac.uk

8.1 Preamble

The assembly of small molecules that possess important biological functions[1] can be seen either as a craft or an art. At its most basic level, simple, designed molecular architectures may be delivered using established methods based on current mechanistic chemistry knowledge. At its most elegant, synthesis is a true form of art, requiring a complex set of skills and the ability to challenge dogmas in a highly innovative and creative fashion.[2]

There is, however, a need to go beyond where we are today to achieve a more predictive understanding of synthesis outcomes in a more sustainable manner.[3] It is not easy to provide universal tools to automate the components of a synthesis process. This is particularly true when trying to automate tasks more commonly associated with human behaviour, such as co-ordinating feedback from an experimental observation *via* the brain into a subsequent manipulative operation at the bench. Furthermore, biologically-active substances rarely are prepared in a single step and it is

Chemical Biology No. 10
Chemical and Biological Synthesis: Enabling Approaches for Understanding Biology
Edited by Nick J. Westwood and Adam Nelson

Published by the Royal Society of Chemistry, www.rsc.org

not uncommon that over ten synthetic steps are required (and often many more during complex natural product syntheses). Even an apparently straightforward process can require extensive optimisation to produce pure product with an acceptable yield. Often forgotten too is the need for multiple downstream operations such as chromatography, pH adjustment, crystallization, distillation and water washes, which add to this technologically challenging scenario. In summary, synthesis is a not trivial activity, and the discovery of a robust route is rarely perfect first time round.[4]

These labour-intensive activities suggests a need for change in the practice of organic molecule synthesis, particularly if the goal is pragmatic and focussed on the biological outcomes rather than demonstrating the prowess of the particular synthetic approach. Here, automation has an important role to play: not to trivialise the science, but to expedite the development of practical syntheses. In most cases, it is wise to adopt a simplified machine-assisted approach to synthesis in the early phase of discovery as opposed to using over-engineered expert-user systems that are better suited to scale-up. It is tempting, however, to seek out "stand-alone" equipment that performs certain isolated, repetitive tasks such as those employed in the automated synthesis of oligomers such as peptides and carbohydrates (see Chapters 9 and 10). However, being able to use the full armoury of synthesis is an altogether more demanding task and necessitates a more flexible modular approach, which harnesses all the different practical elements in a co-ordinated holistic systems fashion to molecular assembly.[5]

In this chapter, we will not attempt to cover the whole field of automated synthesis, since this has been ongoing for many years and has been discussed in many reviews;[6] particularly during the era of combinatorial chemistry where the main focus was on "on-bead" synthesis methods.[7] It is clear that the area has changed significantly from those early days, whereby large numbers of simple molecules were made using only robust and straightforward methods employing rudimentary equipment. Today, there is a much greater emphasis on up-front *in silico* molecular design and there are many new tools and much specialist equipment becoming available to accommodate much higher levels of sophistication and molecular complexity. While speed is still a key driver of discovery, other considerations feature more prominently[8] such as safety, improved audit trails, machine learning and big-data management. The integration of classical batch methods with the development of flow chemistry leads to improved reaction engineering and the discovery of new chemical reactivity.[9] This integration inevitably leads to greater monitoring and control, thus achieving overall better molecular function by design.

It seems appropriate to begin, therefore, with a discussion of some general enabling tools for automated synthesis (Section 8.2), prior to commenting on their engagement in more challenging situations leading to multi-step chemical synthesis processes (Sections 8.3, 8.4 and 8.5).

8.2 Technologies for Machine-assisted Synthesis

8.2.1 Machine Vision

Crucial to many machine-assisted synthesis programs is the need to provide a 'sense' of sight to the monitoring and controlling equipment. This topic forms the basis of an extensive review on the use of camera-enabled technologies for organic synthesis".[10] Consequently, here we restrict our discussion to a few examples that have relevance to monitoring and autonomous feedback control using inexpensive webcams and open-source technologies.

An early example that has importance to an everyday routine synthesis operation involved the automation of liquid–liquid extraction[11] as a downstream process, where a prototype device rapidly mixed organic and aqueous phases which were then pumped into a glass separation chamber where gravity effected the separation of each phase. The position of a small coloured plastic float, which was installed to sit at the interface of the two phases, was monitored using an inexpensive webcam connected to a microprocessor chip using Python open Computer Vision (CV) software. The rate of removal of the lighter upper layer could then be adjusted readily in order to maintain a steady-state volume of the heavy phase, preventing the over- or under-flow of one phase into the outlet of the other.

By way of a simple application of this system (Scheme 8.1), hydrazones were formed from aldehydes and phenylhydrazine in the presence of pyridinium toluenesulfonate, which produced a product stream that was quenched by addition of phosphoric acid. The output was rapidly mixed by an in-line magnetic stirrer, for effective extraction, and subsequently progressed through the webcam-monitored separator for long periods of time at steady state (up to 24 hours). More advanced computer and webcam arrangements and multiple extractor systems are also possible and have been particularly useful when product partition coefficients can cause extraction difficulties (with amino acids, for example).[12]

The use of focussed microwave equipment is commonly employed in organic synthesis where high reaction temperatures and pressures are involved. While it is easy to record reaction times, temperatures and pressures with in-built sensors, some aspects (*e.g.* colour changes, stirring, precipitation and short-circuiting events through arcing) are beyond direct observation, yet can be crucial to a successful reaction outcome. However, by illuminating the reaction cavity using an LED, and placing a small digital camera with access through a port, good images of the reaction in real-time can provide the necessary additional data to further improve safety and reaction knowledge through a video recording.[13]

Although increasing camera monitoring of chemical processes, including thermal imagery and high-speed digital technologies, are with us today, many current applications relate only to observation. We can anticipate in the future, especially with face recognition software and modern communication technology, that much great use of the information to *control* events can be expected.

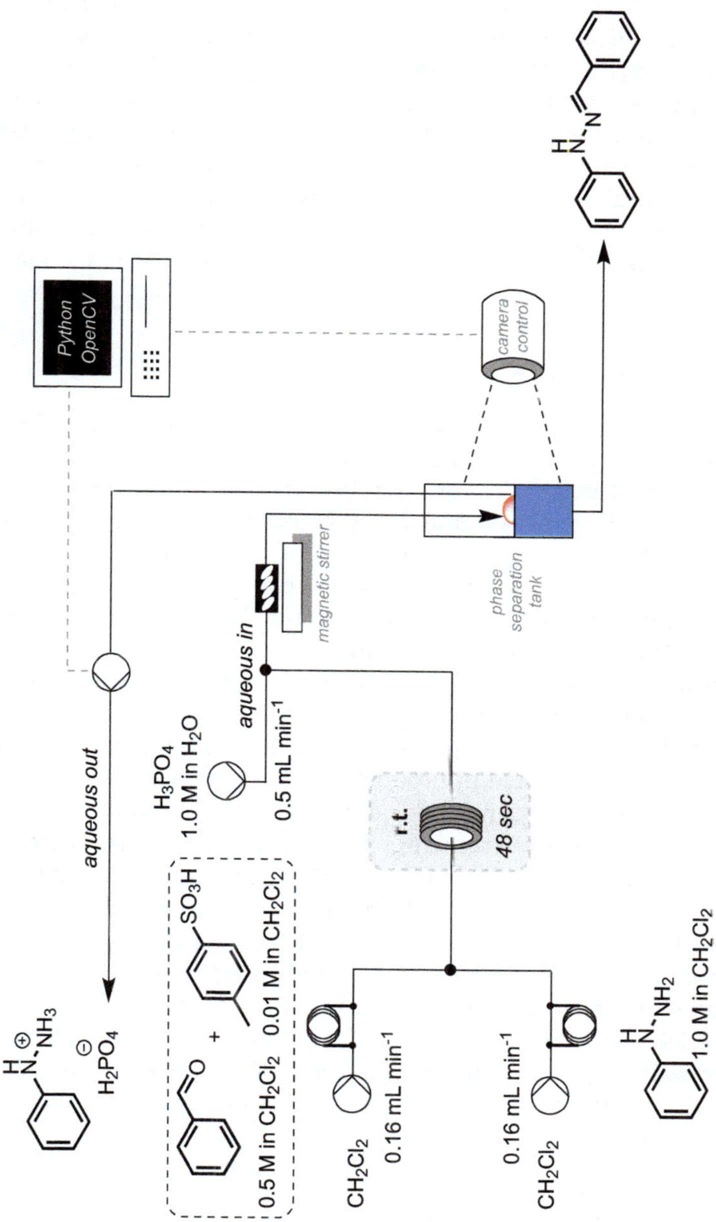

Scheme 8.1 Work-up monitoring during hydrazine preparation.

8.2.2 Process Analytical Technology (PAT)

During any synthesis, in batch or flow, whether single or multiple steps, early intervention to determine structure and product purity facilitates autonomous operation by generating usable analytical data. These data points can indicate real-time success (or failure) of a process, and provide safety information, kinetic understanding and aid optimisation. Indeed, high quality process analytical technology (PAT) is at the heart of any machine-assisted synthesis, off-line, on-line or in-line, depending upon specific needs. While space in this chapter precludes an exhaustive survey of the area,[14] it is useful to be aware of the best methods available to assist decision-making in route planning and during the synthesis process. We focus on applications in discovery research rather than larger-scale applications.

8.2.2.1 Mass Spectrometry

The field of mass spectrometry for in-lab experimental monitoring has been greatly assisted in recent years by the availability of *compact* mass spectrometer devices. Two publications that best represent these improvements, particularly to on-line measurement, are highlighted here. Firstly, a recent report described rapid adaptive optimization using an automated 3 ml coil flow reactor,[15] for the self-optimization of the amidation of nicotinate methyl ester with methylamine to give the corresponding amide. Here Design of Experiment (DoE) software and the use of Stable Noisy Optimization by Branch and Fit (SNOBFIT) self-optimizing algorithm were employed (Scheme 8.2). In this example, the signal from the small mass spectrometer was calibrated using linear relative response values to the HPLC trace. At the end of the cycle, a 93% yield of product could be realized.

In a second example, a continuous-flow reaction was monitored using an in-line miniature mass spectrometer. Of particular interest was that, during a study of a classical benzyne reaction, involving generation *via* diazotisation of anthranilic acid and trapping with furan, a range of primary by-products was detected by the mass spectrometer that had not been observed previously.[16] Moreover, it was also possible to rapidly determine optimum reactor temperatures and residence times for the formulation of the main Diels–Alder product (Scheme 8.3). This study highlights how the mass spectrometer rapidly captures a useful snapshot of the process and provides useful data that aids greater reaction understanding and, eventually, facilitates automation.

8.2.2.2 Infrared Spectroscopy

One of the most useful analytical tools for monitoring the dynamics of product formation and yielding crucial structural information is infrared spectroscopy. While the technique and equipment have been available for

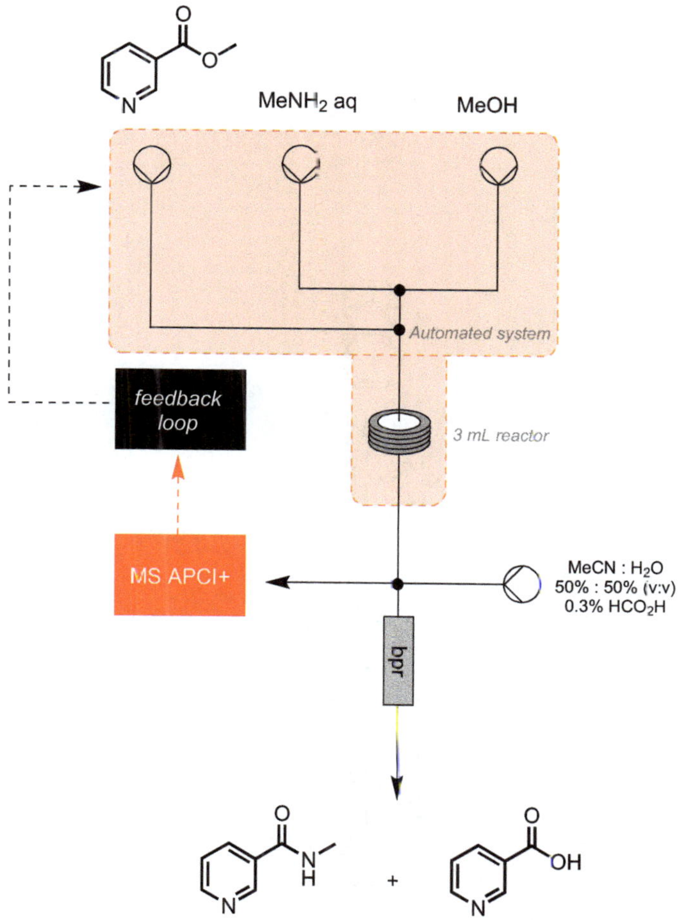

Scheme 8.2 Mass spectrometry monitoring of a self-optimization process.

some time, using fibre-optic connection to attenuated total reflectance IR devices, it has only fairly recently become a truly practical option with the invention of new flow-through ReactIR cells. These small units enable real-time information capture to monitor reagent consumption and product formation to aid rapid reaction optimization *via* data feedback.

Short-lived reaction intermediates can be observed to provide further mechanistic insight. An important paper describing a number of analytical opportunities reports the use of a Fourier transform infra-red (FT-IR) diamond flow head probe to directly monitor batch reactions or interrogate in-line flow processes.[17] Improvements to this technology have resulted in a miniaturized version capable of measuring sub-millimolar concentrations that is highly portable and can be integrated into a multitude of reaction set-ups, most notably when early structural information is of crucial importance, for example in the generation and use of Grignard reagents.[18]

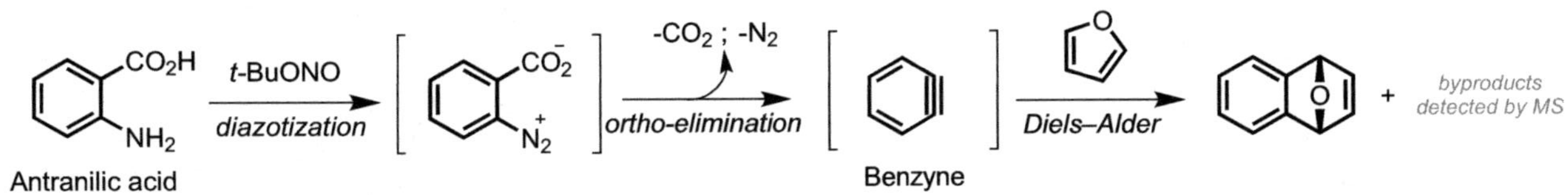

Scheme 8.3 Secondary by-product detection.

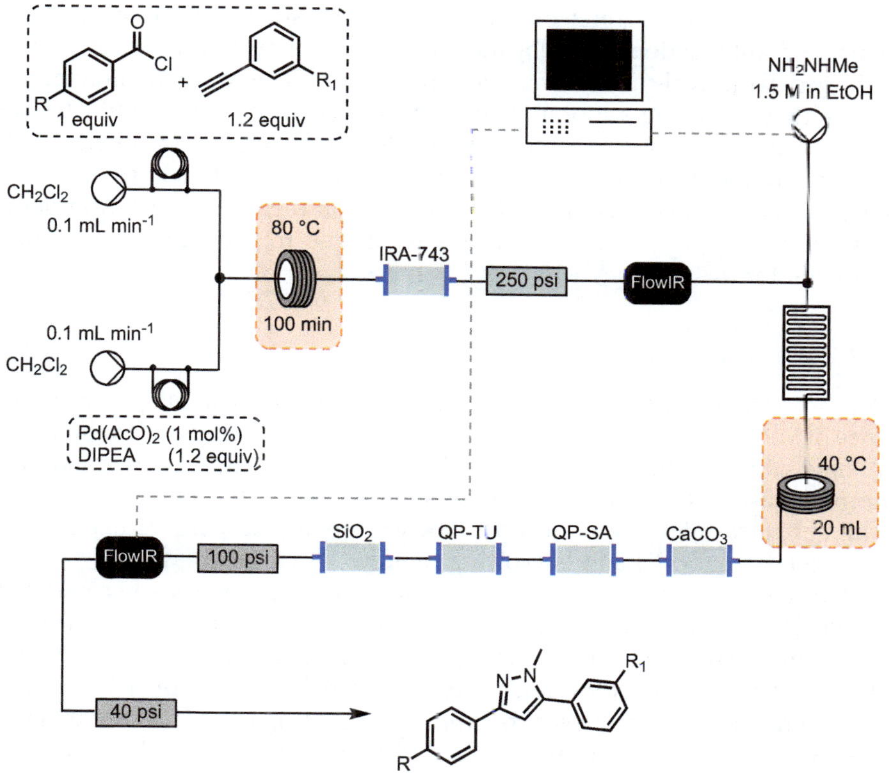

Scheme 8.4 IR monitoring during a pyrazole synthesis.

Using these in-line infrared monitoring devices with a new LabVIEW software application, it was possible to feed back information to accurately control additional pumps in multi-step segmented flow syntheses. In this way, precise mixing and perfect reaction timing greatly improved product quality and minimized the use of chemical feedstocks.[19]

A sophisticated scheme, where the use of this in-line monitoring and control of a multi-step synthetic sequence using both diamond and silicon flow head devices led to full automation and the preparation of a pyrazole library (Scheme 8.4). Notably, the combined use of reagent and scavenger cartridges guarantees the product purity of the flowing reaction stream. Further discussion of these solid-supported materials is included in Section 8.4.

8.2.2.3 Raman Spectroscopy

Analysis of solid-state parameters, for example during crystallization, benefits from the use of non-invasive in-line Raman spectroscopy. In two reviews, much of the background and recent advances have been well documented.[20,21] One of the most useful articles, however, and particularly appropriate to this chapter on automated synthesis, reports four mechanically relevant reactions whereby mesoscale continuous-flow monitoring was achieved with real-time Raman spectroscopy.[22] The work demonstrated nicely that, by using calibration curves, product conversion could be accurately determined. These data were also corroborated by using NMR spectroscopy as a complementary analytical technique. While it is fairly early days for Raman methods to be widely adopted in the research environment, it should be regarded as an additive technique to the suite of methods available for PAT.

8.2.2.4 NMR Spectroscopy

The primary technique of all synthetic chemists–nuclear magnetic resonance (NMR) spectroscopy–is fundamental to product analysis and validation. To date, owing to the physical size and cost of equipment, these methods are normally used as off-line techniques. However, through innovation and the commercial availability of low-field bench-top devices, the situation has changed dramatically and is evolving rapidly.[23,24]

One useful advance towards on-line detection and analysis necessitated the design of a new flow cell placed in the coil region of the spectrometer.[25] This could then be used in conjunction with a reaction sampling, splitter and HPLC system to produce a viable process-development platform (Scheme 8.5).

In an application of a new benchtop 43 MHz compact permanent magnet spectrometer, self-optimization *via* LabView software provided useful data feedback to allow advanced spectral analysis in real-time reaction situations, thereby expediting discovery.[26] Other studies recording ultra-fast 2D NMR spectra to overcome issues of peak overlap on low-field devices have been reported[27] and applied to a synthetically significant Heck–Matsuda process.

Finally, benchtop NMR has been employed as an *in situ* reaction monitoring unit featuring a hypervalent iodine(III)-mediated cyclopropanation reaction to prepare nine different analogues.[28] An important aspect of this work was the development of an in-line solvent switching system that allowed deuterated solvent to be introduced prior to entry into the spectrometer. Again, in-line work-up employed scavenging cartridges to provide clean product streams (Scheme 8.6).

8.2.2.5 UV/Vis Spectroscopy

Although UV monitoring of chemical reactions was one of the most popular in-line techniques and is still useful for dilute product detection, its use as a mainline analysis method has faded, since it provides little in the way of direct structural information. Nevertheless, it does have a role to play in, for example, high-pressure reaction studies where on-line monitoring of potentially hazardous processes can be an issue during product sampling.[29] Using microreactors at up to 600 bar, reactions involving a high pressure nucleophilic aromatic substitution (S_NAr) displacement and second an aza

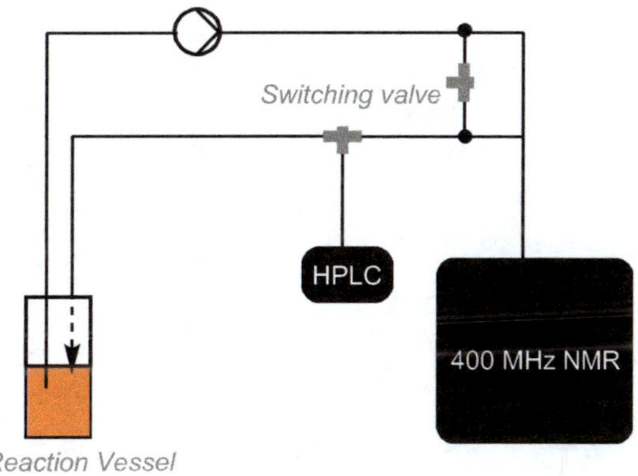

Scheme 8.5 Flow NMR detection.

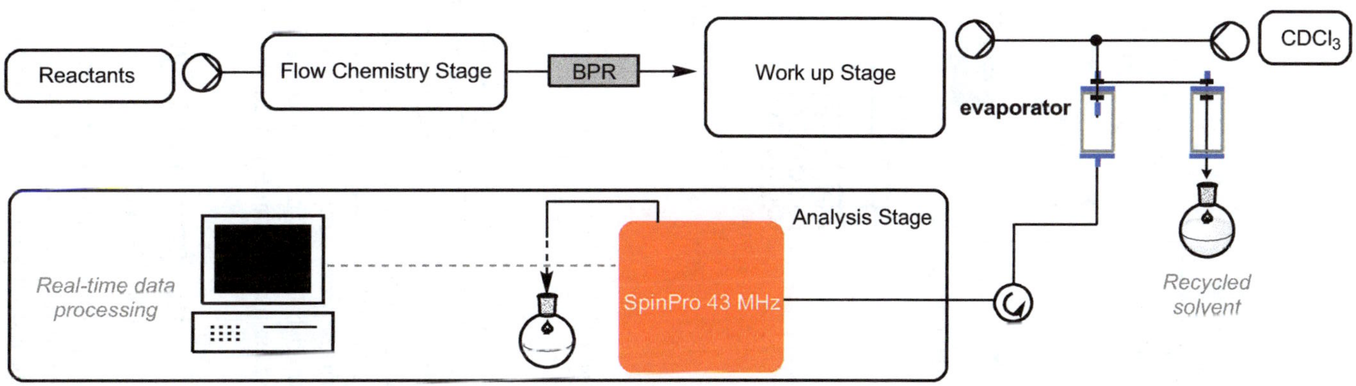

Scheme 8.6 Benchtop NMR monitoring.

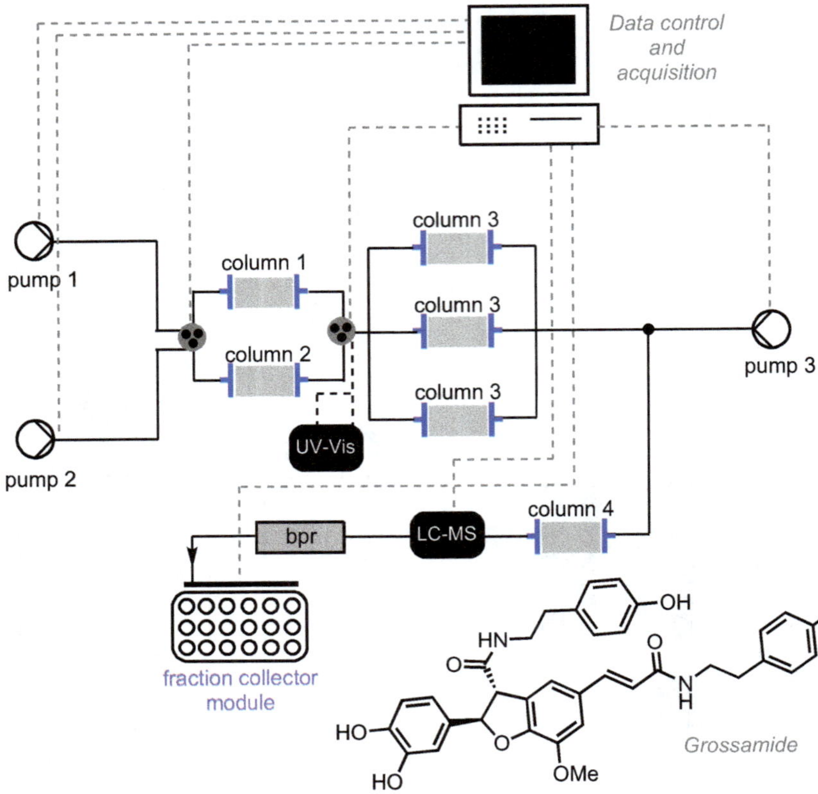

Scheme 8.7 UV/Vis monitoring has been used during the preparation of Grossamide.

Diels–Alder reaction have been reported using UV/Vis detection. Also, a simple in-line UV/Vis technique proved useful to trigger automated fraction collection during a modular machine-assisted synthesis of the natural product grossamide and some analogue precursors[30] (Scheme 8.7). Here the UV/Vis detector was linked with a computer to control valve switching to various processing cartridges and eventual final sample collection. As new, low-cost UV/Vis units become commercially available, we can anticipate greater usage in automated reactor assemblies.

8.3 Integration of Synthesis with Biological Evaluation

The primary purpose of small-molecule design and synthesis is often to determine biological activity. It makes sense therefore to capture this function as soon as possible after initial synthesis, especially if the information is required to feed back into the compound design cycle to improve or alter its activity. Not presented here, however, is emerging technology that relies on microfluidic systems[31] or droplet manipulation methods to link synthesis to screening, as this is discussed elsewhere. Instead, we will focus on the production of clean products arising from batch- or flow-chemistry methods and their automated dilution and diversion to a suitable screening platform.

In an interesting early example of a flow-chemistry synthesis of zolpidem, alpidem and other gamma-aminobutyric acid receptor ($GABA_A$) agonists, biological evaluation was achieved through coupling of an integrated in-line application of a frontal affinity chromatography (FAC) technique.[32] The flow synthesis scheme involves multi-step condensation coupling processes to produce the corresponding library of imidazo[1,2-a]pyridines (in total, 22 analogues), and combines in-line reaction coils and product purification scavenger cartridges, further examples of which are described later in this chapter. The final products are transferred in appropriate autosamplers and fraction collectors and, through a Uniqsis FlowSyn device, directly into the FAC assay following suitable dilution. In this case, binding to human serum albumin (HSA) was evaluated and compared with isoniazid and sodium diclofenac to provide a ranking order of activity (Scheme 8.8).

In a second application searching for modulators of the bromodomain histone reader BRD9, a small FAC cartridge containing the immobilized protein was constructed. This device was evaluated using flow- and microwave-synthesized bromosporine analogues for their affinity to the protein, which was measured accurately using a new compact in-line mass spectrometer.[33] In this technique, the products are flowed through a column containing the immobilized target biomolecule and retained according to their affinity for the target.

An attractive closed-loop approach has been pioneered by researchers at Cyclofluidic. Here, reagents are selected for participation *via* a microfluidic synthesis, which includes integrated purification and analysis, without manual intervention; the products then continue onto a reformatting platform for an on-line biological assay. Generated data on the concentration giving 50% of maximum inhibition (IC_{50}) is then entered into the compound design algorithm which implements the next set of reagent choices to repeat the cycle of discovery.[34] In this way, with just 21 compounds, the fully automated process identified a novel template and hinge binding motif against the ABL kinase. This machine-learning approach has the potential to significantly reduce cycle times in drug-discovery programmes. The versatility of the concept was further demonstrated with integrated synthesis and testing of xanthine dipeptidyl peptidase DPP4 derivatives. Interestingly, each design and assay loop was complete within a two hour cycle, generating useful structure–activity data.[35]

Also of note is the use of discrete enabling technologies to design a 20-membered 1-aryl-4-aminopiperidine library using computational methods which was then synthesized using high-throughput methodology (in flow or batch), purified and screened.[36] Recently reported was a sophisticated integrated platform for the rapid synthesis, purification and

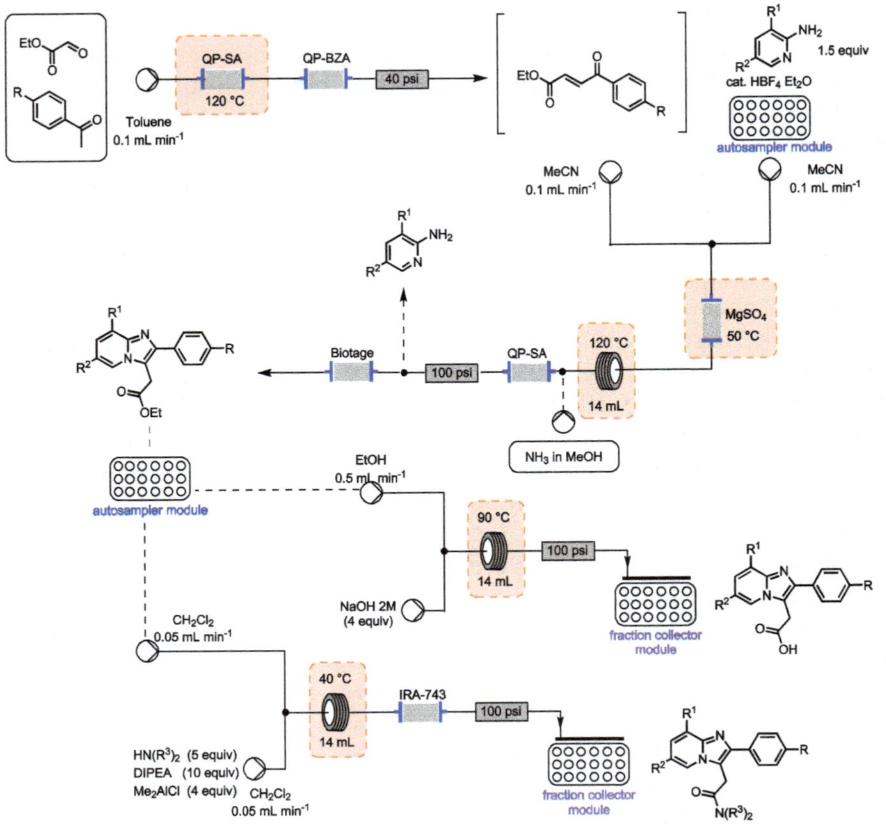

Scheme 8.8 Synthesis of imidazo[1,2-a]pyridines prior to biological evaluation using a frontal affinity chromatography (FAC) system.

testing of small-molecule collections. This work amply demonstrates the complexity that arises with the use of machine-based approaches even when using relatively well-established chemistries. Nevertheless, this integrated approach for two libraries (containing 22 and 33 members) was complete within 15 and 30 hours, respectively, compared with more conventional techniques that required up to a 10-day turnaround.[37]

8.4 Technologies for the Automated Synthesis of Bioactive Compounds

As explained in Section 8.1, the automation of synthesis is non-trivial. Complexity arises at all stages and there is no one solution to the multiplicity of problems that accrue during the preparation of a typical compound. Fortunately, there are approaches that can help, such immobilization[38,39] (Section 8.4.1) and flow chemistry (Section 8.4.2).

8.4.1 Immobilized Systems

While synthesis on beads, in which immobilization occurs through chemical attachment to a functionalized polymer support, is attractive for simple, repetitive and iterative transformations, such as automated peptide synthesis, it is not suitable in more challenging scenarios. Consequently, it makes more sense to immobilize the reagents and the chemical scavenging or quenching agents since it is these, and their decomposition residues, that cause so many separation and purification issues, both upstream and downstream, which are difficult to address through automation.[40]

In a detailed technical review, we have discussed the concept of using these polymer-supported reagents in multi-step organic synthesis programmes.[41] At its most basic level, one can envisage carrying out a chemical transformation in a reaction vessel by adding an appropriate reagent supported on an inert material, most commonly polystyrene beads. Once the desired reaction is complete, the spent reagent is removed by filtration to leave a clean solution containing only pure product that can then be used in the next chemical step (either in the same solvent or after a solvent switch); again, the addition of a suitable polymer-supported reagent enables the next desired transformation. The sequence is repeated until the full synthesis has been achieved, and spent reagents can be easily recovered and reused. Furthermore, convergent syntheses are also possible. Also, to facilitate robotic handling (and therefore automation), the polymer-immobilized reagents can be deployed in porous pouches as a form of containment ("Teabags"). Therefore, overall workload is greatly minimized and the technically-demanding manual steps of a synthesis, such as chromatography, crystallization, distillation, quenching and phase separation, are all avoided.

In reality, in all multi-step syntheses, the situation is more complex than indicated above and often requires further conceptual adjustments. Commonly in a synthetic transformation, especially where components are being coupled to form a new carbon–carbon bond, immobilized reagents can be used, but by-products are formed or there is incomplete consumption of starting materials. This situation leads to work-up difficulties, often causing an increase in the number of downstream unit operations. Here, the problem may be solved by adding polymer-supported scavenging or quenching agents, to specifically target the impurities, which may also be removed by filtration, leading to pure product solutions for the next synthesis step.

Other useful techniques using immobilized systems can also be employed to aid product work-up such as "catch and release" protocols. Here, the product of a reaction is selectively trapped using a functionally-matched polymeric bead to "catch" the material of value, while the impurities and by-products are washed away on filtration. The filtered-trapped product can then be induced to "release" its cargo, often by an acid/base exchange process, to generate a pure product in solution that is suitable for the next step in the assembly process. Rather than discuss a long compilation of schemes that individually showcase particular aspects of these immobilization methods, we feel it is more instructive to consider challenging scenarios that can be addressed using a combination of techniques.

In the first example,[42] immobilized reagents and scavengers were exploited in a multi-step synthesis of the relatively complex amaryllidaceae alkaloid (+)-plicamine (Scheme 8.9). The approach was designed to also

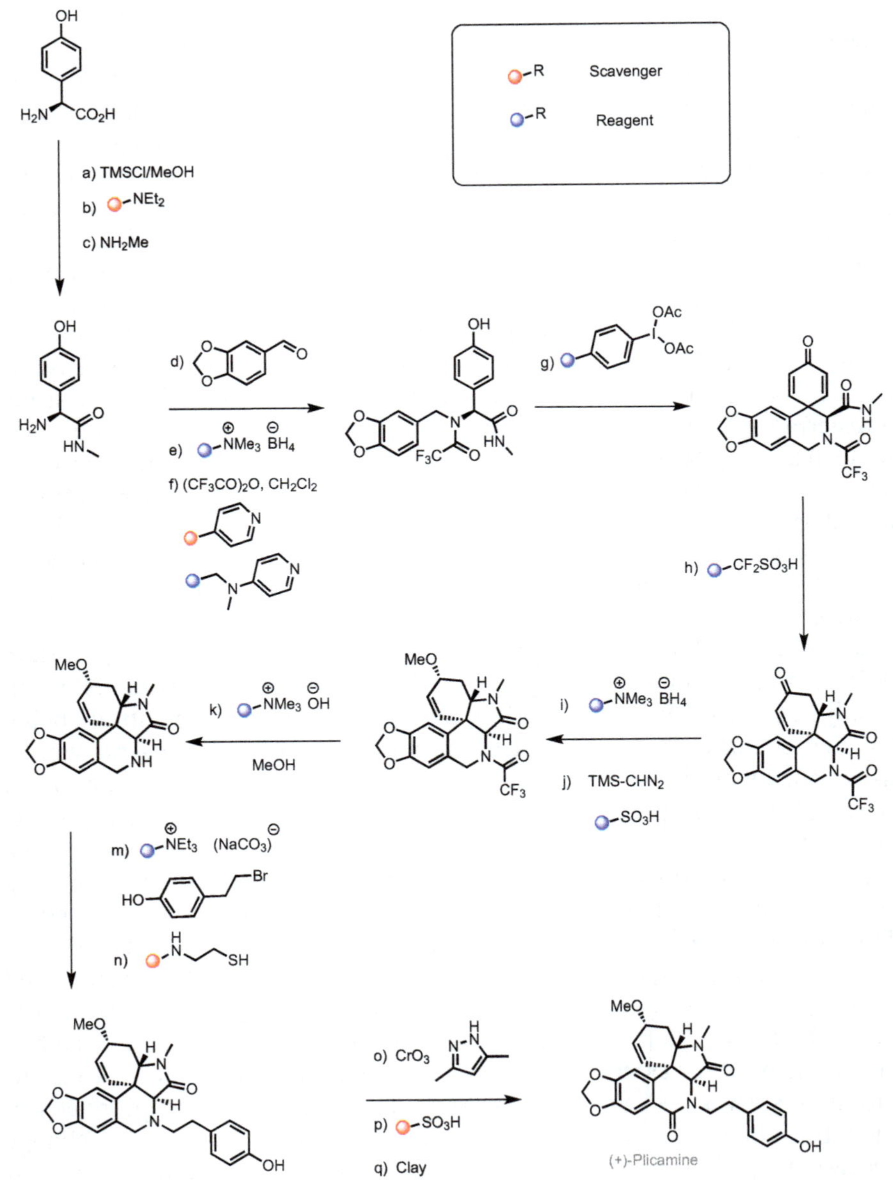

Scheme 8.9 (+)-Plicamine synthesis using immobilized reagents.

produce other related natural products and, *en route*, generate useful model compounds and analogues for biological evaluation. Most notably, the orchestrated reaction sequence that was conceived expedites the synthesis process by avoiding (or, at least, minimizing) time-consuming conventional work-up steps, such as chromatography, distillation, crystallization, water washes and pH adjustments. Here, work-up simply involves the physical removal of spent immobilised reagents by filtration [or even using tweezers when used in porous pouches ("Tea-bags")]. In this way, material necessary for progression of the synthesis route is readily processed to the next reaction in sequence following only a solvent change and minimal operator intervention. The concept accommodates linear and convergent synthesis pathways, owing to its straightforward modularity, and rapidly builds diversity and molecular complexity. Furthermore, each step can be independently optimised and material diverted to alternative reaction processing methods or to other molecular alterations.

While the specific synthesis reported here (Scheme 8.9) may initially appear challenging, all steps follow conventional retrosynthetic planning. The synthesis nicely illustrates how the full scope of immobilized-reagent technology can be deployed, and how further chemical innovation can be stimulated. For example, the interesting use of an immobilized hypervalent iodine reagent was specifically invented to form a hindered C–C bond which led to a novel spiro-dienone intermediate (step g). Also of note was a new application using trimethylsilyl diazomethane and a sulfonic acid resin to methylate a particularly hindered alcohol intermediate (step j).

Many variations of these immobilization strategies in synthesis are now known, enabling the synthesis of complex natural products[43] and active pharmaceutical ingredients (APIs).[44] The importance of these methods extends beyond one of convenience or simple facilitation of chemical reactions. Indeed, their ability to enable recycling, minimise solvent use, generate less waste and reduce human intervention all play a role in delivering more sustainable chemistries. Furthermore, the technically-challenging downstream processing tools and skills common in conventional synthesis programmes can be better managed and automated.

Indeed, given the availability of greatly improved automated software, robotic systems, liquid handlers, in-line analysis and fraction collectors, automation now becomes the next obvious development. Again, space here does not allow a full discussion of all the advances that have been made in this area; rather the focus is on illustrative applications of multi-step polymer-assisted methods. Firstly, a synthesis of a library of histone deacetylase inhibitors provides a convincing example of ease of automation (Scheme 8.10).[45] The approach here utilizes a commercially-available robotic platform that is able to dispense substrates into reaction vessels of different sizes, vortex mix and filter resin-bound reagents to deliver supernatant solutions to reaction blocks in a defined sequence of events to produce an array of sulfonamide hydroxamic acids. Each chemical step includes sulfonylation, alkylation, Heck reactions, catch-and-release amination and

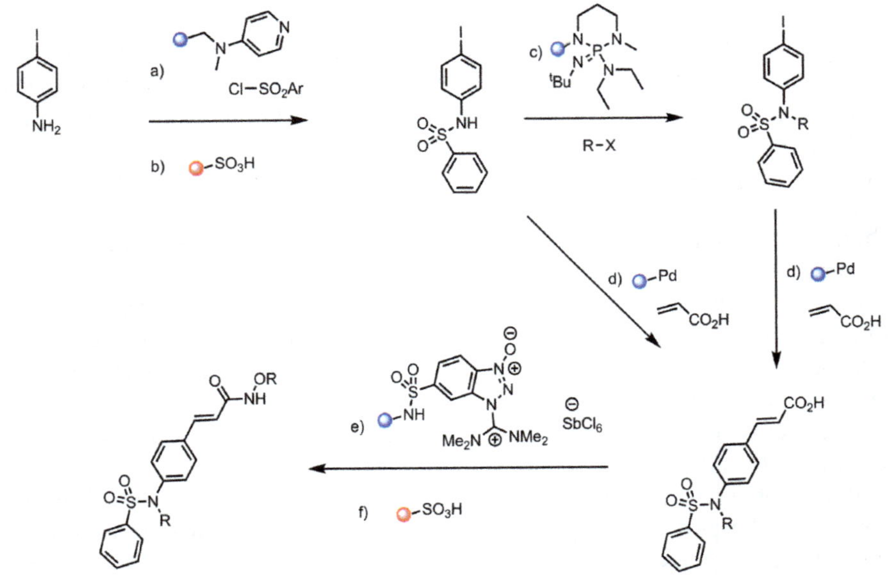

Scheme 8.10 Histone deacetylase inhibitor synthesis.

deprotection to generate the final sample collection plate that can then be processed ahead of biological evaluation.

The whole sequence was automated and led to an unattended four to five step reaction scheme, yielding a library of compounds in just over 3 days compared with the normal manual batch approach that required 5 working days. This approach exemplifies the power of using immobilized chemistries to expedite synthetic chemistry *via* automation.

8.4.2 Flow Chemistry

By far the biggest change to the practice of organic synthesis in recent years has been the introduction of flow chemistry and its corresponding ease of automation. Much has now been written on this topic,[46] consequently, there is need here to distil this down to what constitutes a practical approach and user guide to the area and not to unnecessarily over-sophisticate the application of the devices.

To enter the area, one can begin with the purchase of standard syringe pumps to deliver substrates through inert polymeric or steel flow tubing to various reactor coils or microfluidic chips of different lengths that can be heated or cooled depending upon the perceived reaction processes. Further improvements to this basic system could be the incorporation of commercial flow equipment, back-pressure regulators, in-line immobilized reagents and scavenger cartridges. The next advance would be to integrate in-line analysis through PAT and use computer feedback control algorithms that eventually lead to full automation, self-optimization and multi-step telescoped processing (Scheme 8.11).

An early example of a fully automated continuous flow synthesis of 4,5-disubstituted oxazoles, nicely demonstrated how a chemical library of 31 derivatives was assembled rapidly in individual quantities of up to 10 g.[47] This work clearly showed how a routine heterocyclic synthesis procedure could be readily imported onto a flow chemistry platform consisting of just two chemical inputs: an acid chloride and an alkylisocyanoacetate. The system employed a bespoke reactor design whereby two independent pumps drove the flow stream through a static microfluidic mixer chip and onto two glass cartridges packed firstly with an immobilized Schwesinger base (PS-PEMP) to initiate the base-catalysed coupling and cyclization process, followed by an immobilized high-loading Quadrapure benzylamine resin (QP-BZA) to effect a flow stream clean-up to provide the 4,5-disubstituted oxazoles in high yields and purities usually over 98%. By way of a specific example of this reaction (Scheme 8.12), the reader should not be deluded, however, into thinking this deceptively simple cartoon fully captures the process. We would therefore recommend reading in detail the supplementary information provided by the authors to appreciate the full connectivity of computer-controlled analysis, monitoring, switching and sample collection.

This said, not all syntheses need these high levels of in-line monitoring and control. Much simpler reactor arrangements can be constructed using

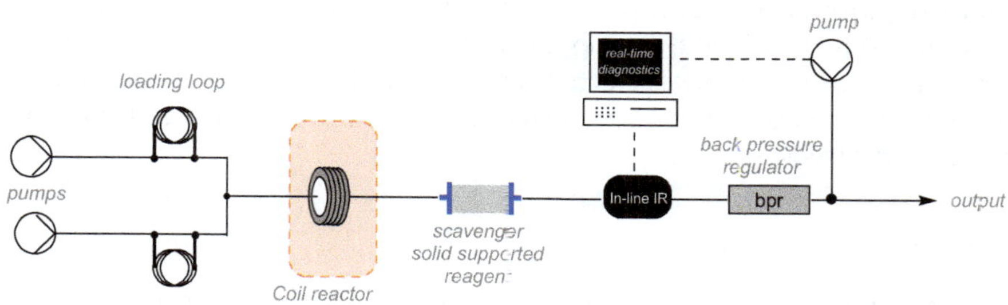

Scheme 8.11 General flow chemistry experiment layout.

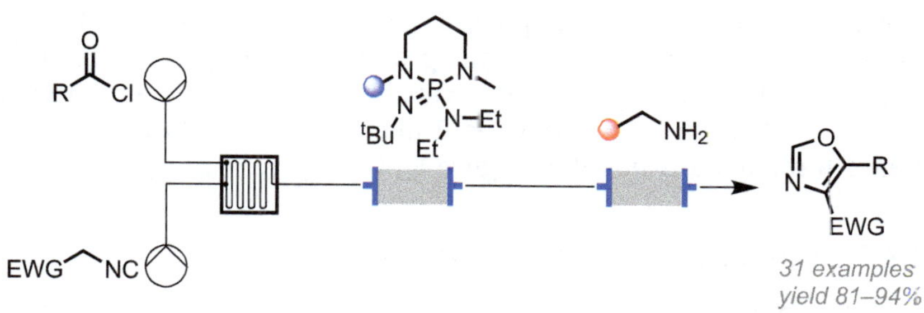

Scheme 8.12 Flow oxazole synthesis.

cheap syringe pumps to deliver chemical substrates to reactor coils, which, when combined with immobilized reagent cartridges and designed sets of scavengers and work-up cartridges, can bring about multi-step transformations that can run unaided for significant periods of time. These modular systems are very attractive for scale-up of routine tasks or for building novel compound collections for screening or in the generation of chemical building blocks to create further molecular complexity.

By way of an example, a multi-step synthesis flow scheme can be quickly assembled for the convergent preparation of triazoles (Scheme 8.13).[48] In this configuration, mixed input streams are used to bring about oxidative conversion of benzylic alcohols to aldehydes, which are intercepted by the Bestmann–Ohira reagent to produce acetylenes which, in turn, undergo coupling *via* copper-catalysed cycloaddition with *in-situ*-generated azides to give the triazole products. This flow stream process also incorporates both reagents to bring about specific transformations and scavengers to ensure in-line product purification. The initial design of these flow schemes uses a classical retrosynthesis analysis approach based on normal mechanistic principles, along with an understanding of anticipated product characteristics, such as solubility. Following this planning stage, a reactor arrangement can be assembled within minutes to produce a basic working platform. Simple optimization, *via* flow rate changes, temperatures and general reaction improvements and some retro-engineering, usually results in a productive device.

These typical flow reactor set-ups can be used beyond single reaction scale-up or novel compound library or building block preparation to effect multi-step, telescoped syntheses of agrochemicals, APIs and even complex natural products. Since these opportunities have been well reviewed in the literature,[49] we will choose selected sequences to best exemplify the concepts of how the engineering of chemistry can enable functional-molecule synthesis and discovery.

8.4.2.1 Tamoxifen Synthesis

For the synthesis of tamoxifen, a front-line product for the treatment of breast cancer, a very short route can be envisaged, which utilizes flow-generated organometallic species.[50] Although many organometallics are known to be pyrophoric and sensitive to oxygen and moisture, they are readily manipulated in multi-step closed-flow systems (Scheme 8.14).

Using new pumping technology along with flow coil precooling devices, a very efficient multi-step synthesis capable of delivering an impressive 223 g day^{-1} of active ingredient (equivalent to 10 000 doses day^{-1}) in a standard fume-hood is possible (Figure 8.1).

8.4.2.2 Casein Kinase 1 Inhibitor Synthesis

During the preparation of 20 diverse casein kinase 1 analogues, a flow synthesis was developed that introduced a number of new technical

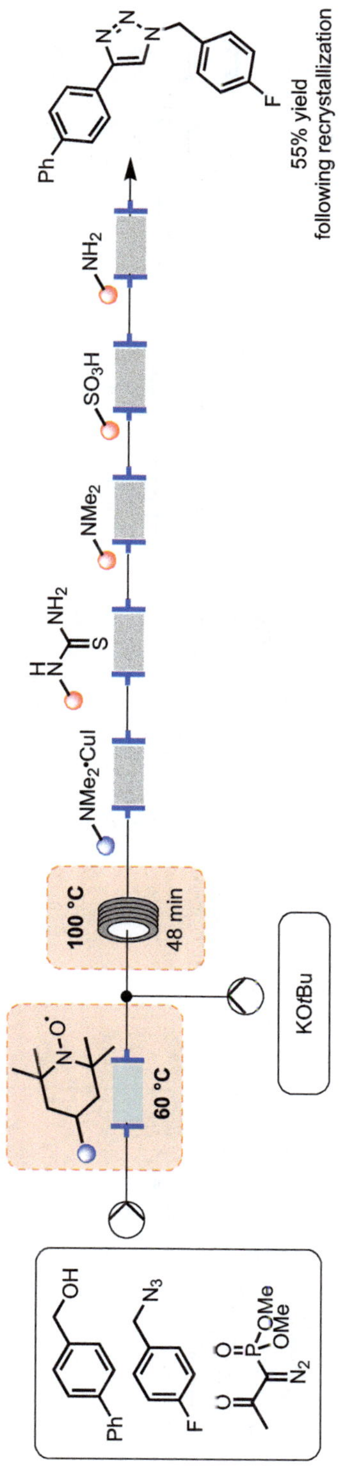

Scheme 8.13 Flow triazole synthesis.

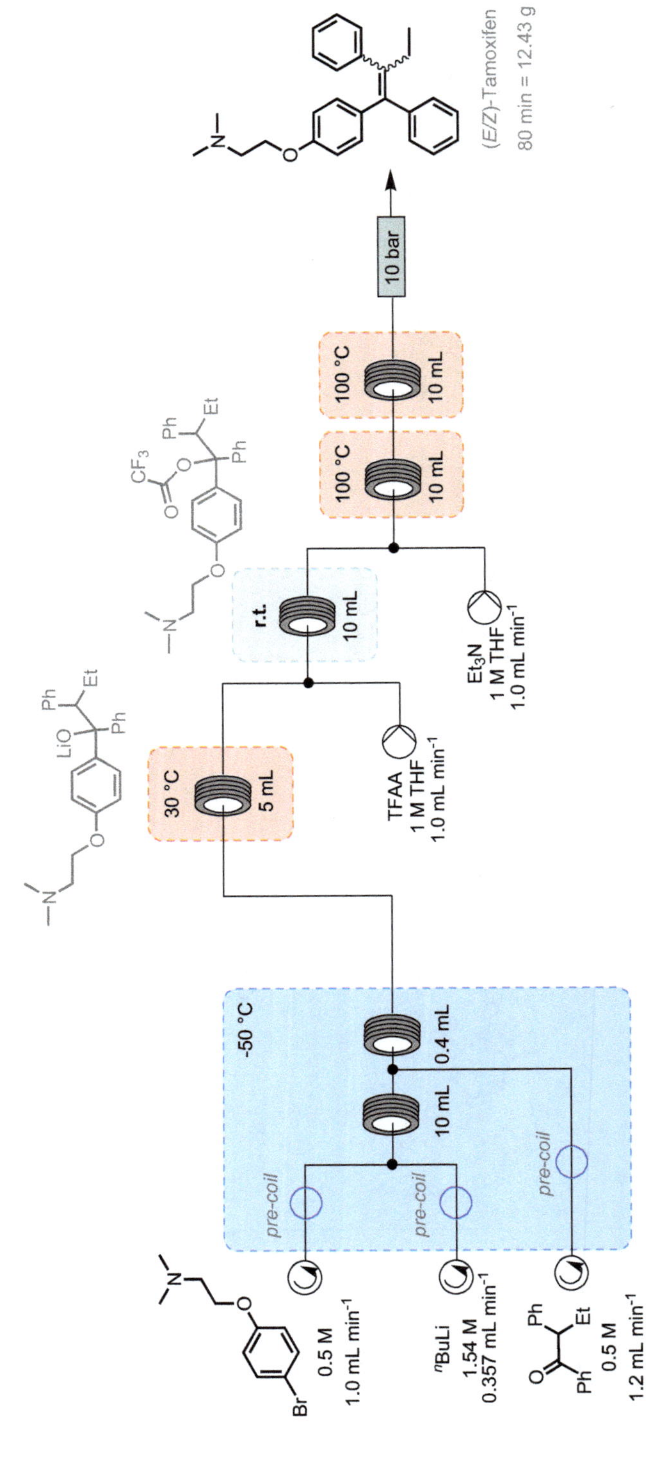

Scheme 8.14 Synthesis of Tamoxifen.

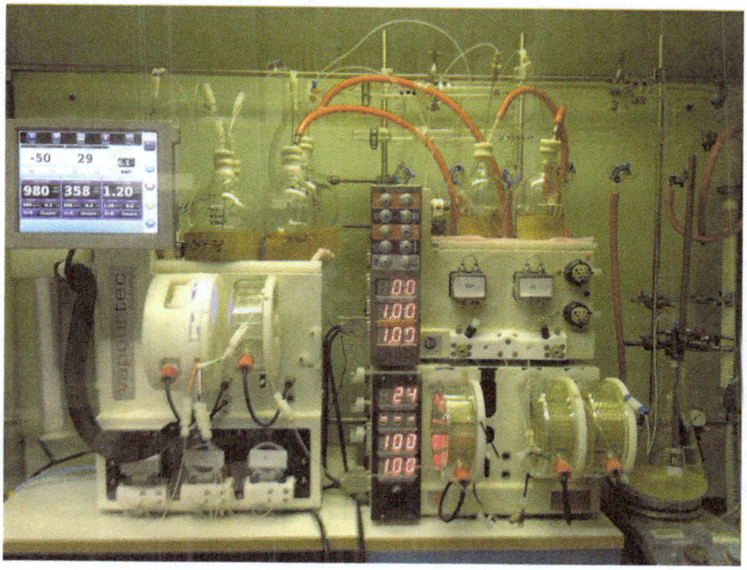

Figure 8.1 The equipment used for the synthesis of Tamoxifen could fit easily into a standard-sized fume-hood.

aspects.[51] All analogues contained a imidazo[1.2b] pyridazine core that was decorated at three sites to provide structural diversity. While the level of automation was not as advanced as in later examples, useful developments arose. In the specific example reported below (Scheme 8.15), the convenience of the approach can be clearly seen from its linear fragment coupling and inherent modularity.

The first problem that needed solving however was how to continually introduce a cooled supply of base (LiHMDS) to effect deprotonation of 4-methylpyridine prior to interception of the resulting anion *in-situ* with 4-fluoroethylbenzoate. This was eventually accomplished by using an automated cooled-loop switching valve system that yielded a single acylated product (no double addition was observed). The next innovation was the use of an immobilized perbromide reagent to effect monobromination; in contrast, most typical batch conditions for this transformation led to dibromination. For the final fragment coupling step, in this specific case with *N*-methylpiperazine, it should be noticed that the reaction is conducted in a steel flow tube at 170 °C using ethanol as solvent. These are conditions that, while suitable for high-pressure and high-temperature in flow-mode, are very difficult to reproduce in batch mode at scale without specialized equipment and extensive safety precautions in place.

8.4.2.3 Efavirenz Synthesis

Efavirenz is an antiretroviral drug used in the treatment of HIV/AIDS. Its flow synthesis employed low-temperature deprotonation of dichlorobenzene and

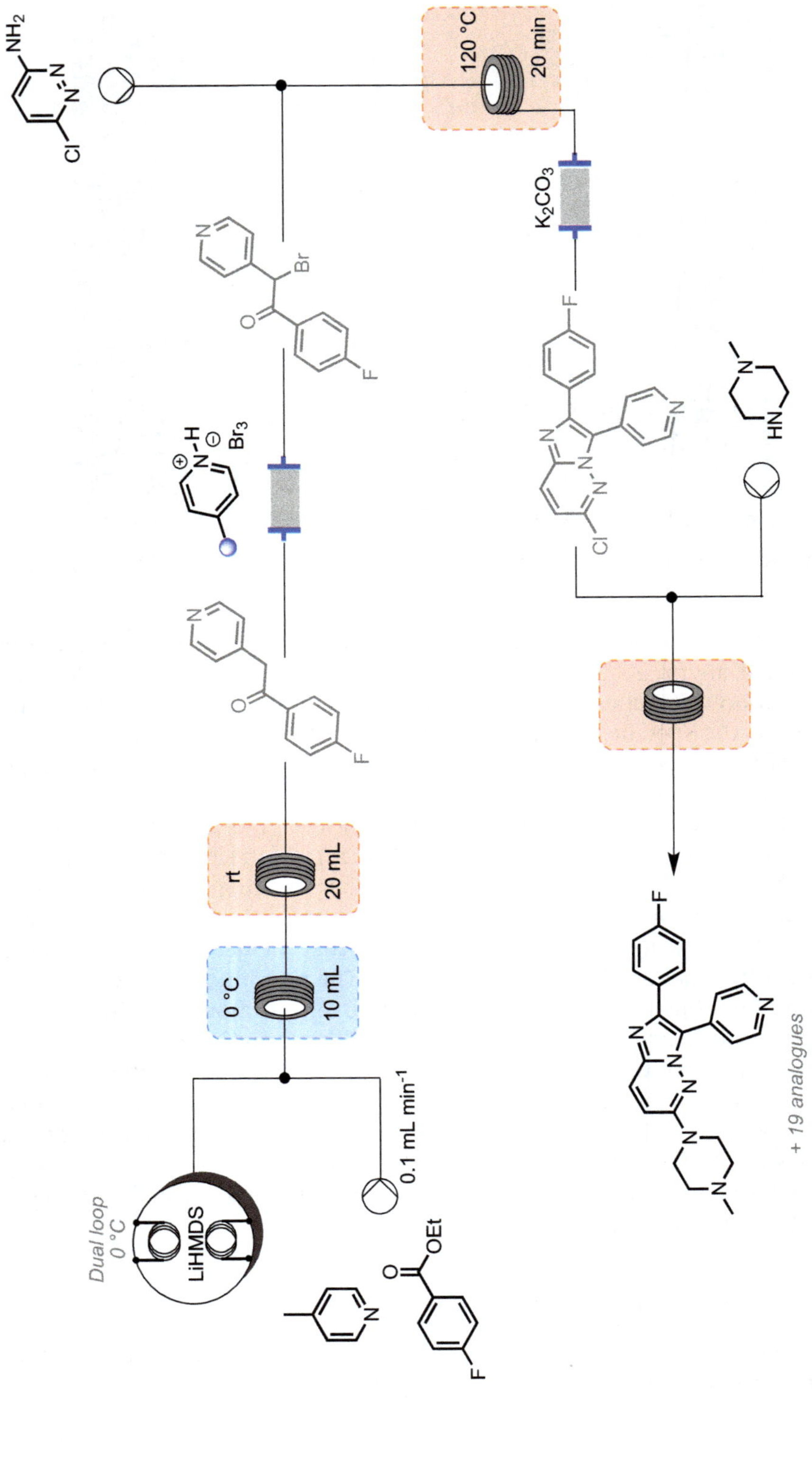

Scheme 8.15 Synthesis of casein kinase 1 inhibitors.

subsequent acylation with trifluoro acetylmorpholine. Passage of the flow stream through a silica gel cartridge, followed by quenching then scavenging of morpholine and salts, yielded a clean product that entered the next coupling step with cyclopropylethyne. Final Cu-mediated chlorine displacement by cyanate and cyclization, generated the API (Scheme 8.16).[52]

8.4.2.4 Imatinib Synthesis

Imatinib, the API of Gleevec, is an important drug developed by Novartis for the treatment of chronic myelogeneous leukaemia and gastrointestinal tumours. Its synthesis using flow-chemistry methods again highlights the power of immobilized reagents, scavenger and catch-and-release techniques. The whole synthesis can be telescoped to produce final material or other analogues to order (Scheme 8.17).[53] Most notable in this work is the use of a single in-line device to switch solvents from CH_2Cl_2 to DMF. Again a modular approach rapidly builds molecular complexity. Another trick of the trade should also be noticed: prior to the back-pressure regulator (bpr), a brief injection of water to dissolve any formed inorganic salts (which can lead to blockages) was very beneficial. This work also features the attractive use of a precatalyst in the Buchwald amination during the final coupling step to assemble the API Gleevec.

8.5 Advanced Automation

While we have described many instances where automation has found use in synthesis laboratories, we have not yet detailed *how* this automation has been achieved. Here we will highlight some of the practical tools that can be used to incorporate automation into our procedures as well as some of the more advanced applications that have been described in the literature.

Many early reports harnessing automation relied on LabVIEW[54] to power computer control. LabVIEW is a graphic programming method, where logical components can be added, connected and repositioned on a workspace to create a sequence of steps to be followed. Given its visual nature and ease of use, it has been popular within the scientific research community, finding excellent use across multiple disciplines. Indeed, it continues to appear frequently in literature reports in the area of organic synthesis.[55]

While LabVIEW can be effective for conducting reactions using repetitive procedures on the same equipment, in many instances this will not suit the typical workflow for a chemist, especially in a discovery environment where multiple different reactions are to be carried out in one day. This is largely owing to the time taken to develop an automation routine using the graphical interface. LabVIEW programs are also somewhat rigid in their construction, often requiring external libraries and software to drive more advanced processes, such as those involving self-optimization as described in more detail below.

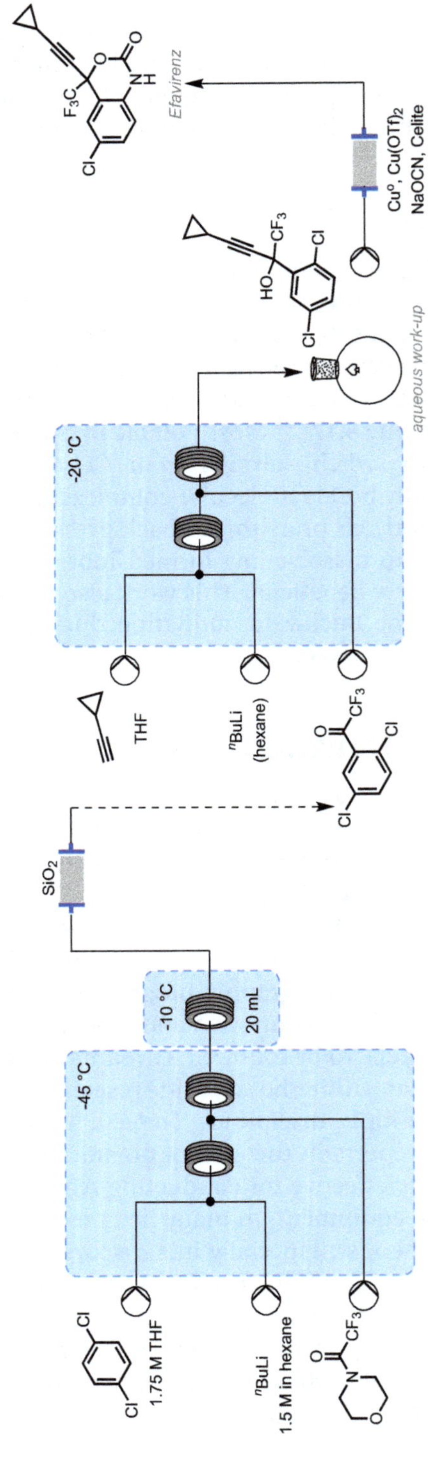

Scheme 8.16 Flow synthesis of Efavirenz.

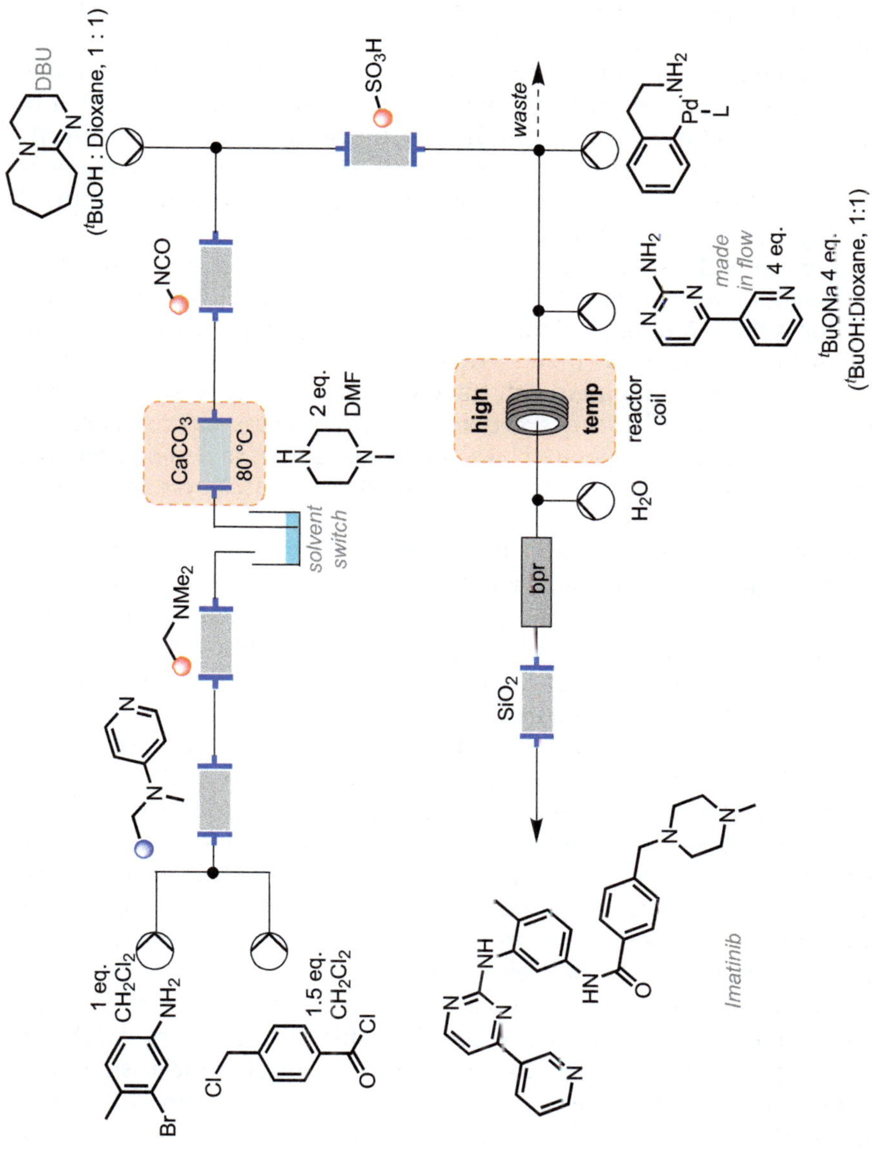

Scheme 8.17 Flow synthesis of Imatinib.

To overcome some of these limitations, new automation systems have been reported recently which are fully modular to ensure that equipment and logical controls can be 'hot-swapped' at any time.[56] These tools do not have a graphical interface, instead operating using either Python or PHP-based scripts. While this does lead to a small learning curve to construct automation sequences, especially for researchers who may not have experience in computer programming, the benefits are numerous.

For example, the multistep synthesis of (*R*,*S*)-piperidine-2-carboxamide has been described,[57] consisting of two reaction steps telescoped together in which crude product mixture from the first transformation was fed directly into the inlet of the next step (Scheme 8.18). Here the control system was used to conduct a set of design of experiment trials to identify optimal conditions for the hydrogenation step.

Subsequently the same automation system was used for the preparation of a precursor to 2-aminoadamantane-2-carboxylic acid,[58] a useful fragment for APIs containing adamantane. The complexity of the set up in this case was significantly increased (Scheme 8.19), where three transformation steps were combined with three downstream processing stages to produce 9.0 mmol h^{-1} of the target product. The process was run at steady state for six hours, and operated by a single researcher, to demonstrate the operational complexity that can be handled by control systems of this nature.

The integration of batch and flow processes has been facilitated with programmable automation software. In one example, a key fragment in the synthesis of AZ82 was synthesised using a series of reactions and downstream processing operations on a single reactor platform.[59] This included a continuous, camera-controlled liquid–liquid quench and separation and a solvent switch from DCM to ethanol.

Other multistep telescoped sequences involving the extensive use of control principles have also been reported, including stand-alone systems that produce APIs suitable for immediate consumption. In one case,[60] lactone aminolysis and Boc deprotection reactions were telescoped with five downstream processing steps, including continuous crystallisation and formulation operations, to produce aliskiren hemifumarate tablets. The overall process was capable of synthesising 45 g h^{-1} of the target API, and was run at steady state for over eight hours.

A second example demonstrated a further extension of the 'end-to-end' synthesis and control principle. A reconfigurable unit has been developed that can produce four different APIs to US Pharmacopeia standards (Scheme 8.20).[61] One of the key characteristics of this system is its small footprint: the entire system fits within a consumer-grade refrigerator. Here a controlled approach was crucial to ensuring the stable operation of multiple units, especially the accurate direction of fluid flow.

These examples show an application of control for synthesis after suitable operation conditions have been identified. One of the advanced applications of automation, however, arises during the optimisation of chemical

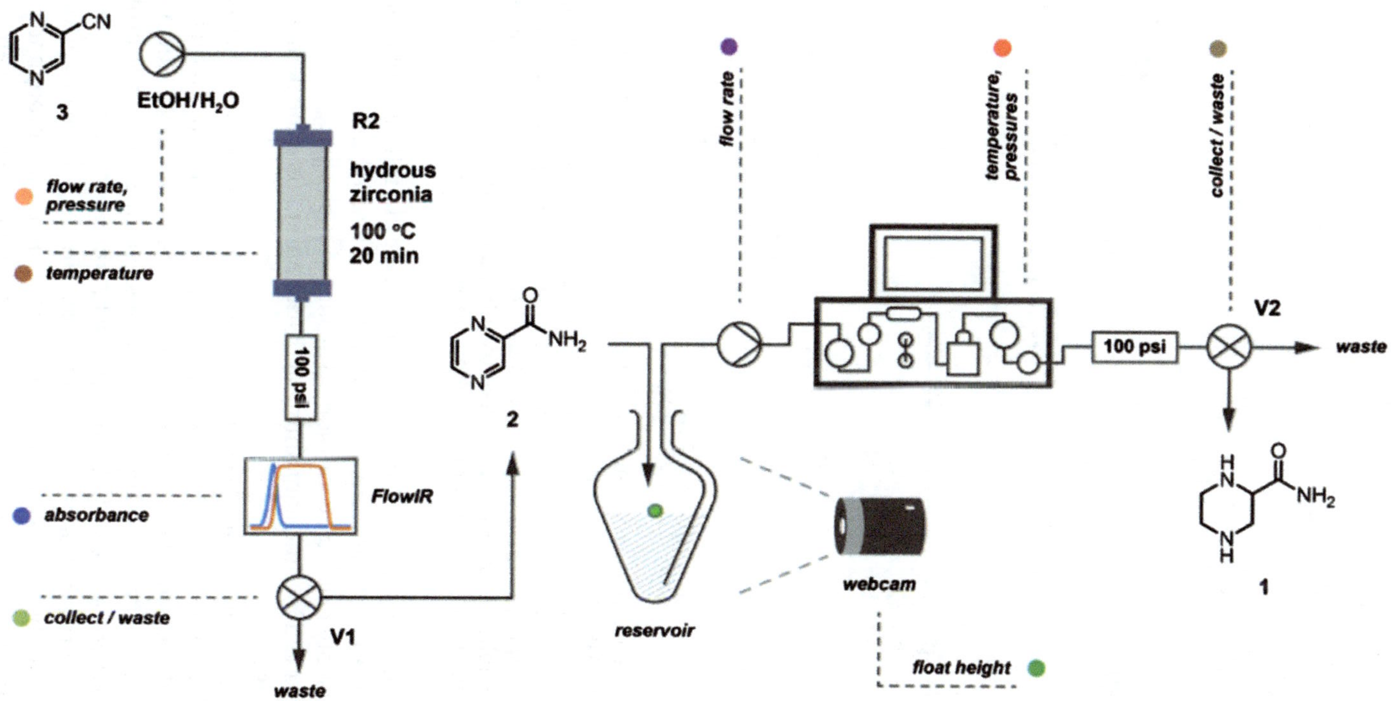

Scheme 8.18 The two-step synthesis of (*R*,*S*)-piperidine-2-carboxamide incorporated a number of automation controls. Reproduced with permission from R. J. Ingham, C. Battilocchio, J. M. Hawkins and S. V. Ley, *Beilstein J. Org. Chem.*, 2014, **10**, 641.[57] Under a Creative Commons Attribution License.

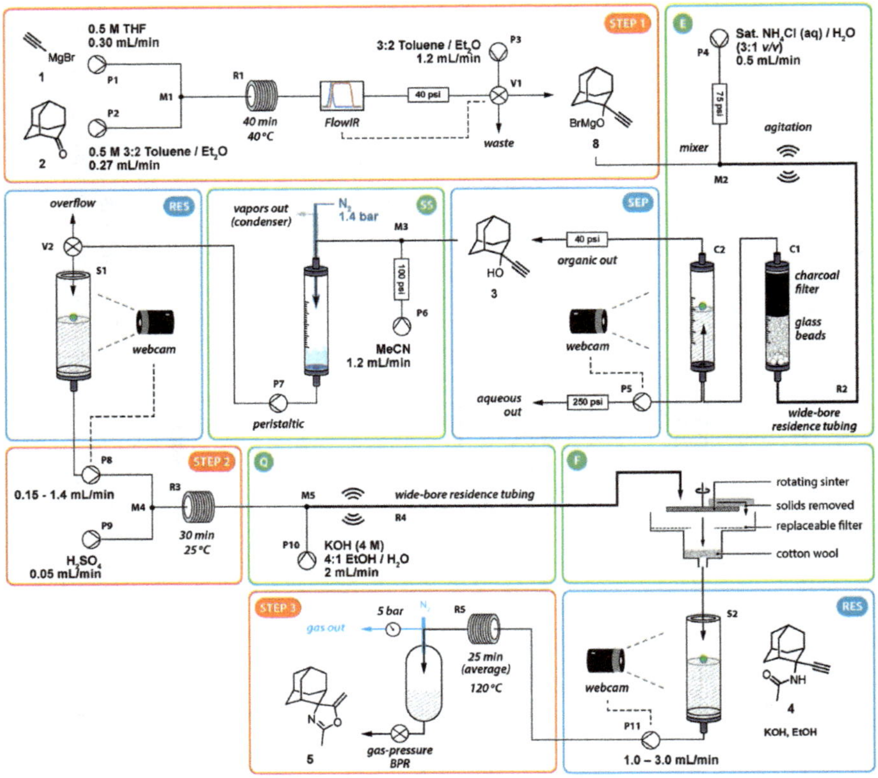

Scheme 8.19 The telescoped synthesis consisted of three reaction steps and three downstream work-up steps.
Reproduced with permission from R. J. Ingham, C. Battilocchio, D. E. Fitzpatrick, E. Sliwinski, J. M. Hawkins and S. V. Ley, *Angew. Chem. Int. Ed.*, 2015, **54**, 144–148.[58] Copyright © 2014 The Authors. Published by Wiley-VCH Verlag GmbH & Co. KGaA Published under the terms of a Creative Commons Attribution License.

reactions. Such self-optimization techniques generally follow evolutionary algorithms, where different sets of conditions are trialled and fed back to equations to create new trial conditions. The aim of such a process is to both accelerate what can be time-consuming activities and reduce the demand on research labour.

As numerous applications of self-optimization have been described in this chapter, and extensive literature reviews have been written in this area previously,[62] discussion here will be kept to a minimum. We did wish, however, to emphasize that approaches to self-optimization have evolved themselves to incorporate a 'human' approach, where important characteristics in addition to reaction yield are optimised. Indeed, a five-dimensional optimisation has been reported that took both material throughput and reagent consumption into account in addition to conversion for an Appel transformation.[56b]

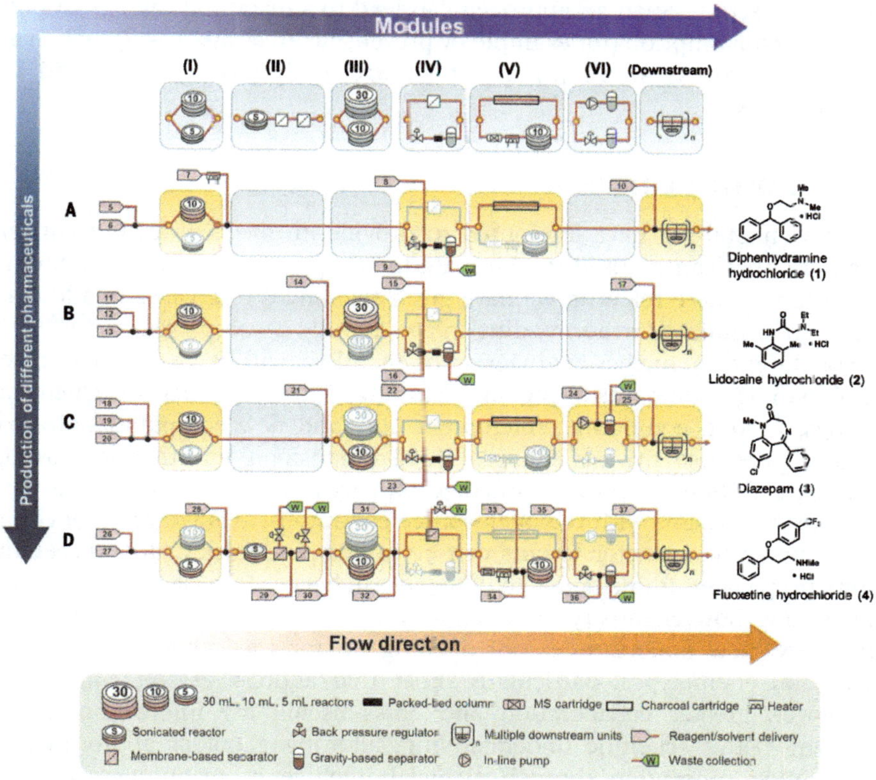

Scheme 8.20 Multiproduct flow sequence.
From A. Adamo, R. L. Beingessner, M. Behnam, J. Chen, T. F. Jamison, K. F. Jensen, J-C. M. Monbaliu, A. S. Myerson, E. M. Revalor, D. R. Snead, T. Stelzer, N. Weeranoppanant, S. Y. Wong and P. Zhang, *Science*, 2016, **352** (6281), 61–67. Reprinted with permission from AAAS.[61]

Some of the most interesting applications of automation and control in a synthetic environment have arisen in the area of discovery. 'Directed serendipity' techniques haven been deployed to accelerate what was previously random discovery of new reactivity patterns, where automated systems combine different materials under different conditions with the aim of finding unexpected products or compositional changes. Such an approach is analogous to high-throughput screening methods common in drug discovery. New photocatalysed transformations,[63] extreme-condition reactions[64] and reactivity pathways[65] have all been identified using directed processes.

Automation has great potential to aid in the safe and extended operation of continuous telescoped processes. For example, by applying machine learning for pattern recognition, automation scripts can analyse detector feedback in real-time to identify failing reactions and take appropriate corrective action. Such an approach can lead to substantial savings in both material consumption (for example by preventing excessive wastage when an inevitable process shutdown and restart/re-equilibration arises) and researcher labour.

8.6 Future Outlook

In this chapter we have attempted to provide an overview of the current methods that can be used to automate the preparation of small functional molecules. Where relevant, we have given references to reviews to add content and stimulate further reading in this rapidly developing area.[66,67]

For the future, further engineering of chemistry will be necessary to meet sustainability standards. Machine learning,[68] the Internet of Chemical Things,[69] cloud computing, microprocessing and new computational algorithms[70] are also driving change. Three-dimensional printing is growing rapidly, and its potential in synthetic chemistry is clear: combining inexpensive reactor designs with the open-source community has led to an environment where new reactor layouts can be prototyped quickly, and successful units propagated immediately to other researchers.[71] It is clear that through integrating flow chemistry methods and a greater focus on novel reaction discovery, new opportunities are opening up for the generation of novel structural chemotypes, particularly those that address greater complexity. Moreover, greater collaboration and cross-disciplinary interaction is developing better scientific understanding, allowing a bolder approach and stimulating innovation through artificial intelligence techniques. Laboratories of the future are evolving whereby the human contribution needs to be maximized in terms of intellectual input while repetitious, routine or scale-up techniques are better relegated to machinery.[4]

References

1. G. M. Whitesides, *Angew. Chem., Int. Ed.*, 2015, **54**, 3196.
2. S. V. Ley, *Tetrahedron*, 2010, **66**, 6270.
3. P. T. Anastas and J. C. Warner, *Green Chemistry: Theory and Practice*, Oxford University Press, New York, 1998.
4. S. V. Ley, D. E. Fitzpatrick, R. J. Ingham and R. M. Myers, *Angew. Chem., Int. Ed.*, 2015, **54**, 3449.
5. R. J. Ingham, C. Battilocchio, D. E. Fitzpatrick, E. Sliwinski, J. M. Hawkins and S. V. Ley, *Angew. Chem., Int. Ed.*, 2015, **54**, 144.
6. For example see (a) D. C. Hill, *Curr. Opin. Drug Discovery Dev.*, 1998, **1**, 92; (b) N. W. Hird, *Drug Discovery Today*, 1999, **4**, 265; (c) W. C. Ripka, G. Barker and J. Krakover, *Drug Discovery Today*, 2001, **6**, 471; (d) R. E. Dolle, *J. Comb. Chem.*, 2000, **2**, 383; (e) R. E. Dolle, *J. Comb. Chem.*, 2001, **3**, 477; (f) H. An and R. D. Cook, *Chem. Rev.*, 2000, **100**, 3311; (g) D. G. Powers and D. L. Cotton, *Drug Discovery Today*, 1999, **4**, 377.

7. S. V. Ley and I. R. Baxendale, *Nat. Rev. Drug Discov.*, 2002, **1**, 573.
8. S. V. Ley, D. E. Fitzpatrick, R. M. Myers, C. Battilocchio and R. J. Ingham, *Angew. Chem., Int. Ed.*, 2015, **54**, 10122.
9. (a) J. I. Yoshida, *Flash Chemistry: Fast Organic Synthesis in Microsystems*, Wiley, Chichester, 2008; (b) J. Hartwig, J. B. Metternich, N. Nikbin, A. Kirschning and S. V. Ley, *Org. Biomol. Chem.*, 2014, **12**, 3611.
10. S. V. Ley, R. J. Ingham, M. O'Brien and D. L. Browne, *Beilstein J. Org. Chem.*, 2013, **9**, 1051.
11. M. O'Brien, P. Koos, D. L. Browne and S. V. Ley, *Org. Biomol. Chem.*, 2012, **10**, 7031.
12. D. X. Hu, M. O'Brien and S. V. Ley, *Org. Lett.*, 2012, **14**, 4246.
13. M. D. Bowman, N. E. Leadbeater and T. M. Barnard, *Tetrahedron Lett.*, 2008, **49**, 195.
14. A. Chanda, A. M. Daly, D. A. Foley, M. A. LaPack, S. Mukherjee, J. D. Orr, G. L. Reid, III, D. R. Thompson and H. W. Ward II, *Org. Process Res. Dev.*, 2015, **19**, 63.
15. N. Holmes, G. R. Akien, R. J. D. Savage, C. Stanetty, I. R. Baxendale, A. J. Blacker, B. A. Taylor, R. L. Woodward, R. E. Meadows and R. A. Bourne, *React. Chem. Eng.*, 2016, **1**, 96.
16. D. L. Browne, S. Wright, B. J. Deadman, S. Dunnage, I. R. Baxendale, R. M. Turner and S. V. Ley, *Rapid Commun. Mass Spectrom.*, 2012, **26**, 1999.
17. C. F. Carter, H. Lange, S. V. Ley, I. R. Baxendale, B. Wittkamp, J. G. Goode and N. L. Gaunt, *Org. Process Res. Dev.*, 2010, **14**, 393.
18. T. Brodmann, P. Koos, A. Metzger, P. Knochel and S. V. Ley, *Org. Process Res. Dev.*, 2012, **16**, 1102.
19. H. Lange, C. F. Carter, M. D. Hopkin, A. Burke, J. G. Goode, I. R. Baxendale and S. V. Ley, *Chem. Sci.*, 2011, **2**, 765.
20. R. S. Das and Y. K. Agrawal, *Vib. Spectrosc.*, 2011, **57**, 163.
21. G. Févotte, *Trans. IchemE, Part A*, 2007, **85**, 906.
22. T. A. Hamlin and N. E. Leadbeater, *Beilstein J. Org. Chem.*, 2013, **9**, 1843.
23. F. Dalitz, M. Cudaj, M. Maiwald and G. Guthausen, *Prog. Nucl. Magn. Reson. Spectrosc.*, 2012, **60**, 52.
24. (a) M. V. Gomez and A. de la Hoz, *Beilstein J. Org. Chem.*, 2017, **13**, 285; (b) B. Blümich and K. Singh, *Angew. Chem., Int. Ed.*, 2018, 57, 6996.
25. D. A. Foley, J. Wang, B. Maranzano, M. T. Zell, B. L. Marquez, Y. Xiang and G. L. Reid, *Anal. Chem.*, 2013, **85**, 8928.
26. V. Sans, L. Porwol, V. Dragone and L. Cronin, *Chem. Sci.*, 2015, **6**, 1258.
27. B. Gouilleux, B. Charrier, E. Danieli, J.-N. Dumez, S. Akoka, F.-X. Felpin, M. Rodriguez-Zubiri and P. Giraudeau, *Analyst*, 2015, **140**, 7854.
28. B. Ahmed-Omer, E. Sliwinski, J. P. Cerroti and S. V. Ley, *Org. Process Res. Dev.*, 2016, **20**, 1603–1614.
29. F. Benito-Lopez, W. Verboom, M. Kakuta, J. G. E. Gardeniers, R. J. M. Egberink, E. R. Ooseterbroek, A. van den Berg and D. N. Reinhoudt, *Chem. Commun.*, 2005, 2857.

30. I. R. Baxendale, C. M. Griffith-Jones, S. V. Ley and G. K. Tranmer, *Synlett*, 2006, 427.
31. L. Shang, Y. Cheng and Y. Zhao, *Chem. Rev.*, 2017, **117**, 7964.
32. L. Guetzoyan, N. Nikbin, I. R. Baxendale and S. V. Ley, *Chem. Sci.*, 2013, **4**, 764.
33. L. Guetzoyan, R. J. Ingham, N. Nikbin, J. Rossignol, M. Wolling, M. Baumert, N. A. Burgess-Brown, C. M. Strain-Damerell, L. Shrestha, P. E. Brennan, O. Fedorov, S. Knapp and S. V. Ley, *Med. Chem. Commun.*, 2014, **5**, 540.
34. B. Desai, K. Dixon, E. Farrant, Q. Feng, K. R. Gibson, W. P. van Hoorn, J. Mills, T. Morgan, D. M. Parry, M. K. Ramjee, C. N. Selway, G. J. Tarver, G. Whitlock and A. G. Wright, *J. Med. Chem.*, 2013, **56**, 3033.
35. W. Czechtizky, J. Dedio, B. Desai, K. Dixon, E. Farrant, Q. Feng, T. Morgan, D. M. Parry, M. K. Ramjee, C. N. Selway, T. Schmidt, G. J. Tarver and A. G. Wright, *ACS Med. Chem. Lett.*, 2013, **4**, 768.
36. M. C. Bryan, C. D. Hein, H. Gao, X. Xia, H. Eastwood, B. A. Bruenner, S. W. Louie and E. M. Doherty, *ACS Comb. Sci.*, 2013, **15**, 503.
37. A. Baranczak, N. P. Tu, J. Marjavovic, P. A. Searle, A. Vasudevan and S. W. Djuric, *ACS Med. Chem. Lett.*, 2017, **8**, 461.
38. (a) A. Akelah and D. C. Sherrington, *Chem. Rev.*, 1981, **81**, 557; (b) A. Kirschning, H. Monenschein and R. Wittenberg, *Angew. Chem., Int. Ed.*, 2001, **40**, 650; (c) I. R. Baxendale, R. I. Storer and S. V. Ley, *Polymeric Materials in Organic Synthesis and Catalysis*, ed. M. R. Buchmeiser, VCH, Berlin, ISBN 3-527-30630-7, 2003, p. 53.
39. (a) R. J. Booth and J. C. Hodges, *Acc. Chem. Res.*, 1999, **32**, 18; (b) S. J. Shuttleworth, S. M. Allin, R. D. Wilson and D. Nasturica, *Synthesis*, 2000, **2000**, 1035; (c) S. J. Shuttleworth, S. M. Allin and P. K. Sharma, *Synthesis*, 1997, **1997**, 1217; (d) S. V. Ley and I. R. Baxendale, *Supported Catalysts and Their Applications*, ed. D. C. Sherrington and A. P. Kybett, *The Royal Society of Chemistry Proceedings*, 2001, p. 9.
40. S. V. Ley and I. R. Baxendale, *Chem. Rec.*, 2002, **2**, 377.
41. S. V. Ley, I. R. Baxendale, R. N. Bream, P. S. Jackson, A. G. Leach, D. A. Longbottom, M. Nesi, J. S. Scott, R. I. Storer and S. J. Taylor, *J. Chem. Soc., Perkin Trans. 1*, 2000, 3815.
42. (a) I. R. Baxendale, S. V. Ley, M. Nesi and C. Piutti, *Tetrahedron*, 2002, **58**, 6285; (b) I. R. Baxendale and S. V. Ley, *Ind. Eng. Chem. Res.*, 2005, **44**, 8588.
43. (a) S. V. Ley, I. R. Baxendale, D. A. Longbottom, R. M. Myers, *Drug Discovery and Development*, ed. M. S. Chorghade, J. Wiley and Sons, 2007, vol. 2, p. 51, ISBN: 0-471-39847-9; (b) S. V. Ley, O. Schmidt, A. W. Thomas and P. J. Murray, *J. Chem. Soc., Perkin Trans. 1*, 1999, 1251; (c) J. Habermann, S. V. Ley and J. S. Scott, *J. Chem. Soc., Perkin Trans. 1*, 1999, 1253; (d) R. I. Storer, T. Takemoto, P. S. Jackson and S. V. Ley, *Angew. Chem., Int. Ed.*, 2003, **42**, 2321; (e) I. R. Baxendale, A.-L. Lee and S. V. Ley, *J. Chem Soc., Perkin Trans. 1*, 2002, 1850.

44. (a) I. R. Baxendale and S. V. Ley, *Bioorg. Med. Chem. Lett.*, 2000, **10**, 1983; (b) R. N. Bream, S. V. Ley and P. A. Procopiou, *Org. Lett.*, 2002, **4**, 3793; (c) A. Bapna, E. Vickerstaffe, B. H. Warrington, M. Ladlow, T.-P. D. Fan and S. V. Ley, *Org. Biomol. Chem.*, 2004, **2**, 611.
45. E. Vickerstaff, B. H. Warrington, M. Ladlow and S. V. Ley, *Org. Biomol. Chem.*, 2003, **1**, 2419.
46. (a) D. Webb and T. F. Jamison, *Chem. Sci.*, 2010, **1**, 675; (b) J. Britton and C. L. Raston, *Chem. Soc. Rev.*, 2017, **46**, 1250; (c) J. Wegner, S. Ceylan and A. Kirschning, *Adv. Synth. Catal.*, 2012, **354**, 17; (d) K. Geyer, J. D. C. Codée and P. H. Seeberger, *Chem. – Eur. J.*, 2006, **12**, 8434; (e) H. R. Sahoo, J. G. Kralj and K. F. Jensen, *Angew. Chem. Int. Ed.*, 2007, **46**, 5704; (f) I. R. Baxendale, *J. Chem. Technol. Biotechnol.*, 2013, **88**, 519; (g) V. Hessel, B. Cortese and M. H. J. M. de Croon, *Chem. Eng. Sci.*, 2011, **66**, 1426; (h) T. N. Glasnov and C. O. Kappe, *J. Heterocycl. Chem.*, 2011, **48**, 11; T. Tsubogo, T. Ishiwata and S. Kobayashi, *Angew. Chem. Int. Ed.*, 2013, **52**, 6590.
47. M. Baumann, I. R. Baxendale, S. V. Ley, C. D. Smith and G. K. Tranmer, *Org. Lett.*, 2006, **8**, 5231.
48. I. R. Baxendale, S. V. Ley, A. C. Mansfield and C. D. Smith, *Angew. Chem., Int. Ed.*, 2009, **48**, 4017.
49. (a) M. Baumann and I. R. Baxendale, *Beilstein J. Org. Chem.*, 2015, **11**, 1194; (b) R. Porta, M. Benaglia and A. Puglisi, *Org. Process Res. Dev.*, 2016, **20**, 2; (c) J. C. Pastre, D. L. Browne and S. V. Ley, *Chem. Soc. Rev.*, 2013, **42**, 8849; (d) D. Ghislieri, K. Gilmore and P. H. Seeberger, *Angew. Chem., Int. Ed.*, 2015, **54**, 678; (e) P. Bana, R. Örkényi, K. Lövei, A. Lakó, G. I. Túrós, J. Éles, F. Faigl and I. Greiner, *Bioorg. Med. Chem.*, 2017, **23**, 6180.
50. P. R. D. Murray, D. L. Browne, J. C. Pastre, C. Butters, D. Guthrie and S. V. Ley, *Org. Process Res. Dev.*, 2013, **17**, 1192.
51. F. Venturoni, N. Nikbin, S. V. Ley and I. R. Baxendale, *Org. Biomol. Chem.*, 2010, **8**, 1798.
52. C. A. Correia, K. Gilmore, D. T. McQuade and P. H. Seeberger, *Angew. Chem., Int. Ed.*, 2015, **54**, 4945.
53. M. D. Hopkin, I. R. Baxendale and S. V. Ley, *Org. Biomol. Chem.*, 2013, **11**, 1822.
54. For more information about the LabVIEW software system, refer to http://www.ni.com/en-gb/shop/labview.html. (Accessed 29 September 2017).
55. (a) C. A. Hone, N. Holmes, G. R. Akien, R. A. Bourne and F. L. Muller, *React. Chem. Eng.*, 2017, **2**, 103; (b) N. Holmes, G. R. Akien, R. J. D. Savage, C. Stanetty, I. R. Baxendale, A. J. Blacker, B. A. Taylor, R. L. Woodward, R. E. Meadows and R. A. Bourne, *React. Chem. Eng.*, 2016, **1**, 96.
56. (a) R. J. Ingham, C. Battilocchio, J. M. Hawkins and S. V. Ley, *Beilstein J. Org. Chem.*, 2014, **10**, 641; (b) D. E. Fitzpatrick, C. Battilocchio and S. V. Ley, *Org. Process Res. Dev.*, 2016, **20**, 386.

57. R. J. Ingham, C. Battilocchio, J. M. Hawkins and S. V. Ley, *Beilstein J. Org. Chem.*, 2014, **10**, 641.
58. R. J. Ingham, C. Battiloccio, D. E. Fitzpatrick, E. Sliwinski, J. M. Hawkins and S. V. Ley, *Angew. Chem., Int. Ed.*, 2015, **54**, 144.
59. D. E. Fitzpatrick and S. V. Ley, *React. Chem. Eng.*, 2016, **1**, 629.
60. S. Mascia, P. L. Heider, H. Zhang, R. Lakerveld, B. Benyahia, P. I. Barton, R. D. Braatz, C. L. Cooney, J. M. B. Evans, T. F. Jamison, K. F. Jensen, A. S. Myerson and B. L. Trout, *Angew. Chem., Int. Ed.*, 2013, **52**, 12359.
61. A. Adamo, R. L. Beingessner, M. Behnam, J. Chen, T. F. Jamison, K. F. Jensen, J.-C. M. Monbaliu, A. S. Myerson, E. M. Revalor, D. R. Snead, T. Stelzer, N. Weeranoppanant, S. Y. Wong and P. Zhang, *Science*, 2016, **352**, 61.
62. (a) D. C. Fabry, E. Sugiono and M. Rueping, *Isr. J. Chem.*, 2014, **54**, 341; (b) V. Sans and L. Cronin, *Chem. Soc. Rev.*, 2016, **45**, 2032.
63. A. McNally, C. K. Prier and D. W. C. MacMillan, *Science*, 2011, **334**, 1114.
64. Z. Amara, E. S. Streng, R. A. Skilton, J. Jin, M. W. George and M. Poliakoff, *Eur. J. Org. Chem.*, 2015, 6141.
65. V. Dragone, V. Sans, A. B. Henson, J. M. Granda and L. Cronin, *Nat. Commun.*, 2017, **8**, 15733.
66. C. A. Shukla and A. A. Kulkarni, *Beilstein J. Org. Chem.*, 2017, **13**, 960.
67. D. E. Fitzpatrick, C. Battilocchio and S. V. Ley, *ACS Cent. Sci.*, 2016, **2**, 131.
68. (a) L. Wang, L. Sun, J. Yu, R. Zhang and F. Gao, *Ind. Chem. Eng. Chem. Res.*, 2017, **56**, 10099; (b) J. N. Wei, D. Duvenaud and A. Aspuru-Guzik, *ACS Cent. Sci.*, 2016, **2**, 725; (c) M. Pardakhti, E. Moharreri, D. Wanik, S. L. Suib and R. Srivastava, *ACS Comb. Sci.*, 2017, **19**, 640.
69. S. V. Ley and D. E. Fitzpatrick, *Beilstein Mag.*, 2015, **1**, 11.
70. (a) M. A. Kabashov, E. Sliwinski, D. E. Fitzpatrick, B. Musio, J. A. Newby, W. D. W. Blaylock and S. V. Ley, *Chem. Commun.*, 2015, **51**, 7172; (b) V. Dragone, V. Sans, A. B. Henson, J. M. Granda and L. Cronin, *Nat. Commun.*, 2017, **8**, 15733.
71. M. O'Brien, L. Konings, M. Martin and J. Heap, *Tetrahedron Lett.*, 2017, **58**, 2409.

Section 2:
Synthetic Approaches to Classes of Modified Biomolecule

CHAPTER 9

Genetically Encoded Cyclic Peptide Libraries

CATRIN SOHRABI AND ALI TAVASSOLI*

Department of Chemistry, University of Southampton, Southampton SO17 1 BJ, UK
*Email: ali1@soton.ac.uk

9.1 Introduction

Cyclic peptides have emerged as promising scaffolds when employed against historically challenging targets such as protein–protein interactions (PPIs). By virtue of their conformationally constrained frameworks and stereochemical complexity, these compounds often exhibit increased biological activity when compared against their more flexible linear counterparts, as well as increased affinity and target selectivity (resulting from the reduced entropic costs upon binding) when employed against challenging biomolecular structures.[1] In addition, the application of chemical backbone modifications, such as peptide stapling and head-to-tail peptide cyclization, has enabled the development of peptide-based molecules with increased membrane permeability, oral bioavailability and proteolytic stability.[2] Not surprisingly, there are several examples of therapeutically relevant and biologically active natural cyclic peptides, including cyclosporin A (immunosuppressant),[3] polymyxin E (antibacterial),[4] and oxytocin (an endogenous hormone released *via* the posterior pituitary to stimulate the termination of pregnancy, Figure 9.1).[5]

In the light of the potential advantages of this structural class, the use of cyclic peptide libraries in high-throughput screening for the discovery of novel PPI inhibitors has emerged as an attractive therapeutic strategy for

Chemical Biology No. 10
Chemical and Biological Synthesis: Enabling Approaches for Understanding Biology
Edited by Nick J. Westwood and Adam Nelson

Published by the Royal Society of Chemistry, www.rsc.org

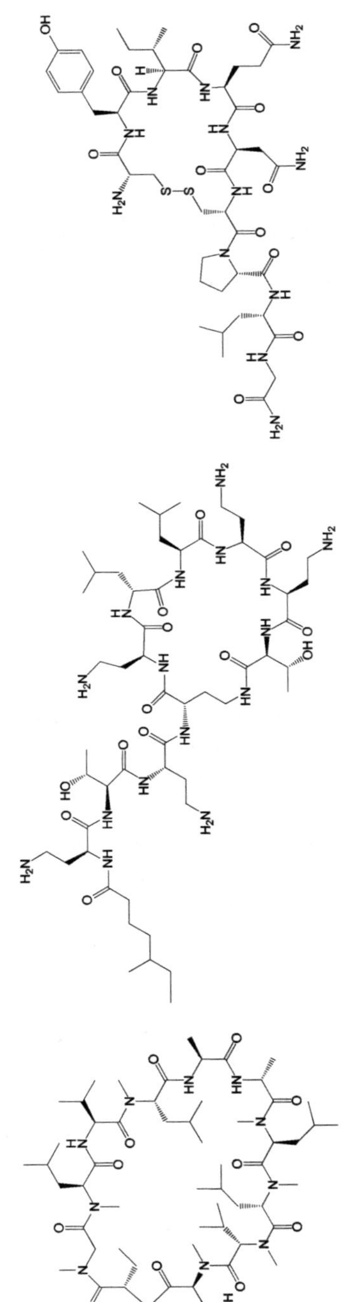

Cyclosporin A
Immunosuppresant

Polymixin E
Antibacterial

Oxytocin
Pregnancy termination

Figure 9.1 Examples of natural head-to-tail cyclic peptides and their respective bioactivities.

early-stage drug discovery in both academic and industrial settings. Whilst a variety of efficient and versatile methods exist for the generation of large and diversified cyclic peptide libraries (10^7–10^{12} members), genetic encoding retains a significant advantage over chemical synthesis; namely, being more readily accessible, enabling coupling to various high-throughput screening platforms (*e.g.* phage, yeast or mRNA display),[6] and allowing for the straightforward deconvolution of hits *via* sequencing of the peptide-encoding gene. In particular, the advent of Sanger sequencing (chain-termination or dideoxy technique) has since established a robust and facile method for DNA sequencing that now lies at the heart of deconvoluting hits from genetically encoded libraries.[7] Another enabling technology behind the rapid expansion of modern peptide-based libraries, and a solution to the limited capacity of chemically competent cells, emerged with the advent of high-efficiency electroporation.[8]

Whilst the majority of biologically synthesized cyclic peptide libraries are formed using phage/phagemid display,[9] split-intein-mediated cyclization of peptides and proteins (SICLOPPS) represents a simple and readily accessible method that uses a selectively randomized library of split inteins for the production of genetically encoded, backbone-cyclized peptide libraries. As such, for the purpose of this chapter, we describe a range of genetically encoded methods for the generation and screening of head-to-tail cyclized peptide libraries and detail examples of their successful implementation.

9.2 SICLOPPS

SICLOPPS represents a simple and accessible method that uses split inteins to direct the intracellular cyclization of peptide backbones for the rapid biosynthesis of cyclic peptide libraries ranging from 10^6 to 10^8 peptides in size.[6] In particular, inteins (internal proteins) represent naturally occurring in-frame protein domains that facilitate their self-excision from flanking extein (external protein) segments at the post-translational level, with concomitant extein ligation by a native peptide bond, *via* an autocatalytic mechanism termed protein splicing (analogous to mRNA splicing).[10,11] In exploiting this natural process in forming a N-to-C association, SICLOPPS uses rearranged intein segments derived from the *Synechocystis* sp. (*Ssp*) PCC6803 *trans*-splicing split DnaE intein in which the C-terminal intein (I_C, 36 amino acids) precedes the N-terminal domain (I_N, 123 amino acids) whilst flanking an extein sequence of choice (I_C–extein–I_N, Figure 9.2A). Upon intein association, the resulting active *cis*-intein cyclizes the intervening extein sequence to liberate a head-to-tail cyclized polypeptide product, provided that a nucleophilic cysteine or serine residue is present as the first amino acid of the extein (Figure 9.2B).[12–14] More recently, the faster splicing *Nostoc punctiforme* (*Npu*) DnaE split inteins have been used in SICLOPPS for the rapid generation of cyclic peptide libraries and of cyclic proteins that have the added advantage of being more tolerant than *Ssp* inteins to amino acid variation near the splice junction, a critical requirement for library production.[15]

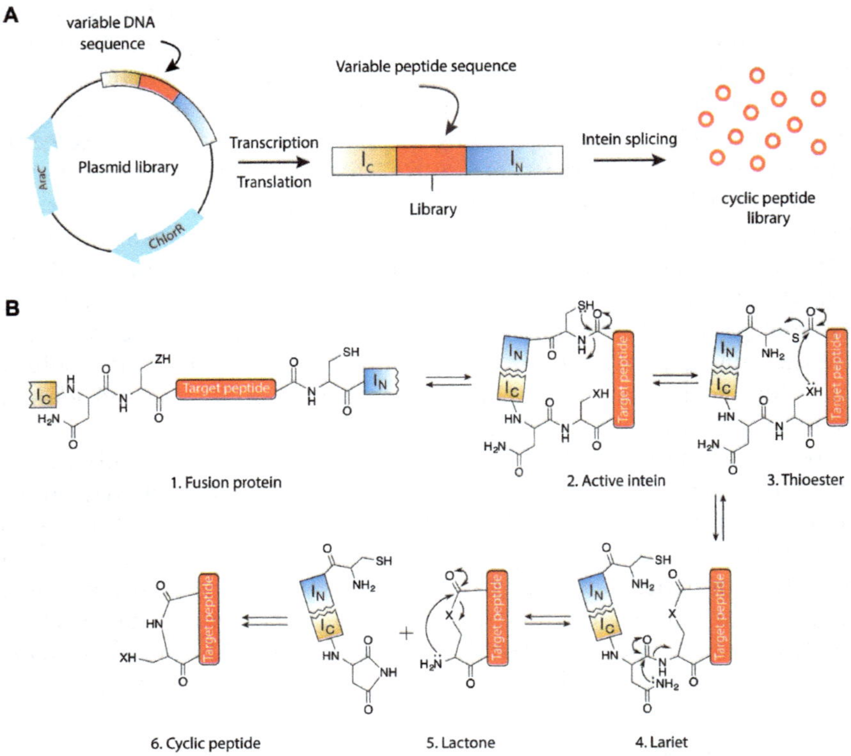

Figure 9.2 (A) Illustration of SICLOPPS library construction for the intracellular generation of cyclic peptide libraries. Typically *Ssp* or *Npu* inteins are used in SICLOPPS, although other split-inteins may also be used. (B) Illustrative mechanism of SICLOPPS intein splicing, whereby the expression target (or fusion) peptide folds following the association and splicing of I_C and I_N components to form an active cyclized peptide. X = O or S.

The ease, simplicity and speed with which SICLOPPS plasmid libraries comprising hundreds of millions of members may be created attests to a key advantage of this platform when employed against challenging targets such as PPIs.[14,16] As such, the construction of a SICLOPPS cyclic peptide library is mediated by the polymerase chain reaction (PCR)-based incorporation of degenerate oligonucleotides encoding a randomized extein segment of NNS or NNB mixed base repeats, where N represents any of the four canonical DNA bases (A, C, G or T), S represents either C or G, and B represents C, G, or T (Figure 9.3A).[14] The NNS and NNB motifs provide coverage of 32 and 48 codons, respectively, coding for all 20 peptidogenic amino acids whilst eliminating both ochre (UAA) and opal (UGA) stop codons. Since there is no theoretical limitation to the number of incorporated degenerate oligonucleotide repeats and hence to the number of amino acids within each finalized peptide variant, SICLOPPS may be used for the controlled assembly

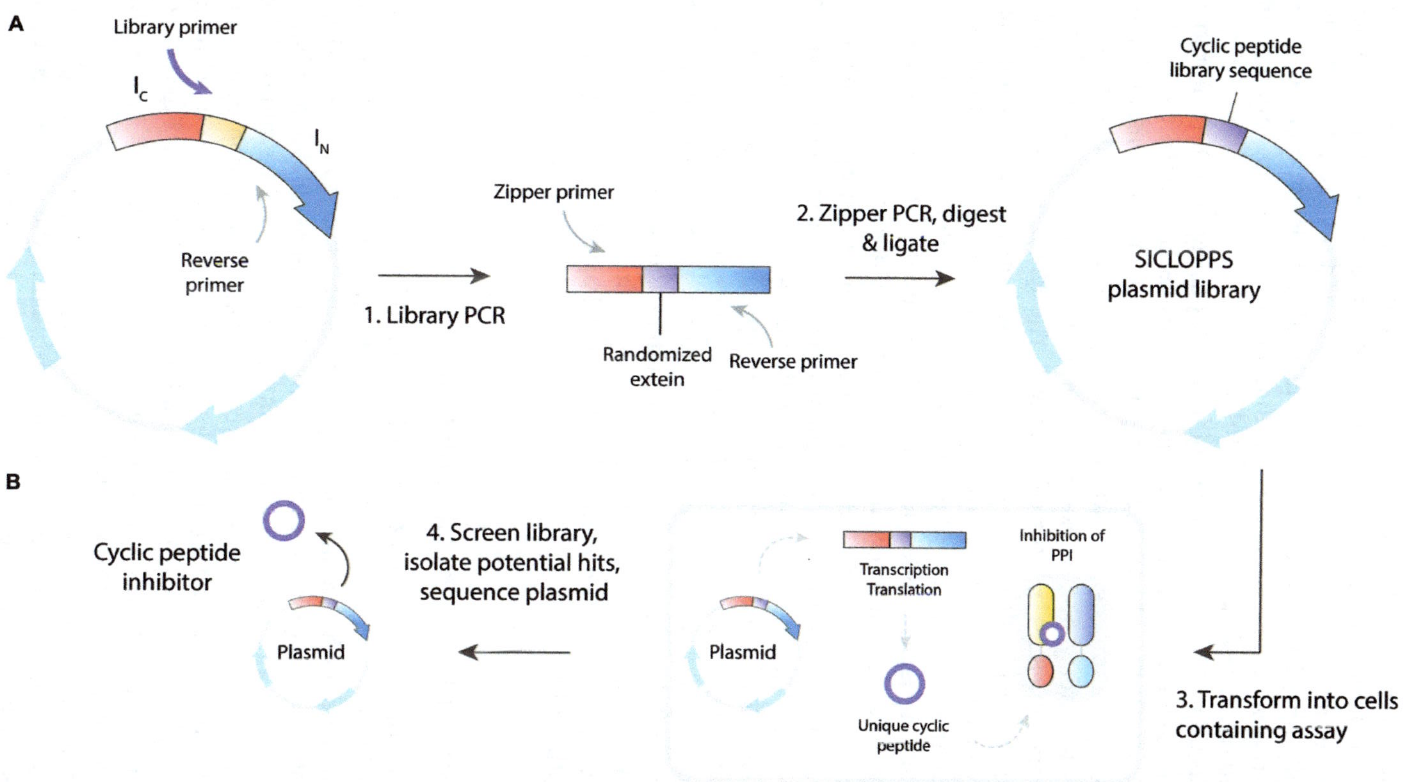

Figure 9.3 (A) Representative workflow for the generation of a SICLOPPS library using a two-step PCR-based methodology. (B) Subsequent transformation of the resulting plasmid library into cells comprising an assay (*e.g.* FRET, life/death) enables screening for the identification of active cyclic peptide hits.

of macrocyclic products of virtually any size and sequence. Although limited by the transformation efficiency of its cellular host and therefore to the formation of larger cyclic peptide rings when interfaced with cell-based assays, five or six variable amino acids (3.2×10^6 and 6.4×10^7, respectively) are typically introduced to ensure that the total number of cyclic peptide library variants remains lower than that of the maximum number of *Escherichia coli* transformations (10^8–10^{10}).

The selection of unique modulatory molecules using SICLOPPS libraries is typically performed in conjunction with a bacterial reverse two-hybrid system (RTHS).[13] In a classical reverse genetics approach, the disruption of protein complexation is systematically linked to the downstream expression of various reporter genes, whose regulation can be monitored through use of chromogenic assays or host cell survival. The RTHS typically employed in conjunction with SICLOPPS libraries is based on the λ bacteriophage regulatory system, whereby the disruption of a homodimer or heterodimer protein complex is linked to the expression of three downstream reporter genes including *HIS*3 (imidazoleglycerol-phosphate dehydratase) and Kan^R (aminoglycoside 3′-phoshopytansgerase for kanamycin resistance), to amplify selection stringency and allow for chemical tenability, respectively, and *LacZ* (β-galactosidase), for quantitative β-galactosidase-mediated measurements. Thereafter, homodimerization or heterodimerization of the targeted protein(s) reconstitutes the bacteriophage functional repressor, thereby blocking the transcription of reporter genes and thus leading to cell death upon selective media (Figure 9.4). As a result, members of a SICLOPPS library that inhibit the targeted protein–protein interaction will, in turn, prevent the formation of a functional repressor to thereby enable its survival and growth upon selective media for the rapid identification of potential cyclic peptide hits.[17–19]

9.2.1 SICLOPPS in Prokaryotic Cells

The systematic coupling of SICLOPPS library production with the RTHS in *E. coli* for inhibitor identification was first reported through monitoring of the iron-dependent heterodimeric enzyme ribonucleotide reductase (RR),[17] responsible for catalysing the rate-limiting formation of 2′-deoxyribonucleotides *via* ribonucleotide reduction during *de novo* DNA synthesis.[9] To mediate the functional discovery of protein–protein interaction inhibitors, SICLOPPS libraries encoding a randomized hexapeptide insert with an invariable cysteine residue as the cyclization nucleophile were exposed *via* transformation to a RTHS *E. coli* strain expressing murine RR fusions. Of the approximately 10^8 transformants created, a final cohort of four systematically selected hexapeptides displaying a strong preference and high level of target selectivity for RR were identified, exhibiting the greatest degree of target specificity and potency whilst possessing a suitable residue composition for chemical-mediated peptide synthesis (RR93, RR127, RR130 and

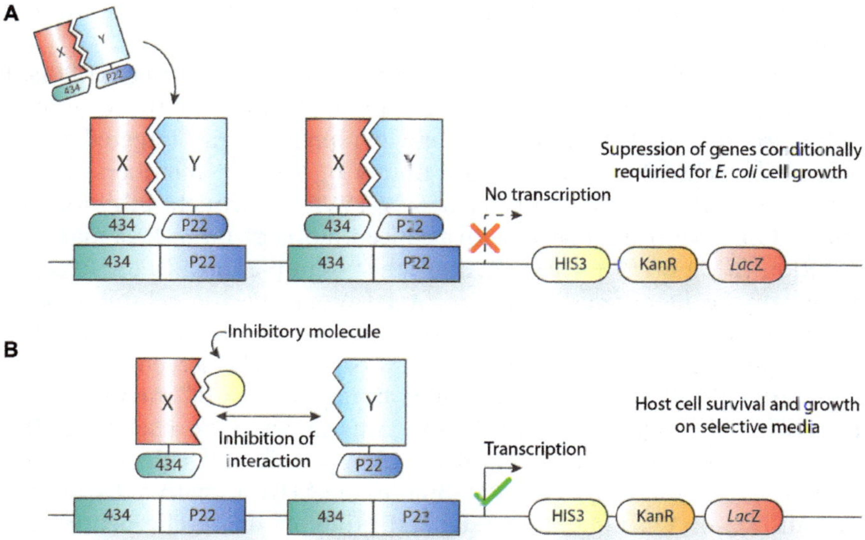

Figure 9.4 Illustration of the SICLOPPS reverse two-hybrid system. (A) Interaction of target proteins X and Y leads to the formation of a functional 434:P22 repressor complex with the inherent ability to bind to operators engineered onto the chromosome of the host *E. coli* strain upstream of three key reporter genes (HIS3, Kan and *LacZ*). The binding and interaction of X and Y inhibits the expression of downstream genes, leading to cellular death upon selective media. (B) The presence of a PPI inhibitor with the ability to selectively disrupt the targeted interaction between X and Y, thereby disrupting the downstream repressor complex, permits transcription and translation of the resulting reporter genes and therefore host cell survival on selective media.

RR133) alongside an exhibited approximately twofold enhanced activity over their corresponding linear forms.

In another example, the SICLOPPS RTHS was successfully employed in the selection of a unique cyclic peptide inhibitor, *cyclo*-CRYFNV with an inhibition constant (K_i) of 17 ± 4 μM when compared against its linear counterpart (K_i of 142 ± 22.5 μM), from a 3.2×10^6 member *cyclo*-CX_5 library of the previously undruggable and difficult-to-target highly conserved 64 kDa bifunctional homodimeric enzyme aminoimidazole carboxamide ribonucleotide transformylase/inosine monophosphate cyclohydrolase (ATIC, with a dimeric interface spanning approximately 5000 Å^2 [19–21]), which catalyses the final two steps of *de novo* purine biosynthesis.[19] Since rapidly dividing cells are reliant upon *de novo* purine biosynthesis for nucleotide production, the selective inhibition of enzymes within this pathway for use as potential antineoplastic agents represents an attractive therapeutic approach.[21] In a later experiment, a cell-permeable small-molecule ATIC inhibitor with *in vivo* activity was subsequently identified (K_i 685 ± 35 nM, representing a 25-fold improvement in activity) from a library of backbone *N*-methylated and non-natural analogues based upon the active arginine–tyrosine dipeptide motif of the parent cyclic peptide.[21] This compound has since been shown to be functional *in vivo*.[22]

Similarly, the above approach has been used to target a variety of PPIs. In one example, the NADH-dependent homodimerization of the C-terminal binding protein (CtBP) transcriptional co-repressor was targeted with a nonamer *cyclo*-SGWX$_6$ cyclic peptide library of 6.4×10^7 members, leading to the identification of CP61 (*cyclo*-SGWTVVRMY) which inhibits CtBP dimerization *in vitro* and in cells with a concentration giving 50% of maximum inhibition (IC_{50}) of 19 ± 4 μM.[23] In another example, SICLOPPS screening has led to the identification of the first isotope-specific inhibitor of the heterodimeric transcription factor HIF-1, which alters the expression of over 300 genes in response to hypoxia (low oxygen). The screen targeted the HIF-1α/HIF-1β PPI with a CX$_5$ SICLOPPS library of 3.2×10^6 members, from which the lead compound (*cyclo*-CLLFVY) was shown to be a selective HIF-1 inhibitor *in vitro* and in cells. Correspondingly, this molecule was demonstrated to act by binding to the PAS-B domain of HIF-1α with a dissociation constant (K_d) of 124 nM.[24] The extended utility of this cyclic peptide as a chemical tool was further illustrated through the discovery of a feedback loop by which the HIF-1 transcription factor upregulates the expression of HIF-1α in response to hypoxia.[25] The inteins encoding this peptide have also been incorporated into human cells to allow for the selective production of this cyclic peptide and inhibition of HIF-1 assembly in response to hypoxia.[26] Alongside this, other examples include the development of a SICLOPPS RTHS for the selection of two cyclic octapeptides (with the potential for future development into novel antibiotic classes) that interfere with the dimerization of the *Staphylococcus aureus* β-sliding clamp of the replisome, leading to the inhibition of DNA replication, altered cell morphology and, ultimately, cell death.[27]

Interestingly, not all SICLOPPS screens result in the identification of a cyclic peptide hit. The codon set used to encode the randomised region of SICLOPPS includes a stop codon (TAG), which results in a short peptide sequence (of one to five amino acids, depending on the location of TAG) attached to the I_c. There is therefore the potential to identify linear peptide hits when using this codon set. This was observed in a recent example using a SICLOPPS library for the systematic selection of inhibitors of the *Bacillus anthracis* protective antigen with human CMG2. The screen identified a series of hits with TAG stop codons within the randomised extein region. This was taken to indicate the requirement for a linear backbone to target this interaction, and indeed, the leading linear peptide was shown to be an inhibitor of this PPI *in vitro* and to block anthrax toxin entry into cells.[28]

9.2.2 SICLOPPS in Eukaryotic Cells

Alongside *E. coli*, SICLOPPS has likewise been employed in various eukaryotic-cell-based assays for the systematic identification of cyclic peptide modulators of various PPIs. In a landmark example, SICLOPPS libraries comprising variants of the DnaB intein from *Ssp* PCC6083 were engineered into a retroviral expression system for the generation of cyclic peptide libraries in human BJAB lymphocytes to target the interleukin-4 (IL-4, a T-cell-derived multifunctional cytokine) signalling pathway, whose overstimulation is linked to atopic conditions including asthma, allergic rhinitis and atopic dermatitis.[29,30] Accordingly, the construction of a GFP-tagged retroviral vector encoding a SX_4 library of 1.60×10^5 variants led to the selection of 13 sequences for inhibitors of IL-4 induction of sterile ε-germ line transcription. Although inadvertently disadvantaged by its low actual diversity and lengthy week-long selection process, the above study is noted for having attained the first demonstration of functional screening of SICLOPPS-derived cyclic peptide libraries in human cells. In another example, a modified SICLOPPS library with a C-terminal intein mutation (N36A) incapable of splicing beyond the lariat intermediate, was employed for the identification of cyclic peptide modulators using a yeast two-hybrid system of the bacterial SOS response protein LexA[31] as well as to Abl kinase (H396P).[32]

More recently, SICLOPPS libraries have been employed in *Saccharomyces cerevisiae* to identify novel structurally stable inhibitors of α-synuclein,[33] a presynaptic neuronal protein whose abnormal aggregation to form insoluble fibrils characterized by the presence of Lewy bodies is linked genetically and neuropathologically to Parkinson's disease (PD).[34] Here, a yeast-based selection-compatible, high-diversity octamer SICLOPPS library was interfaced with a yeast synucleinopathy model screen that resulted in the isolation of two unique cyclic peptides that suppressed α-synuclein toxicity whilst preventing dopaminergic neuron loss in a *Caenorhabditis elegans* Parkinson's model. It should be noted, however, that a major limitation to screening in yeast is its relatively low transformation efficiency when compared with *E. coli* (5 million transformants *versus* 1.28 billion). Together, the above examples illustrate the utility, flexibility and overall ease of use of SICLOPPS libraries that can be both readily prepared using standard molecular biology techniques and interfaced with a variety of cell-based assays.

9.3 Phage Display

Phage-display represents an established technique for the selection of unique polypeptide ligands from highly diverse combinatorial libraries of up to 10^{10} variants that are presented (or displayed) on the surface of phage.[35] In particular, a major advantage of phage display lies in its ability to establish a direct physical linkage between the phage phenotype and its encapsulated genotype, whereby exogenous DNA encoding the molecule to be displayed (*i.e.* peptide library) is cloned into the phage genome such that its expression is achieved in fusion with phage coat proteins (typically viral coat protein pIII, the most exploited for efficient N-terminal surface accessibility) for display upon the surface of modified phage virions.[36–40] In practice, a variety of proteins and peptides have been displayed using lysogenic filamentous

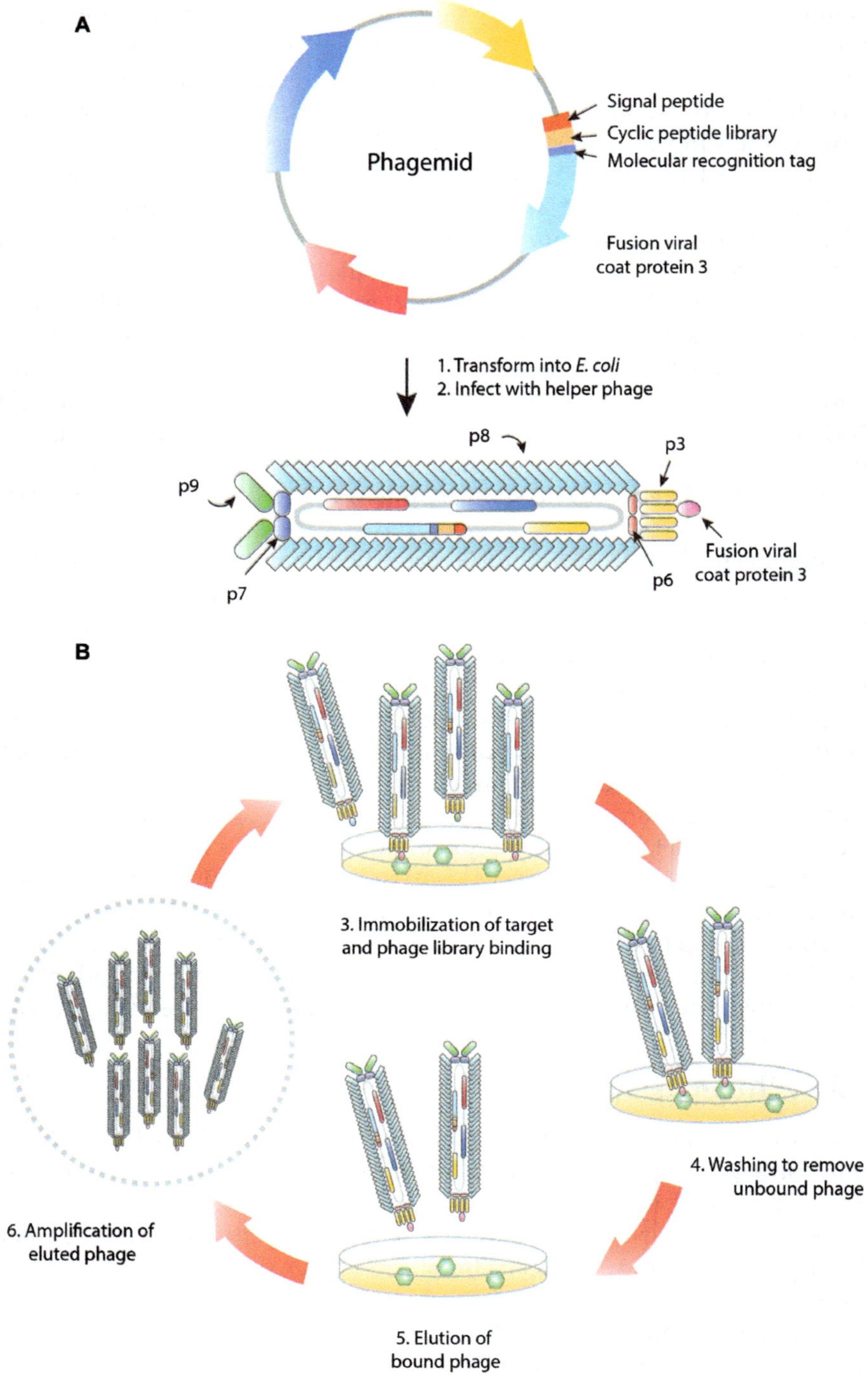
A
Phagemid
Signal peptide
Cyclic peptide library
Molecular recognition tag
Fusion viral coat protein 3
1. Transform into E. coli
2. Infect with helper phage
p8
p3
p9
Fusion viral coat protein 3
p7
p6
B
3. Immobilization of target and phage library binding
4. Washing to remove unbound phage
5. Elution of bound phage
6. Amplification of eluted phage

phage, including alkaline phosphatase (60 kDa), mustard trypsin inhibitor (7 kDa), Src homology 3 (SH3) (6.5 kDa) and cytochrome b562 (11 kDa),[6] for which the *E. coli* filamentous bacteriophage (Ff, referred to as phage) typically employed in phage display include members of the *Inoviridae* family (genus *Inovirus*) such as M13, f1 and fd comprising circular single-stranded DNA enclosed within phage coat proteins (approximately 98 % identity).[36]

For selections, the identification of unique modulatory molecules from combinatorial phage-displayed libraries is typically achieved through use of an *in vitro* high-throughput affinity-selection-based strategy, termed biopanning, in which the target molecule for which selections will be performed against is immobilized upon a solid support or surface (*e.g.* a multi-well microtitre plate) and later exposed to a phage population comprising the displayed peptide library (Figure 9.5).[41] Thereafter, the enrichment of favourable binding affinities is achieved through exposing the target–phage complex to successive and stringent wash steps to eliminate non-specific or weakly binding phage. Finally, propagation of desirable phage *via* bacterial-host-cell infection followed by the identification of target-specific binders of the selected modulatory molecule is achieved through DNA sequencing of individual phage clones.[41,42]

Although other advantages of phage display over traditional randomized screening techniques include the simplicity, reduced cost, effectiveness and speed of screening, another major strength lies in its ability to generate enormously diverse exogenous peptide or protein libraries of up to 10^{10} variants using standard molecular biology techniques, as opposed to the use of individual genetically engineered variants.[35] In addition, several possibilities exist for cyclising the displayed peptides, *e.g.* disulfide bonds (discussed further below). As a result, phage display has emerged as a universally adopted and extensively used method for the systematic identification of specific proteinaceous ligands for various macromolecular targets.

9.3.1 Applications of Phage Display in Drug Discovery

The creation of cyclic peptide libraries using phage display for the selection of unique modulatory molecules against challenging PPIs can be achieved

Figure 9.5 Illustration of the creation of phage-displayed cyclic peptide libraries with phagemid. (A) A plasmid encoding the viral coat protein 3 (p3) fused to a randomized peptide sequence is transformed into *E. coli* and subsequently infected with a helper phage that encodes the remaining proteins for correct phage assembly. The resulting members of the newly created phage-displayed library are next assessed for binding to the desired biological target *in vitro via* biopanning. (B) Schematic of biopanning in which the target molecule is immobilized against a solid support and exposed to a phage population displaying the desired library. Non-specific binding or weakly binding phage are eliminated through stringent washing steps, after which elution and retrieval of target-bound phage with the desired binding activity is performed.

by disulphide cysteine oxidation, whereby a randomized linear sequence is flanked by a pair of cysteine residues that form a spontaneous disulfide bond under non-reducing conditions.[43–45] In particular, M13-based phages are particularly suited to the construction of such libraries, since their resilience and added tolerance to extremes of pH, DNase and/or proteolytic enzymes, in addition to non-aqueous solutions, permits the cyclization of sequences including $-CX_5C$ and $-CX_6C$ (where X = any peptidogenic amino acid).[6] As such, this methodology has been widely employed for the creation of vast disulphide-bonded cyclic peptide libraries for the identification of cyclized polypeptide ligands of various biological macromolecules.

In one example, disulphide cysteine oxidation was used for the display of a random disulphide-constrained heptapeptide library for the selection of two high affinity ligands (C-WSFFSNI-C and C-WPFWGPW-C) against full-length hepatitis B core antigen (HBcAg, a Hepatitis B viral protein found on the surface of the nucleocapsid core) with K_d values of 13.5 ± 2.1 nM and 22.1 ± 1.7 nM, respectively.[46] In another example, phage display was used to discover a novel NeutrAvidin and Avidin-binding cyclic peptide motif ($DX_aAX_bPX_c$ where $X_a = R$ or L; $X_b = S$ or T; and $X_c = Y$ or W) distinct from the well-studied streptavidin-binding His–Pro–Gln peptide motif, with two highly selective six-residue disulphide-constrained cyclic peptide ligands, CDRATPYC and CDRASPYC, for NeutrAvidin (a chemically deglycosylated version of Avidin) and Avidin binding.[47]

A cysteine-constrained cyclic nonapeptide-bearing phage-display library was similarly utilized to identify several novel peptide inhibitors including *cyclo*-CPSNVNNIC, *cyclo*-CPMSQNPTC and *cyclo*-CPKLHPGGC, against density-gradient-purified and UV-treated Andes virus (ANDV) strain CHI-7913, a National Institute of Allergy and Infectious Diseases (NIAID) category A agent linked to hantavirus cardiopulmonary syndrome (HCPS) and characterized by pulmonary oedema caused by capillary leak, with death often resulting from cardiogenic shock.[48]

Phage-display screening has also lead to the identification of ligands for the proinflammatory cytokine tumour necrosis factor-α (TNF-α),[49] an abundant cytokine generated by immunological cell types, including macrophages and monocytes, but implicated in various inflammatory and autoimmune conditions, such as rheumatoid arthritis, Crohn's disease, multiple sclerosis and chronic hepatitis C.[49] Specifically, screening of a cyclic heptamer peptide library enabled the identification of CPATLTSLC from 28 determined peptide sequences with a predicted half-life ($T_{1/2}$) of 20 hours and estimated apparent dissociation constant of 1.12×10^{-4} M when compared against its linear form (1.79×10^{-4} M). In the same year, two cyclic peptide inhibitors of the lens epithelium-derived growth factor (LEDGF)/p75 PPI were identified. The targeted PPI is a transcriptional co-activator and HIV-integrase (IN) co-factor required for the tethering and correct integration of the viral genome into the host chromatin, to prevent human immunodeficiency virus (HIV) replication.[50] Here, selectivity was observed for two cyclic peptides, CVSGHPLWCGGGK and CILGHSDWCGGGK, that preferentially inhibit the LEDGF/p75-IN interaction *in vitro*, and over successive days blocked HIV replication at nontoxic levels in cells, thereby serving as proof-of-concept for the direct targeting of LEDGF/p75 as a novel therapeutic strategy for HIV.

A more recent development of phage-display technology has been the ability to produce and screen bicyclic peptide libraries. This approach uses a chemical linker that reacts with three free cysteines in the displayed peptide to form two rings that constrain the linear peptide and thereby increase its overall rigidity.[51] In one example, a displayed protein containing a CX_nCX_nC (X = three to six variable residues) terminus may be cyclized *via* reaction with a tris-(bromomethyl)benzene derivative to permit the formation of bicyclic peptide libraries.[52] This methodology enables the production of highly structured libraries, which has been proposed to increase the target selectivity of these peptides. In another example a library of constrained bicyclic peptides was screened for molecules that bind inhibitors against the trypsin-like serine protease, urokinase-type plasminogen activator (uPA).[53] With a combinatorial repertoire of 4.0×10^9 peptide variants cyclized by the trifunctional tris(bromomethyl)benzene on the surface of filamentous bacteriophage, a potent 17-mer bicyclic competitive inhibitor (UK18, ACSRYEVDCRGRGSACG) of human uPA activity ($K_i = 53 \pm 4$ nM) was isolated, in which the conformational constraint imposed by bicyclization lead to improved potency of inhibition against human uPA when compared with linear ($K_i = 17.5 \pm 0.7$ nM) and monocyclic forms ($K_i = 383 \pm 37$ nM). More recently, bicyclic phage-displayed libraries have been used for the identification of six bicyclic peptides binding β-catenin, a multifunctional intracellular signal transducer of the Wnt growth factor signalling pathway involved in the over-proliferation and onset of tumorigenesis from *e.g.* gain-of-function mutations, with micromolar K_d *via* three library cyclization strategies.[54] Here, phages displaying over 4×10^9 randomized peptide variants adopting a ACX_6CX_6CG format were cyclized using three chemical linkers, 1,3,5-tris(bromomethyl)benzene (TBMB), 1,3,5-triacryloyl-1,3,5-triazinane (TATA) and N,N',N''-(benzene-1,3,5-triyl)-tris(2-bromoacetamide) (TBAB) to yield three separate combinatorial bicyclic peptide libraries yielding a total diversity of over 12 billion peptide macrocycles (Figure 9.6).

In another example, a phage-display-based selection strategy has been used for the isolation of a potent thioether-linked bicyclic macrocycle inhibitor (PK15), with two peptide loops anchored to a mesitylene core following tris-(bromomethyl)benzene conjugation *via* three reactive cysteine residues, of human plasma kallikrein (a subgroup of serine proteases). The most potent hit had a K_i of 1.5 nM and was shown to interrupt the intrinsic coagulation pathway in human plasma tested *ex vivo*.[51]

In summary, phage display represents an accessible method for the generation of disulphide-bonded cyclic peptides and for their use in affinity-based assays. These cyclic peptides are, however, limited by the fact that they are disulphide-bonded, making them susceptible to reducing

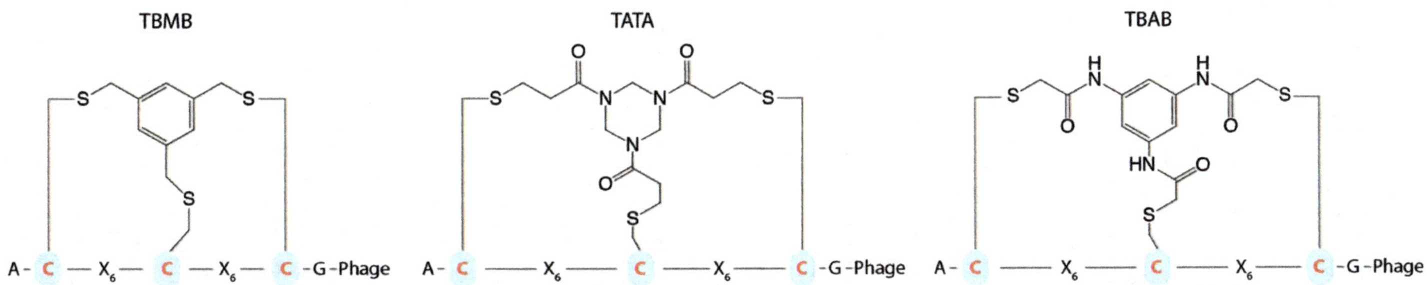

Figure 9.6 Cyclization of phage-displayed peptides of the format ACX_6CX_6CG in separate rate reactions with the three chemical linkers: 1,3,5-tris(bromomethyl)benzene (TBMB), 1,3,5-triacryloyl-1,3,5-triazinane (TATA) and *N*,*N*′,*N*″-(benzene-1,3,5-triyl)-tris(2-bromoacetamide) (TBAB).[54]

environments.[55] Like all display-based technologies, it is only compatible with affinity-based selection systems, and therefore additional assays are required to determine whether a potent binder of a target also inhibits its function. Nevertheless, the development of bicycle peptide libraries has breathed new life into phage display and promises to provide a series of highly potent binders; however, due to the challenges of cell permeability for these large compounds, they are best employed against extracellular protein targets.

9.4 mRNA Display

The decoding of hits from genetically encoded cyclic peptide libraries requires co-localization of the cyclic peptide with the oligonucleotide encoding it. In SICLOPPS and phage display the container is a cell or a phage. An alternative approach for keeping the code with the resulting molecule is to covalently link the two molecules. In mRNA display, the C-terminus of the encoded polypeptide is linked to the encoding mRNA sequence by a puromycin linkage.[55,57] Following puromycin attachment, reverse transcription [used to generate complementary DNA (cDNA) from an RNA template] is utilized to prevent the inadvertent interference of mRNA with peptide binding; thereafter, *in vitro*-based affinity selection for an immobilized target protein with subsequent amplification of desirable affinities by PCR to generate an enriched library of initial hits can be performed. Cyclization of the resulting peptide chain requires additional steps, which necessitates the addition of reactive groups during translation or *via* post-translational modification. Examples of these include dibromoxylene-mediated bridging of the sulfhydryl group of two cysteines *via* a reconstituted bacterial translation system,[58] and disuccinimidyl glutarate cross-linking between the *N*-terminus of a rabbit reticulocyte-lysate-produced peptide and lysine side-chain.[59] A key advantage of such methods is their applicability to the standard proteinogenic amino acids, but the presence of three or more reactive residues within the randomized region of a library can lead to scrambling of crosslinking patterns, making it difficult to deconvolute hits.[55]

A more technically challenging yet consistent method for the *in vitro* generation of macrocyclic-based peptide libraries uses custom-made reconstituted translational systems that omit certain aminoacyl-tRNA synthases and tRNAs that are reassigned to nonproteinogenic amino acids. One methodology using this approach uses lanthionine-like macrocyclic peptides through mischarging of aminoacyl-tRNA synthetases to incorporate 4-selenalysine; selective oxidation and elimination of the seleno group forms a dehydroalanine residue that in turn reacts *via* Michael addition with the sulfhydryl group of an existing cysteine to yield a thioether bond.[60] An alternative approach uses 'flexible' tRNA acylation ribozymes capable of charging a diverse range of amino acids onto tRNA, referred to as flexizymes, to facilitate genetic code reprogramming with

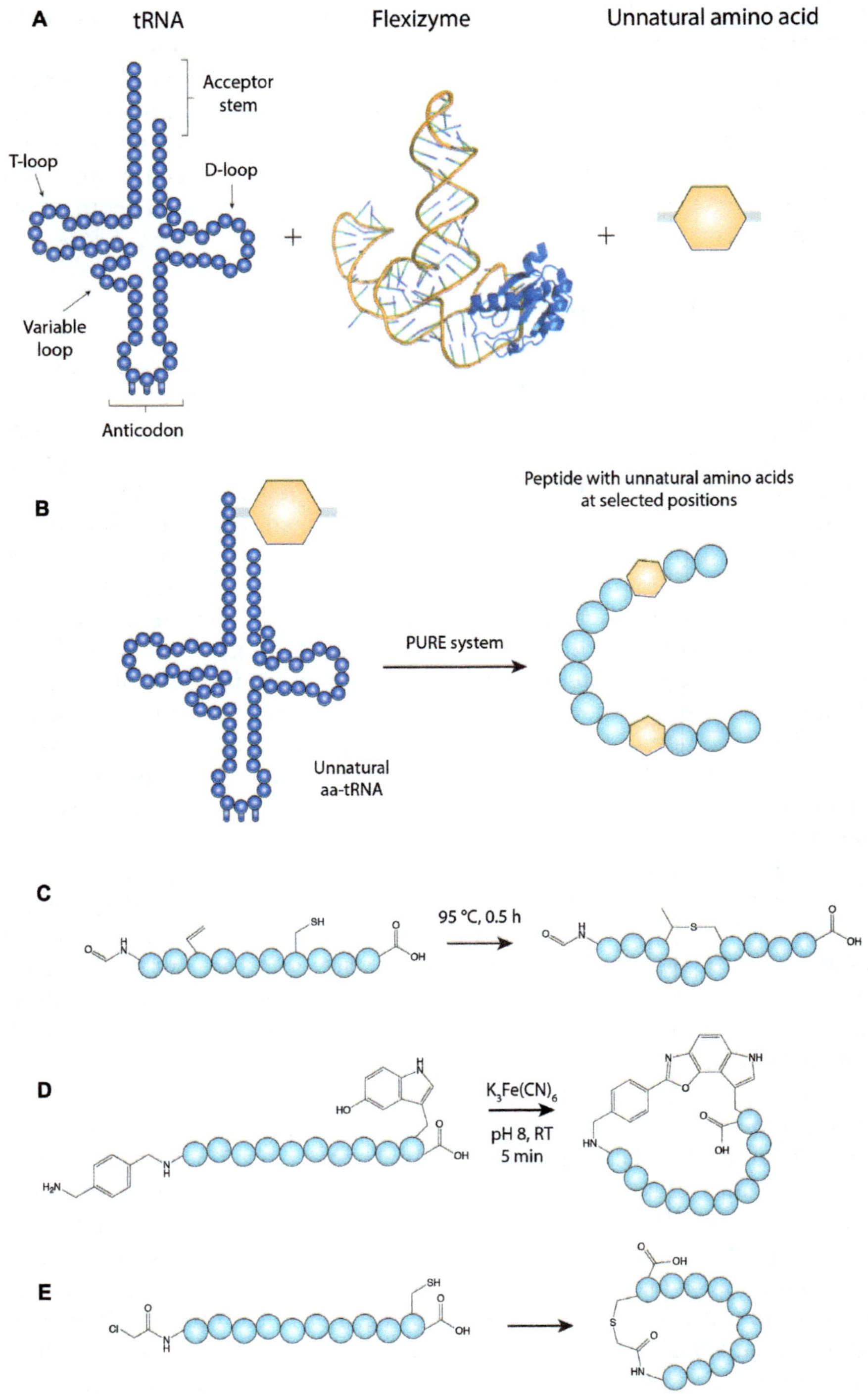
A
tRNA
Flexizyme
Unnatural amino acid
Acceptor stem
T-loop
D-loop
Variable loop
Anticodon
B
Peptide with unnatural amino acids at selected positions
PURE system
Unnatural aa-tRNA
C
95 °C, 0.5 h
D
K3Fe(CN)6
pH 8, RT
5 min
E

facile preparation of a diverse selection of nonproteinogenic aminoacyl tRNAs with the capacity to read vacant codons created *via* a flexible *in vitro* translation (FIT) system.[61–64] Correspondingly, flexizymes have been widely utilized for the creation of nonstandard peptide libraries to facilitate the translation of exotic structures comprising *e.g.* *N*-alkylglycines,[65] *N*-methyl,[66] and D-amino acids[67] with physiologically stable macrocyclic scaffolds.[6]

The FIT system, based upon the protein-synthesizing PURE *in vitro* transcription/translation (IVTT) system reconstituted from recombinant tagged protein factors purified to homogeneity,[68] achieves the incorporation of unnatural amino acids at defined positions through the removal of aminoacyl tRNA synthetases (aaRSs) specific to the targeted codon for reprogramming. The removal of aaRSs alongside their related amino acids generates an empty codon that can be targeted by the addition of flexizyme-generated unnatural aminoacyl-tRNA displaying the relevant anti-codon (Figure 9.7A,B). In general, the FIT system allows for the cyclization of linear peptide sequences *via* a plethora of cyclization techniques through exploitation of the ribosomal synthesis of macrocyclic nonproteinogenic peptides capable of crosslinking with other proteinogenic or non-proteinogenic residues.[55] For example, thermal isomerization of a systematically incorporated vinylglycine to dehydrobutyrine with concurrent thioether bond formation with cysteine residues (thereby generating methyllanthionine-like macrocyclic peptides[69]) in addition to the creation of a fluorogenic indole benzoxazole bridge *via* mild oxidative cyclization resulting from the translation of peptide sequences initiated with *N*-benzylamino and the incorporation of a downstream 5-hydroxytryptophan,[70] have utilized the FIT system for successful macrocycle formation (Figure 9.7C,D). In particular, the initiation of peptide translation *via* an *N*-chloroacetyl-amino acid for linear precursor peptide synthesis with selective thioether bond formation (and hence ring closure) resulting from a spontaneous attack of the sulfhydryl group on the terminal cysteine residue to the α-carbon of the *N*-terminal chloroacetyl group, remains one of the most convenient and reliable methods for FIT-system-mediated peptide

Figure 9.7 Overview of the flexible *in vitro* translation system. (A) Components required for the flexizyme-mediated incorporation of unnatural amino acids into a polypeptide sequence for cyclization. (B) Generation of an unnatural aa-tRNA with the desired anticodon and subsequent incorporation into a cyclic peptide *via* coupling with a specifically modified PURE translational system lacking the corresponding native aa-tRNA [the flexible *in vitro* translation (FIT) system]. (C) The incorporation of vinylglycine, forming a methyllanthionine bridge with a cysteine residue. (D) Initiation of peptide translation *via* *N*-benzylamine and the incorporation of hydroxytryptophan to create a fluorogenic benzoxazole bridge, as well as (E) an *N*-chloroacetyl amino acid to form thioether bonds with cysteine residues.

cyclization (Figure 9.7E).[71] The potential diversity of mRNA display libraries is several orders of magnitude larger than those from phage display and SICLOPPS that are limited by transformation efficiency into the host organism (approximately 10^9 in *E. coli*).[57] To date, various mRNA-based display selections of macrocyclic peptides have been reported, including the preparation of over 10^{11} unique lanthionine-containing peptides from which a macrocyclic inhibitor with high specific affinity against Sortase A ($K_d = 3 \pm 0.2$ μM), a transamidase required for *Staphylococcus aureus* virulence, was selected.[72]

Combination of the FIT system with mRNA display, termed the random non-standard peptides integrated discovery (RaPID) system, has enabled the generation of libraries of cyclic peptides containing multiple non-natural amino acids.[73–76] For instance, this system was used for the selection of "natural-product-like" thioether-macrocyclic *N*-methyl-peptide inhibitors (CM_{11}-1) of ubiquitin ligase E6AP-catalyzing polyubiquitination on the target proteins Prx1 and p53.[77] Likewise, *N*-chloroacetyl-L-amino acid- or D-amino acid-mediated initiation of peptide translation, yielding thioether macrocyclized peptide libraries that are covalently displayed upon their corresponding mRNAs, has been used to achieve the selection of serine/threonine (S/T) kinase Akt-selective and Akt2-isoform-selective thioether-macrocyclic peptide inhibitors from a library of 10^{12} variants to control the misregulation of Akt signalling and of its associated downstream pathways, including apoptosis and cellular proliferation.[73]

In addition to the latter, the RaPID system has been successfully employed for the identification of isoform-specific inhibitors of human deacetylase SIRT2 (NAD-dependent deacetylase sirtuin-2) with an IC_{50} of 3 nM, 10- and 100-fold greater than those against SIRT1 and SIRT3, respectively.[75,78] In another example, three inhibitory macrocyclic peptide inhibitors (MaL6, MaD3S and MaD5) of PfMATE (*Pyrococcus furiosus* multidrug and toxic compound extrusion, a transmembrane drug–sodium or proton antiporter protein homologous to NorM-VC) serving also as co-crystalization ligands of selenomethionine-labelled PfMATE at 3.0–2.4 Å resolutions, revealed the first instance of three-dimensional structure determination of selected macrocyclic inhibitors isolated *via* the RaPID system.[74,79] More recently, a novel macrocyclic peptide inhibitor (aCAP) against a eukaryotic ATP-binding cassette (ABC) P-glycoprotein homolog-drug transporter (CmABCB1 from *Cyanidioschyzon merolae*) with an $IC_{50} = 65$ nM was used to obtain CmABCB1 crystal structures in unbound and unique allosteric bound inhibitor forms at 2.6 and 2.4 Å resolutions, respectively.[80]

Arguably, mRNA display has been the most commercially successful method for the generation and screening of cyclic peptide libraries. Although the major advantage of this method lies in its ability to incorporate multiple non-natural amino acids, it suffers from the same limitation as phage display, in that it is only compatible with affinity-based assays. Nonetheless, this method has proven to be robust for the identification of high-affinity binders to a wide variety of protein targets.[55]

9.5 Concluding Remarks

There is increasing and significant interest in using cyclic peptides as molecular scaffolds in drug discovery for the development of novel therapeutic compounds against challenging biomolecular targets, including PPIs.[14] This progression has been forthcoming, driven partly by the diminishing returns from current small-molecule pipelines. As a result, increasing attention has been drawn to cyclic peptide backbones as potential scaffolds that bridge the gap between the pharmacokinetic properties of small organic molecules and biologics whilst maintaining high potency and target selectivity.[81] The improvement of their pharmacological properties in conjunction with current advances in rational drug design, genetically encoded methods for generation, which have been detailed herein, peptide synthesis and structure determination have, in turn, resulted in the development of an increasing number of cyclic peptide inhibitors against a variety of challenging targets. As such, cyclic-peptide-based biologics have emerged as privileged scaffolds in drug discovery that, alongside their promising efficacy, hold much promise for use in early-stage drug discovery and beyond.

References

1. A. R. Horswill and S. J. Benkovic, *Cell Cycle*, 2005, **4**(4), 552–555.
2. A. Zorzi, K. Deyle and C. Heinis, *Curr. Opin. Chem. Biol.*, 2017, **38**, 24–29.
3. A. Laupacis, P. A. Keown, R. A. Ulan, N. McKenzie and C. R. Stiller, *Can. Med. Assoc. J.*, 1982, **126**(9), 1041–1046.
4. D. Landman, C. Georgescu, D. A. Martin and J. Quale, *Clin. Microbiol. Rev.*, 2008, **21**(3), 449–465.
5. C. S. Carter, *Annu. Rev. Psychol.*, 2014, **65**, 17–39.
6. A. D. Foster, J. D. Ingram, E. K. Leitch, K. R. Lennard, E. L. Osher and A. Tavassoli, 2015, **20**(5), 563–576.
7. F. Sanger, S. Nicklen and A. R. Coulson, *Proc. Natl. Acad. Sci. U. S. A.*, 1977, **74**(12), 5463–5467.
8. W. J. Dower, J. F. Miller and C. W. Ragsdale, *Nucleic Acids Res.*, 1988, **16**(13), 6127–6145.
9. C. Heinis, T. Rutherford, S. Freund and G. Winter, *Nat. Chem. Biol.*, 2009, **5**(7), 502–507.
10. S. Chong, K. S. Williams, C. Wotkowicz and M.-Q. Xu, *J. Biol. Chem.*, 1998, **273**(17), 10567–10577.
11. H. Paulus, *Chem. Soc. Rev.*, 1998, **27**, 375.
12. C. P. Scott, E. Abel-Santos, A. D. Jones and S. J. Benkovic, *Chem. Biol.*, 2001, **8**(8), 801–815.
13. A. Tavassoli and S. J. Benkovic, *Nat. Protoc.*, 2007, **2**(5), 1126–1133.
14. A. Tavassoli, *Curr. Opin. Chem. Biol.*, 2017, **38**, 30–35.
15. J. E. Townend and A. Tavassoli, *ACS Chem. Biol.*, 2016, acschembio.6b00095.
16. K. R. Lennard and A. Tavassoli, *Chemistry*, 2014, **20**(34), 10608–10614.

17. A. R. Horswill, S. N. Savinov and S. J. Benkovic, *Proc. Natl. Acad. Sci. U. S. A.*, 2004, **101**, 15591–15596.
18. M. Akoachere, R. C. Squires, A. M. Nour, L. Angelov, J. Brojatsch and E. Abel-Santos, *J. Biol. Chem.*, 2007, **282**(16), 12112–12118.
19. A. Tavassoli and S. J. Benkovic, *Angew. Chem., Int. Ed.*, 2005, **44**, 2760–2763.
20. R. J. Giordano, M. Cardó-Vila, R. L. Sidman, L. F. Bronk, Z. Fan, J. Mendelsohn, W. Arap and R. Pasqualini, *Proc. Natl. Acad. Sci. U. S. A.*, 2010, **107**(11), 5112–5117.
21. I. B. Spurr, C. N. Birts, F. Cuda, S. J. Benkovic, J. P. Blaydes and A. Tavassoli, *ChemBioChem*, 2012, **13**(11), 1628–1634.
22. D. J. Asby, F. Cuda, M. Beyaert, F. D. Houghton, F. R. Cagampang and A. Tavassoli, *Chem. Biol.*, 2015, **22**(7), 838–848.
23. C. N. Birts, S. K. Nijjar, C. A. Mardle, F. Hoakwie, P. J. Duriez, J. P. Blaydes and A. Tavassoli, *Chem. Sci.*, 2013, **4**(8), 3046–3057.
24. E. Miranda, I. K. Nordgren, A. L. Male, C. E. Lawrence, F. Hoakwie, F. Cuda, W. Court, K. R. Fox, P. A. Townsend, G. K. Packham, S. A. Eccles, A. Tavassoli, 2013, **4**(8), 3046–3057.
25. D. J. Asby, F. Cuda, F. Hoakwie, E. Miranda and A. Tavassoli, *Mol. BioSyst.*, 2014, **10**(10), 2505–2508.
26. I. N. Mistry and A. Tavassoli, *ACS Synth. Biol.*, 2017, **6**(3), 518–527.
27. S. Kjelstrup, P. M. P. Hansen, L. E. Thomsen, P. R. Hansen and A. Løbner-Olesen, *PLoS One*, 2013, **8**(9), e72273.
28. A. L. Male, F. Forafonov, F. Cuda, G. Zhang, S. Zheng, P. C. F. Oyston, P. R. Chen, E. D. Williamson and A. Tavassoli, *Sci. Rep.*, 2017, **7**(1), 3104.
29. H. C. Oettgen and R. S. Geha, *J. Allergy Clin. Immunol.*, 2001, **107**(3), 429–440.
30. T. M. Kinsella, C. T. Ohashi, A. G. Harder, G. C. Yam, W. Li, B. Peelle, E. S. Pali, M. K. Bennett, S. M. Molineaux, D. A. Anderson, E. S. Masuda and D. G. Payan, *J. Biol. Chem.*, 2002, **277**(40), 37512–37518.
31. K. Barreto, V. M. Bharathikumar, A. Ricardo, J. F. DeCoteau, Y. Luo and C. R. Geyer, *Chem. Biol.*, 2009, **16**(11), 1148–1157.
32. V. M. Bharathikumar, K. Barreto, J. F. Decoteau and C. R. Geyer, *ChemBioChem*, 2013, **14**(16), 2119–2125.
33. J. A. Kritzer, S. Hamamichi, J. M. McCaffery, S. Santagata, T. A. Naumann, K. A. Caldwell, G. A. Caldwell and S. Lindquist, *Nat. Chem. Biol.*, 2009, **5**(9), 655–663.
34. L. Stefanis, *Cold Spring Harbor Perspect. Med.*, 2012, **2**(2), 1–23.
35. M. Hamzeh-Mivehroud, A. A. Alizadeh, M. B. Morris, W. Bret Church and S. Dastmalchi, *Drug Discovery Today*, 2013, **18**(23–24), 1144–1157.
36. J. Bazan, I. Całkosiñski and A. Gamian, *Hum. Vaccines Immunother.*, 2012, **8**(12), 1817–1828.
37. G. Smith, *Science*, 1985, **228**(4705), 1315–1317.
38. S. F. Parmley and G. P. Smith, *Gene*, 1988, **73**(2), 305–318.
39. A. N. Zacher, C. A. Stock, J. W. Golden and G. P. Smith, *Gene*, 1980, **9**(1–2), 127–140.

40. J. K. Scott and G. P. Smith, *Science*, 1990, **249**(4967), 386–390.
41. M. A. Barry, W. J. Dower and S. A. Johnston, *Nat. Med.*, 1996, **2**(3), 299–305.
42. L. Aghebati-Maleki, B. Bakhshinejad, B. Baradaran, M. Motallebnezhad, A. Aghebati-Maleki, H. Nickho, M. Yousefi and J. Majidi, *J. Biomed. Sci.*, 2016, **23**(1), 66.
43. K. T. O'Neil, R. H. Hoess, S. A. Jackson, N. S. Ramachandran, S. A. Mousa and W. F. DeGrado, *Proteins: Struct., Funct., Genet.*, 1992, **14**(4), 509–515.
44. A. Luzzago, F. Felici, A. Tramontano, A. Pessi and R. Cortese, *Gene*, 1993, **128**(1), 51–57.
45. M. A. McLafferty, R. B. Kent, R. C. Ladner and W. Markland, *Gene*, 1993, **128**(1), 29–36.
46. K. L. Ho, K. Yusoff, H. F. Seow and W. S. Tan, *J. Med. Virol.*, 2003, **69**(1), 27–32.
47. S. C. Meyer, T. Gaj and I. Ghosh, *Chem. Biol. Drug Des.*, 2006, **68**(1), 3–10.
48. P. R. Hall, B. Hjelle, H. Njus, C. Ye, V. Bondu-Hawkins, D. C. Brown, K. A. Kilpatrick and R. S. Larson, *J. Virol.*, 2009, **83**(17), 8965–8969.
49. C. Sclavons, C. Burtea, S. Boutry, S. Laurent, L. Vander Elst and R. N. Muller, *Int. J. Pept.*, 2013, **2013**, 348409.
50. B. A. Desimmie, M. Humbert, E. Lescrinier, J. Hendrix, S. Vets, R. Gijsbers, R. M. Ruprecht, U. Dietrich, Z. Debyser and F. Christ, *Mol. Ther.*, 2012, **20**(11), 2064–2075.
51. C. Heinis, T. Rutherford, S. Freund and G. Winter, *Nat. Chem. Biol.*, 2009, **5**(7), 502–507.
52. P. Timmerman, J. Beld, W. C. Puijk and R. H. Meloen, *ChemBioChem*, 2005, **6**(5), 821–824.
53. A. Angelini, L. Cendron, S. Chen, J. Touati, G. Winter, G. Zanotti and C. Heinis, *ACS Chem. Biol.*, 2012, **7**(5), 817–821.
54. D. Bertoldo, M. M. G. Khan, P. Dessen, W. Held, J. Huelsken and C. Heinis, *ChemMedChem*, 2016, **11**(8), 834–839.
55. N. K. Bashiruddin and H. Suga, *Curr. Opin. Chem. Biol.*, 2015, **24**, 131–138.
56. N. Nemoto, E. Miyamoto-Sato, Y. Husimi and H. Yanagawa, *FEBS Lett.*, 1997, **414**(2), 405–408.
57. R. W. Roberts and J. W. Szostak, *Proc. Natl. Acad. Sci. U. S. A.*, 1997, **94**(23), 12297–12302.
58. Y. V. Guillen Schlippe, M. C. T. Hartman, K. Josephson and J. W. Szostak, *J. Am. Chem. Soc.*, 2012, **134**(25), 10469–10477.
59. S. W. Millward, T. T. Takahashi and R. W. Roberts, *J. Am. Chem. Soc.*, 2005, **127**(41), 14142–14143.
60. F. P. Seebeck, A. Ricardo and J. W. Szostak, *Chem. Commun.*, 2011, **47**(21), 6141–6143.
61. Y. Goto, T. Katoh and H. Suga, *Nat. Protoc.*, 2011, **6**(6), 779–790.
62. H. Murakami, A. Ohta, Y. Goto, Y. Sako and H. Suga, *Nucleic Acids Symp. Ser.*, 2006, **50**, 35–36.
63. M. Ohuchi, H. Murakami and H. Suga, *Curr. Opin. Chem. Biol.*, 2007, **11**(5), 537–542.
64. T. Passioura and H. Suga, *Chem. – Eur. J.*, 2013, **19**(21), 6530–6536.

65. T. Kawakami, H. Murakami and H. Suga, *J. Am. Chem. Soc.*, 2008, 16861–16863.
66. T. Kawakami, H. Murakami and H. Suga, *Chem. Biol.*, 2008, **15**(1), 32–42.
67. T. Fujino, Y. Goto, H. Suga and H. Murakami, *J. Am. Chem. Soc.*, 2013, **135**(5), 1830–1837.
68. Y. Shimizu, A. Inoue, Y. Tomari, T. Suzuki, T. Yokogawa, K. Nishikawa and T. Ueda, *Nat. Biotechnol.*, 2001, **19**(8), 751–755.
69. Y. Goto, K. Iwasaki, K. Torikai, H. Murakami and H. Suga, *Chem. Commun.*, 2009, **23**, 3419.
70. Y. Yamagishi, H. Ashigai, Y. Goto, H. Murakami and H. Suga, *ChemBioChem*, 2009, **10**(9), 1469–1472.
71. Y. Goto, A. Ohta, Y. Sako, Y. Yamagishi, H. Murakami and H. Suga, *ACS Chem. Biol.*, 2008, **3**(2), 120–129.
72. F. T. Hofmann, J. W. Szostak and F. P. Seebeck, *J. Am. Chem. Soc.*, 2012, **134**(19), 8038–8041.
73. Y. Hayashi, J. Morimoto and H. Suga, *ACS Chem. Biol.*, 2012, **7**(3), 607–613.
74. C. J. Hipolito, Y. Tanaka, T. Katoh, O. Nureki and H. Suga, *Molecules*, 2013, **18**(9), 10514–10530.
75. J. Morimoto, Y. Hayashi and H. Suga, *Angew. Chem., Int. Ed.*, 2012, **51**(14), 3423–3427.
76. Y. Yamagishi, I. Shoji, S. Miyagawa, T. Kawakami, T. Katoh, Y. Goto and H. Suga, *Chem. Biol.*, 2011, **18**(12), 1562–1570.
77. Y. Yamagishi, I. Shoji, S. Miyagawa, T. Kawakami, T. Katoh, Y. Goto and H. Suga, Chem. Biol., 2011, **18**(12), 1562–1570.
78. K. Yamagata, Y. Goto, H. Nishimasu, J. Morimoto, R. Ishitani, N. Dohmae, N. Takeda, R. Nagai, I. Komuro, H. Suga and O. Nureki, *Structure*, 2014, **22**(2), 345–352.
79. Y. Tanaka, C. J. Hipolito, A. D. Maturana, K. Ito, T. Kuroda, T. Higuchi, T. Katoh, H. E. Kato, M. Hattori, K. Kumazaki, T. Tsukazaki, R. Ishitani, H. Suga and O. Nureki, *Nature*, 2013, **496**(7444), 247–251.
80. A. Kodan, T. Yamaguchi, T. Nakatsu, K. Sakiyama, C. J. Hipolito, A. Fujioka, R. Hirokane, K. Ikeguchi, B. Watanabe, J. Hiratake, Y. Kimura, H. Suga, K. Ueda and H. Kato, *Proc. Natl. Acad. Sci. U. S. A.*, 2014, **111**(11), 4049–4054.
81. J.-P. Bingham, *Med. Chem.*, 2014, **4**(5), 469–472.

CHAPTER 10

Modern Methods for the Synthesis of Carbohydrates

DARSHITA BUDHADEV,[a] RALF SCHWOERER[b] AND MARTIN A. FASCIONE*[a]

[a] Department of Chemistry, University of York, Heslington, York YO10 5DD, UK; [b] Ferrier Research Institute, Victoria University of Wellington, P.O. Box 33436, Petone 5046, New Zealand
*Email: martin.fascione@york.ac.uk

10.1 Introduction

Carbohydrates, or sugars as they are commonly known, are Nature's most prevalent and diverse biopolymer.[1] In their monomeric and polymeric forms they play important roles in a number of biological processes, including tumour metastasis,[2,3] bacterial and viral recognition[4–6] and the immunological response.[7–9] In order to obtain pure samples of oligosaccharides or polysaccharides for biological studies, synthetic carbohydrate chemists have to overcome the myriad challenges presented by their complex synthesis. These include establishing control over the stereoselectivity, chemoselectivity and regioselectivity of glycosidic-bond-forming glycosylation reactions, which differ for nearly every monosaccharide building block. In addition, it is necessary to master the practical challenges associated with the multistep synthesis and purification of target oligo- or polysaccharides. Such is the breadth of research in this field, the aim of this chapter is to provide only a brief introduction to the fundamentals of glycosylation reactions, which have been summarised in great detail on a number of occasions previously, and instead highlight the most recent advances there have been in "modern

Chemical Biology No. 10
Chemical and Biological Synthesis: Enabling Approaches for Understanding Biology
Edited by Nick J. Westwood and Adam Nelson

Published by the Royal Society of Chemistry, www.rsc.org

chemical carbohydrate synthesis" with a focus on synthetic methods that have begun to yield carbohydrates with greater stereocontrol, of longer lengths and in a faster more practical manner.

10.2 Glycosylation: The Basics

A wide variety of monosaccharides (single-carbohydrate units) are commonly found in nature, with different abundances and biological roles across all kingdoms of life. The most common monosaccharide is D-glucose (Glc) (Figure 10.1) which is the component of the energy storage vehicle starch, and in which all alcohol substituents are equatorial. Other common monosaccharides include glucose epimers (diastereomers) such as D-galactose (Gal) and D-mannose (Man), nitrogen-containing *N*-acetyl-D-glucosamine (GlcNAc), the deoxy-sugar L-fucose (Fuc) and the negatively charged nine-carbon sugar *N*-acetyl neuraminic acid (Neu5Ac) (Figure 10.1, showing both the chemical and glycobiology representations of the monosaccharides). All of these monosaccharides (and more) are components of larger oligosaccharides known as *N*-glycans or *O*-glycans, presented on the surface of glycoproteins. The roles these carbohydrates play in sustaining life and in the etiology of the disease is complex and still vastly understudied, which makes the synthesis of pure carbohydrates, in all forms, of paramount importance for chemical biologists.

Glycosylation is the fundamental process in synthetic carbohydrate chemistry and involves the formation of glycosidic bonds in the union of monosaccharide building blocks.[10–12] The most reliable strategy for glycosylations entails the use of a glycosyl donor **1** which presents a reactive anomeric centre (C1) to a suitably protected glycosyl alcohol, or glycosyl acceptor **2** (Scheme 10.1). This process invariably requires chemoselective activation of a leaving group (X) on the glycosyl donor, followed by reaction of the glycosyl acceptor with the newly generated oxocarbenium ion **3** to produce a glycoside **4**.

A testament to the wealth of studies in this area is the wide variety of glycosyl donors and activators that have been developed for use in glycosylation reactions. Glycosyl halides (bromides, chlorides, fluorides and iodides)[13,14] thioglycosides[15] and glycosyl trichloroacetimidates[16] are the most commonly used classes of donors,[17] but the development of glycosyl sulfoxides[18] for challenging glycoside linkages and *n*-pentenyl glycosides[19] for use in armed/disarmed strategies also deserve special mention.

10.3 Stereoselectivity in Glycosylation Reactions

An area of fervent study in synthetic carbohydrate chemistry is the control of the stereochemical outcome of glycosylation reactions.[20] Traditionally the synthesis of 1,2-*trans* β-glycosides (wherein a simplified view β = equatorial) can be achieved with complete stereoselectivity by virtue of neighbouring group participation (NGP) (Scheme 10.2).[21] For example, after activation of

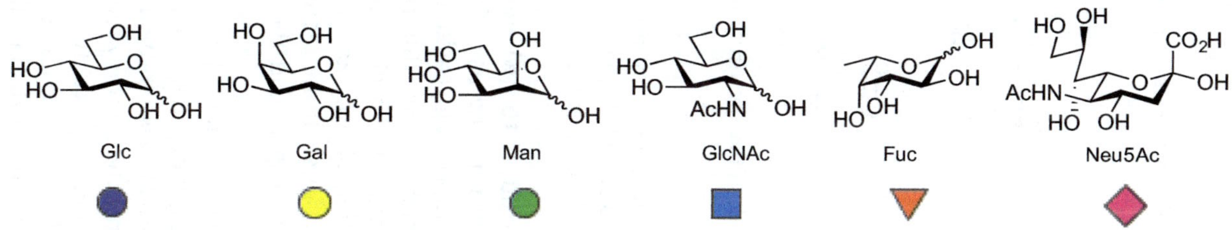

Figure 10.1 Chemical structures of common monosaccharide building blocks, and glycobiology symbol representations.

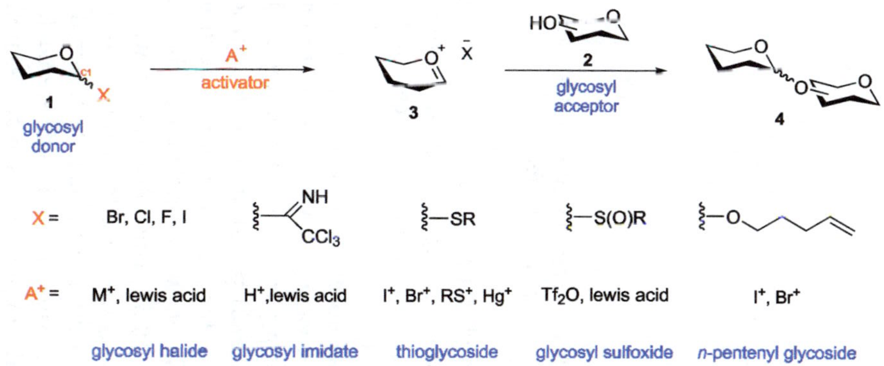

Scheme 10.1 Schematic representation of a glycosylation reaction, and combinations of anomeric leaving groups and activating agents.

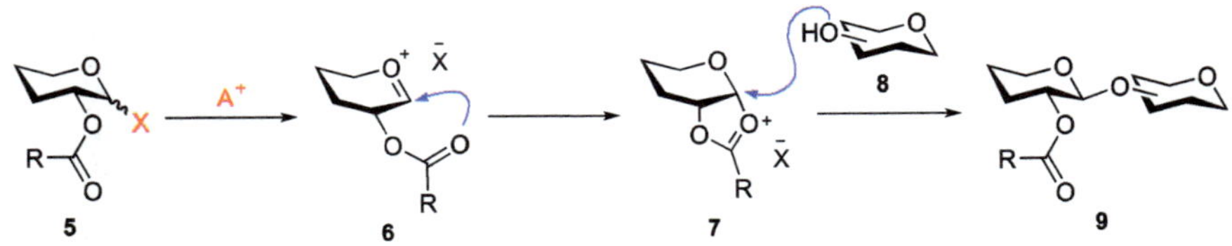

Scheme 10.2 Neighbouring group participation (NGP) in the synthesis of 1,2-*trans* β-glycosides **9**.

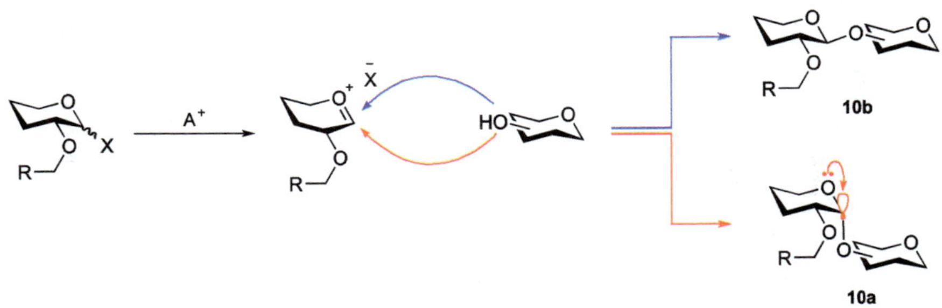

Scheme 10.3 Stereoselective synthesis of 1,2-*cis* α-glycosides (**10a**) by virtue of the anomeric effect.

the glycosyl donor 5, the oxocarbenium ion **6** is intercepted by an ester participating group at *O*-2 to produce a dioxolenium ion 7, which can then react with a glycosyl acceptor **8** stereoselectively to yield a 1,2-*trans* β-glycoside **9**.

The stereoselective synthesis of 1,2-*cis* α-glycosides **10** (wherein a simplified view α = axial) is less straightforward however. The use of non-participating groups at *O*-2 is required to produce 1,2-*cis* α-glycosides **10a** (Scheme 10.3) predominantly by virtue of the anomeric effect; this is the preference for electronegative substituents to orientate themselves in an axial position when attached to the anomeric position of a pyranose ring (Scheme 10.3). This observation was postulated to arise from preferred hyperconjugative overlap between the ring oxygen lone pair and the anti-bonding orbital of the axially oriented substituent.[22]

Despite this established stereoelectronic preference for the 1,2-*cis* α-glycoside, in practice when an *O*-2 non-participating group (such as a benzyl group) is used, quantities of 1,2-*trans* β-glycosides **10b** are also formed, and separation of these stereoisomers at the anomeric centre (commonly referred to as "anomers") is often very difficult, if not impossible (Scheme 10.3).

This problematic purification and implicit inefficiency in the stereo-selective formation of glycosides is a challenge which, in particular, limits the application of developing one-pot and automated synthetic methods for the synthesis of oligosaccharides (see Section 10.6) as consecutive glyco-sylations need to proceed with complete stereoselectivity to prevent the formation of complex inseparable anomeric mixtures on these platforms.

Many elegant strategies have therefore been pioneered to solve this stereoselectivity issue,[23] including the use of glycosyl donors bearing chiral auxiliaries,[24] directing protecting groups or novel leaving groups.[25] One of the earliest approaches to the problem, however, was the use of temporary tethering of glycosyl acceptors (defined as an "aglycon" in this context) to the *O*-2 position of a glycosyl donor **11**; this method is widely known as intramolecular aglycon delivery (IAD, Scheme 10.4). A successful IAD strategy requires an efficient tethering reaction to produce a mixed acetal **12** on either side of the α or β face of the sugar, followed by activation of a glycosyl donor and subsequent stereoselective intramolecular glycosylation to generate 1,2-*cis* α-glycosides **13α** or 1,2-*cis* β-glycosides **13β** (Scheme 10.4 A and B).

IAD was first conceived simultaneously by Barresi and Hindsgaul[26] and by Stork and colleagues[27] as a solution to the problematic synthesis of β-mannosides, widely considered the most challenging linkage to synthesise in carbohydrate chemistry due to steric hindrance arising from the axially positioned *O*-2 during β-face attack, and the absence of any anomeric stereoelectronic stabilisation, which as mentioned favours the α-anomer. Hindsgaul and Barresi utilized vinyl ether **14** as their thioglycoside donor,[28] and tethered a glycosyl acceptor **15** under acidic conditions to generate mixed ketal **16**. Standard I^+ activation of the thioethyl leaving group resulted in an excellent yield of β-mannoside **17** as the sole product, when the primary acceptor was used (Scheme 10.5). Although a high yield of β-mannoside was isolated from the glycosylation reaction, the less efficient tethering reaction limited the general applicability of this method.

Stork and La Clair utilized a simple temporary silicon tether in their version of IAD,[29] where combination with the use of a glycosyl sulfoxide donor allowed β-mannosides to be synthesised intramolecularly at low temperatures.

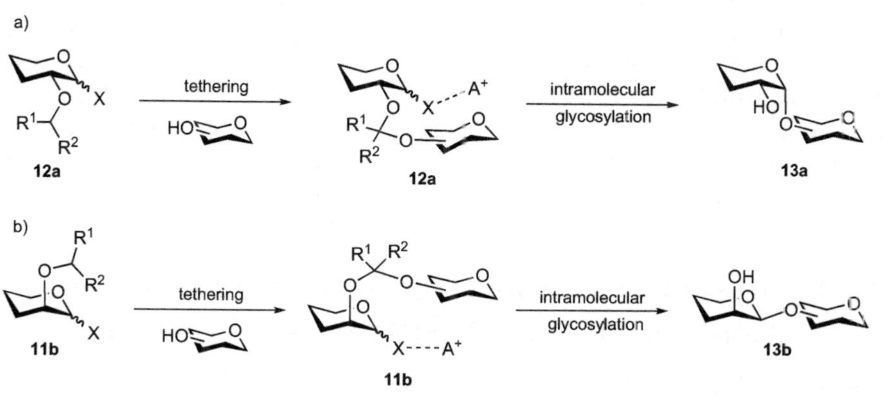

Scheme 10.4 Schematic representation of intramolecular aglycon delivery (IAD) for the synthesis of A. 1,2-*cis* α-glycosides **13a** and B. 1,2-*cis* β-glycosides **13b**.

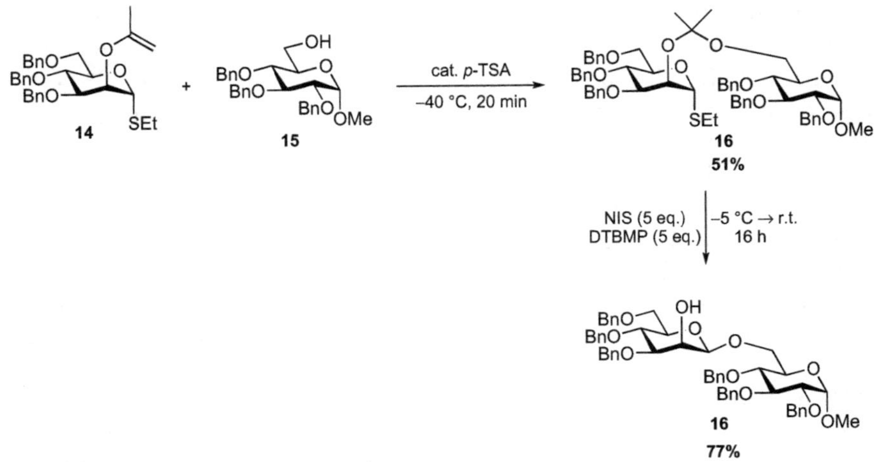

Scheme 10.5 Hindsgaul's vinyl ether-IAD approach to β-mannoside synthesis.

One of the most effective alternative IAD strategies for β-mannoside synthesis was developed by Ogawa and co-workers[30,31] and improved on both these methods by only requiring the use of the common *para*-methoxybenzyl (PMB) protecting group at *O*-2. Highest yields were observed when a cyclic 4,6-*O* protecting group was used, with 4,6-di-*O*-cyclohexylidene acetal being optimal. The increased rigidity imposed by the cyclic protecting group was postulated to promote a more 'S_N2 like' glycosylation and suppress side reactions.[32,33] Tethering reactions were facilitated by the mild oxidising agent 2,3-dichloro-5,6-dicyano-1,4-benzoquinone (DDQ), and methyl triflate activated a thioglycoside leaving group with impressive efficiency.[33]

Ito and co-workers[34,35] subsequently disclosed an analogous IAD strategy whereby napthyl ethers (NAP) at *O*-2 were also oxidised in a tethering reaction to generate mixed ketal derivatives. The yields of NAP-mediated IAD reactions in β-mannoside formation were even superior to those of the original PMB-based strategy, although the 4,6-di-*O*-cyclohexylidene acetal was still necessary for high yields.[35] Alternative IAD methods have also been used for the stereoselective synthesis of α-glucosides,[36] most notably by Fairbanks and co-workers, who developed a plethora of *N*-iodosuccinimide (NIS) mediated IAD variations.[37–43] These include the use of 2-*O*-propargyl, 2-*O*-vinyl and most successfully 2-*O*-allyl protecting groups.

10.4 Solvent and Protecting Group Strategies for Controlling Stereoselective Glycosylation

Alongside these highly engineered methods for controlling stereoselectivity in glycosylations, there also exist more conventional approaches. For example, a variation in solvent mixtures is a tactic that has been routinely used to tune selectivity in glycosylation reactions. The electron-donating effect of ethereal solvents has long been thought to help stabilise any oxocarbenium ion intermediates in glycosylation reactions.[44] A classic example focussed on solvent screening for glycosylations using a benzylated thioglucoside donor and identified 3 : 1 dioxane : toluene (v : v) as the optimum mixture for increasing 1,2-*cis*-α-selectivity.[45] Interestingly, the choice of thioglycoside activator also seems to have an effect on α-selectivity. When iodinium di-collidine perchlorate (IDCP) is used to activate a thioethyl glycoside in 3 : 1 dioxane : toluene (v : v), excellent selectivities ranging from α : β 15 : 1 to 28 : 1 with primary and secondary alcohols have been achieved.[45] It has also been shown that haloalkane and ethereal solvent mixtures can effectively promote α-selective glycosylations.[46]

Stereocontrol in glycosylations of the benzylated galactose thioglycosides could be achieved by tuning protecting groups as well as solvents.[47] Following activation of a completely benzylated galactose donor with *N*-iodosuccinimide/trimethylsilyl trifluoromethanesulfonate (NIS/TMSOTf), only a target disaccharide with modest selectivity of α : β 2.2 : 1 was synthesised. However on switching from the *O*-4 benzyl ether to an *O*-4 acetate protecting group, then to the *O*-4 benzoyl group, the α-selectivity of the reaction dramatically improved to α : β 17 : 1. To explain these results the authors proposed that NGP from the *O*-4 position shields the β-face during glycosylation, leading to the predominant formation of 1,2-*cis* glycosides. Evidence in support of this theory was garnered when α-selectivity increased with the increasing electron-donating potential of the *O*-4 ester group. The more electron-rich the ester, the more effectively it can participate in NGP, with an impressive selectivity of α : β 33 : 1 achieved when an electron-rich *O*-4 PMB protecting group was utilized.

Despite this evidence, studies probing the existence of NGP from various ring positions still refute the possibility of galactose *O*-4 ester participation.[48] However, significant precedent for the use of *O*-6 protecting groups in aiding α-selectivities in glycosylation reactions also exists.[49] Principal examples using the mild activation of thioglycosides generated glycoside products with excellent α-selectivities when bulky trichloroethoxycarbonyl (Troc), or triphenylmethyl (Trityl) protecting groups were used at the *O*-6 position.[50] The bulky *O*-6 protecting groups were thought to hinder attack from the β-face, thus increasing α-selectivity.

Overall, the use of protecting groups to aid selectivity, be it by shielding a reactive face or imposing conformational constraints on the glycosyl donor,[51,52] has been shown to be an effective and synthetically attractive option in stereoselective glycosylations. However, many of these methods are limited to individual monosaccharide configurations, and therefore do not offer a general solution to the aforementioned selectivity challenges. As such, one of the standing aims of synthetic carbohydrate chemistry still remains to design "plug and play" methods that can provide precise control over selectivity of glycosylations across the whole gamut of monosaccharide donors. Where possible, this will limit the need to tailor the glycosyl donor for each new synthetic target. One approach to this lofty goal of acquiring greater stereocontrol over the synthesis of carbohydrates which has shown recent promise, is the use of external organocatalysts in glycosylation reactions.

10.5 Organocatalyst-mediated Glycosylations

Hydrogen bonding catalysis studies which enabled enantioselective additions to oxocarbenium ions were reported by Jacobsen and co-workers in 2008.[53] These studies initially highlighted the possibility that stereocontrol over glycosylations may not always have to be substrate-controlled and could possibly be modulated by organocatalyst activators. An anion-binding, hydrogen-bonding donor thiourea **18** was used for activation of chloride leaving groups attached to simplified substrates, akin to glycosyl chloride scaffolds (Scheme 10.6). In the presence of carbon centred nucleophiles, enantioselective addition to oxocarbenium ions **19** generated new stereocentres with enantiomeric excess (ee) ranging from 85% to 94%, under conditions potentially compatible with glycosylation. The report of only carbon-centred nucleophilic attack on simplified oxocarbenium ion substrates indicated that in 2008 this methodology may have been less compatible with traditional glycosyl donors and acceptors, but rightly still piqued the interest of others in the field to the possibility of catalyst-derived stereocontrol in glycosylation reactions.

Fairbanks and co-workers[54] were one of the first to employ such a strategy in the activation of classical trichloroacetimidate glycosyl donors using chiral 1,1′-bi-2-naphthol (BINOL)-derived phosphoric acid catalysts, which had previously been employed in asymmetric Bronsted acid catalysis strategies, and achieved diastereoselective glycosylation of galactose donors **20** (Scheme 10.7). The authors noted that although activation with chiral

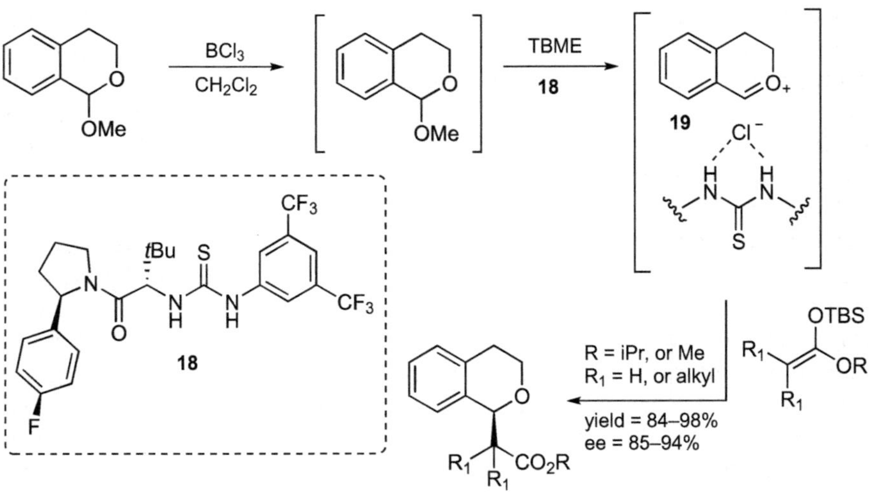

Scheme 10.6 Jacobsen's enantioselective addition to oxocarbenium ions.

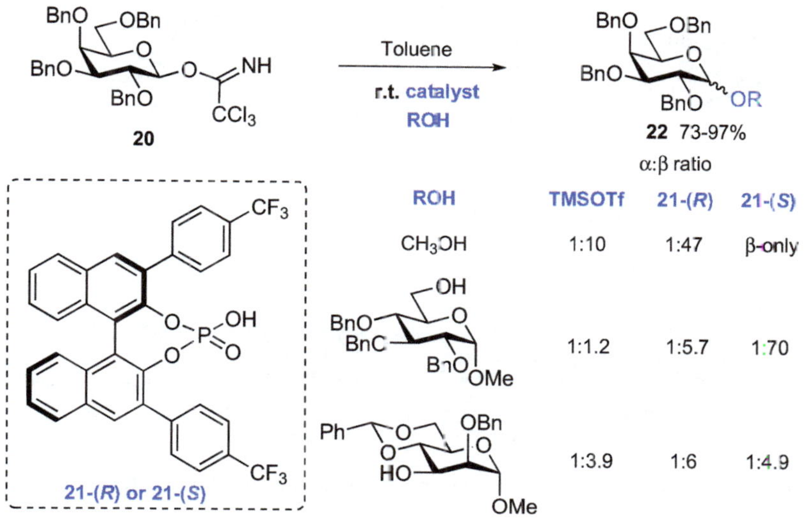

Scheme 10.7 Using chiral phosphoric acids for the activation of glycosyl imidate donors.

phosphoric acids **21** yielded β-galactosides **22** stereoselectively, the use of different enantiomers could not change the selectivity of glycosylation completely from one enantiomer to another and the use of the (*R*)-enantiomer of the catalyst resulted in the highest β-selectivities with a range of glycosyl acceptor substrates. These results when placed in the context of the stereoselectivity achieved with traditional TMSOTf activation demonstrated that the chirality of the catalyst does result in an increased level of stereocontrol.

Thiourea organocatalysis was next applied to the stereoselective synthesis of challenging 2-deoxy sugars. McGarrigle, Galan and co-workers demonstrated activation using an achiral thiourea organocatalyst **23** after 24 h at reflux.[55] Notably, using a range of glycosyl acceptors, with benzyl protected galactal **24**, only α-glycosides **25** were formed (Scheme 10.8). The stereoselectivity of this reaction was proposed to arise through proton delivery to the least hindered α-face in **24** *via* an alcohol/glycosyl acceptor–thiourea complex, with formation of an ion pair, which collapses to yield only the α-product. This procedure highlights both the opportunities and challenges of using organocatalysts for stereoselective glycosylations. In this case, the exquisite control is achieved by using an achiral catalyst, which accentuates the galactosyl donor's innate preference to react in a stereoselective manner under the appropriate conditions. However to achieve 'switchable' access to both α and β-anomers in the presence of different enantiomers of a chiral version of the catalyst, this innate preference would need to be overcome solely by the stereochemical environment created by the catalyst. Galan and co-workers further emphasised the extent of this challenge by demonstrating that activation of galactal **24** using both enantiomers of a chiral Bronsted

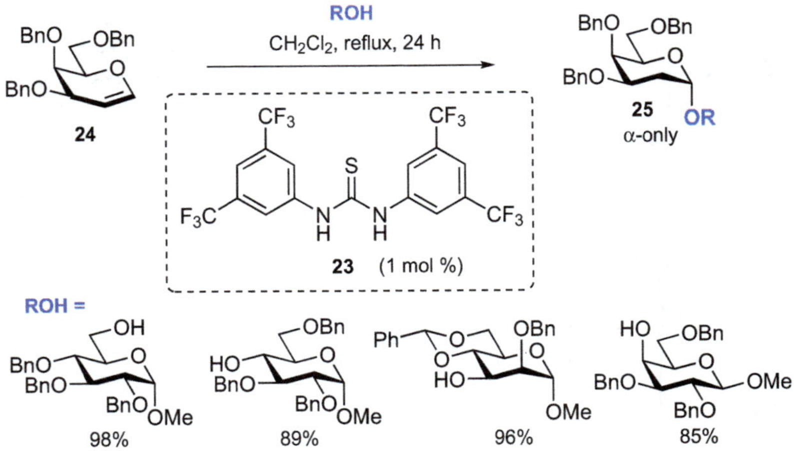

Scheme 10.8 Achiral thiourea activation of galactals in the synthesis of 2-deoxy-α-galactosides.

acid, generated α-glycosides with high stereoselectivity.[56] Similarly, 2-nitro galactals also yielded α-glycosides when activated with bifunctional thiourea–cinchona alkaloid organocatalysts, with cooperative catalysis proving essential for glycosylation by activating the nitro group (thiourea), and directing the incoming nucleophilic glycosyl acceptor (cinchona-alkaloid) simultaneously.[57] Schmidt and co-workers also explored this concept of cooperative catalysis using thiourea organocatalysts, in the presence of acid co-catalysts. The authors noted that thiourea alone was unable to activate trichloroacetimidate donors under conventional glycosylation conditions, but when combined with a range of Bronsted acid catalysts, β-glycosides could be synthesised with high stereoselectivity from α-imidate donors.[58] Although it is clear that the thiourea and acid catalyst work co-operatively in this procedure, and glycosylations proceed with the level of stereospecificity expected from an 'S_N2-like' glycosylation, an unequivocal mechanistic rationale for the stereocontrol has yet to be established.

Nearly a decade after their first foray into abstraction of chloride leaving groups using thiourea organocatalysts, Jacobsen and co-workers also proposed an 'S_N2-like' stereospecific mechanism to account for high β-stereoselectivities in a return to glycosylation reactions.[59] On this occasion they employed glycosyl chlorides as donors and demonstrated that in the presence of macrocyclic bis-thioureas (monomeric chiral, achiral thioureas and ureas were less effective)[60] stereospecific 'S_N2-like' glycosylations took place to generate a range of β-glycosides with high yield and stereoselectivities (α : β 15 : 85 to 1 : 99) (Scheme 10.9). The authors reported that this stereospecificity could be achieved across a wide range of acceptors and donors, including α-chlorides of glucose, galactose, xylose and 2-azido-galactose among others. Perhaps most impressively, β-mannose glycosides **26** could also be synthesised with high yield and stereoselectivity, using both

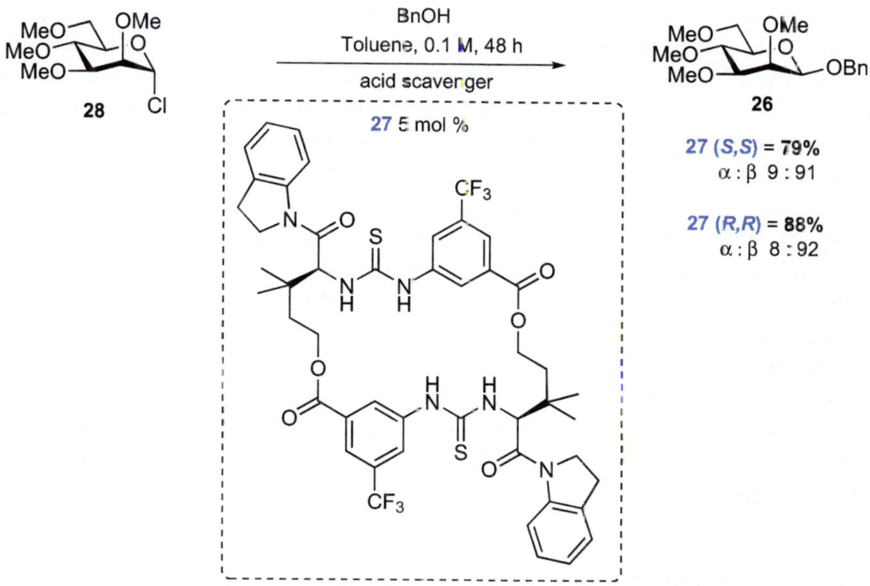

Scheme 10.9 Stereospecific synthesis of β-mannosides using macrocyclic bis-thioureas.

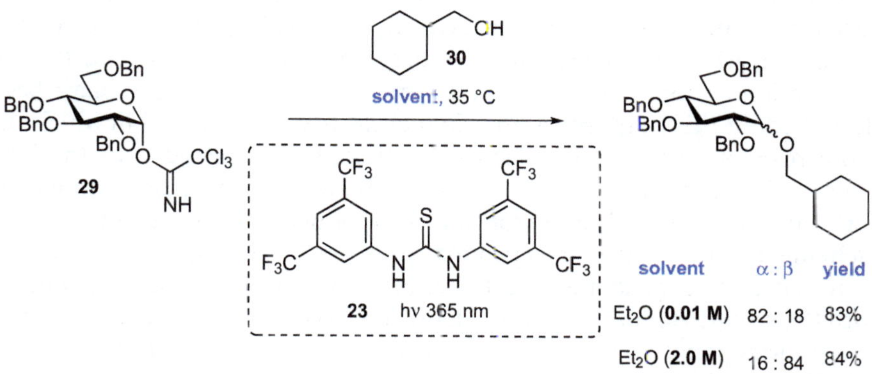

Scheme 10.10 Toshima's use of a thiourea organo photoacid enables the stereoselective synthesis of both α and β-glycosides.

enantiomers of macrocyclic bis-thiourea **27** for activation of a simple ether-protected glycosyl donor **28**.

Finally, Toshima and co-workers highlighted the potential influence of reaction concentration on the stereochemical outcome of glycosylation reactions using trichloroacetimidate donors in the presence of thiourea 23 (Scheme 10.10).[61] As previously noted[58] this catalyst alone is not acidic enough to activate the imidate-leaving group, however upon photoirradiation at 365 nm an organo photoacid is formed from the thiourea, and in this excited state proved acidic enough to initiate glycosylation. Again using α-imidate donor **29**, stereoselective glycosylation proceeded with a range of glycosyl acceptors. Using the specific case of primary alcohol **30** as an example, the effects of reaction concentration on stereoselectivity are clear, at low concentrations in diethyl ether α-glycosides were formed, and at higher concentrations β-glycosides dominated. The authors proposed that at high concentrations the reaction also proceeds through an 'S_N2-like' mechanism, where the activation of α-imidate is followed by stereospecific β-attack by the acceptor. Whereas at low concentrations formation of an oxocarbenium ion intermediate is followed by stereoelectronically preferred attack from the α-face. This methodology was further showcased with a range of glycosyl acceptors and galactosyl and glucosyl donors, with the relationship between stereoselectivity and reaction concentration conserved across all examples. Therefore, the combination of α-imidate donors and thiourea organo photoacids could be considered one of the rare examples where access to both α and β-stereoisomers can be achieved from a single glycosyl donor, and the state-of-the-art in terms of stereoselective glycosylation.

10.6 Automated Oligosaccharide Synthesis

As well as progression in the methods for stereoselective glycosylation, recent advances have also greatly reduced the labour required in preparing carbohydrate materials and focussed on synthesising complex oligosaccharides in a faster more practical manner. This includes one-pot glycosylation, wherein oligosaccharides are synthesized without isolation or purification of intermediates. Three general one-pot strategies have been described for this purpose: (a) the chemoselective strategy,[62–65] in which the more reactive donor **31** is selectively activated and treated with the less-reactive donor **32** to provide a new glycoside, which can subsequently react with another glycosyl acceptor **33** to produce an oligosaccharide **34** (Scheme 10.11); (b) the preactivation strategy,[66–68] in which the donor is activated alone to generate a reactive species, which is subsequently coupled with the second donor bearing an identical aglycon at the reducing end, and (c) the orthogonal strategy,[69,70] in which the leaving group of one donor is selectively activated over another. Several excellent reviews describe these one-pot glycosylation methods in more detail.[71–74]

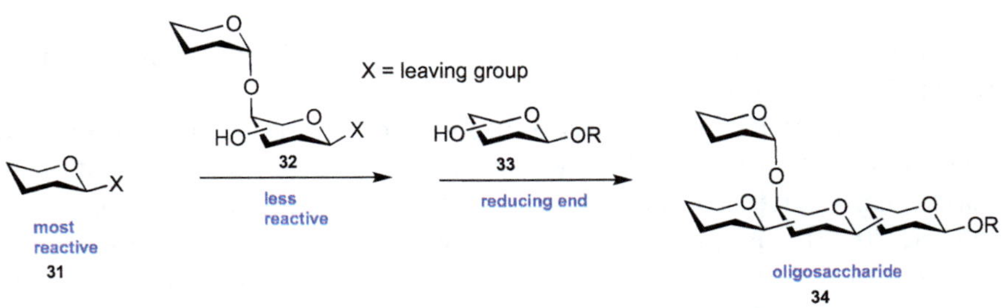

Scheme 10.11 One strategy for one-pot synthesis of oligosaccharides.

A computer database, OptiMer, was created by Wong and co-workers to store the relative reactivity values (RRVs) of many donor and donor/acceptor compounds (*e.g.* building blocks with one hydroxy group unprotected).[75] The programmable one-pot synthesis of oligosaccharides has the potential to affect many areas of drug discovery, as it provides rapid access to complex carbohydrate structures. By using the OptiMer database, oligosaccharides containing three to six monosaccharides could be rapidly assembled in minutes to hours without intermediate workup or purification procedures. The carbohydrate-associated cancer antigen Lewis^y **35** expressed on the surfaces of colorectal adenocarcinoma and hepatocellular carcinomas was prepared through the programmable one-pot approach (Scheme 10.12).

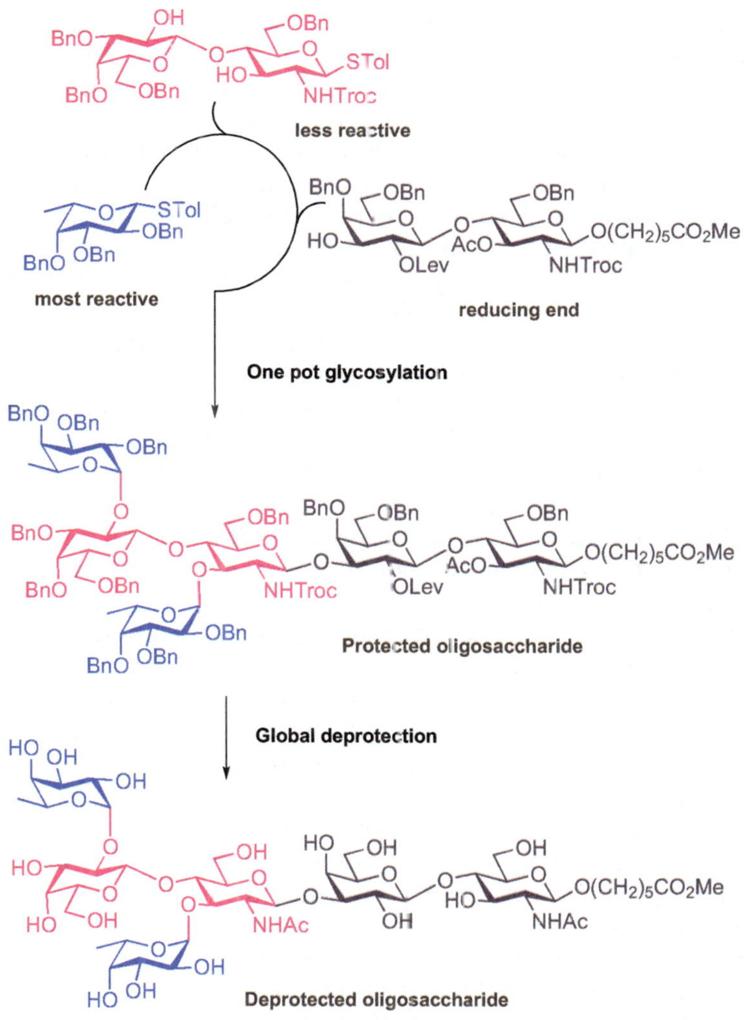

35

Scheme 10.12 Programmable one-pot synthesis of Lewis^y hapten.

The evolution of a solid-phase paradigm for the construction of oligosaccharides was initiated with Schuerch and Frechet's synthesis of di- and trisaccharides on a polymer support in 1971.[76] Since then, solid-phase synthesis, which eliminates the need for purifying intermediates and simplifies the removal of excess reagents, has seen much advancement. In 2001 Seeberger *et al.* reported the first automated oligosaccharide synthesis using a modified peptide synthesizer.[77–80] In 2012, Seeberger *et al.* also reported "the first fully automated solid-phase oligosaccharide synthesizer," initially in its experimental form,[81] and in 2013 it was marketed as the Glyconeer 2.1.

The synthesis of various long-chain carbohydrates has been accomplished on the Glyconeer platform, including synthesis of a 30mer[82] and 50mer[83] α-(1,6)-oligomannoside **36**, using a mannosyl phosphate building block **37** and Merrifield resin functionalised with a photocleavable linker **38** (Scheme 10.13). Removal of the temporary Fmoc protecting group with piperidine revealed the C6 hydroxy group for chain elongation. A capping step was included in the synthesis cycle to block any unreacted hydroxy groups prior to the following glycosylation cycle, and the coupling efficiency was assessed by UV/Vis measurement of the piperidine–dibenzofulvene adduct released during Fmoc cleavage.[84] A capture–release technique was also adopted to facilitate polysaccharide purification, relying on the covalent attachment of the labelled target molecule to a solid support to separate the desired oligosaccharide from any deletion sequences. The iterative automated synthesis of conjugation-ready linear and branched β-(1,3)-glucan oligosaccharides[85] was also reported by the same group. The linear structure was assembled first, followed by the introduction of β-(1,6)-branching **39–42** (Scheme 10.14). These glucan oligosaccharides were subsequently immobilised for microarray-based human antibody epitope evaluation.

Mixed-linkage glucan (MLG) oligosaccharides[86] were also recently synthesized, and implemented as tools to determine the substrate specificity of lichenase. These cello-oligosaccharide sequences are composed of short stretches of β-(1,4)-linked oligosaccharides connected through β-(1,3)-linkages. The 1,3-linkages in the 1,4-linked glucan chain form molecular kinks that prevent an intermolecular alignment into microfibrils as observed in the case of cellulose. The MLG-specific hydrolase lichenase cleaves every 1,4-linkage following a 1,3-linkage in a mixed-linkage glucan backbone. Synthetic MLG-oligosaccharides of varying connectivity would permit a simple liquid chromatography–mass spectrometry (LC–MS) analysis of their digestion products after hydrolysis, providing a toolkit for the determination of substrate specificities of the lichenase and other MLG endoglucanases.

A conjugation-ready *Streptococcus pneumoniae* serotype 3 (ST3) capsular trisaccharide[87] was also assembled using a novel glucuronic acid building block to circumvent the need for a late-stage oxidation. The building blocks were synthesized at high yields using standard protecting group chemistry and the automated glycosylation protocol was employed three times with three equivalents of the building block to ensure complete glycosylation of

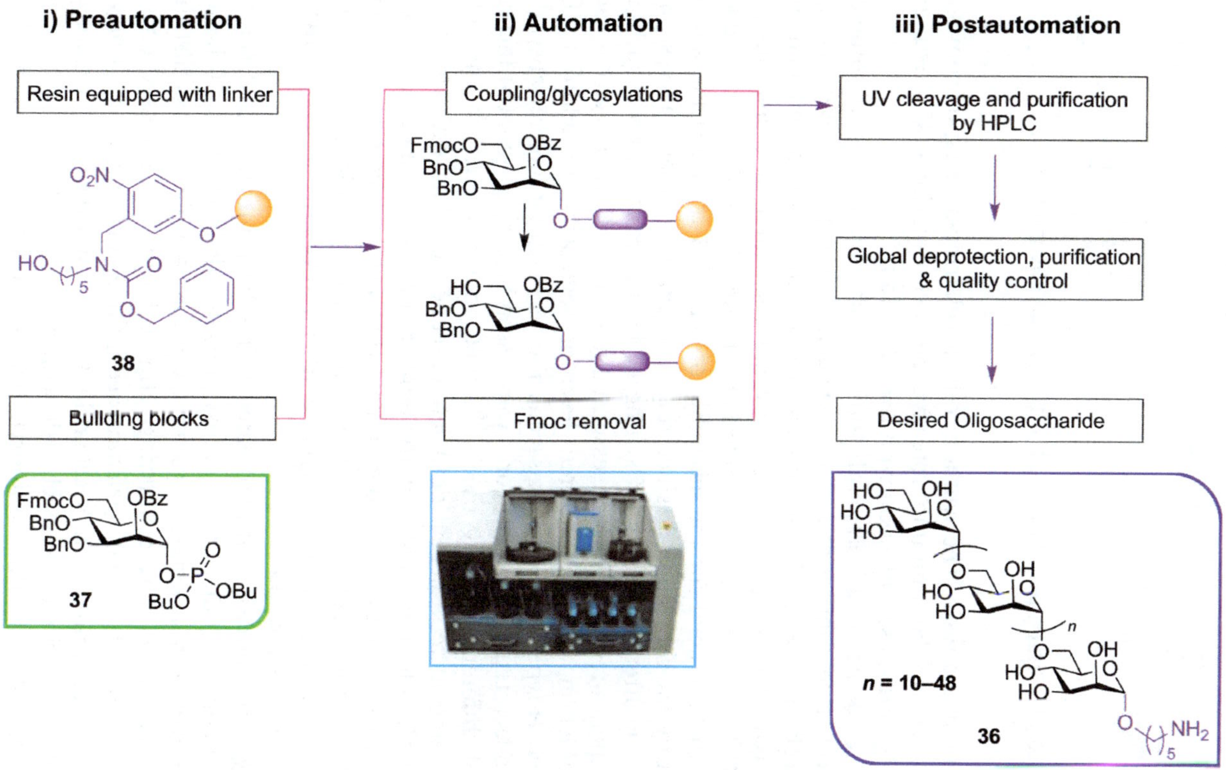

Scheme 10.13 Schematic overview of automated glycan assembly using Glyconeer 2.1.

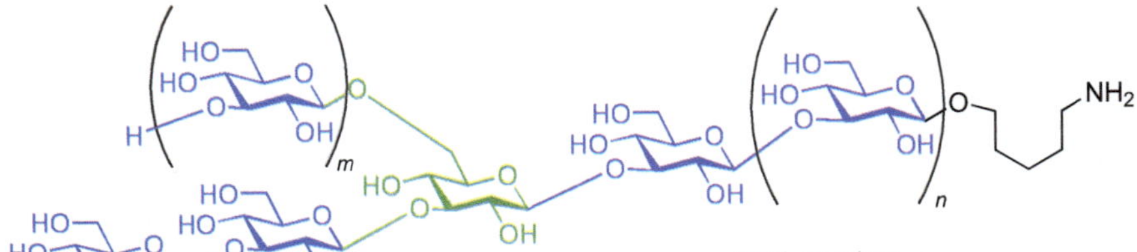

Scheme 10.14 Structure of branched β-(1,3)-glucan oligosaccharides.

the nucleophilic resin. The synthetic-oligosaccharide-based glycoconjugates were shown to be immunoprotective against experimental pneumonia caused by transnasal infection with the ST3 strain.[88]

Despite being relatively embryonic in its development, automated oligosaccharide synthesis has also yielded a number of other complex oligosaccharides with diverse applications, including the automated solid-phase assembly of the protected tumour-associated antigens Gb-3 and Globo-H.[89] Stereoselective installation of the α-1,4-galactosidic linkage proved to be crucial for the assembly of the Gb-3 trisaccharide and was mandatory for the coupling since the solid-phase approach allows purification only after completion of synthesis. Longer reaction times (3 h) and low temperatures (−50 °C) were essential in order to drive this glycosylation to completion with the desired level of selectivity. In addition, fully protected H-type I and II pentasaccharides containing lactotetraosyl (Lc4) and neo-lactotetraosyl (nLc4) cores were assembled by sequential incorporation of different building blocks.[90] HPLC indicated the presence of an Fmoc-protected tetrasaccharide deletion sequence along with the desired pentasaccharide, but increasing the temperature of the Fmoc deprotection module to 30 °C was found to eliminate this deletion sequence. Changing the glycosylation solvent for the fucose donor building block from DCM : Et_2O (1 : 3) to DCM also significantly improved the stereoselectivity of the fucose glycosylation.

Using a combination of automated solid-phase methods and solution-phase fragment coupling, a glycosylphosphatidylinositol-based hexasaccharide, which is a potential malarial vaccine candidate,[91] has also been synthesised. A mannose containing tetrasaccharide was accessed on solid phase using readily available trichloroacetimidate building blocks. Coupling of trichloroacetimidate tetrasaccharide donor with a disaccharide acceptor produced the desired hexasaccharide with modest yield.

A library of 11 plant arabinoxylan oligosaccharides[92] have also been constructed using automated synthesis, and the binding specificities of xylan-directed antibodies determined on microarrays using this library. In this case, the required β-selectivity in the glycosylation steps was ensured through the installation of benzoate esters on the C-2 hydroxyls. With arabinoxylan fragments in hand, microarray slides were printed and used as tools to probe the binding preferences of 31 anti-xylan monoclonal antibodies.[93] The synthesis of the more complex linear and branched arabinofuranose oligosaccharides were also accessible using automated solid-phase synthesis.[94]

Furthermore, a general approach providing access to 14 type-II arabinogalactan glycans[95] of different size and complexity has also been described using four monosaccharide building blocks. In the synthesized arabinogalactan oligosaccharides, the arabinose unit is either attached to the β-(1,3)-linked galactan backbone or to the β-(1,6)-linked galactan side chain. In this synthesis a standard capping protocol employing acetic anhydride and pyridine was modified to obtain benzoyl esters as capping moieties due to their stability towards the hydrazine used for Lev ($ROCO(CH_2)_2COCH_3$) deprotection.

The potential of synthetic lipopolysaccharide (LPS) inner core structures as protective antigens against *Neisseria meningitidis* has also been studied.[96] A highly conserved LPS core tetrasaccharide was selected based on the specific reactivity of human serum antibodies to the synthetic LPS cores. The distal trisaccharide of the LPS tetrasaccharide was recognised as a crucial epitope in inducing a robust IgG response in mice when formulated as an immunogenic glycoconjugate. While a proximal 3-deoxy-α-D-*manno*-oct-2-ulosonic acid (Kdo) moiety was immunodominant and induced mainly nonprotective antibodies.

Sialylated oligosaccharides with α-selectivity were also synthesised by automated glycan assembly using sialic acid glycosyl phosphate building blocks.[97] Wherein a cyclic 4*O*,5*N*-oxazolidine protecting group was utilized for efficient chemical sialylation. In order to achieve better α-selectivity, *O*-chloroacetyl groups were installed on the C-7 and C-8 positions of the sialic acid donor.

A general strategy for the diversity-oriented synthesis of a library of inner core oligosaccharides of the LPS of pathogenic Gram-negative bacteria[93] has also been reported using the automated oligosaccharide synthesizer, where flexible building blocks are required to install the highly branched sugar residues of LPS included derivatives of Kdo, L-*glycero*-D-*manno*-heptose (Hep) and 4-amino-4-deoxy-β-L-arabinose (Ara4N). A regioselective and stereoselective extension of the Kdo 7,8-diol with Ara4N proved challenging but was accomplished by identification of appropriate reaction conditions. This synthetic strategy was modular and enabled rapid access to other LPS oligosaccharides consisting of Hep and Kdo.

Recently, an automated approach to synthesize negatively charged chondroitin sulphate glycosaminoglycans (GAGs)[99] was also reported using a stable supply of tailor-made differentially protected building blocks and a robust but easily-cleaved linker to connect the first monosaccharide of the nascent oligosaccharide to the solid support. Linker selection posed a challenge because of the breadth and variety of chemical reactions that must be withstood during GAG synthesis.

In a union of classical and modern carbohydrate chemistry, oligosaccharides containing multiple challenging *cis*-linkages[100] were also efficiently synthesized by automated glycan assembly using monosaccharide building blocks equipped with remote participating groups. Targets bearing α-galactosidic linkages and α-glucans were selected for the development of automated methods for stereoselective *cis*-glycosidic bond formation. Indeed, glycosylation of thiogalactose bearing acetyl or benzoyl esters at C4 and C6 with disaccharide acceptors proceeded with complete stereoselectivity, but still with significant amounts of the disaccharide deletion sequence remaining after the coupling. However the same glycosylation with C3,C4-bis-acetylated thiogalactose donors furnished better yields of the trisaccharide with a good α-selectivity (39.5 : 1). Glucose building blocks with remote participating acetyl groups at C3 and C6 positions, in analogy to the galactose series, resulted in the best selectivity. A further benefit was achieved from the addition of diethyl ether as a solvent to promote the desired α-glucoside formation. Despite these successes, the field of automated glycan assembly still requires further methodology advancements, and new practical developments continue to be published apace, including the important recent disclosure of a traceless photocleavable linker which generates product carbohydrates in their reducing form.[101]

An alternative solid-phase strategy for oligosaccharide synthesis has also been reported by Jaipuri and Pohl,[102] who used fluorous solid-phase extraction (FSPE) protocols to synthesize both linear **43** and branched **44** mannose oligomers (Scheme 10.15). The fluorous tag proved essential for maintaining the solubility of the growing oligosaccharide chain and was shown to be stable under all the reaction conditions required for the

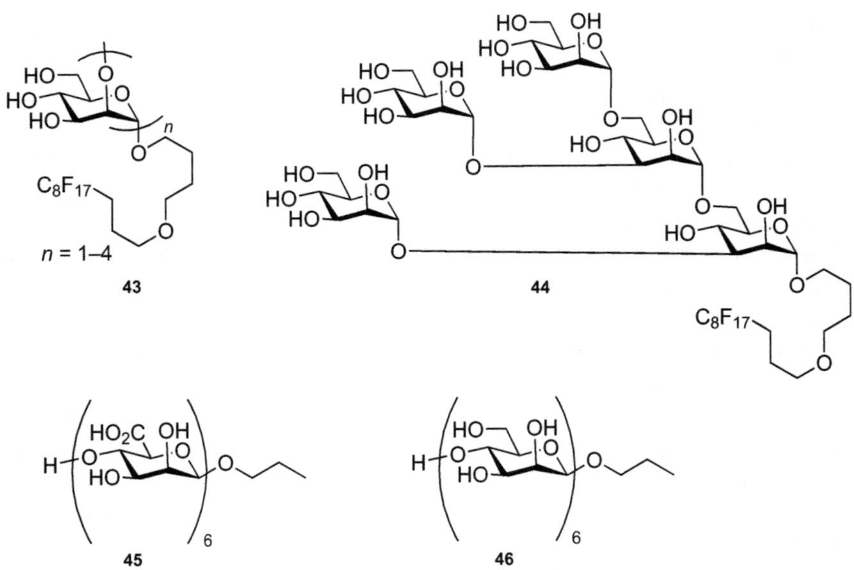

Scheme 10.15 Programmable one-pot synthesis of Lewis^y hapten.

requisite glycosylation and deprotection conditions in a manner amenable to automation. By utilizing fluorous-tag-assisted purification after repeated reaction cycles, β-1,4-mannuronate and β-1,4-mannan oligomers (**45** and **46** in Scheme 10.15) were synthesized[103] in a new approach to β-mannan synthesis. An α-linked trimannose, one of the major components in both the outer lipophosphoglycan (LPG) of *Leishmania* parasites[104] and the lipoarabinomannan (LAM) of the *Mycobacterium tuberculosis* cell wall,[105] was also synthesised using the FSPE protocol with excellent yields, and a carboxylate handle was incorporated at the reducing end of the sugar for attachment to a polymeric bead for multivalent display.[106]

Demchenko and co-workers have also developed their own approach to automated oligosaccharide assembly, surface-tethered iterative carbohydrate synthesis (STICS) technology[107] in which a surface-functionalised 'stick' made of nanoporous gold plates[108] simplified transformation of the solid-support-bound molecules between the reaction vessels (Scheme 10.16). At the end of the synthesis, the oligosaccharide could be cleaved off and the 'stick' reused for subsequent synthesis; allowing cost-efficient and simple synthesis of oligosaccharide chains. In 2016, the same group also developed a HPLC-based platform[109] for simple, scalable and transformative

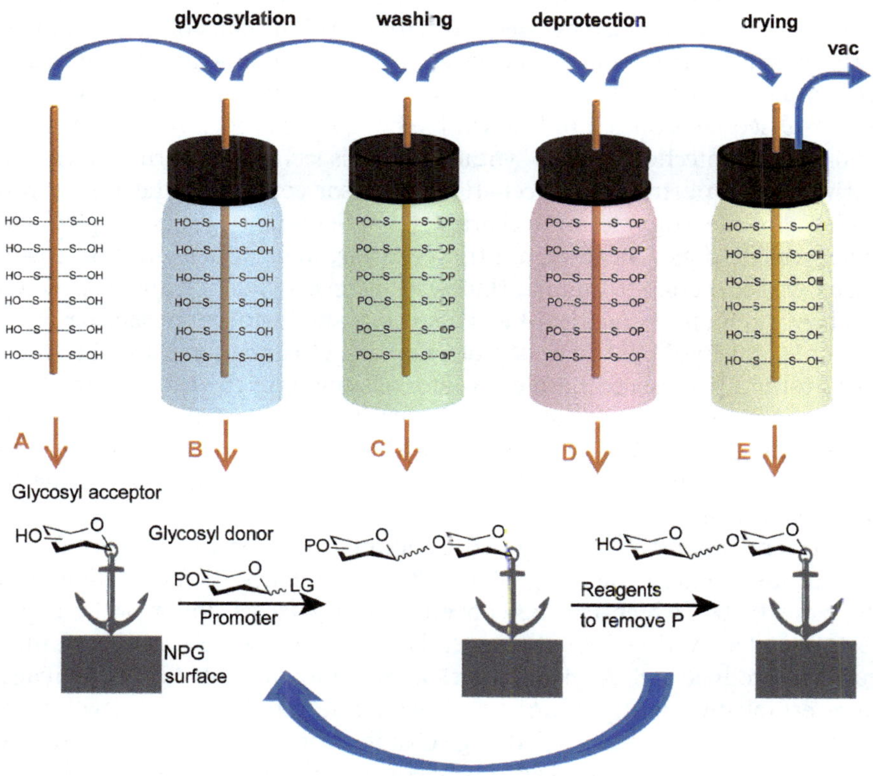

Scheme 10.16 Outline of the STICS concept.

automation of oligosaccharide synthesis that does not rely on specialized equipment. Using the acceptor-bound approach,[110] preloaded Tentagel resin was packed in an Omnifit column and integrated into a HPLC system.[111] This platform allows real-time UV monitoring of all steps, including glycosylation, which, in turn, helps to reduce the reaction time and the amount of reagents and solvent needed.

10.7 Pushing the Limits of Solution-phase Synthesis

Although automated methods for carbohydrate synthesis are undoubtedly of great potential utility, the longest and most complex carbohydrates whose synthesis has been described in the literature have still been conquered through the use of traditional solution-phase chemistry, sometimes in combination with enzymatic approaches.

Heparin (HP) and heparan sulfate (HS), a class of heterogenic, linear polysaccharides are part of the glycosaminoglycan (GAG) family, and their interactions with a plethora of proteins[112,113] and involvement in important physiological processes[114] makes them attractive synthetic targets. Whereas earlier synthetic efforts focussed on heparin and analogues for their anticoagulant activity and on shorter sequences of HS,[115] In recent years syntheses of longer HS oligosaccharides have also been reported. Hung and co-workers[116] synthesised two octasaccharides as inhibitors of herpes simplex virus–host-cell interaction, by stepwise addition of mono- and disaccharides. The authors masked the iduronic acid functionality as a 1,6-anhydro sugar and oxidised later on the fully assembled oligosaccharide, followed by sulfation and deprotection. In this synthesis five disaccharide intermediates were derived from a common disaccharide precursor containing the 1,6-anhydro moiety. In one example, disaccharide **47** (Scheme 10.17) was coupled with thioglycoside donor **48** to yield a trisaccharide, upon which the 1,6-anhydro group was converted to another thioglycoside glycosyl donor. In a convergent strategy, trichloroacetamide **49** was also combined with monosaccharide acceptor **50**, followed by *O*-4-Lev deprotection and glycosylation with S-Tol donor **51** to form a pentasaccacharide acceptor. Coupling of the two intermediates led to the desired heparin sulfated octasaccharide **52**. A similar approach, but based on only one 1,6-anhydro disaccharide building block, was recently used by the same group to synthesise a ^{15}N- and ^{13}C-labelled octasaccharide for protein NMR studies.[117]

Tyler and his team[118] developed a modular synthesis to HS based on a central disaccharide building block **53**, in the *gluco-* or *ido*-configuration (Scheme 10.18), which can be converted to a glycosyl donor **54** by deprotecting and activating the reducing end or by Fmoc cleavage on the terminal sugar to produce a glycosyl acceptor **55**. The more difficult 1,2-*cis* α-linkages were established first, within the disaccharide building blocks, while the 1,2-*trans* stereochemistry of the glycosidic bond in oligosaccharide **56**, formed during chain extension, was controlled by anchimeric assistance provided by benzoyl protecting groups. Amino groups were protected as

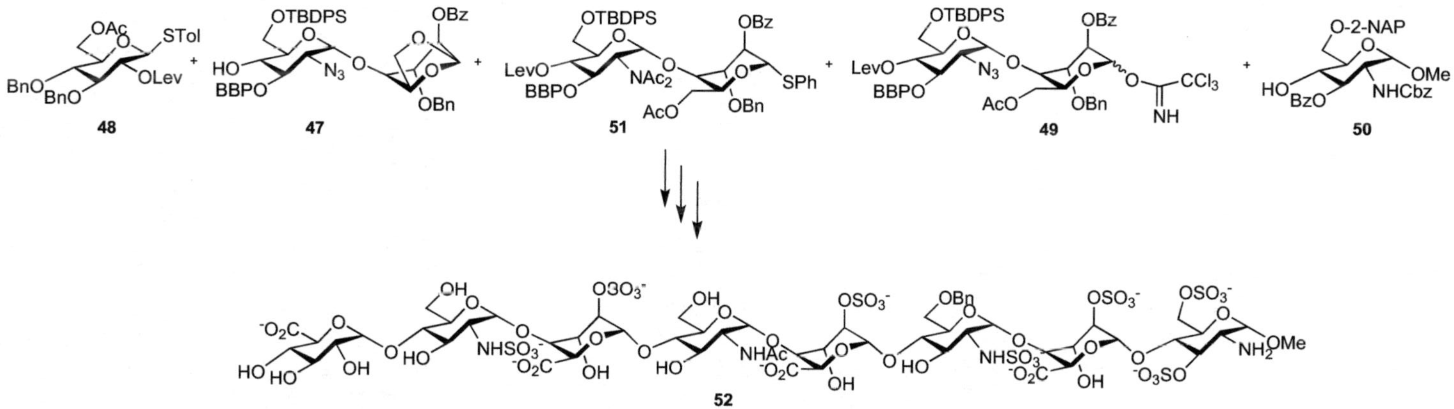

Scheme 10.17 Hung's synthesis of irregular 3-*O*-sulfonated heparan sulfate oligosaccharides.

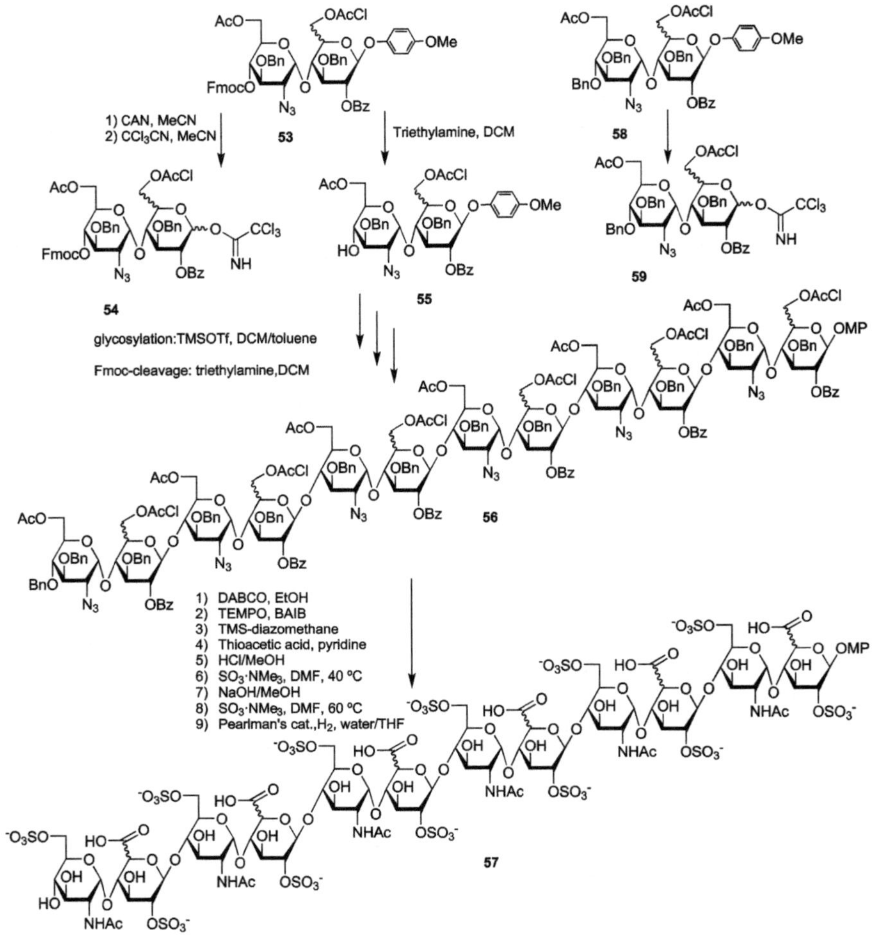

Scheme 10.18 Tyler's synthesis of a heparan sulfate oligosaccharide library.

azides and the uronic acids, both *ido* and *gluco,* were accessed by late-stage oxidation of the assembled oligosaccharides after selective removal of chloroacetyl protecting groups. Selective sulfation of either only the 6-*O* or both the 2-*O* and 6-*O* positions and catalytic hydrogenolysis for global deprotection yielded the final products **57**. Furthermore, this methodology was applied in the synthesis of a 16-compound library of hexasaccharides to dodecasaccharides starting from only two different central building blocks: one containing a *gluco*- the other an *ido*-configured uronic acid precursor, plus the respective chain-terminating building blocks bearing benzyl instead of Fmoc protection (**58** and **59** in Scheme 10.18). The resulting library was tested against beta secretase 1 (BACE1), an enzyme involved in the formation of toxic A-peptides associated with Alzheimer's disease, and shown to contain very potent inhibitors. Although there was no obvious decline in yields with increasing chain length, the authors did not pursue larger oligosaccharides of this type, as the increase in inhibitory activity was only moderate for fragments bigger than octasaccharides.

Miller *et al.* further expanded the size of accessible HS fragments[119] by using a thiophenyl tetrasaccharide **60** (Scheme 10.19), they had developed for an earlier synthesis of a dodecasaccharide,[120] as a glycosyl donor to rapidly assemble larger oligosaccharides. The iduronates were protected as methyl esters, avoiding oxidation at a later stage, and the chain extension was based on 4-*O*-trichloroacetate protection at the non-reducing end. The 1,2-*cis* stereochemistry was again pre-established in the tetrasaccharide donor, and the formation of the 1,2-*trans* glycosidic bonds assisted by neighbouring benzoates. Starting from a known dodecasaccharide alcohol **61**, repeated coupling cycles (NIS, AgOTf) of the growing acceptor **62** with thiophenyl donor **60** and deprotection cycles (methanol/pyridine) yielded a fully protected 40-mer **63**. After deprotection and sulphation an *N*/*O*-2-sulfated heparin-like 20-mer **64** was formed, making these the longest chemically synthesised heparin-like oligosaccharides. From a more general perspective, this synthesis begins to approach a block strategy, which has also been successfully used in the synthesis of large branched oligosaccharides, mainly arabinomannans and arabinogalactans.[121]

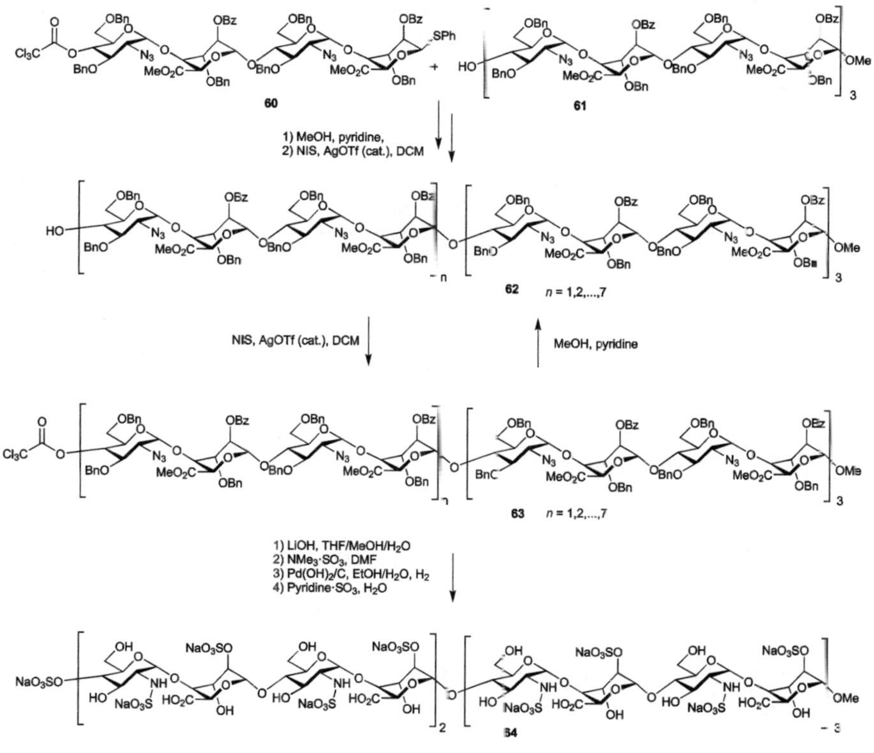

Scheme 10.19 Miller's synthesis of a fully protected 40-mer and a sulphated heparin-like 20-mer

The largest chemically synthesised oligosaccharide described to date is a 92-mer mycobacterial arabinogalactan, synthesised by Ye and co-workers in such a block fashion.[122] Mycobacterial cell walls contain arabinogalactan as an essential structural element and this polysaccharide plays a role in the infection and pathogenicity of *M. tuberculosis*,[123] making it an attractive, if not formidable, synthetic target. A preactivation-based one-pot glycosylation protocol was used throughout the synthesis, starting with the quick assembly of a variety of pentasaccharide, hexasaccharide and heptasaccharide blocks, which provide access to the two larger main building blocks. A 31-mer branched arabinan donor **65** (Scheme 10.20) was constructed *via* a

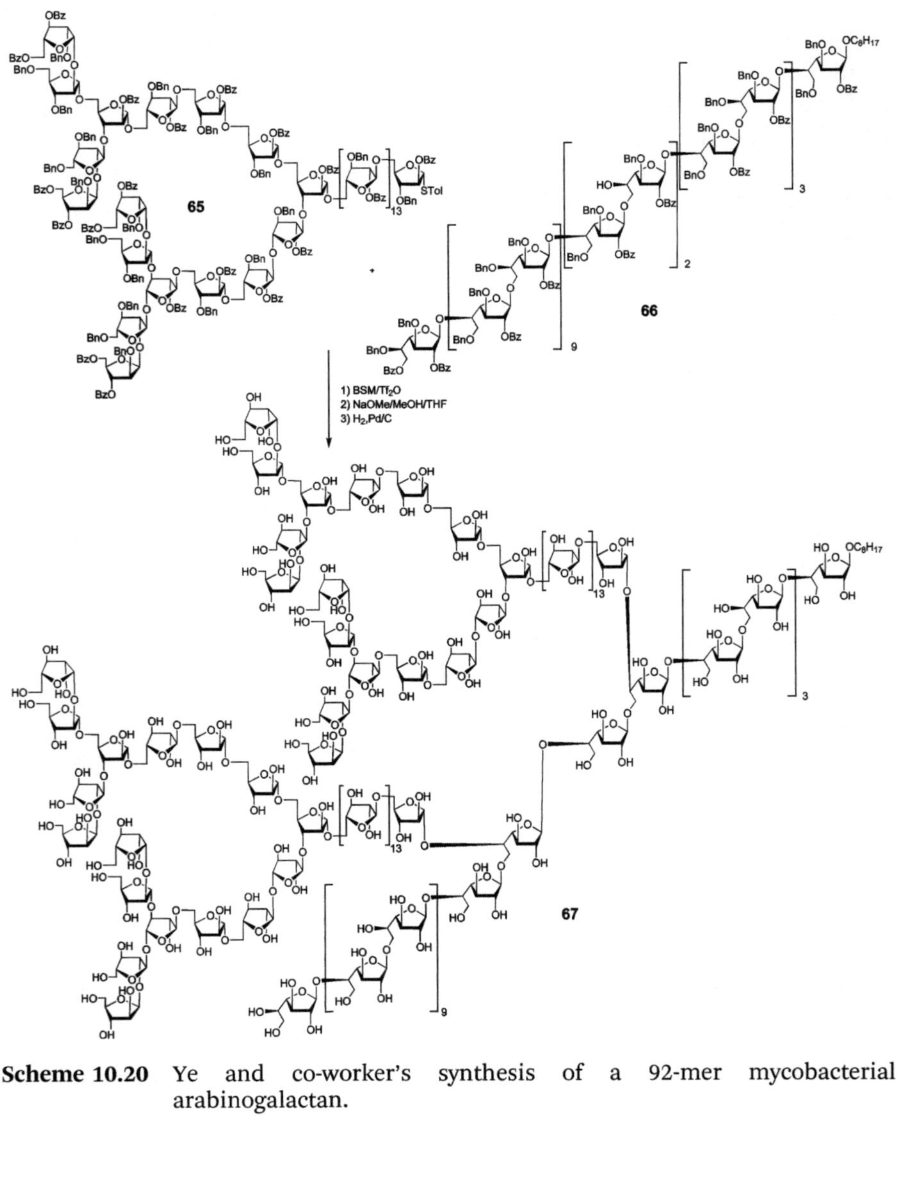

Scheme 10.20 Ye and co-worker's synthesis of a 92-mer mycobacterial arabinogalactan.

sequential [7 + 7 + 5 + 6 + 6] one-pot glycosylation. Similarly a linear 30-mer galactan acceptor **66** was synthesised in an iterative one-pot [6 + 6 + 6 + 6 + 6] glycosylation. Two of the 31 sugar arabinan donors were then simultaneously coupled with the 30 sugar units containing galactan. While attempting to achieve this [31 + 31 + 30] glycosylation, a variety of promotor systems were tried unsuccessfully, until the benzenesulfinyl morpholine–triflic anhydride couple[124] enabled the double glycosylation to the protected target molecule with a remarkable 84% yield. Finally, global deprotection by Zemplén deacetylation and catalytic hydrogenolysis yielded the final 92mer arabinogalactan **67**.

Another challenging area of carbohydrate synthesis where the development of better and more efficient methods and the combination with enzymatic synthesis has resulted in considerable progress is the synthesis of *N*-glycans. Although smaller in size compared with the aforementioned cell-wall components, *N*-glycans are involved in diverse and highly complex post-translational modification of proteins. The issues inherent in the synthesis of *N*-glycans include the stereochemical challenge of installing the crucial β-mannoside and the steric demands around the branching points, especially when installing a bisecting *N*-acetylglucosamine. The chemical synthesis of various types of *N*-glycans has been achieved over the years by several research groups, among them bi-and triantennary *N*-glycans by Danishefsky and co-workers,[125] complex high-mannose-type glycans by Ito and co-workers,[126] complex core-fucosylated *N*-glycans by Fukase and co-workers,[127] and multiantennary bisected glycans by Unverzagt and co-workers.[128] However, due to their diversity in structure and biological function, rapid access to libraries of complex structures is still highly desirable. Recently, a number of practical chemoenzymatic approaches for the production of libraries of *N*-glycans have been published.[129–132] In these strategies a core oligosaccharide is first made by chemical synthesis and then selectively modified by either trimming or extending enzymes. Ito and co-workers[130] first demonstrated access to multiple structures by enzymatic trimming of a chemically synthesised high-mannose-type tetradecasaccharide (Scheme 10.21A). In this non-natural precursor the three oligomannose branches are capped by selected monosaccharides other than mannose, to enable selective enzymatic deprotection with the respective glycosidase. Once the mannose sequence is exposed, further trimming by mannosidases enables the construction of a high-mannose-glycan library with variable branch lengths on three different branches. Boons and co-workers also proposed a general strategy for the synthesis of asymmetrically branched glycans by enzymatic extension of a chemically derived decasaccharide (Scheme 10.21B).[133] Here an acetylated galactose on the terminus of one arm of the glycan is not recognised by a sialyltransferase and therefore unmodified during the sialylation of a neighbouring arm, the resulting oligosaccharide is then further extended by a selection of glycosyltransferases. This strategy was recently expanded to include a combination of enzymatic trimming and extension, with the goal of providing access to bi-, tri, and tetraantennary *N*-glycans from a common

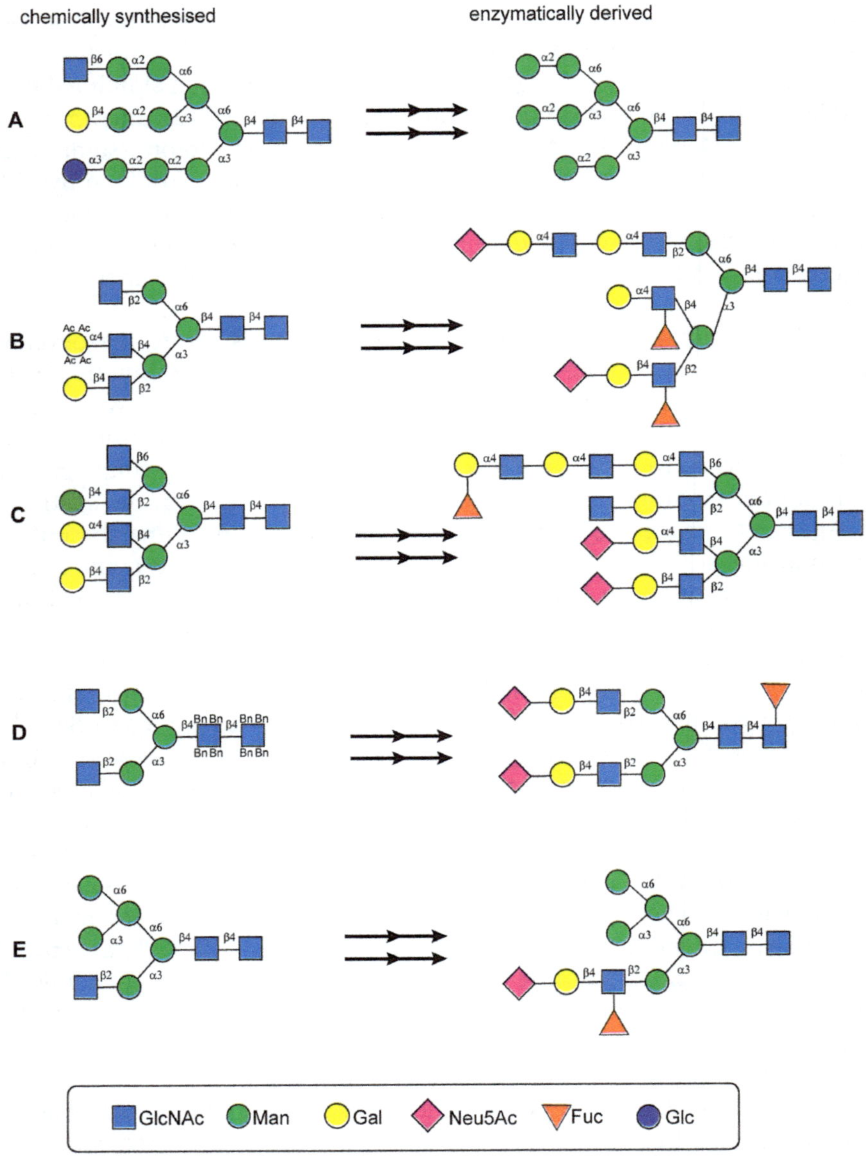

Scheme 10.21 Chemoenzymatic approaches to *N*-glycan synthesis.

precursor (Scheme 10.21C), this time blocking branches with unnatural hexoses to control enzymatic extension.[134] Reichardt and co-workers also recently chemoenzymatically prepared a ^{13}C-labelled library of complex *N*-glycans as standards for absolute glycan quantification by enzymatically extending a labelled heptasaccharide (Scheme 10.21D).[132] The *N*-acetylglucosamine residues are isotopically labelled by acylation with ^{13}C-substituted acetic anhydride. Deblocking, either by catalytic hydrogenolysis of benzyl protecting groups or enzymatic removal of terminal *N*-acetylglucosamine residues, followed by enzymatic extension with fucosyltransferases, galactosyltransferases and sialyltransferases leads to a library of standards for absolute glycan quantification by mass spectrometry. Finally a library consisting of 73 *N*-glycans was also recently synthesised by Wang and coworkers[131] through extension of eight chemically synthesised core structures, pentasaccharides to octasaccharides (Scheme 10.21E), with the help of four glycosyltransferases to produce 15-mer *N*-glycans.

10.8 Conclusion

It is clear that in the 21st century the improvements in synthetic methods for carbohydrate chemistry, as well as the increasing availability of advanced technical equipment, has begun to provide chemical biologists and glycobiologists with access to carbohydrates of unprecedented purity, complexity and size. The next challenge in the field is to further refine and exploit these new methods so they yield carbohydrates in a quantity, and with a level of practicality, to enable the next generation of glycoscientists to answer the ever more complex questions that continue to arise about the roles carbohydrates play in life and disease.

References

1. D. Peters, *Biotechnol. J.*, 2006, **1**, 806.
2. F. T. Liu and G. A. Rabinovich, *Nat. Rev. Cancer*, 2005, **5**, 29–41.
3. R. Sasisekharan, Z. Shriver, G. Venkataraman and U. Narayanasami, *Nat. Rev. Cancer*, 2002, **2**, 521–528.
4. P. I. Kitov, J. M. Sadowska, G. Mulvey, G. D. Armstrong, H. Ling, N. S. Pannu, R. J. Read and D. R. Bundle, *Nature*, 2000, **403**, 669–672.
5. W. I. Weis and K. Drickamer, *Annu. Rev. Biochem.*, 1996, **65**, 441–473.
6. Y. Kinjo, D. Wu, G. S. Kim, G. W. Xing, M. A. Poles, D. D. Ho, M. Tsuji, K. Kawahara, C. H. Wong and M. Kronenberg, *Nature*, 2005, **434**, 520–525.
7. G. Ragupathi, *Cancer Immunol. Immunother.*, 1996, **43**, 152–157.
8. G. Ragupathi, T. K. Park, S. L. Zhang, I. J. Kim, L. Graber, S. Adluri, K. O. Lloyd, S. J. Danishefsky and P. O. Livingston, *Angew. Chem., Int. Ed. Engl.*, 1997, **36**, 125–128.
9. J. N. Reitter, R. E. Means and R. C. Desrosiers, *Nat. Med.*, 1998, **4**, 679–684.
10. K. Toshima and K. Tatsuta, *Chem. Rev.*, 1993, **93**, 1503–1531.
11. B. G. Davis, *J. Chem. Soc., Perkin Trans. 1*, 2000, **1**, 2137–2160.
12. P. Sinay, *Pure Appl. Chem.*, 1991, **63**, 519–528.
13. T. Mukaiyama, Y. Murai and S. Shoda, *Chem. Lett.*, 1981, 431–434.
14. W. Koenigs and E. Knorr, *Ber. Dtsch. Chem. Ges.*, 1901, **34**, 957–981.
15. P. J. Garegg, *Adv. Carbohydr. Chem. Biochem.*, 1997, **52**, 179–205.

16. R. R. Schmidt and W. Kinzy, *Adv. Carbohydr. Chem. Biochem.*, 1994, **50**, 21–123.
17. F. Barresi and O. Hindsgaul, *J. Carbohydr. Chem.*, 1995, **14**, 1043–1087.
18. D. Kahne, S. Walker, Y. Cheng and D. Vanengen, *J. Am. Chem. Soc.*, 1989, **111**, 6881–6882.
19. B. Fraser Reid, U. E. Udodong, Z. F. Wu, H. Ottosson, J. R. Merritt, C. S. Rao, C. Roberts and R. Madsen, *Synlett*, 1992, 927–942.
20. H. Paulsen, *Angew. Chem., Int. Ed. Engl.*, 1982, **21**, 155–224.
21. A. V. Demchenko, *Handbook of Chemical Glycosylation; Advances in Stereoselectivity and Therapeutic Relevance*, Wiley-VCH, Weinheim, 2008.
22. R. U. Lemieux, *Pure Appl. Chem.*, 1971, **25**, 527–548.
23. S. S. Nigudkar and A. V. Demchenko, *Chem. Sci.*, 2015, **6**, 2687–2704.
24. R. Brabham and M. A. Fascione, in *Selective Glycosylations: Synthetic Methods and Catalysts*, ed. C. S. Bennet, Wiley-VCH, Weinheim, Germany, 2017, ch. 5, pp. 97–112.
25. R. Das and B. Mukhopadhyay, *ChemistryOpen*, 2016, **5**, 401–433.
26. F. Barresi and O. Hindsgaul, *J. Am. Chem. Soc.*, 1991, **113**, 9376–9377.
27. G. Stork, H. S. Suh and G. Kim, *J. Am. Chem. Soc.*, 1991, **113**, 7054–7056.
28. F. Barresi and O. Hindsgaul, *Synlett*, 1992, 759–761.
29. G. Stork and J. J. La Clair, *J. Am. Chem. Soc.*, 1996, **118**, 247–248.
30. A. Dan, Y. Ito and T. Ogawa, *J. Org. Chem.*, 1995, **60**, 4680–4681.
31. Y. Ito and T. Ogawa, *Angew. Chem., Int. Ed. Engl.*, 1994, **33**, 1765–1767.
32. A. Dan, M. Lergenmuller, M. Amano, Y. Nakahara, T. Ogawa and Y. Ito, *Chem. – Eur. J.*, 1998, **4**, 2182–2190.
33. Y. Ito, Y. Ohnishi, T. Ogawa and Y. Nakahara, *Synlett*, 1998, 1102–1104.
34. Y. J. Lee, A. Ishiwata and Y. Ito, *J. Am. Chem. Soc.*, 2008, **130**, 6330–6331.
35. A. Ishiwata, Y. Munemura and Y. Ito, *Eur. J. Org. Chem.*, 2008, 4250–4263.
36. M. Bols, *J. Chem. Soc., Chem. Commun.*, 1992, 913–914.
37. S. C. Ennis, A. J. Fairbanks, R. J. Tennant-Eyles and H. S. Yeates, *Synlett*, 1999, 1387–1390.
38. C. M. P. Seward, I. Cumpstey, M. Aloui, S. C. Ennis, A. J. Fairbanks and A. J. Redgrave, *Chem. Commun.*, 2000, 1409–1410.
39. I. Cumpstey, A. J. Fairbanks and A. J. Redgrave, *Org. Lett.*, 2001, **3**, 2371–2374.
40. S. C. Ennis, A. J. Fairbanks, C. A. Slinn, R. J. Tennant-Eyles and H. S. Yeates, *Tetrahedron*, 2001, **57**, 4221–4230.
41. K. Chayajarus, D. J. Chambers, M. J. Chughtai and A. J. Fairbanks, *Org. Lett.*, 2004, **6**, 3797–3800.
42. E. Attolino, T. W. D. F. Rising, C. D. Heidecke and A. J. Fairbanks, *Tetrahedron: Asymmetry*, 2007, **18**, 1721–1734.
43. M. Aloui, D. J. Chambers, I. Cumpstey, A. J. Fairbanks, A. J. Redgrave and C. M. P. Seward, *Chem. – Eur. J.*, 2002, **8**, 2608–2621.
44. G. H. Veeneman, *Carbohydrate Chemistry*, Blackie Academic & Professional, London, 1999.

45. A. Demchenko, T. Stauch and G. J. Boons, *Synlett*, 1997, 818–820.
46. A. Ishiwata, Y. Munemura and Y. Ito, *Tetrahedron*, 2008, **64**, 92–102.
47. A. V. Demchenko, E. Rousson and G.-J. Boons, *Tetrahedron Lett.*, 1999, **40**, 6523–6526.
48. D. Crich, T. S. Hu and F. Cai, *J. Org. Chem.*, 2008, **73**, 8942–8953.
49. G. J. Boons, S. Bowers and D. M. Coe, *Tetrahedron Lett.*, 1997, **38**, 3773–3776.
50. K. Fukase, Y. Nakai, T. Kanoh and S. Kusumoto, *Synlett*, 1998, 84–86.
51. T. Zhu and G. J. Boons, *Org. Lett.*, 2001, **3**, 4201–4203.
52. D. Crich and W. L. Cai, *J. Org. Chem.*, 1999, **64**, 4926–4930.
53. S. E. Reisman, A. G. Doyle and E. N. Jacobsen, *J. Am. Chem. Soc.*, 2008, **130**, 7198–7199.
54. D. J. Cox, M. D. Smith and A. J. Fairbanks, *Org. Lett.*, 2010, **12**, 1452–1455.
55. E. I. Balmond, D. M. Coe, M. C. Galan and E. M. McGarrigle, *Angew. Chem., Int. Ed.*, 2012, **51**, 9152–9155.
56. C. Palo-Nieto, A. Sau, R. Williams and M. C. Galan, *J. Org. Chem.*, 2017, **82**, 407–414.
57. S. Medina, M. J. Harper, E. I. Balmond, S. Miranda, G. E. M. Crisenza, D. M. Coe, E. M. McGarrigle and M. C. Galan, *Org. Lett.*, 2016, **18**, 4222–4225.
58. Y. Geng, A. Kumar, H. M. Faidallah, H. A. Albar, I. A. Mhkalid and R. R. Schmidt, *Angew. Chem., Int. Ed.*, 2013, **52**, 10089–10092.
59. Y. Park, K. C. Harper, N. Kuhl, E. E. Kwan, R. Y. Liu and E. N. Jacobsen, *Science*, 2017, **355**, 162–166.
60. L. Sun, X. Wu, D.-C. Xiong and X.-S. Ye, *Angew. Chem., Int. Ed.*, 2016, **55**, 8041–8044.
61. T. Kimura, T. Eto, D. Takahashi and K. Toshima, *Org. Lett.*, 2016, **18**, 3190–3193.
62. S. Raghavan and D. Kahne, *J. Am. Chem. Soc.*, 1993, **115**, 1580–1581.
63. S. V. Ley and H. W. Priepke, *Angew. Chem.*, 1994, **106**, 2412–2414.
64. S. V. Ley and H. W. Priepke, *Angew. Chem., Int. Ed.*, 1994, **33**, 2292–2294.
65. H. Yamada, T. Harada and T. Takahashi, *J. Am. Chem. Soc.*, 1994, **116**, 7919–7920.
66. D. Crich and S. Sun, *J. Am. Chem. Soc.*, 1998, **120**, 435–436.
67. J. D. Codée, R. E. Litjens, R. den Heeten, H. S. Overkleeft, J. H. van Boom and G. A. van der Marel, *Org. Lett.*, 2003, **5**, 1519–1522.
68. X. Huang, L. Huang, H. Wang and X. S. Ye, *Angew. Chem., Int. Ed. Engl.*, 2004, **116**, 5333–5336.
69. O. Kanie, Y. Ito and T. Ogawa, *J. Am. Chem. Soc.*, 1994, **116**, 12073–12074.
70. H. Yamada, T. Harada, H. Miyazaki and T. Takahashi, *Tetrahedron Lett.*, 1994, **35**, 3979–3982.
71. H. Tanaka, H. Yamada and T. Takahashi, *Trends Glycosci. Glycotechnol.*, 2007, **19**, 183.
72. Y. Wang, X.-S. Ye and L.-H. Zhang, *Org. Biomol. Chem.*, 2007, **5**, 2189–2200.

73. Y. Wang, L.-H. Zhang and X.-S. Ye, *Comb. Chem. High Throughput Screening*, 2006, **9**, 63–75.
74. B. Yu, Z. Yang and H. Cao, *Curr. Org. Chem.*, 2005, **9**, 179–194.
75. Z. Zhang, I. R. Ollmann, X.-S. Ye, R. Wischnat, T. Baasov and C.-H. Wong, *J. Am. Chem. Soc.*, 1999, **121**, 734–753.
76. C. Schuerch and J. M. Frechet, *J. Am. Chem. Soc.*, 1971, **93**, 492–496.
77. O. J. Plante, E. R. Palmacci and P. H. Seeberger, *Science*, 2001, **291**, 1523–1527.
78. P. H. Seeberger, *Chem. Soc. Rev.*, 2008, **37**, 19–28.
79. O. J. Plante, E. R. Palmacci and P. H. Seeberger, *Adv. Carbohydr. Chem. Biochem.*, 2003, **58**, 35–54.
80. P. H. Seeberger, *Acc. Chem. Res.*, 2015, **48**, 1450–1463.
81. L. Kröck, D. Esposito, B. Castagner, C.-C. Wang, P. Bindschädler and P. H. Seeberger, *Chem. Sci.*, 2012, **3**, 1617–1622.
82. O. Calin, S. Eller and P. H. Seeberger, *Angew. Chem., Int. Ed.*, 2013, **52**, 5862–5865.
83. K. Naresh, F. Schumacher, H. S. Hahm and P. Seeberger, *Chem. Commun.*, 2017, **53**, 9085–9088.
84. J. Niggemann, K. Michaelis, R. Frank, N. Zander and G. Höfle, J. Chem. Soc., Perkin Trans. *1*, 2002, 2490–2503.
85. M. Weishaupt, H. Hahm, A. Geissner and P. Seeberger, *Chem. Commun.*, 2017, **53**, 3591–3594.
86. P. Dallabernardina, F. Schuhmacher, P. H. Seeberger and F. Pfrengle, *Chem. – Eur. J.*, 2017, **23**, 3191–3196.
87. M. W. Weishaupt, S. Matthies, M. Hurevich, C. L. Pereira, H. S. Hahm and P. H. Seeberger, *Beilstein J. Org. Chem.*, 2016, **12**, 1440.
88. S. G. Parameswarappa, K. Reppe, A. Geissner, P. Ménová, S. Govindan, A. D. Calow, A. Wahlbrink, M. W. Weishaupt, B. P. Monnanda and R. L. Bell, *Cell Chem. Biol.*, 2016, **23**, 1407–1416.
89. D. B. Werz, B. Castagner and P. H. Seeberger, *J. Am. Chem. Soc.*, 2007, **129**, 2770–2771.
90. H. S. Hahm, C.-F. Liang, C.-H. Lai, R. J. Fair, F. Schuhmacher and P. H. Seeberger, *J. Org. Chem.*, 2016, **81**, 5866–5877.
91. M. C. Hewitt, D. A. Snyder and P. H. Seeberger, *J. Am. Chem. Soc.*, 2002, **124**, 13434–13436.
92. D. Schmidt, F. Schuhmacher, A. Geissner, P. H. Seeberger and F. Pfrengle, *Chem. – Eur. J.*, 2015, **21**, 5709–5713.
93. S. Pattathil, U. Avci, D. Baldwin, A. G. Swennes, J. A. McGill, Z. Popper, T. Bootten, A. Albert, R. H. Davis and C. Chennareddy, *Plant Phys.*, 2010, **153**, 514–525.
94. J. Kandasamy, M. Hurevich and P. H. Seeberger, *Chem. Commun.*, 2013, **49**, 4453–4455.
95. M. P. Bartetzko, F. Schuhmacher, H. S. Hahm, P. H. Seeberger and F. Pfrengle, *Org. Lett.*, 2015, **17**, 4344–4347.
96. A. Reinhardt, Y. Yang, H. Claus, C. L. Pereira, A. D. Cox, U. Vogel, C. Anish and P. H. Seeberger, *Chem. Biol.*, 2015, **22**, 38–49.

97. C.-H. Lai, H. S. Hahm, C.-F. Liang and P. H. Seeberger, *Beilstein J. Org. Chem.*, 2015, **11**, 617.
98. Y. Yang, S. Oishi, C. E. Martin and P. H. Seeberger, *J. Am. Chem. Soc.*, 2013, **135**, 6262–6271.
99. S. Eller, M. Collot, J. Yin, H. S. Hahm and P. H. Seeberger, *Angew. Chem., Intl. Ed.*, 2013, **52**, 5858–5861.
100. H. S. Hahm, M. Hurevich and P. H. Seeberger, *Nat. Commun.*, 2016, **7**, 12482.
101. M. Wilsdorf, D. Schmidt, M. Bartetzko, P. Dallabernardina, F. Schuhmacher, P. Seeberger and F. Pfrengle, *Chem. Commun.*, 2016, **52**, 10187–10189.
102. F. A. Jaipuri and N. L. Pohl, *Org. Biomol. Chem.*, 2008, **6**, 2686–2691.
103. S.-L. Tang and N. L. Pohl, *Org. Lett.*, 2015, **17**, 2642–2645.
104. S. J. Turco and A. Descoteaux, *Annu. Rev. Microbiol.*, 1992, **46**, 65–92.
105. A. Vercellone, J. Nigou and G. Puzo, *Front. Biosci.*, 1998, **3**, e149–e163.
106. E.-H. Song, A. O. Osanya, C. A. Petersen and N. L. Pohl, *J. Am. Chem. Soc.*, 2010, **132**, 11428–11430.
107. P. Pornsuriyasak, S. C. Ranade, A. Li, M. C. Parlato, C. R. Sims, O. V. Shulga, K. J. Stine and A. V. Demchenko, *Chem. Commun.*, 2009, 1834–1836.
108. N. V. Ganesh, K. Fujikawa, Y. H. Tan, S. S. Nigudkar, K. J. Stine and A. V. Demchenko, *J. Org. Chem.*, 2013, **78**, 6849–6857.
109. S. G. Pistorio, S. S. Nigudkar, K. J. Stine and A. V. Demchenko, *J. Org. Chem.*, 2016, **81**, 8796–8805.
110. N. V. Ganesh, K. Fujikawa, Y. H. Tan, K. J. Stine and A. V. Demchenko, *Org. Lett.*, 2012, **14**, 3036–3039.
111. M. C. Parlato, M. N. Kamat, H. Wang, K. J. Stine and A. V. Demchenko, *J. Org. Chem.*, 2008, **73**, 1716–1725.
112. D. Xu and J. D. Esko, *Annu. Rev. Biochem.*, 2014, **83**, 129–157.
113. A. Ori, M. C. Wilkinson and D. G. Fernig, *J. Biol. Chem.*, 2011, **286**, 19892–19904.
114. J. R. Bishop, M. Schuksz and J. D. Esko, *Nature*, 2007, **446**, 1030–1037.
115. M. Mende, C. Bednarek, M. Wawryszyn, P. Sauter, M. B. Biskup, U. Schepers and S. Bräse, *Chem. Rev.*, 2016, **116**, 8193–8255.
116. Y.-P. Hu, S.-Y. Lin, C.-Y. Huang, M. M. L. Zulueta, J.-Y. Liu, W. Chang and S.-C. Hung, *Nat. Chem.*, 2011, **3**, 557–563.
117. T.-Y. Huang, D. Irene, M. M. L. Zulueta, T.-J. Tai, S.-H. Lain, C.-P. Cheng, P.-X. Tsai, S.-Y. Lin, Z.-G. Chen, C.-C. Ku, C.-D. Hsiao, C.-L. Chyan and S.-C. Hung, *Angew. Chem., Int. Ed.*, 2017, **56**, 4192–4196.
118. R. Schwörer, O. V. Zubkova, J. E. Turnbull and P. C. Tyler, *Chem. – Eur. J.*, 2013, **19**, 6817–6823.
119. S. U. Hansen, G. J. Miller, M. J. Cliff, G. C. Jayson and J. M. Gardiner, *Chem. Sci.*, 2015, **6**, 6158–6164.
120. S. U. Hansen, G. J. Miller, C. Cole, G. Rushton, E. Avizienyte, G. C. Jayson and J. M. Gardiner, *Nat. Commun.*, 2013, **4**, 2016.

121. T. L. Lowary, *Acc. Chem. Res.*, 2016, **49**, 1379–1388.
122. Y. Wu, D.-C. Xiong, S.-C. Chen, Y.-S. Wang and X.-S. Ye, *Nat. Commun.*, 2017, **8**, 14851.
123. P. J. Brennan, *Tuberculosis*, 2003, **83**, 91–97.
124. C. Wang, H. Wang, X. Huang, L.-H. Zhang and X.-S. Ye, *Synlett*, 2006, **2006**, 2846–2850.
125. M. A. Walczak, J. Hayashida and S. J. Danishefsky, *J. Am. Chem. Soc.*, 2013, **135**, 4700–4703.
126. M. Hagiwara, M. Dohi, Y. Nakahara, K. Komatsu, Y. Asahina, A. Ueki, H. Hojo, Y. Nakahara and Y. Ito, *J. Org. Chem.*, 2011, **76**, 5229–5239.
127. M. Nagasaki, Y. Manabe, N. Minamoto, K. Tanaka, A. Silipo, A. Molinaro and K. Fukase, *J. Org. Chem.*, 2016, **81**, 10600–10616.
128. M. Mönnich, S. Eller, T. Karagiannis, L. Perkams, T. Luber, D. Ott, M. Niemietz, J. Hoffman, J. Walcher, L. Berger, M. Pischl, M. Weishaupt, C. Wirkner, R. G. Lichtenstein and C. Unverzagt, *Angew. Chem., Int. Ed.*, 2016, **55**, 10487–10492.
129. S. S. Shivatare, S.-H. Chang, T.-I. Tsai, S. Y. Tseng, V. S. Shivatare, Y.-S. Lin, Y.-Y. Cheng, C.-T. Ren, C.-C. D. Lee, S. Pawar, C.-S. Tsai, H.-W. Shih, Y.-F. Zeng, C.-H. Liang, P. D. Kwong, D. R. Burton, C.-Y. Wu and C.-H. Wong, *Nat. Chem.*, 2016, **8**, 338–346.
130. A. Koizumi, I. Matsuo, M. Takatani, A. Seko, M. Hachisu, Y. Takeda and Y. Ito, *Angew. Chem., Int. Ed.*, 2013, **52**, 7426–7431.
131. L. Li, Y. Liu, C. Ma, J. Qu, A. D. Calderon, B. Wu, N. Wei, X. Wang, Y. Guo, Z. Xiao, J. Song, G. Sugiarto, Y. Li, H. Yu, X. Chen and P. G. Wang, *Chem. Sci.*, 2015, **6**, 5652–5661.
132. B. Echeverria, J. Etxebarria, N. Ruiz, Á. Hernandez, J. Calvo, M. Haberger, D. Reusch and N.-C. Reichardt, *Anal. Chem.*, 2015, **87**, 11460–11467.
133. Z. Wang, Z. S. Chinoy, S. G. Ambre, W. Peng, R. McBride, R. P. de Vries, J. Glushka, J. C. Paulson and G.-J. Boons, *Science*, 2013, **341**, 379–383.
134. I. A. Gagarinov, T. Li, J. S. Toraño, T. Caval, A. D. Srivastava, J. A. W. Kruijtzer, A. J. R. Heck and G.-J. Boons, *J. Am. Chem. Soc.*, 2017, **139**, 1011–1018.

CHAPTER 11

Precursor-directed Biosynthesis and Semi-synthesis of Natural Products

EDWARD KALKREUTER,[a,b] SAMANTHA M. CARPENTER[a] AND GAVIN J. WILLIAMS*[a,b]

[a] Department of Chemistry, NC State University, Raleigh, NC 27695-8204, USA; [b] Comparative Medicine Institute, NC State University, Raleigh, NC, USA
*Email: gjwillia@ncsu.edu

11.1 Introduction

Natural products comprise just approximately 0.7% of the commercially available chemical space but constitute a significantly larger portion of the biologically active space.[1,2] As such, accessing these molecules and their analogues has been of great interest to chemists and biologists, especially over the last 60 years. In order to address limitations in scope, efficiency and scale of purely biological or chemical approaches to access natural products and their analogues, strategies have been developed to bridge the gap between traditional synthetic chemistry and biological syntheses. The discovery of taxol (paclitaxel) in 1971 provided a test case for this philosophy. Within ten years of its discovery, taxol was identified as a potential blockbuster anti-cancer drug candidate and grossed $1.6 billion annually by the year 2000. It became apparent that continued isolation of

Chemical Biology No. 10
Chemical and Biological Synthesis: Enabling Approaches for Understanding Biology
Edited by Nick J. Westwood and Adam Nelson

Published by the Royal Society of Chemistry, www.rsc.org

taxol from the bark of the Pacific yew (*Taxus brevifolia*) was not a sustainable route to large scale production (killing 360 000 trees annually in 1988).[3] Six total syntheses of taxol were published between 1994 to 1999, highlighted by Robert Holton and co-workers 1982–1994 46-step *tour de force*.[4–6] Though an impressive feat, the total synthesis of taxol in 46-steps could not solve the product shortfall. Notably, even though yew trees contain very little taxol, plantations around the world have been used for the isolation of precursors of taxol that can be used as starting materials for the semi-synthesis of taxol and its analogues (*e.g.* docetaxel).[7] Additionally, biological engineering approaches have yielded precursors suitable for semi-synthesis, such as the overexpression of a β-xylosidase in yeast to remove an undesired xylose from 7-β-xylosyl-10-deacetyltaxol to yield 10-deacetyltaxol.[8] Similarly, taxol has been produced *via* cell cultures from various *Taxus* species.[9] Other production approaches that might meet the required demand include the production of taxol in a heterologous host. However, the biosynthetic pathway for taxol is expected to consist of 19 steps but has not been fully elucidated. Subsequently, partial pathways have been transferred into *Escherichia coli*, yeast and plants, but a range of issues (especially oxidation steps) have thwarted the advancement of this approach.[10–12] Nevertheless, several intermediates suitable for semi-synthesis have already been produced.[7,8]

Another classic, though less monumental example of partial or semi-synthesis is the conversion of rifamycin, an ansamycin antibiotic from *Amycolatopsis rifamycinica*, to rifampicin, a leading anti-tuberculosis drug. The addition of the methyl-piperazinyl moiety does not alter the bioactivity in any measurable way but improves its bioavailability.[13] While this was a very early example of semi-synthesis, 21st century chemists have continued to develop novel processes to couple biosynthesis and chemical synthesis.[14–16] While elegant, the use of non-enzymatic catalysis to facilitate semi-synthetic modification of natural products will not be included in this chapter, but the reader is directed to a recent comprehensive review of this area.[17]

11.2 Strategies for Coupling of Biosynthesis and Chemical Synthesis

Biological and chemical synthesis can be coupled in a number of ways that vary according to the number of total steps involved, the order of biological *vs.* chemical transformations and the degree of biological manipulations or chemical synthesis required. In general, such approaches usually capitalize on the stepwise and often modular biosynthetic assembly of natural products (Figure 11.1A). Most commonly, these strategies can be divided into the four approaches described below and in Figure 11.1B–E.

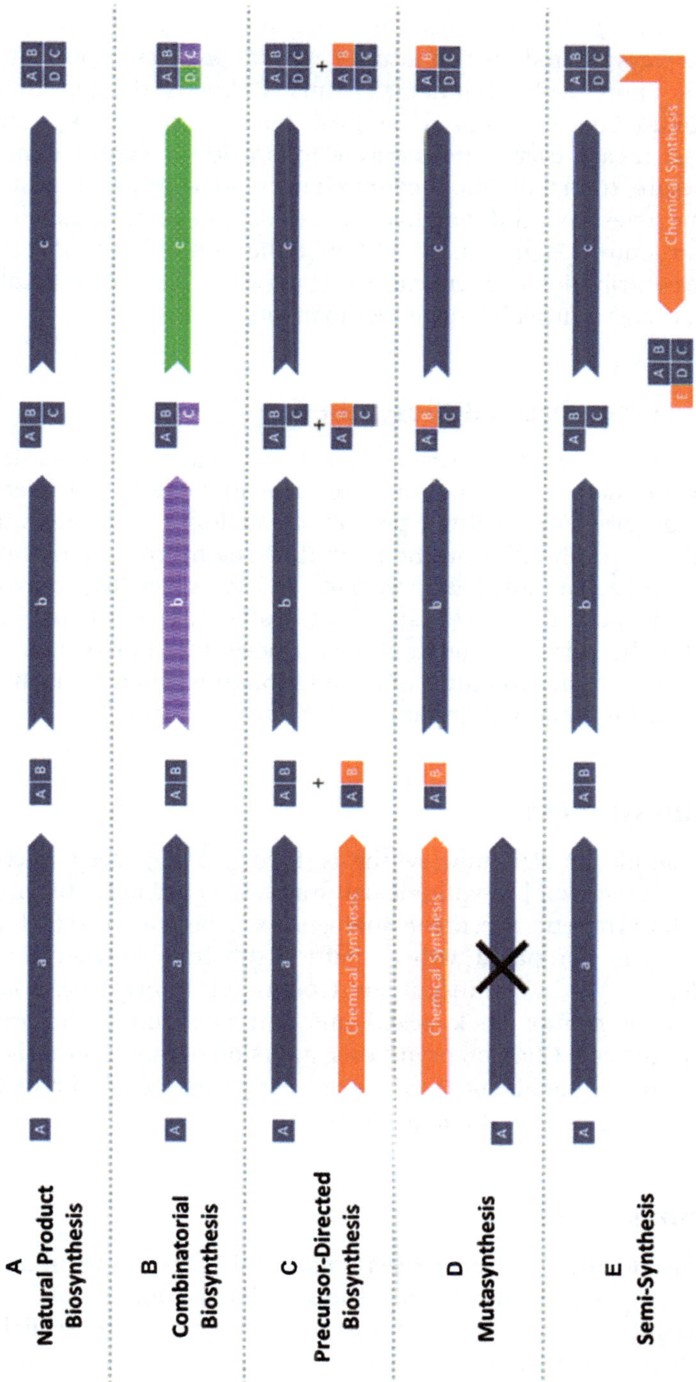

Figure 11.1 (A) In the natural pathway, the native machinery synthesizes various intermediates and final products from native precursors (individual blocks). (B)–(E). Different chemical or biological approaches to introduce synthetic elements or additional non-native substrates (shaded blocks).

11.2.1 Combinatorial Biosynthesis

Broadly, combinatorial biosynthesis is the genetic manipulation of biosynthetic pathways to produce "unnatural" natural products. Formally, this approach does not involve synthetic chemistry, but is the platform for strategies that do. Combinatorial biosynthesis ranges from making relatively small changes to a biosynthetic pathway to provide a modified natural product structure, to making more comprehensive changes, such as mixing and matching genes from different pathways or organisms to make libraries of chimeric structures (Figure 11.1B). Although the focus of this chapter is to highlight the contribution of chemical synthesis, we will include examples of combinatorial biosynthesis by way of comparison.

11.2.2 Precursor-directed Biosynthesis

Precursor-directed biosynthesis (PDB) (Figure 11.1C) leverages organic synthesis to provide native or unnatural substrates that are fed to microbial strains and converted to a natural product or analogue by a biosynthetic pathway. This approach relies on the natural or engineered promiscuity of enzymes that utilize the artificial precursor and is especially prevalent in biosynthetic pathways that use easily diversifiable substrates like amino acids or carboxylic acids. Although this is a powerful approach to introducing diversity into the structures of complex natural products, PDB often produces a mixture of natural and unnatural final products.[18]

11.2.3 Mutasynthesis

While highly similar to PDB, mutasynthesis (Figure 11.1D) uses a microbial host in which a precursor biosynthetic pathway is knocked out. This process is useful if the synthetic precursor analogue is a poor substrate for the intercepted enzyme compared with its natural substrate or when a single product is desired.[19–23] A significant shortcoming of mutasynthesis is that often a precursor cannot be knocked out if it is required for primary metabolism in the cell. Chemobiosynthesis, a version of mutasynthesis used mostly for assembly line-like pathways, skips early-stage steps of a pathway through the feeding of a synthetic precursor.[24]

11.2.4 Semi-synthesis

Semi-synthesis (Figure 11.1E) accesses small molecules using traditional organic chemistry using starting materials isolated from nature. These synthetic modifications may be useful to improve efficacy, bioavailability, stability or other properties. In many cases, production of a natural product by semi-synthesis is much more practical than a total synthesis.[25]

11.3 Leveraging Assembly Line Biosynthetic Systems for Polyketide and Non-ribosomal Peptide Diversification

11.3.1 Overview of Assembly Line Biosynthesis

Nature has organized some of the most complex enzyme systems into pathways akin to assembly lines. These biosynthetic pathways are organized in a modular fashion, whereby each module acts as a 'station' on the assembly line to add and tailor a portion of the final product. Type I polyketide synthases (PKSs) and non-ribosomal peptide synthetases (NRPSs) are assembly lines that produce secondary metabolites that are diverse in structure and biological activities.[26,27] A significant portion of these products are pharmaceutically relevant, as seen with many classes of natural products, and are considered to be privileged scaffolds.[28,29] As with synthetic drugs, there is a great interest in diversifying or making analogues of these natural products in the hopes of producing molecules with improved activity, decreased toxicity, *etc.* A variety of approaches, including semi-synthesis, total synthesis and biosynthesis, have been applied to these products with varying levels of success.[30,31] As more is discovered about the structure and enzymology of these assembly lines, targeted and rational engineering has become more feasible. Many of the first engineering successes were the results of precursor-directed biosynthesis and combinatorial biosynthesis, which have now been complemented with mutasynthesis and other hybrid strategies.

Type I PKSs are modular enzymatic systems that stitch together 'ketide' units—most commonly derived from malonyl- and methylmalonyl-CoA—in a stepwise fashion to create complex and stereo-rich partially reduced polyketides. The canonical and arguably most well-studied example of a type I PKS is 6-deoxyerythronolide B synthase (DEBS), which synthesizes the macrolactone core of the antibiotic erythromycin A (Figure 11.2A).[32] Other notable examples of polyketides produced by type I PKSs include rapamycin (immunosuppressant), epothilone A (anti-tumour) and lovastatin (anti-cholesterol, Figure 11.2B). This small pool of examples reveals the remarkable diversity in structure and biological activity of this natural product class. Indeed, these privileged structures have an exceptionally high hit rate when compared with other natural product classes. For example, a 2005 study calculated a 0.3% hit rate from naturally occurring polyketides, compared with less than 0.001% hit rate from a standard pharmaceutical collection.[33]

Despite the diversity of these natural products, they are biosynthesized from a modest pool of simple small-molecule building blocks (Figure 11.2C).[34] The intricacy of the natural product is derived from their templated biosynthesis; the proteins are organized as modules, which are further divided into domains with distinct enzymatic activities. Each module is responsible for the installation of a building block and consists minimally of an acyltransferase (AT), ketosynthase (KS) and an acyl carrier protein (ACP). The AT initiates the

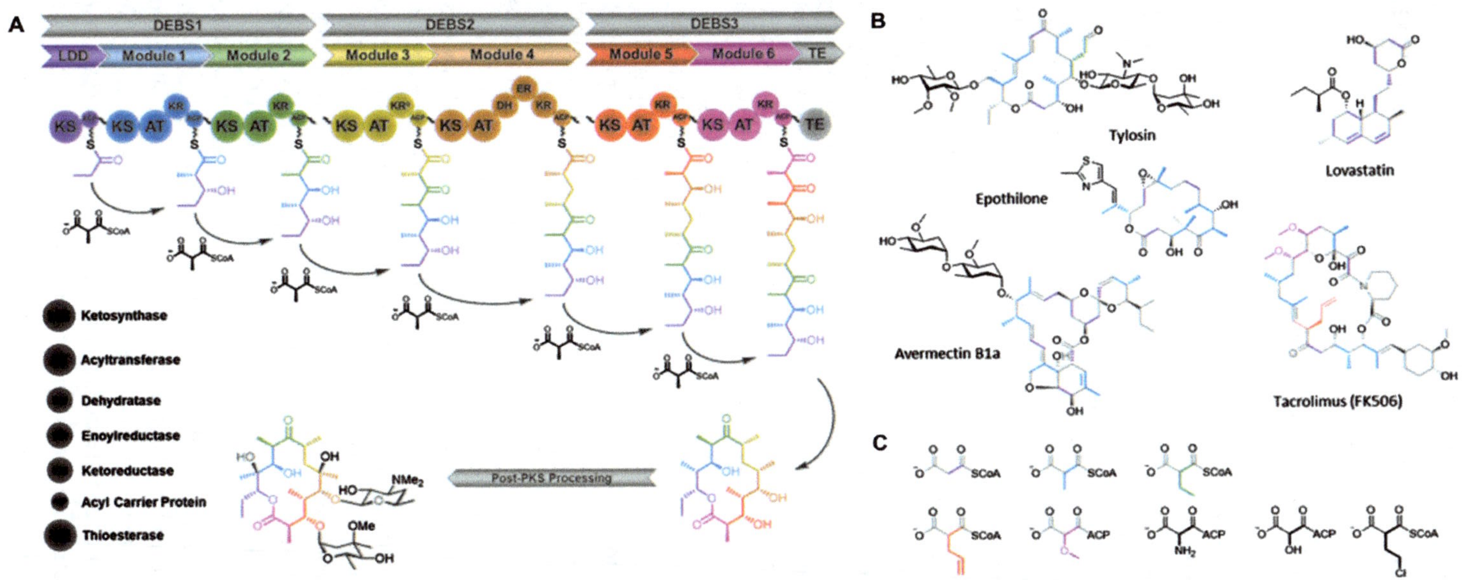

Figure 11.2 Type I PKSs. (A) Organization of DEBS. (B) Examples of polyketides produced by type I PKSs. (C) Building blocks utilized by type I PKSs. Common extender units are shown in the top row.

incorporation of the thioester-activated building block at each module by acylating its active site serine residue. In most cases, the AT is specific for a single substrate and discriminates against other similar substrates available in the cell. The malonyl-derived building block is then transferred to the phosphopantetheine (Ppant) arm of the ACP. While tethered to the ACP, the KS catalyses a Claisen-like condensation with the precursor from the upstream module to yield the newly extended product. Modules can also include up to three reductive domains: a ketoreductase (KR), enoylreductase (ER) and dehydratase (DH). These reduce the β-carbonyl to an alcohol, alkene or fully saturated methylene respectively. Further complexity can be introduced by modules that contains in-line tailoring domains, such as *C*- or *O*-methyltransferases (MTs). Finally, at the terminal module, the product is either hydrolysed or cyclized by the thioesterase (TE) to yield the final polyketide product. The polyketide product is then usually further decorated by post-PKS tailoring enzymes such as glycosyltransferases (GTs), P450s, or MTs.

Non-ribosomal peptide synthetases (NRPS) follow a thio-templated biosynthesis, similar to PKSs. The building blocks for these modular enzymatic systems are amino acids. The adenylation domain (A) is analogous to the AT domain from PKSs and the thiolation domain (T) or the peptidyl carrier protein domain (PCP) is analogous to the ACP. The amino acid is first activated by the A domain before being transferred to the Ppant of the T domain. The downstream condensation domain (C) catalyses peptide bond formation between the newly introduced amino acid and the growing peptide from the upstream domain. As with PKSs, NRPS can contain other tailoring domains, including methylation (MT), epimerization (E), oxidase (Ox) and cyclization domains (Cy). The terminal module contains a thioesterase domain, which is responsible for either hydrolysing off the linear peptide or cyclizing and offloading the product. Because of their modular organization and known domain function, NRPSs are excellent candidates for PDB and mutasynthesis to produce novel analogues.

Shortly after PKS pathways were first discovered, the concept of producing 'hybrid' antibiotics by genetic engineering was quickly translated into practice by leveraging actinorhodin biosynthesis to produce analogues, though little was known about the mechanism and specificity of these pathways.[35] A few years later, avermectin analogues were biosynthesized using precursor-directed biosynthesis approaches to yield compounds with anthelmintic and insecticidal activity.[36] These successes, among others, served as inspiration for the future of polyketide synthase engineering for the production of analogues.

11.3.2 A Model for Biosynthetic Engineering: 6-Deoxyerythronolide B Synthase (DEBS)

The discovery and decoding of the modular organization and domain functions of type I PKSs[32,37–39] coupled with the advent of modern molecular biology techniques allowed for construction and manipulation of hybrid PKS pathways. This was limited only by the available DNA segments, resulting in the rapid acceleration of engineering efforts to produce non-natural polyketide analogues. Many of the most notable early successes were achieved with DEBS.

Included in these engineering efforts were module deletions or swaps, in which an entire module or modules were either deleted to produce a smaller product or replaced with a module from another pathway to introduce an alternative building block. Cortes and co-workers discovered that the terminal thioesterase domain at the terminus of DEBS3 could be appended to the terminus of DEBS1 (which includes the loading module and modules 1 and 2, Figure 11.2A) to create a truncated system capable of producing triketide lactones, providing evidence that the thioesterase was capable of cyclizing alternative substrates and allowing for a smaller model system to explore combinatorial biosynthesis.[40] The creation of this minimal model system combined with the newfound ability to express these enzymes in a heterologous host[41] and to use them *in vitro*[42,43] greatly accelerated combinatorial biosynthesis efforts. Khosla and co-workers determined that the TE was also active when fused to the C-terminus of module 5 to biosynthesize the predicted 12-membered macrolactone[44] and to module 3 to produce tetraketides,[45] further supporting the modularity of these systems.

The DEBS1-TE system was modified by Leadlay and co-workers to create a hybrid system, replacing the natural propionate-accepting loading module with the known promiscuous loading module from the avermectin PKS, producing a panel of triketide lactones. This activity successfully transferred when the promiscuous loading module was inserted into the complete DEBS pathway in the producing strain, to synthesize a panel of erythromycin analogues.[46] This advance provided additional evidence that unnatural precursors could be carried and processed through downstream enzymatic transformations when introduced at early biosynthetic steps.

Rowe *et al.* constructed a hybrid DEBS pathway capable of synthesizing expanded macrolactones by inserting modules 2 and 5 from the rapamycin synthase (Rap). The hybrid PKSs incorporated a malonyl unit at the location predicted by the unnatural module insertion.[47] However, as was seen with many hybrid systems, these suffered substantially from decreased yields of the desired product, and in some cases the major product was the result of the inserted module being skipped. The failings of these hybrid systems are multifaceted and still more complex than we fully understand, though considerable progress has been made to improve hybrid PKSs like these. For example, by assembling a panel of DEBS1-TE–Rap hybrids and determining the productivity of the construct, it was determined that maintaining the natural ACP–KS domain partners between successive modules resulted in productive constructs, as it allows for the native protein–protein interactions to be maintained.[48] A panel of 154 bimodular constructs containing modules from eight different PKSs reveal flexibility in the linker region between the loading module and module 1 and that DEBS module 6, which has no natural downstream KS partner, was a universal donor in tested constructs.[49] In some cases, previously unproductive hybrid didomains could be restored by transplanting the natural ketosynthase domain for the donor module into the acceptor module, further supporting the importance of ACP–KS protein–protein interactions.[50]

In addition to success with module swapping, specific domains have been excised and replaced within a module to create a hybrid module. The truncated DEBS1 + TE system was used in a domain-swapping experiment in which the AT from module 1 was replaced with rapamycin AT2 to produce a triketide lactone.[51] This concept was later expanded to synthesize full desmethyl-erythromycin analogues, demonstrating that the change in structure could be processed by downstream modules and post-PKS enzymes, albeit at low efficiencies.[52] In a similar fashion, AT4 from DEBS (the AT from module 4) was replaced with a predicted ethylmalonyl-CoA-specific AT from the niddamycin PKS to successfully produce an ethyl-erythromycin derivative.[53] Such approaches have been applied to several other type I PKSs.[54]

Domain swapping has also been applied to the reductive domains, including substitution of an active KR with an inactive KR to yield the corresponding keto-analogue[55] and a swap of an active KR with an active DH-KR cassette to produce an α,β-unsaturated ketone.[56] Ketoreductase inactivation was also achieved through a targeted deletion of a portion of KR in module 5 of DEBS to produce a 6-dEB analogue.[57] High homology among reductive loops domains combined with downstream module tolerance to changes in oxidation have led to many successes in ketoreductase and dehydratase domain engineering.

The construction of hybrid PKSs often yields unproductive or inefficient enzymes. In a landmark demonstration, multiple genetic modifications of the erythromycin PKS DEBS yielded a library of 61 6dEB analogues through the combinatorial replacement and insertion of various AT, KR, DH and ER domains.[58] Yet, the yields of most of the compounds were less than 1% of that of the natural product, 6dEB, and could not be accurately determined. In recent years improvements have been made in predicting the appropriate boundaries for domain replacement and improving faulty protein–protein interactions. *In vitro* analysis of wild-type and hybrid modules revealed that the introduction of heterologous ATs significantly disrupts protein interactions between other domains within the module.[59] To address this, an efficient chimeragenesis strategy was employed to create hybrid DEBS–pikromycin modules for the production of engineered macrolactones and successfully discovered a significant portion of active hybrids.[60] The throughput of this platform, although improved over previous efforts, is still limited by the need to carry out a chromatographic separation for each chimera tested. Further investigations have reinforced the importance of the ACP–KS interaction epitope in chimeric systems,[54] as poor activities could be improved through mutagenesis.[61,62] More recently, Keasling and co-workers aimed to more systematically design boundaries for AT replacement, which may reinvigorate PKS engineering through AT swaps.[63] Other strategies have focused on identifying amino acid mutations that can reprogram the extender unit selectivity of a given AT, thus enabling production of regioselectively modified polyketides if the target extender unit is available to the host strain.[64]

The power of combinatorial biosynthesis to tailor the structures of polyketides is evident from the examples described above. Biosynthetic pathways can be predictably modified through inactivation, swapping and deletions to yield non-natural analogues of natural products. Most of these examples take advantage of natural enzymatic and small-molecule components, thus limiting the chemical space that can be accessed by combinatorial biosynthesis. For example, there is no known organism that natively produces propargylmalonyl-CoA, a potential extender unit that could be used to facilitate semi-synthesis *via* "click" chemistry. Accordingly, there is also no known PKS that natively utilizes propargylmalonyl-CoA as an extender unit. Subsequently, alternative methods to access non-natural chemical functionalities need to be developed. This has been the primary driving force behind the emergence of precursor-directed biosynthesis and mutasynthesis.

11.3.3 Precursor-directed Biosynthesis and Mutasynthesis of Unnatural Complex Polyketides

Precursor-directed biosynthesis and mutasynthesis take advantage of synthesis to probe and interrogate biosynthetic pathways with unnatural precursors to produce natural product analogues. Notably, these approaches differ from combinatorial biosynthesis by using unnatural small molecules instead of rearranging and recombining nature-derived components.[65–68] Successful application of precursor-directed biosynthesis requires that (1) the precursor to be introduced to intact cells, crude extracts or purified enzymes must be chemically stable; (2) the precursor must be installed and processed by the PKS domain at the targeted step and (3) the intermediates generated by the processing of the precursor must also be utilized by the entire downstream PKS domains and post-PKS tailoring steps.

The increased understanding of type I PKSs has greatly accelerated the influence of PDB and mutasynthesis.[69–71,93] For example, the loading module of DEBS1 was shown to have relaxed specificity *in vitro*, accepting acetyl-CoA[42,72] and butyryl-CoA[72] in addition to the natural propionyl-CoA started unit. Additionally, the DEBS1 + TE construct could accept *N*-acetylcysteamine (NAC) thioesters as precursor mimics *in vitro*.[43] *In vivo*, these unnatural substrates would probably be outcompeted by the natural substrate, propionyl-CoA. To overcome this, Khosla and co-workers knocked out

KS1 in an engineered heterologous strain (*Streptomyces coelicolor*) and supplemented the growth media with cell-permeable diketide NAC-thioesters. In this way, biosynthesis *via* the KS1 and upstream domains is completely intercepted by the fed diketide to create the corresponding 6-dEB analogues.[21] Notably, this strategy was also demonstrated in an engineered *E. coli* host.[22] The overall yield of 6-dEB analogues has been improved up to tenfold with some precursors by truncating the entire DEBS module.[23] A variety of unnatural functional groups have been installed into the starter unit portion of erythromycin by extending this approach, yielding fluoro-[73–75] and azido-erythromycin A[75] analogues, amongst others. Also targeting replacements of the propionyl-CoA starter unit, Khosla and coworkers screened 19 diketide–NAC thioesters as substrates for a truncated DEBS1 by using a colony bioassay. An alkynyl-erythromycin derivative with similar antibiotic potency that can further be modified through orthogonal chemistry.[76] The ease of intercepting PKS starter units with unnatural precursors has encouraged success with other systems. For example, analogues of the angiogenesis inhibitor borrelidin were generated by mutasynthesis. Biosynthesis of the native starter unit was knocked-out by mutagenesis, and a series of analogues were provided by simply feeding in a panel of starter unit replacements.[77] Although the library of borrelidin analogues was quite modest, several displayed cell line selectivities different from the parent compound.[77,78] Similarly, mutasynthesis was successfully employed to generate a series of ansamycin and geldanamycin analogues *via* mutation of the native starter unit biosynthetic pathway and feeding in benzoic acid derivatives.[79–81]

In some examples of PDB and mutasynthesis, only a few starter unit derivatives lead to generation of the expected analogue. This has been attributed to the narrow substrate specificity of some PKS loading modules (*e.g.* borrelidin[77]), but may also reflect narrow specificity of the acyl-CoA ligase required to activate the free carboxylic acids. Swapping the native loading module for one with flexible start unit specificity can overcome this barrier. Such an approach was used to beneficial effect to generate analogues of the commercial insecticide spinosyn.[82]

Starter units have been frequently and easily targeted for PDB because they are structurally differentiated from other PKS building blocks and are therefore intrinsically regioselectively selected for installation by the single PKS loading module. Extender unit positions of erythromycin account for the majority of the carbon scaffold of the mature natural product and have thereby been the subject of intensive engineering to alter their structure by PDB and mutasynthesis. The terminal DEBS module was the subject of combined computational modelling and mutagenesis in order to identify mutations that could improve the ability of the AT6 to utilize an unnatural extender unit, propargylmalonyl-CoA. Subsequently, a series of single amino acid changes were introduced to the AT5 in the genome of the erythromycin-producing strain, *Saccharopolyspora erythraea*. Similarly to NAC thioesters of starter units, malonyl-SNAC analogues are known to be utilized by native ATs in place of CoAs.[83,84] Thus, instead of building an artificial biosynthetic route to propargylmalonyl-CoA, the authors chemically synthesized 2-propargylmalonyl-SNAC as a mimic of the desired CoA-thioester. Subsequently, the combined approach dubbed "enzyme-directed mutasynthesis" led to the identification of a molecular ion with a mass consistent with the desired propargyl-erythromycin A analogue, although not at yields that could support structural confirmation.[85] Further optimizations have highlighted the need to discover AT mutations that not only improve utilization of the target unnatural extender unit but also reduce the ability of the enzyme to utilize its native extender unit.[86] This particular long-standing challenge was recently addressed by Williams and co-workers by challenging a panel of AT active site mutants with a pair of competing extender units. In this way, a single AT amino acid substitution in the DEBS module 6 was found to be sufficient to completely invert substrate specificity to propargylmalonyl-CoA away from the native substrate. Furthermore, Williams and colleagues have also demonstrated that the potential scope of PDB could be much broader that previously assumed by probing the extender unit promiscuity of DEBS with a large panel of non-native and unnatural extender units. This study revealed remarkable promiscuity of the DEBS terminal module.[87,88]

Sherman and co-workers have leveraged the power of combined chemical and biological synthesis by chemically synthesizing and utilizing advanced biosynthetic intermediates of various macrolide antibiotics. These intermediates are normally shuttled between modules *via* attachment to an ACP's phosphopantetheine arm, but as described previously, the field has traditionally used SNAC compounds for synthetic starter units, extender units and intermediates. Hansen *et al.* tested a panel of alternative electrophilic esters and found that modules can differentiate between esters.[89] In the pikromycin pathway, the terminal TE domain can produce either a 12-membered macrolactone (favoured by a hexaketide-SNAC intermediate) or a 14-membered macrolactone (favoured by a hexaketide-thiophenol). This discovery ultimately led to the chemo-enzymatic synthesis of nine tylactone-based macrolides from a common synthetic hexaketide-thiophenol intermediate fed through two final PKS modules *in vitro* (Figure 11.3).[90] The resulting macrolide, tylactone, is a common polyketide core for 16-membered macrolides. Purified tylactone was fed into a growing *Streptomyces* strain for addition of the desosamine sugar (synthetically difficult to add) before diversification by a variety of chemical and P450-mediated oxidations of the scaffold to yield the final products.

The immunosuppressant rapamycin, produced by *Streptomyces hygroscopicus*, has been the target of many biosynthetic engineering efforts, resulting in a collection of products called 'rapalogs' with activities including antifungal, immunosuppressive, antitumor, neuroprotective and anti-aging.[91–93] The enzyme responsible for incorporating lysine-derived L-pipecolic acid into rapamycin also installs proline and closely related

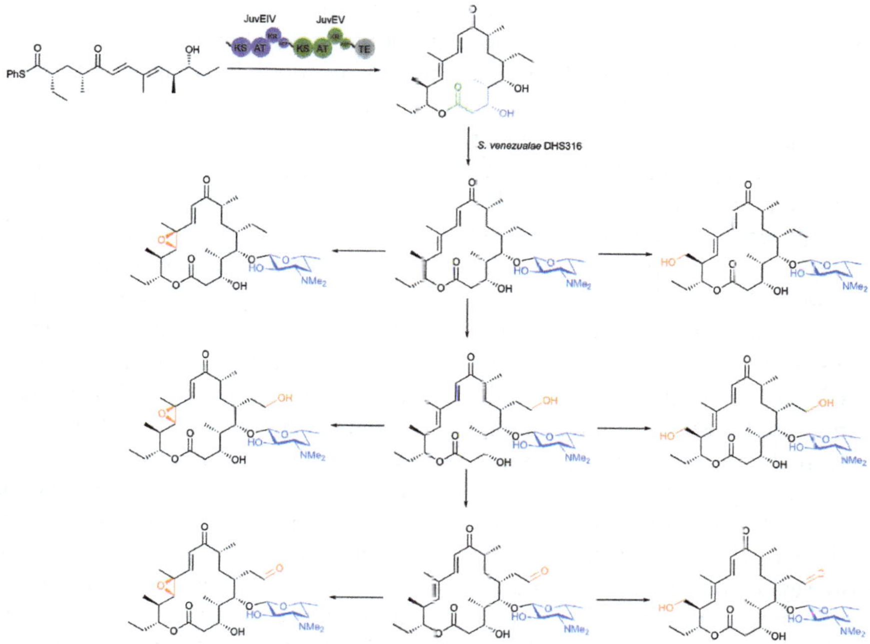

Figure 11.3 Lowell *et al.*'s chemo-enzymatic synthesis of various macrolide antibiotics starting from a thiophenol-linked extended chain followed by extension and cyclization by two PKS modules.
Adapted with permission from A. N. Lowell, M. D. DeMars, S. T. Slocum, F. Yu, K. Anand, J. A. Chemler, N. Korakavi, J. K. Priessnitz, S. R. Park, A. A. Koch, P. J. Schultz and D. H. Sherman, *J. Am. Chem. Soc.*, 2017, **139**, 7913–7920. Copyright 2017 American Chemical Society.[90]

analogues when L-pipecolic biosynthesis is inhibited.[94,95] Aside from leveraging these post-PKS enzymes to modify the rapamycin scaffold, substrate flexibility of the loading module and KS1 of the rapamycin PKS was also utilized to yield pre-rapamycin analogues containing alternative starter units, including a seven-membered ring.[96] A technological *tour de force* by Wilkinson and colleagues used molecular modelling to guide the identification of a rapalog that lacked *O*- and *C*-linked methyl groups at positions 16 and 17. The authors predicted that this analogue would be a potent inhibitor of cancer cell line proliferation. A combination of AT-swapping and mutasynthesis led to the production of the target compound 16-*O*-desmethyl-17-desmethylrapamycin with good titers.[97]

Structurally similar to rapamycin, the immunosuppresant FK506 contains a unique allylmalonyl-CoA extender unit. Yoon and co-workers were able to characterize the full pathway for allylmalonyl-CoA biosynthesis and through this generated a knockout production strain that was unable to synthesize the natural allylmaonyl-CoA.[98] This knockout strain was used to produce three analogues generated through feeding of synthetic carboxylic acids. It is important to note that this was done successfully without altering the AT domain. The analogue created from isobutenylmalonyl-CoA, 36-methyl-FK506, showed an approximately 20% improvement in neurite outgrowth activity and is under further exploration. Moore and co-workers were then able to produce this analogue by creating a hybrid pathway to biosynthesize isobutenylmalonyl-CoA *in vivo* from isobutyryl-CoA, increasing the titres more than sixfold compared with feeding of 4-methylpentanoic acid.[99]

An alternative method for introducing non-natural extender units is through *trans*-complementation. This involves knocking out an in-line *cis*-AT at a specific module and complementing with a naturally discrete AT that possesses a desired natural or engineered specificity. Chang and colleagues harnessed a naturally occurring pathway to produce fluoromalonyl-CoA, which was introduced into triketide lactones through an AT-null mutant of a truncated DEBS system complemented with the *trans*-AT, DszsAT.[100] Although DszsAT natively uses malonyl-CoA, it is able to transfer the unnatural fluoromalonyl-CoA to the DEBS ACP for incorporation. *Trans*-ATs can also be used to produce analogues of their own natural product. This was accomplished in the kirromycin PKS–NRPS hybrid system to produce allyl- and propargyl-kirromycin analogues *in vivo* through feeding of allylmalonic acid and propargylmalonic acid.[101] This demonstrated remarkable plasticity of the kirromycin hybrid PKS–NRPS and post-assembly line enzymatic machinery. Although both of these examples of *trans*-complementation involve harnessing the natural promiscuity of enzymes, engineering the specificity of these discrete enzymes could further increase the scope of *trans*-complementation.

In many cases, enzymes can accept and incorporate unnatural substrates, probably because of the lack of evolutionary pressure to discriminate against them. Mutagenesis can also be combined with feeding of unnatural substrates to improve titres through eliminating competition from the natural substrate or by increasing affinity for the unnatural substrate. By combining the scope of synthetic chemistry with the precision of biosynthetic pathways, the chemical space available is nearly limitless and will only improve as we continue to learn how to manipulate these pathways.

11.3.4 Semi-synthesis of Complex Unnatural Polyketides

The semi-synthesis of the cholesterol-lowering drug simvastatin provides a comprehensive example illustrative of the ability of combined chemical and biosynthetic approaches to out-perform purely chemical methods. Simvastatin is a semi-synthesized derivative of the natural product lovastatin, and differs only in the presence of a C8 α-dimethylbutyrate side chain on the decaline core of simvastatin, *versus* an α-*S*-methylbutyrate in lovastatin. Tang and colleagues explored the use of the free-standing acyltransferase LovD to install the dimethylbutyryl side chain to the lovastatin precursor monacolin J. Indeed, membrane-permeable donors including, α-dimethyl-butyryl-SNAC, could be delivered to *E. coli* and used to produce simvastatin from monacolin J that was fed to the culture.[102] Substrate engineering then yielded an improved acyl donor, α-dimethylbutyryl-*S*-methyl-3-mercaptopropionate, which supported faster reaction rates and is more economical than the original substrate.[103] The rate of simvastatin synthesis *via* whole-cell biocatalysis was further improved by deleting a carboxylesterase that otherwise rapidly hydrolysed the acyl donor.[103] Still not content, the same group then used directed evolution to improve the activity of LovD 11-fold compared with the wild-type enzyme.[104] Next, advanced directed evolution methods were used to identify a mutant LovD with 29 amino acid mutations and a 1000-fold increase in enzyme activity.[105]

The production of second-generation macrolide antibiotics from erythromycin exemplifies the potential of chemo-selective organic transformations to access new and improved drugs using complex fermentation products as the starting materials.[106] The chemical modification of erythromycin yielded roxithromycin, clarithromycin and azithromycin in efforts to produce derivatives that were more acid-stable.[107] Rapid emergence of bacterial resistance spurred the development of additional generations of derivatives. The most promising is Solithromycin, a fourth-generation macrolide antibiotic that is semi-synthesized in nine steps from clarithromycin, itself prepared from erythromycin in four steps.[108] The overall yield on the multi-metric ton scale is expected to be 35–40%.[107] In comparison, a total of 40 chemical transformations are required for the total synthesis of Solithromycin *via* Myer's sophisticated chemistry,[109] and the ability to use this chemistry to manufacture metric tons has yet to be demonstrated.

Perhaps rather conspicuously, the true combinatorial potential of type I PKSs has yet to be fully coupled with the power of synthetic chemistry. Thus, most new macrolide derivatives are accessed *via* chemo-selective modification of natural macrolides or those that require only minimal engineering of the corresponding biosynthetic machinery. For example, the removal of cladinose and installation of the keto group of the ketolide class of macrolides is carried out completely chemically from clarithromycin, rather than modifying the biosynthesis of erythromycin to make these structural changes. Probably the biggest obstacle to overcome in this respect, discussed above, is the rather low activity of most engineered PKSs. One notable exception is the mutasynthesis of a series of 15-*R*-erythromycin analogues by Khosla and co-workers, whereby the natural ethyl side-chain of erythromycin (derived from the starter unit) was replaced with a panel of unnatural side-chains.[110] These unnatural macrolides served as the starting materials for semi-synthesis to produce a series of 6-*O*-arylalkyl ketolides.[106] In addition, the relatively newly discovered ability to introduce unnatural chemical handles into polyketides *via* extender-unit engineering might also facilitate coupling biosynthetic engineering and chemo-selective organic transformations.[111–113]

11.3.5 Precursor-directed Biosynthesis and Mutasynthesis of Unnatural Non-ribosomal Peptides

Targeting NRPS pathways by PDB and mutasynthesis has enjoyed similar successes to that of PKSs. Such chemical approaches can complement the limitations of purely combinatorial biosynthetic approaches, which are usually restricted to the use of naturally occurring building blocks.[114,115] One of the earliest successes of PDB in NRPSs was the production of cyclosporin analogues through the feeding of unnatural amino acids. It was found that position two in the undecapeptide was responsive to increased feeding of natural amino acids, indicating that this was a flexible position within the assembly line. Indeed, this position showed incorporation of unnatural amino acids when the fermentation broth was supplemented.[116]

PDB was also successfully employed by Süssmuth and co-workers to generate a fluorine-containing analogue of vancomycin, the glycopeptide antibiotic used to treat antibiotic-resistant infections. The producing strain was mutated to prohibit the biosynthesis of natural substrate β-hydroxytyrosine and the fermentation broth was supplemented with 3-fluoro-β-hydroxytyrosine, which was incorporated at two positions by the natural NRPS machinery.[117] The producing strain was later mutated to knock-out biosynthesis of (*S*)-3,5-dihydroxyphenylglycine and unnatural methoxy-containing phenylglycine analogues were fed to the culture, leading to the successful production of vancomycin analogues with antibiotic activity.[118] In another example, a thioesterase domain from the decapeptide tyrocidine synthetase was used to catalyse the macrocyclization of alkynyl modified peptide–SNAC thioesters. Subsequent conjugation to 21 azido sugars yielded a library of 247 glycopeptides with improved therapeutic indexes, compared with tyrocidine.[119]

Calcium-dependent antibiotics (CDAs) are a class of lipopeptides that are biosynthesized by NRPSs from *Streptomyces coelicolor* (Figure 11.4). Production of analogues of CDAs has been successful through PDB and mutasynthesis. The Micklefield group produced CDA analogues with modified fatty acid sidechains through feeding of chemically synthesized fatty acid–SNAC analogues in conjunction with inactivation of the PCP of module 1 of the CDA NRPS.[120] They were also able to produce analogues through targeted mutagenesis of the enzymes involved in fatty acid chain length, biosynthesizing two novel CDAs with truncated fatty acid chains without the need to supplement the fermentation with synthetic precursors.[121] The previous two examples involve manipulation of the fatty acid portion of the molecule, but the peptide portion was also altered in subsequent studies through the modification of the A-domain in module 10, allowing for incorporation of unnatural synthetic (2*S*,3*R*)-3-methyl glutamine.[122] The efforts towards biosynthesis of CDA analogues provide an excellent example of combining both synthetic chemistry and biosynthetic pathway engineering to efficiently produce desired complex molecules.

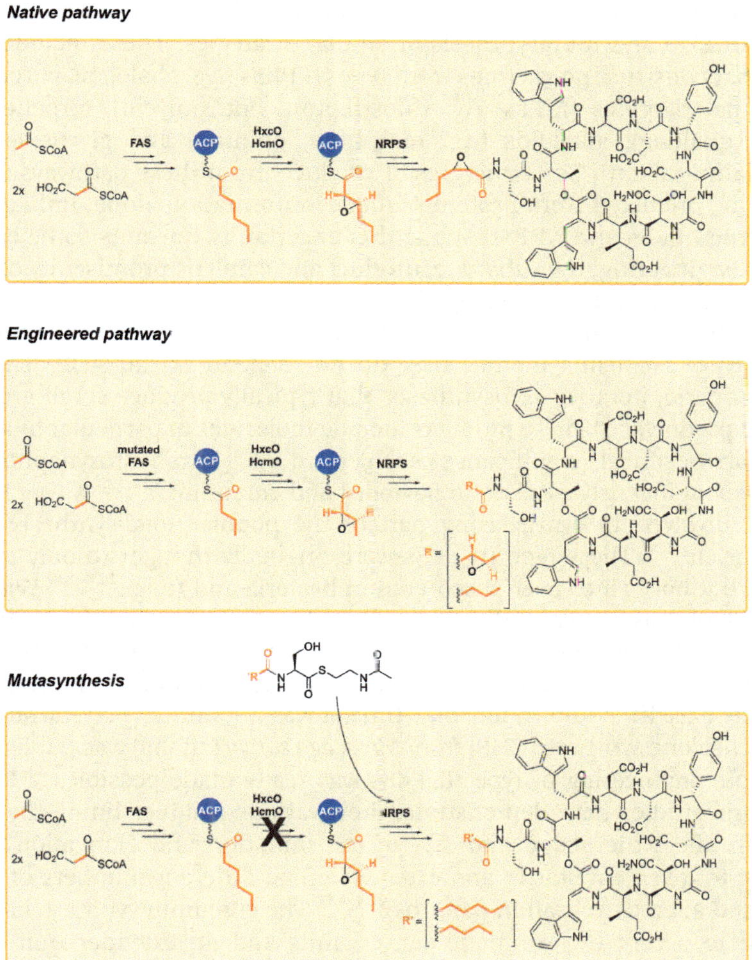

Figure 11.4 Production of CDA analogues through PDB and mutasynthesis by Micklefield and co-workers.[121] The natural pathway is shown in the top panel. The lipid portion of the molecule was modified through FAS mutagenesis, enzyme inactivation (*hxcOΔ*), and synthetic SNAC precursor mimics.

11.4 Leveraging Non-templated Biosynthetic Systems for Natural Product Diversification

11.4.1 Overview of Non-templated Biosynthesis of Natural Products

As highlighted above, the powerful templated logic of type I PKSs and NRPSs is ideally suited to interception by chemically synthesized precursors to generate unnatural complex polyketides and non-ribosomal peptides.

Nevertheless, non-assembly-line biosynthetic pathways have also been the target of PDB and mutasynthesis or similar strategies. These include pathways that construct polyketides from type III PKSs (*e.g.* chalcone, curcumin), ribosomal peptides (RiPPs; *e.g.* thiostrepton, bottromycin), terpenes (*e.g.* taxol, squalene), alkaloids (*e.g.* morphine, quinine) and glycosides (*e.g.* amygdalin, salicin). The structures of products from these pathways cannot always be predicted from protein-sequence information alone, unlike those from type I PKSs and NRPSs. While this uncertainty warrants some caution during engineering, the substrate, product and catalytic promiscuity of these systems can be carefully leveraged.

Type III PKSs, also known as chalcone-like synthases, differ from their type I and type II cousins in that they do not depend on an ACP. They are homodimeric, iterative ketosynthases that typically produce small aromatic natural products.[123] These products include molecules of particular biological or pharmaceutical significance such as 1,3,6,8-tetrahydroxynaphthalene (THN; spontaneously oxidizes to flaviolin) and curcumin.[124,125] A type III PKS is also involved in synthesizing part of the popular total-synthesis target kendomycin.[126] These type III PKSs were originally thought to only exist in plants, but have since been discovered in bacteria and fungi.[127–130] With this diversity of hosts comes diversity of products: curcumenoids, acridones, quinolinones, benzophenones, biphenyls, bibenzyls, alkylresorcinols, alkylpyrones, stilbenes, chalcones and benzalacetones.[131] While the more famous products have been known for some time, it was not until the crystal structure of the chalcone synthase (CHS) from *Medicago sativa* (alfalfa) was published in 1999 that engineering of type III PKSs was really made possible.[132] Further structural studies have demonstrated how various residues lining the active site can affect the shape and size of the initiation and elongation cavity, leading to divergent starter and extender units, different numbers of extensions and altered cyclization patterns.[133–137] The curcumin synthase has even evolved to accept two aromatic starter units and an extender unit simultaneously.[138] As the knowledge base for type III PKSs has grown, inherent promiscuity and structure-guided mutagenesis has been used to produce unnatural products.

Ribosomally-synthesized and post-translationally modified peptides (RiPPs, ribosomal peptides, or ribosomal natural products) are found in all kingdoms of organisms, and they have a wide variety of biological activities.[139] There are more than twenty subgroups of RiPPs. However, those designations are mostly grouped by structure and function rather than by their biosynthetic origins.[140,141] These structures can vary from the very large (polytheonamide A, 5 kDa) to the small (PQQ, 330 Da), though they are often greater than 1000 Da. While similar in structure to many NRPs, RiPPs are synthesized using the ribosome, followed by post-translational modifications. The general biosynthetic pathway for RiPPs starts with a precursor peptide that contains an N-terminal signal peptide, a leader peptide, the core peptide and a C-terminal recognition sequence. Post-translational modifications are made to the core peptide before proteolysis and export leads to the mature RiPP.

Prior to the advent of widespread genome mining, there was much less interest from the global research community in these small, difficult-to-identify gene clusters, but now the field has evolved into one that can utilize combinatorial biosynthesis to expand the already impressive chemical repertoire of RiPPs.[140,142–145] Despite the breadth of chemical scope natively available to RiPPs, the nature of their biosynthesis allows for five main avenues of combinatorial biosynthesis, as described by Sardar and Schmidt and listed below:[144]

1. Introduce variety (by core peptide hypervariability).
2. Pair variety with promiscuity (using broad-substrate pathways).
3. Provide roots to promiscuous partnership (through conserved recognition sequences).
4. Swap partners (by pathway modularity).
5. Achieve novelty (creating unique PTMs).

While some of these methods are accessible to other natural product classes (*e.g.* pathway modularity), others, such as core peptide hypervariability, can provide unique and powerful approaches to diversification of RiPPs.

Terpenes (isoprenoids, terpenoids) are produced primarily by plants and make up the largest and arguably most structurally diverse class of natural products.[146] This claim is evidenced by the wide range of sizes, cyclization patterns and oxidation states, and their uses include pharmaceuticals, fragrances, flavors, agrochemicals and other useful products.[147] Unlike most of the other natural product classes discussed here, terpene biosynthesis only requires two 5-carbon precursor isoprenes: a starter unit, dimethylallyl pyrophosphate (DMAPP), and an extender unit, isopentenyl pyrophosphate (IPP). Terpene synthases will stitch together one or more IPPs onto a single DMAPP unit, resulting in the acyclic geranyl pyrophosphate (GPP, C_{10}), farnesyl pyrophosphate (FPP, C_{15}) or geranylgeranyl pyrophosphate (GGPP, C_{20}). Longer chains can be produced by the additions of GPP, FPP or GGPP. Terpene cyclases can follow and promote cyclization of a myriad of products: monoterpenes (C_{10}, *e.g.* limonene), sesquiterpenes (C_{15}, *e.g.* caryophyllene), diterpenes (C_{20}, *e.g.* taxadiene), triterpenes (C_{30}, *e.g.* lanosterol), tetraterpenes (C_{40}, *e.g.* β-carotene) and polyterpenes (C_{50+}). Even from the same cyclase, dozens of products are possible. Indeed, the γ-humulene synthase has been shown to produce 52 distinct sesquiterpenes.[148] This extreme variability is due to a carbocation-based cyclization mechanism, catalyzed by release of pyrophosphate. Post-cyclization modifications (often oxidations of the hydrocarbon scaffold) lead to the bioactive final products.

With over 55 000 compounds isolated by 2007, terpene structural diversity allows for a range of bioactivities, but the chiral centers and quaternary centers of terpenes render them difficult targets for total syntheses.[146,149] While biomimetic approaches to terpenes have been explored, combinatorial chemistry—especially semi-synthesis—has been particularly useful for accessing new derivatives. Other research has focused on harnessing the inherently promiscuous product specificity of the biosynthetic enzymes. In both cases, the primary role of organic synthesis may be unique among natural product classes due to the extreme homogeneity and lack of functionality of terpene precursor molecules.

Alkaloids, nitrogen-containing natural products of plant origin, are widely represented in ancient and modern pharmaceuticals. A formal definition for alkaloids is lacking, but it is frequently accepted that the nitrogen(s) must be incorporated into a cyclic moiety—thus ruling out amino acids and other terminal amines.[150] Due to this ambiguous classification, there is a wide variety in both precursor and final product structures. Some of the most common precursors (and final products) are niacin (nicotine), tyrosine (morphine), tryptophan (quinine), and xanthosine (caffeine).[150] Currently, over 30 phenylalanine- and tyrosine-derived alkaloids are FDA-approved medicines.[151] Each class of alkaloid has their own biosynthetic similarities, but there is no real overarching mechanism, as with polyketides, RiPPs or terpenes. Predictably however, many alkaloid biosynthetic pathways use Schiff bases or the Mannich reaction.[152]

One of the key challenges of this class of natural products is the difficulty of engineering plant biosynthesis. Much work has been done to combat this issue.[153–155] However, sustainable engineering and production of alkaloids may be best achieved by utilizing microbes.[151] While many alkaloids can be targeted due to their amino acid precursors *via* PDB, much of the successful work with alkaloids has come about by semi-synthesis.

11.4.2 Unnatural Aromatic Polyketides *via* Type III PKSs

Two of the most well-studied of these enzymes, germicidin synthase (Gcs) and SCO7671, are both from *Streptomyces coelicolor* and were shown by Sherman and colleagues to have unexpectedly broad substrate promiscuity.[156] They originally explored the *in vitro* starter unit promiscuity and found products corresponding to acceptance of a wide variety of branched, short-chain, long-chain and aromatic CoA thioesters. Using a malonyl-CoA extender unit, nearly 20 unique pyrone analogues were produced and characterized. Just as with other types of PKSs, the CoA moiety could be replaced with synthetic SNAC. This could have important scale-up ramifications by eliminating the need for the expensive CoA in the production of synthetically-useful pyrones. Nearly a decade later, the same group discovered that the extender units could be varied from malonyl-CoA to methyl- and ethylmalonyl-CoA.[157] Notably, it was concurrently discovered that Gcs preferred ACP-linked starter units, potentially providing another route for further PDB.

In 2007, Abe *et al.* extended PDB to include structure-based engineering of pentaketide chromone synthase (PCS), a type III PKS from *Aloe arborescens* (Liliaceae).[137] PCS, like many divergent type III PKSs, has a sterically altered active site compared with chalcone synthase (CHS) that leads to the condensation of five units of malonyl-CoA. Upon a single mutation (Met→Gly), an additional three malonyl-CoA condensations were observed. A triple mutant (Met→Gly, Tyr→Ala, Phe→Ala) yielded an unnatural nonaketide naphthopyrone, the longest known product from a type III PKS. To account for the expanded number of condensations catalyzed by these mutants, homology modeling of the active site showed an approximately 400% increase in volume.

While the previous example was achieved mostly through steric control of the enzyme, Austin *et al.* demonstrated the importance of electronic control in converting CHS to a stilbene synthase (STS).[158] Structural analysis revealed only minor changes in the overall topology of each enzyme, with no apparent structural reason for the divergent cyclization patterns. To catalogue key residues, the authors identified and mutagenized four regions of CHS *via* combinations of partial or complete conversion to the STS sequence. In variants of CHS that had STS-like activity, a thioesterase-like hydrogen bond network and corresponding set of residues were identified. Additional alanine scanning highlighted two residues as making up the "aldol switch" for STS aldol cyclization. This work is a nice example of engineering a type III PKS without changing the product length.

The polyketide synthase 1 (PKS1) from *Huperzia serrata* (Chinese club moss) has also been shown to be conducive to combinatorial approaches since its discovery in 2007. PKS1 naturally synthesizes naringenin chalcone from 4-coumaroyl-CoA and three units of malonyl-CoA. However, it was subsequently noted that other bulky, non-natural starter units such as *N*-methylanthraniloyl-CoA would lead to the production of 1,3-dihydroxy-*N*-methylacridone by PKS1.[159] While the unnatural starter unit is similar in size to the natural product, the nucleophilic nitrogen allowing for the formation of a fused tricyclic ring system is missing. In 2011, the promiscuity of PKS1 was further demonstrated by the incorporation of 2-carbamoylbenzoyl-CoA and 3-carbamoylpicolinoyl-CoA (one nitrogen difference) with two malonyl-CoA or methylmalonyl-CoA extensions.[160] A much larger naphthalene-containing starter unit was also accepted by the wild-type PKS1. The resulting fused tricyclic and tetracyclic compounds are not found in nature. Morita *et al.* then targeted an active site serine for mutagenesis and a glycine variant resulted in the same 2-carbamoylbenzoyl-CoA starter unit being extended three times to give a fused tricyclic ring with a central seven-membered ring and a novel cyclization mechanism.[160]

11.4.3 Diversification of RiPPs

The power of RiPP combinatorial biosynthesis is exemplified by the diversity-oriented biosynthesis work on the trunkamide cyanobactin (*tru*) pathway, producing more than 300 new compounds by modifying the core peptide.[142] Two of the seven residues were fixed due to the macrocylcase specificity; however, 3.2 million (20^5) compounds were still theoretically possible. The 300+ compounds identified is a relatively high number for a single combinatorial experiment, but with this strategy, it was a small subset selected due to the lack of a high-throughput screen. By careful library design, nearly full saturation was achieved at all five targeted positions, and the data were used to determine some basic rules about amino acid incorporation at each position. The authors did note the major caveat with this approach: that the rules only applied under the specific circumstances screened. However, of the natural product combinatorial biosynthesis approaches, it may be the easiest to rapidly produce large numbers of analogues.

Similar to the work with the *tru* pathway, the core peptide sequence of the natural product cypemycin has been altered to yield various single and double mutants, including one with increased antimicrobial activity.[161] While far fewer variants were produced overall, it was still a notable advancement due to the high level of post-translational modification, with 8 of 21 residues being altered. The work also highlighted the importance of bioinformatics in RiPP engineering, especially regarding conserved residues in this family of compounds.

While RiPPs have a massive number of conceivable permutations just by modifying the core peptide sequence, feeding in unnatural amino acids further increases the possible options. For example, an engineered *E. coli* strain was utilized to produce a bioactive 32-residue lantibiotic by replacing a tryptophan residue with the non-canonical L-β-(thieno[3,2-*b*]pyrrolyl)alanine.[162] The use of non-canonical or unnatural amino acids may not be as feasible for large-scale syntheses, but it provides another useful tool when evaluating RiPPs and their biosynthetic pathways.

While it is not technically difficult to alter the core peptide sequence, protein engineers are still mostly dependent on nature to provide promiscuous post-translational modifying enzymes. One such example of "natural combinatorial biosynthesis" involves the biosynthesis of lanthipeptides.[143] The prochlorosin synthetase ProcM from the cyanobacterium *Prochlorococcus* MIT9313 has been shown to take up to 29 diverse precursor peptides. The mechanism behind the promiscuity is unknown, as it was also discovered that approximately two-thirds of the leader peptides were not required. While shorter leader peptides have been utilized in RiPP biosynthesis before,[163,164] this was the first instance where this property was paired with a broad substrate scope. Homologous enzymes have been discovered with similarly diverse but different substrate profiles.[143] Enzymes such as ProcM could eventually be targeted using chemically synthesized probes for mechanistic or functional studies.

Most combinatorial biosynthesis of RiPPs has focused on making analogues of specific natural products; however, entire pathways can be adapted or engineered to produce novel molecules. To this end, a flexible, modular *in vitro* route to pharmaceutically-valuable cyanobactin RiPPs was developed.[165] Several different azoline-containing macrocycles were produced using substrates and enzymes from five different RiPP pathways, while enzymes from another pathway could be added for conversion to thiazolines. This plug-and-play method with promiscuous post-translational enzymes led to the production of 1–2 mg of 17 novel compounds. The *in vitro* approach to combinatorial biosynthesis removes many of the complications of engineering enzymes to work in a living system (for example with the non-canonical amino acids) and maybe more importantly, allows the production of compounds that would inhibit growth of an organism.

11.4.4 Terpene Diversification

The discovery of the potent antimalarial artemisinin, led to the 2015 Nobel Prize in Physiology or Medicine after Youyou Tu isolated the endoperoxide sesquiterpenoid from the plant *Artemisia annua* (sweet wormwood). Until the past decade, the only source of artemisinin was this herb, leading to supply and price fluctuations. Several combinatorial biosynthetic approaches over the past few years have helped produce artemisinin and its clinically-relevant semi-synthetic derivatives. In 2006, the Keasling group reported the production of the artemisinin precursor, artemisinic acid, in *Saccharomyces cerevisiae* (yeast), providing an efficient biosynthetic route to a semi-synthetic platform for medicinal chemists.[166] This was followed up six years later with production of dihydroartemisinic acid, a more advanced precursor, by the same group.[167] In 2013, the four-step semi-synthesis of artemisinin from microbially-produced artemisinic acid was published with a 40–45% overall yield.[168] These synthetic biology platforms have been used to build a road map for future semi-synthetic approaches for valuable natural products and analogues.[168–170]

In an apparent blow to the combinatorial chemistry community, the new source of artemisinin had only a modest effect on the overall market, and in 2015, just one year after initial production of semi-synthetic artemisinin, Sanofi halted production.[171] Nevertheless, two years later, the Allemann group published an alternative approach to the semi-synthesis of artemisinin.[172,173] Instead of utilizing a series of enzymes to produce artemisinic acid like the work in engineered yeast or *A. annua*, chemically-synthesized 12-hydroxy-FPP was used with amorphadiene cyclase to give dihydroartemisinic aldehyde in a single biosynthetic step.[172] This intermediate bypasses a transition-metal-catalyzed hydrogenation from previous work. A small-scale semi-synthetic route to artemisinin was published using this approach.[173]

Several groups in recent years have explored the use of chemically-synthesized isoprenoid precursor analogues as inhibitors and mechanistic probes for terpene cyclases.[174–181] For some analogues, like 3-bromo-FPP and GPP, the electronics are too different from the natural substrates for cyclization, leading to inhibitors that could be used for structural studies of these enzymes.[179] This is a powerful tool for a class of synthases where the catalytically active conformation has yet to be captured as a crystal. Other synthetic precursor analogues can be cyclized. Of particular note, 6-fluoro-GGPP was incubated with taxadiene synthase to show that the previously accepted stereochemistry of an intermediate was incorrect.[178] The Allemann group and others have used deuterated, methylated and fluorinated precursors to great effect, especially through the production of more stable versions of germacrene, a sesquiterpene that affects the olfactory response of insects.[174–176,182]

Taking advantage of the natural promiscuity of terpene cyclases is a useful combinatorial tool to produce terpene analogues. However, the lack of product control can be a disadvantage as well. To combat this, researchers have mutated many putative plasticity residues in the more promiscuous cyclases, for example in γ-humulene synthase.[148,183,184] Yoshikuni *et al.* displayed impressive control and understanding of the cyclization mechanism through rationally designed mutations.[184] The wild-type enzyme had seven major products, but a series of one to five mutations led to seven enzyme variants producing product profiles significantly shifted towards each of the major products. The authors noted that specific mutations promoted predicted cyclization patterns and that this knowledge, in principle, could be combined with precursor-directed biosynthesis to give even more varied analogues.

Recently, biosynthetic studies of a previously poorly characterized diterpene biosynthetic pathway led to the development of an efficient and effective semi-synthetic platform. Pleuromutilin is an antibiotic produced by *Clitopilus passeckerianus* that has served as a precursor for semi-synthetic derivatives used in veterinary and human medicine. Although several semi-synthetic derivatives of this diterpene antibiotic are being produced, large-scale production of designer semi-synthetic analogues requires an increased understanding of the biosynthetic pathway to pleuromutilin. Genetic characterization of the steps involved in pleuromutilin biosynthesis, through heterologous expression in *Aspergillus oryzae*, led to a panel of mutant strains that accumulate each pathway intermediate. Subsequently, *A. oryzae* was established as a platform for bio-conversion of chemically modified analogues of pleuromutilin intermediates, and was leveraged to generate a semi-synthetic pleuromutilin derivative with improved antibiotic activity.[185]

In addition to the previously described diversity, some natural products, like meroterpenoids (isoprenoid-polyketides), are synthesized by hybrid pathways. These products often involve the prenylation of small aromatic molecules, as in the case of the anti-oxidant naphterpin. Orf2, the promiscuous aromatic prenyltransferase from the naphterpin pathway, has been shown to accept a panel of 1-, 2- and 3-ring aromatic prenyl acceptors.[186] It is notable that a geranyl moiety was transferred to both aromatic carbons and exocyclic oxygens, depending on the substrate. In a similar example, *in vivo* production of an analogue of pyripyropene A was accomplished by feeding in benzoic acid instead of nicotinic acid.[187] Yet another prenyltransferase, CnqP3, has shown not only aromatic prenyl acceptor promiscuity, but also isoprenoid substrate promiscuity (DMAPP and GPP).[188] These examples of natural prenyltransferase promiscuity notwithstanding, the scope and utility of terpene diversification could be further expanded by rational design and site-directed mutagenesis of the component enzymes.[189]

11.4.5 Alkaloid Diversification

Alkaloids have been diversified through a variety of methods, but some of the earliest examples were the precursor-directed biosynthesis of non-natural lysergic derivatives from the ergopeptine class.[190–192] These results were achieved by the relatively simple procedure of feeding in unnatural amino acids such as norvaline and 4-thioproline into cultures of the producing strain. This approach has been utilized in several more cases for alkaloids.[193–196] One such case was the replacement of tryptamine, a common precursor for many alkaloid pathways, with a series of four aza-indole isomers.[193] The simple addition of a nitrogen in the indole ring provides new hydrogen bonding properties and improved water solubility. The aza-tryptamines were shown to be compatible with the Pictet–Spengler reaction, the first committed step in many alkaloid biosynthetic pathways. The ease of analogue production by these approaches highlights the simplicity and power of PDB when promiscuous enzymes or pathways are known and precursors are chemically available.

While promiscuous enzymes are necessary for PDB, in many natural-products-producing hosts, subsequent product diversity can lead to issues in isolating the different analogues. Recently, the utility of mutasynthesis to address this problem was demonstrated using the prodiginine class of alkaloids.[19] The prodiginines are red pigments that demonstrate antitumor, antimalarial and antimicrobial activities. The results of preliminary experiments demonstrated that fed-in precursors were indeed imported into the heterologous host, accepted, and utilized by the native biosynthetic machinery but that the desired analogue was present as a mixture with the natural product. A knock-out strain was subsequently established whereby the natural monopyrrole precursor was no longer produced. This strain allowed for exploration of the native promiscuity of PigC, the final condensing enzyme of the pathway, and production of more than a dozen new analogues that would not have been easily accessed *via* total synthesis (Figure 11.5). These new derivatives included alkenyl moieties for further semi-synthesis. In this example, the substrate specificity of the key bond-forming enzyme (PigC) limited the full scope of PDB. In cases like this, the precursor scope can be manipulated by engineering the substrate specificity of the key enzyme.[197,198]

In the examples described above, the precursors utilized were of synthetic origin. However, there are cases in which a combinatorial strategy can allow for production of the precursors *in vivo*. One particularly powerful method involves the use of halogenases. O'Connor and colleagues have previously used PyrH and RebH in periwinkle to chlorinate (and brominate in high bromide salt concentrations) the 5 and 7 positions of tryptophan, respectively, for incorporation into a monoterpene alkaloid strictosidine.[199] In periwinkle, strictosidine is functionalized into more than 100 different alkaloids, and previous PDB efforts had shown that halogenated indoles could be processed to mature products when fed *in planta*.[200,201]

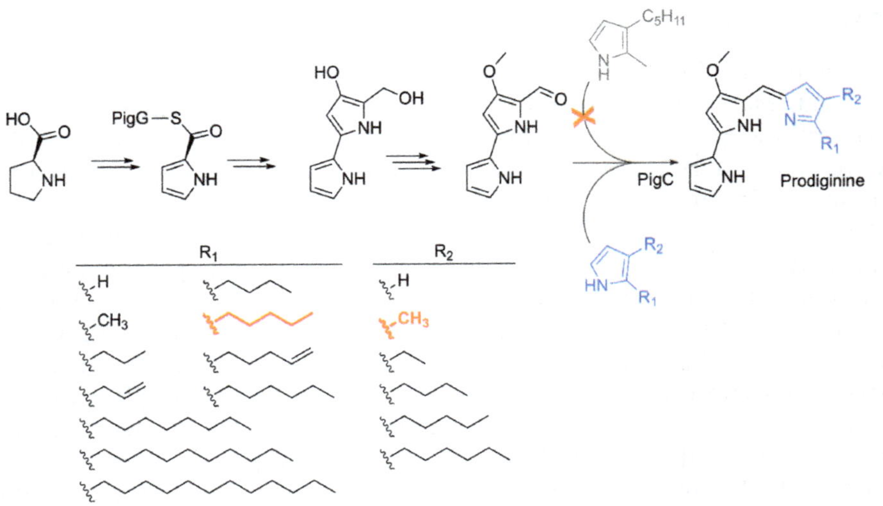

Figure 11.5 One of the two natural precursor pathways was knocked out in prodiginine biosynthesis and a series of synthetic precursors were fed in. The promiscuity of the final enzyme, PigC, allowed for the mutasynthetic production of more than a dozen new analogues. The natural R groups are highlighted.

In such a broad class of natural products like the alkaloids, PDB or mutasynthesis might work well with some pathways and not at all with others. In cases where biological manipulations are possible, semi-synthetic handles may be incorporated. In other cases, semi-synthesis without genetic manipulation might be a preferable alternative to total synthesis for useful analogues.[202] Because of this, dozens of semi-syntheses have been published for alkaloids.[151,203–206] Some of the most famous semi-syntheses for alkaloids have been reported for the opioid family. One unique and notable case in alkaloid production is the *de novo* production of thebaine and the semi-synthetic opioid hydrocodone in yeast and later in *E. coli*.[207–209] Smolke and co-workers introduced and used up to 23 enzymes from plants, mammals, bacteria and yeast in producing these highly valuable alkaloids. This is a significant example demonstrating the power of combined chemical and biological synthesis, as none of the reported approximately 30 chemical syntheses of morphine derivatives have as yet been shown to be robust enough for large-scale production.[210]

11.4.6 Diversification of Natural Product Glycosides

The glycosylation of natural products and synthetic molecules is well-known to affect the pharmacological properties of the parent scaffold. Yet, the regio- and stereo-controlled glycosylation of complex small molecules is particularly challenging, especially given the requirement for activated sugars. To address this problem, approaches that depend on glycosyltransferases have enjoyed much success, while chemo-selective glycosylation chemistries have been developed that facilitate semi-synthesis of novel glycosides.

With respect to glycosyltransferases, many naturally-occurring enzymes that catalyze the transfer of sugars from activated nucleotide-diphosphate (NDP) donors to natural product scaffolds have been utilized in various combinatorial biosynthesis formats, with some enzymes displaying sufficient substrate flexibility to generate small libraries of differentially glycosylated natural products.[211] In addition, Thorson and co-workers developed an artificial chemo-enzymatic "glycorandomization" platform for producing libraries of glycodiversified natural products and small molecules *in vitro* and *in vivo*.[212] This platform depends on the ability of natural or engineered enzymes to process reducing sugars to their corresponding 'activated' NDP-sugars. Furthermore, the specificity of glycosyltransferases has been expanded by directed evolution in order to broaden the scope of glycorandomization.[213–215]

In terms of semi-synthesis, although several approaches meet the requirements for chemoselective conjugation of carbohydrates,[216] one particularly notable advance includes alkoxylamine-based "neoglycorandomization"—one-step sugar ligation that does not need any sugar protection or deprotection steps. Indeed, this method has been used to glycodiversify many complex acceptor molecules isolated from natural sources, including vancomycin, calicheamicin, colchicine and digitoxin.[217] Notably, neoglycorandomization of digitoxin provided a series of lead compounds with improved cytotoxicity activities that served the basis of an entirely new class of drug leads, referred to as extracellular drug conjugates.[218]

11.5 Summary and Outlook

Arguably the most impressive advances in PDB, mutasynthesis and semi-synthesis have focused on accessing unnatural polyketides *via* PKSs, including production of the blockbuster drug simvastatin and chemical diversification of naturally produced macrolide antibiotics. NRPSs have enjoyed success too but perhaps the ease of peptide synthesis, at least those that are not heavily *N*-methylated, has mitigated enthusiasm in this area. At the same time, non-templated biosynthetic pathways have also been used in combination with synthetic chemistry to access unnatural derivatives. Yet, the true potential of both templated and non-templated biosynthetic approaches has yet to be fully leveraged and coupled with synthetic chemistry. For example, most examples of PDB, mutasynthesis and semi-synthesis of polyketides involve minimally engineered PKSs. In addition, the building block repertoire of some biosynthetic machinery, particularly that of terpenes, remains stringent and has yet to be fully explored with unnatural precursors. We predict that the continued combination of protein engineering, metabolic engineering, synthetic biology and chemistry will lead to significant advances in our ability to access unnatural natural products. For example, the development of designer macrolide biosensors now enables directed evolution of chimeric PKS pathways and the associated biosynthetic machinery.[219] Additional improvements in heterologous expression, host strain engineering and the discovery of enzymes with novel specificities and activities will also be required to fully hyphenate chemistry and biosynthesis.[220]

Acknowledgements

We gratefully acknowledge support from the National Institutes of Health (GM104258, G.J.W.), the National Science Foundation (NSF CAREER Award, CHE-1151299, G.J.W.), the North Carolina Biotechnology Center, the Chancellors Innovation Fund at NC State, and the Comparative Medicine Institute at NC State.

References

1. D. J. Newman and G. M. Cragg, *J. Nat. Prod.*, 2016, **79**, 629–661.
2. A. L. Harvey, R. Edrada-Ebel and R. J. Quinn, *Nat. Rev. Drug Discovery*, 2015, **14**, 111–129.
3. J. Goodman and V. Walsh, *The Story of Taxol: Nature and Politics in the Pursuit of an Anti-Cancer Drug*, Cambridge University Press, 2001.
4. R. A. Holton, R. R. Juo, H. B. Kim, A. D. Williams, S. Harusawa, R. E. Lowenthal and S. Yogai, *J. Am. Chem. Soc.*, 1988, **110**, 6558–6560.
5. R. A. Holton, C. Somoza, H. B. Kim, F. Liang, R. J. Biediger, P. D. Boatman, M. Shindo, C. C. Smith and S. Kim, *J. Am. Chem. Soc.*, 1994, **116**, 1597–1598.
6. R. A. Holton, H. B. Kim, C. Somoza, F. Liang, R. J. Biediger, P. D. Boatman, M. Shindo, C. C. Smith and S. Kim, *J. Am. Chem. Soc.*, 1994, **116**, 1599–1600.
7. W. C. Liu, T. Gong and P. Zhu, *RSC Adv.*, 2016, **6**, 48800–48809.
8. W. C. Liu, T. Gong, Q. H. Wang, X. Liang, J. J. Chen and P. Zhu, *Sci. Rep.*, 2016, **6**, 18439.
9. A. Christen, J. Bland and D. Gibson, *Proc. Am. Assoc. Cancer Res.*, 1989, **30**, 566.
10. Q. Huang, C. A. Roessner, R. Croteau and A. I. Scott, *Bioorg. Med. Chem.*, 2001, **9**, 2237–2242.
11. B. Engels, P. Dahm and S. Jennewein, *Metab. Eng.*, 2008, **10**, 201–206.
12. O. Besumbes, S. Sauret-Gueto, M. A. Phillips, S. Imperial, M. Rodriguez-Concepcion and A. Boronat, *Biotechnol. Bioeng.*, 2004, **88**, 168–175.
13. W. Lester, *Annu. Rev. Microbiol.*, 1972, **26**, 85–102.
14. G. Griesgraber, Y. S. Or, D. T. Chu, A. M. Nilius, P. M. Johnson, R. K. Flamm, R. F. Henry and J. J. Plattner, *J. Antibiot.*, 1996, **49**, 465–477.

15. W. R. Baker, J. D. Clark, R. L. Stephens and K. H. Kim, *J. Org. Chem.*, 1988, **53**, 2340–2345.
16. S. Morimoto, Y. Misawa, T. Adachi, T. Nagate, Y. Watanabe and S. Omura, *J. Antibiot.*, 1990, **43**, 286–294.
17. C. R. Shugrue and S. J. Miller, *Chem. Rev.*, 2017, **117**, 11894–11951.
18. J. Kennedy, *Nat. Prod. Rep.*, 2008, **25**, 25–34.
19. A. S. Klein, A. Domröse, P. Bongen, H. U. C. Brass, T. Classen, A. Loeschcke, T. Drepper, L. Laraia, S. Sievers, K.-E. Jaeger and J. Pietruszka, *ACS Synth. Biol.*, 2017, **6**, 1757–1765.
20. L. Chen, Y. Li, Q. Yue, A. Loksztejn, K. Yokoyama, E. A. Felix, X. Liu, N. Zhang, Z. An and G. F. Bills, *ACS Chem. Biol.*, 2016, **11**, 2724–2733.
21. J. R. Jacobsen, C. R. Hutchinson, D. E. Cane and C. Khosla, *Science*, 1997, **277**, 367–369.
22. K. Kinoshita, B. A. Pfeifer, C. Khosla and D. E. Cane, *Bioorg. Med. Chem. Lett.*, 2003, **13**, 3701–3704.
23. S. L. Ward, R. P. Desai, Z. Hu, H. Gramajo and L. Katz, *J. Ind. Microbiol. Biotechnol.*, 2007, **34**, 9–15.
24. S. Murli, K. S. MacMillan, Z. Hu, G. W. Ashley, S. D. Dong, J. T. Kealey, C. D. Reeves and J. Kennedy, *Appl. Environ. Microbiol.*, 2005, **71**, 4503–4509.
25. P. M. Wright, I. B. Seiple and A. G. Myers, *Angew. Chem., Int. Ed. Engl.*, 2014, **53**, 8840–8869.
26. D. E. Cane and C. T. Walsh, *Chem. Biol.*, 1999, **6**, R319–R325.
27. D. E. Cane, C. T. Walsh and C. Khosla, *Science*, 1998, **282**, 63–68.
28. D. J. Newman and G. M. Cragg, *J. Nat. Prod.*, 2016, **79**, 629–661.
29. G. M. Cragg and D. J. Newman, *Biochim. Biophys. Acta*, 2013, **1830**, 3670–3695.
30. M. E. Maier, *Org. Biomol. Chem.*, 2015, **13**, 5302–5343.
31. G. Li, J. Wang and M. T. Reetz, *Bioorg. Med. Chem.*, 2017, **26**, 1241–1251.
32. J. Cortes, S. F. Haydock, G. A. Roberts, D. J. Bevitt and P. F. Leadlay, *Nature*, 1990, **348**, 176–178.
33. K. J. Weissman and P. F. Leadlay, *Nat. Rev. Microbiol.*, 2005, **3**, 925–936.
34. Y. A. Chan, A. M. Podevels, B. M. Kevany and M. G. Thomas, *Nat. Prod. Rep.*, 2009, **26**, 90–114.
35. D. A. Hopwood, F. Malpartida, H. M. Kieser, H. Ikeda, J. Duncan, I. Fujii, B. A. Rudd, H. G. Floss and S. Ōmura, *Nature*, 1985, **314**, 642–644.
36. T. S. Chen, E. S. Inamine, O. D. Hensens, D. Zink and D. A. Ostlind, *Arch. Biochem. Biophys.*, 1989, **269**, 544–547.
37. S. Donadio, M. J. Staver, J. B. McAlpine, S. J. Swanson and L. Katz, *Science*, 1991, **252**, 675–679.
38. S. Donadio and L. Katz, *Gene*, 1992, **111**, 51–60.
39. S. Donadio, J. B. McAlpine, P. J. Sheldon, M. Jackson and L. Katz, *Proc. Natl. Acad. Sci. U. S. A.*, 1993, **90**, 7119–7123.
40. J. Cortes, K. Wiesmann, G. Roberts, M. Brown, J. Staunton and P. Leadlay, *Science*, 1995, **268**, 1487–1489.

41. C. M. Kao, L. Katz and C. Khosla, *Science*, 1994, **265**, 509–512.
42. K. E. Wiesmann, J. Cortes, M. J. Brown, A. L. Cutter, J. Staunton and P. F. Leadlay, *Chem. Biol.*, 1995, **2**, 583–589.
43. R. Pieper, G. Luo, D. E. Cane and C. Khosla, *Nature*, 1995, **378**, 263–266.
44. C. M. Kao, G. Luo, L. Katz, D. E. Cane and C. Khosla, *J. Am. Chem. Soc.*, 1995, **117**, 9105–9106.
45. C. M. Kao, G. Luo, L. Katz, D. E. Cane and C. Khosla, *J. Am. Chem. Soc.*, 1996, **118**, 9184–9185.
46. A. F. Marsden, W. Barrie, J. Cortés, N. J. Dunster, J. Staunton and P. F. Leadlay, *Science*, 1998, **279**, 199–202.
47. C. J. Rowe, I. U. Böhm, I. P. Thomas, B. Wilkinson, B. A. M. Rudd, G. Foster, A. P. Blackaby, P. J. Sidebottom, Y. Roddis, A. D. Buss, J. Staunton and P. F. Leadlay, *Chem. Biol.*, 2001, **8**, 475–485.
48. A. Ranganathan, M. Timoney, M. Bycroft, C. Jesús, I. P. Thomas, B. Wilkinson, L. Kellenberger, U. Hanefeld, I. S. Galloway, J. Staunton and P. F. Leadlay, *Chem. Biol.*, 1999, **6**, 731–741.
49. H. G. Menzella, R. Reid, J. R. Carney, S. S. Chandran, S. J. Reisinger, K. G. Patel, D. A. Hopwood and D. V. Santi, *Nat. Biotechnol.*, 2005, **23**, 1171–1176.
50. S. S. Chandran, H. G. Menzella, J. R. Carney and D. V. Santi, *Chem. Biol.*, 2006, **13**, 469–474.
51. M. Oliynyk, M. J. Brown, J. Cortes, J. Staunton and P. F. Leadlay, *Chem. Biol.*, 1996, **3**, 833–839.
52. X. Ruan, A. Pereda, D. L. Stassi, D. Zeidner, R. G. Summers, M. Jackson, A. Shivakumar, S. Kakavas, M. J. Staver, S. Donadio and L. Katz, *J. Bacteriol.*, 1997, **179**, 6416–6425.
53. D. L. Stassi, S. J. Kakavas, K. A. Reynolds, G. Gunawardana, S. Swanson, D. Zeidner, M. Jackson, H. Liu, A. Buko and L. Katz, *Proc. Natl. Acad. Sci. U. S. A.*, 1998, **95**, 7305–7309.
54. J. Zhang, Y. J. Yan, J. An, S. X. Huang, X. J. Wang and W. S. Xiang, *Microb. Cell Fact.*, 2015, **14**, 1–12.
55. D. Bedford, J. R. Jacobsen, G. Luo, D. E. Cane and C. Khosla, *Chem. Biol.*, 1996, **3**, 827–831.
56. R. McDaniel, C. M. Kao, H. Fu, P. Hevezi, Claes Gustafsson, M. Betlach, G. Ashley, D. E. Cane and C. Khosla, *J. Am. Chem. Soc.*, 1997, **119**, 4309–4310.
57. S. Donadio, M. J. Staver, J. B. McAlpine, S. J. Swanson and L. Katz, *Gene*, 1992, **115**, 97–103.
58. R. McDaniel, A. Thamchaipenet, C. Gustafsson, H. Fu, M. Betlach, M. Betlach and G. Ashley, *Proc. Natl. Acad. Sci. U. S. A.*, 1999, **96**, 1846–1851.
59. M. Hans, A. Hornung, A. Dziarnowski, D. E. Cane and C. Khosla, *J. Am. Chem. Soc.*, 2003, **125**, 5366–5374.
60. J. A. Chemler, A. Tripathi, D. A. Hansen, M. O'Neil-Johnson, R. B. Williams, C. Starks, S. R. Park and D. H. Sherman, *J. Am. Chem. Soc.*, 2015, **137**, 10603–10609.

61. M. Klaus, M. P. Ostrowski, J. Austerjost, T. Robbins, B. Lowry, D. E. Cane and C. Khosla, *J. Biol. Chem.*, 2016, **291**, 16404–16415.
62. M. Klaus, M. P. Ostrowski, J. Austerjost, T. Robbins, B. Lowry, D. E. Cane and C. Khosla, *J. Biol. Chem.*, 2016, **291**, 16404–16415.
63. S. Yuzawa, K. Deng, G. Wang, E. E. K. Baidoo, T. R. Northen, P. D. Adams, L. Katz and J. D. Keasling, *ACS Synth. Biol.*, 2017, **6**, 139–147.
64. C. D. Reeves, M. Sumati, G. W. Ashley, M. Piagentini, C. R. Hutchinson and R. McDaniel, *Biochemistry*, 2001, **40**, 15464–15470.
65. S. Weist and R. D. Sussmuth, *Appl. Microbiol. Biotechnol.*, 2005, **68**, 141–150.
66. J. Kennedy, *Nat. Prod. Rep.*, 2008, **25**, 25–34.
67. K. J. Weissman, *Trends Biotechnol.*, 2007, **25**, 139–142.
68. A. Kirschning and F. Hahn, *Angew. Chem., Int. Ed.*, 2012, **51**, 4012–4022.
69. A. J. Birch, *Pure Appl. Chem.*, 1963, **7**, 527–538.
70. W. T. Shier, K. L. Rinehart and D. Gottlieb, *Proc. Natl. Acad. Sci. U. S. A.*, 1969, **63**, 198–204.
71. K. L. Rinehart, *Pure Appl. Chem.*, 1977, **49**, 1361–1384.
72. R. Pieper, G. Luo, D. E. Cane and C. Khosla, *J. Am. Chem. Soc.*, 1995, **117**, 11373–11374.
73. R. J. M. Goss and H. Hong, *Chem. Commun.*, 2005, 3983–3985.
74. R. P. Desai, E. Rodriguez, J. L. Galazzo and P. Licari, *Biotechnol. Prog.*, 2004, **20**, 1660–1665.
75. G. W. Ashley, M. Burlingame, R. Desai, H. Fu, T. Leaf, P. J. Licari, C. Tran, D. Abbanat, K. Bush and M. Macielag, *J. Antibiot.*, 2006, **59**, 392–401.
76. C. J. B. Harvey, J. D. Puglisi, V. S. Pande, D. E. Cane and C. Khosla, *J. Am. Chem. Soc.*, 2012, **134**, 12259–12265.
77. S. J. Moss, I. Carletti, C. Olano, R. M. Sheridan, M. Ward, V. Math, E. A. M. Nur, A. F. Brana, M. Q. Zhang, P. F. Leadlay, C. Mendez, J. A. Salas and B. Wilkinson, *Chem. Commun.*, 2006, 2341–2343.
78. B. Wilkinson, M. A. Gregory, S. J. Moss, I. Carletti, R. M. Sheridan, A. Kaja, M. Ward, C. Olano, C. Mendez, J. A. Salas, P. F. Leadlay, R. vanGinckel and M. Q. Zhang, *Bioorg. Med. Chem. Lett.*, 2006, **16**, 5814–5817.
79. Y. N. Song, R. H. Jiao, W. J. Zhang, G. Y. Zhao, H. Dou, R. Jiang, A. H. Zhang, Y. Y. Hou, S. F. Bi, H. M. Ge and R. X. Tan, *Org. Lett.*, 2015, **17**, 556–559.
80. F. Taft, M. Brünjes, H. G. Floss, N. Czempinski, S. Grond, F. Sasse and A. Kirschning, *ChemBioChem*, 2008, **9**, 1057–1060.
81. S. Eichner, H. G. Floss, F. Sasse and A. Kirschning, *ChemBioChem*, 2009, **10**, 1801–1805.
82. L. S. Sheehan, R. E. Lill, B. Wilkinson, R. M. Sheridan, W. A. Vousden, A. L. Kaja, G. D. Crouse, J. Gifford, P. R. Graupner, L. Karr, P. Lewer, T. C. Sparks, P. F. Leadlay, C. Waldron and C. J. Martin, *J. Nat. Prod.*, 2006, **69**, 1702–1710.

83. B. J. Carroll, S. J. Moss, L. Bai, Y. Kato, S. Toelzer, T.-W. Yu and H. G. Floss, *J. Am.Chem. Soc.*, 2002, **124**, 4176–4177.
84. N. L. Pohl, R. S. Gokhale, D. E. Cane and C. Khosla, *J. Am. Chem. Soc.*, 1998, **120**, 11206–11207.
85. U. Sundermann, K. Bravo-Rodriguez, S. Klopries, S. Kushnir, H. Gomez, E. Sanchez-Garcia and F. Schulz, *ACS Chem. Biol.*, 2013, **8**, 443–450.
86. K. Bravo-Rodriguez, S. Klopries, K. R. M. Koopmans, U. Sundermann, S. Yahiaoui, J. Arens, S. Kushnir, F. Schulz and E. Sanchez-Garcia, *Chem. Biol.*, 2015, **22**, 1425–1430.
87. I. Koryakina, J. B. McArthur, M. M. Draelos and G. J. Williams, *Org. Biomol. Chem.*, 2013, **11**, 4449–4458.
88. I. Koryakina and G. J. Williams, *ChemBioChem*, 2011, **12**, 2289–2293.
89. D. A. Hansen, A. A. Koch and D. H. Sherman, *J. Am. Chem. Soc.*, 2015, **137**, 3735–3738.
90. A. N. Lowell, M. D. DeMars, 2nd, S. T. Slocum, F. Yu, K. Anand, J. A. Chemler, N. Korakavi, J. K. Priessnitz, S. R. Park, A. A. Koch, P. J. Schultz and D. H. Sherman, *J. Am. Chem. Soc.*, 2017, **139**, 7913–7920.
91. J. Staunton and B. Wilkinson, *Chem. Rev.*, 1997, **97**, 2611–2630.
92. S. R. Park, Y. J. Yoo, Y.-H. Ban and Y. J. Yoon, *J. Antibiot.*, 2010, **63**, 434–441.
93. D. Benjamin, M. Colombi, C. Moroni and M. N. Hall, *Nat. Rev. Drug Discovery*, 2011, **10**, 868–880.
94. L. E. Khaw, G. A. Bohm, S. Metcalfe, J. Staunton and P. F. Leadlay, *J. Bacteriol.*, 1998, **180**, 809–814.
95. F. V. Ritacco, E. I. Graziani, M. Y. Summers, T. M. Zabriskie, K. Yu, V. S. Bernan, G. T. Carter and M. Greenstein, *Appl. Environ. Microbiol.*, 2005, **71**, 1971–1976.
96. M. A. Gregory, H. Petkovic, R. E. Lill, S. J. Moss, B. Wilkinson, S. Gaisser, P. F. Leadlay and R. M. Sheridan, *Angew. Chem., Int. Ed.*, 2005, **44**, 4757–4760.
97. M. A. Gregory, A. L. Kaja, S. G. Kendrew, N. J. Coates, T. Warneck, M. Nur-e-Alam, R. E. Lill, L. S. Sheehan, L. Chudley, S. J. Moss, R. M. Sheridan, M. Quimpere, M.-Q. Zhang, C. J. Martin and B. Wilkinson, *Chem. Sci.*, 2013, **4**, 1046–1052.
98. S. Mo, D. H. Kim, J. H. Lee, J. W. Park, D. B. Basnet, Y. H. Ban, Y. J. Yoo, S. W. Chen, S. R. Park, E. A. Choi, E. Kim, Y. Y. Jin, S. K. Lee, J. Y. Park, Y. Liu, M. O. Lee, K. S. Lee, S. J. Kim, D. Kim, B. C. Park, S.-g. Lee, H. J. Kwon, J.-W. Suh, B. S. Moore, S.-K. Lim and Y. J. Yoon, *J. Am. Chem. Soc.*, 2011, **133**, 976–985.
99. A. Lechner, M. C. Wilson, Y. H. Ban, J.-Y. Hwang, Y. J. Yoon and B. S. Moore, *ACS Synth. Biol.*, 2013, **2**, 379–383.
100. M. C. Walker, B. W. Thuronyi, L. K. Charkoudian, B. Lowry, C. Khosla and M. C. Y. Chang, *Science*, 2013, **341**, 1089–1094.
101. E. M. Musiol-Kroll, F. Zubeil, T. Schafhauser, T. Härtner, A. Kulik, J. McArthur, I. Koryakina, W. Wohlleben, S. Grond, G. J. Williams, S. Y. Lee and T. Weber, *ACS Synth. Biol.*, 2017, **6**, 421–427.

102. X. Xie, K. Watanabe, W. A. Wojcicki, C. C. Wang and Y. Tang, *Chem. Biol.*, 2006, **13**, 1161–1169.
103. X. Xie, W. W. Wong and Y. Tang, *Metab. Eng.*, 2007, **9**, 379–386.
104. X. Gao, X. Xie, I. Pashkov, M. R. Sawaya, J. Laidman, W. Zhang, R. Cacho, T. O. Yeates and Y. Tang, *Chem. Biol.*, 2009, **16**, 1064–1074.
105. G. Jimenez-Oses, S. Osuna, X. Gao, M. R. Sawaya, L. Gilson, S. J. Collier, G. W. Huisman, T. O. Yeates, Y. Tang and K. N. Houk, *Nat. Chem. Biol.*, 2014, **10**, 431–436.
106. L. Katz and G. W. Ashley, *Chem. Rev.*, 2005, **105**, 499–528.
107. P. Fernandes, E. Martens and D. Pereira, *J. Antibiot.*, 2017, **70**, 527–533.
108. P. Fernandes, E. Martens, D. Bertrand and D. Pereira, *Bioorg. Med. Chem.*, 2016, **24**, 6420–6428.
109. I. B. Seiple, Z. Zhang, P. Jakubec, A. Langlois-Mercier, P. M. Wright, D. T. Hog, K. Yabu, S. R. Allu, T. Fukuzaki, P. N. Carlsen, Y. Kitamura, X. Zhou, M. L. Condakes, F. T. Szczypinski, W. D. Green and A. G. Myers, *Nature*, 2016, **533**, 338–345.
110. J. R. Jacobsen, C. R. Hutchinson, D. E. Cane and C. Khosla, *Science*, 1997, **277**, 367–369.
111. X. Zhu and W. Zhang, *Front. Chem.*, 2015, **3**, 11.
112. E. M. Musiol-Kroll, F. Zubeil, T. Schafhauser, T. Hartner, A. Kulik, J. McArthur, I. Koryakina, W. Wohlleben, S. Grond, G. J. Williams, S. Y. Lee and T. Weber, *ACS Synth. Biol.*, 2017, **6**, 421–427.
113. I. Koryakina, C. Kasey, J. B. McArthur, A. N. Lowell, J. A. Chemler, S. Li, D. A. Hansen, D. H. Sherman and G. J. Williams, *ACS Chem. Biol.*, 2017, **12**, 114–123.
114. K. A. J. Bozhüyük, F. Fleischhacker, A. Linck, F. Wesche, A. Tietze, C.-P. Niesert and H. B. Bode, *Nat. Chem.*, 2017, **10**, 275–281.
115. D. L. Niquille, D. A. Hansen, T. Mori, D. Fercher, H. Kries and D. Hilvert, *Nat. Chem.*, 2017, **10**, 282–287.
116. R. Traber, H. Hofmann and H. Kobel, *J. Antibiot.*, 1989, **42**, 591–597.
117. S. Weist, B. Bister, O. Puk, D. Bischoff, S. Pelzer, G. J. Nicholson, W. Wohlleben, G. Jung and R. D. Süssmuth, *Angew. Chem., Int. Ed.*, 2002, **41**, 3383–3385.
118. S. Weist, C. Kittel, D. Bischoff, B. Bister, V. Pfeifer, G. J. Nicholson, W. Wohlleben and R. D. Süssmuth, *J. Am. Chem. Soc.*, 2004, **126**, 5942–5943.
119. H. Lin and C. T. Walsh, *J. Am. Chem. Soc.*, 2004, **126**, 13998–14003.
120. A. Powell, M. Borg, B. Amir-Heidari, J. M. Neary, J. Thirlway, B. Wilkinson, C. P. Smith and J. Micklefield, *J. Am. Chem. Soc.*, 2007, **129**, 15182–15191.
121. R. A. Lewis, L. Nunns, J. Thirlway, K. Carroll, C. P. Smith and J. Micklefield, *Chem. Commun.*, 2011, **47**, 1860–1862.
122. J. Thirlway, R. Lewis, L. Nunns, M. Al Nakeeb, M. Styles, A.-W. Struck, C. P. Smith and J. Micklefield, *Angew. Chem., Int. Ed.*, 2012, **51**, 7181–7184.

123. C. Stewart, Jr., C. R. Vickery, M. D. Burkart and J. P. Noel, *Curr. Opin. Plant Biol.*, 2013, **16**, 365–372.
124. N. Funa, Y. Ohnishi, Y. Ebizuka and S. Horinouchi, *J. Biol. Chem.*, 2002, **277**, 4628–4635.
125. Y. Katsuyama, T. Kita, N. Funa and S. Horinouchi, *J. Biol. Chem.*, 2009, **284**, 11160–11170.
126. S. C. Wenzel, H. B. Bode, I. Kochems and R. Muller, *ChemBioChem*, 2008, **9**, 2711–2721.
127. M. G. Bangera and L. S. Thomashow, *J. Bacteriol.*, 1999, **181**, 3155–3163.
128. N. Funa, T. Awakawa and S. Horinouchi, *J. Biol. Chem.*, 2007, **282**, 14476–14481.
129. J. Li, Y. Luo, J. K. Lee and H. Zhao, *Bioorg. Med. Chem. Lett.*, 2011, **21**, 6085–6089.
130. N. Funa, Y. Ohnishi, I. Fujii, M. Shibuya, Y. Ebizuka and S. Horinouchi, *Nature*, 1999, **400**, 897–899.
131. I. Abe and H. Morita, *Nat. Prod. Rep.*, 2010, **27**, 809–838.
132. J. L. Ferrer, J. M. Jez, M. E. Bowman, R. A. Dixon and J. P. Noel, *Nat. Struct. Biol.*, 1999, **6**, 775–784.
133. J. M. Jez, M. B. Austin, J. Ferrer, M. E. Bowman, J. Schroder and J. P. Noel, *Chem. Biol.*, 2000, **7**, 919–930.
134. H. Morita, S. Kondo, S. Oguro, H. Noguchi, S. Sugio, I. Abe and T. Kohno, *Chem. Biol.*, 2007, **14**, 359–369.
135. I. Abe, T. Watanabe, W. Lou and H. Noguchi, *FEBS J.*, 2006, **273**, 208–218.
136. I. Abe, S. Oguro, Y. Utsumi, Y. Sano and H. Noguchi, *J. Am. Chem. Soc.*, 2005, **127**, 12709–12716.
137. I. Abe, H. Morita, S. Oguro, H. Noma, K. Wanibuchi, N. Kawahara, Y. Goda, H. Noguchi and T. Kohno, *J. Am. Chem. Soc.*, 2007, **129**, 5976–5980.
138. H. Morita, K. Wanibuchi, H. Nii, R. Kato, S. Sugio and I. Abe, *Proc. Natl. Acad. Sci. U. S. A.*, 2010, **107**, 19778–19783.
139. J. A. McIntosh, M. S. Donia and E. W. Schmidt, *Nat. Prod. Rep.*, 2009, **26**, 537–559.
140. P. G. Arnison, M. J. Bibb, G. Bierbaum, A. A. Bowers, T. S. Bugni, G. Bulaj, J. A. Camarero, D. J. Campopiano, G. L. Challis, J. Clardy, P. D. Cotter, D. J. Craik, M. Dawson, E. Dittmann, S. Donadio, P. C. Dorrestein, K.-D. Entian, M. A. Fischbach, J. S. Garavelli, U. Goransson, C. W. Gruber, D. H. Haft, T. K. Hemscheidt, C. Hertweck, C. Hill, A. R. Horswill, M. Jaspars, W. L. Kelly, J. P. Klinman, O. P. Kuipers, A. J. Link, W. Liu, M. A. Marahiel, D. A. Mitchell, G. N. Moll, B. S. Moore, R. Muller, S. K. Nair, I. F. Nes, G. E. Norris, B. M. Olivera, H. Onaka, M. L. Patchett, J. Piel, M. J. T. Reaney, S. Rebuffat, R. P. Ross, H.-G. Sahl, E. W. Schmidt, M. E. Selsted, K. Severinov, B. Shen, K. Sivonen, L. Smith, T. Stein, R. D. Sussmuth, J. R. Tagg, G.-L. Tang, A. W. Truman, J. C. Vederas, C. T. Walsh,

J. D. Walton, S. C. Wenzel, J. M. Willey and W. A. van der Donk, *Nat. Prod. Rep.*, 2013, **30**, 108–160.

141. C. L. Cox, J. R. Doroghazi and D. A. Mitchell, *BMC Genom.*, 2015, **16**, 778.
142. D. E. Ruffner, E. W. Schmidt and J. R. Heemstra, *ACS Synth. Biol.*, 2015, **4**, 482–492.
143. Q. Zhang, X. Yang, H. Wang and W. A. van der Donk, *ACS Chem. Biol.*, 2014, **9**, 2686–2694.
144. D. Sardar and E. W. Schmidt, *Curr. Opin. Chem. Biol.*, 2016, **31**, 15–21.
145. J. I. Tietz, C. J. Schwalen, P. S. Patel, T. Maxson, P. M. Blair, H. C. Tai, U. I. Zakai and D. A. Mitchell, *Nat. Chem. Biol.*, 2017, **13**, 470–478.
146. T. J. Maimone and P. S. Baran, *Nat. Chem. Biol.*, 2007, **3**, 396–407.
147. J. L. F. Monteiro and C. O. Veloso, *Top. Catal.*, 2004, **27**, 169–180.
148. C. L. Steele, J. Crock, J. Bohlmann and R. Croteau, *J. Biol. Chem.*, 1998, **273**, 2078–2089.
149. Z. G. Brill, M. L. Condakes, C. P. Ting and T. J. Maimone, *Chem. Rev.*, 2017, **117**, 11753–11795.
150. J. Ziegler and P. J. Facchini, *Annu. Rev. Plant Biol.*, 2008, **59**, 735–769.
151. A. M. Ehrenworth and P. Peralta-Yahya, *Nat. Chem. Biol.*, 2017, **13**, 249–258.
152. S. Bunsupa, M. Yamazaki and K. Saito, *Front. Plant Sci.*, 2012, **3**, 239.
153. W. S. Glenn, W. Runguphan and S. E. O'Connor, *Curr. Opin. Biotechnol*, 2013, **24**, 354–365.
154. A. Diamond and I. Desgagné-Penix, *Plant Biotechnol. J.*, 2016, **14**, 1319–1328.
155. R. N. Kulkarni, K. Baskaran and T. Jhang, *Plant Genet. Resour.*, 2016, **14**, 283–302.
156. S. Gruschow, T. J. Buchholz, W. Seufert, J. S. Dordick and D. H. Sherman, *ChemBioChem*, 2007, **8**, 863–868.
157. J. A. Chemler, T. J. Buchholz, T. W. Geders, D. L. Akey, C. M. Rath, G. E. Chlipala, J. L. Smith and D. H. Sherman, *J. Am. Chem. Soc.*, 2012, **134**, 7359–7366.
158. M. B. Austin, M. E. Bowman, J. L. Ferrer, J. Schroder and J. P. Noel, *Chem. Biol.*, 2004, **11**, 1179–1194.
159. K. Wanibuchi, P. Zhang, T. Abe, H. Morita, T. Kohno, G. Chen, H. Noguchi and I. Abe, *FEBS J.*, 2007, **274**, 1073–1082.
160. H. Morita, M. Yamashita, S. P. Shi, T. Wakimoto, S. Kondo, R. Kato, S. Sugio, T. Kohno and I. Abe, *Proc. Natl. Acad. Sci. U. S. A.*, 2011, **108**, 13504–13509.
161. T. Mo, W.-Q. Liu, W. Ji, J. Zhao, T. Chen, W. Ding, S. Yu and Q. Zhang, *ACS Chem. Biol.*, 2017, **12**, 1484–1488.
162. A. Kuthning, P. Durkin, S. Oehm, M. G. Hoesl, N. Budisa and R. D. Sussmuth, *Sci. Rep.*, 2016, **6**, 33447.
163. W. L. Cheung, S. J. Pan and A. J. Link, *J. Am. Chem. Soc.*, 2010, **132**, 2514–2515.

164. M. R. Levengood, G. C. Patton and W. A. van der Donk, *J. Am. Chem. Soc.*, 2007, **129**, 10314–10315.
165. W. E. Houssen, A. F. Bent, A. R. McEwan, N. Pieiller, J. Tabudravu, J. Koehnke, G. Mann, R. I. Adaba, L. Thomas, U. W. Hawas, H. Liu, U. Schwarz-Linek, M. C. Smith, J. H. Naismith and M. Jaspars, *Angew. Chem., Int. Ed. Engl.*, 2014, **53**, 14171–14174.
166. D.-K. Ro, E. M. Paradise, M. Ouellet, K. J. Fisher, K. L. Newman, J. M. Ndungu, K. A. Ho, R. A. Eachus, T. S. Ham, J. Kirby, M. C. Y. Chang, S. T. Withers, Y. Shiba, R. Sarpong and J. D. Keasling, *Nature*, 2006, **440**, 940–943.
167. P. J. Westfall, D. J. Pitera, J. R. Lenihan, D. Eng, F. X. Woolard, R. Regentin, T. Horning, H. Tsuruta, D. J. Melis, A. Owens, S. Fickes, D. Diola, K. R. Benjamin, J. D. Keasling, M. D. Leavell, D. J. McPhee, N. S. Renninger, J. D. Newman and C. J. Paddon, *Proc. Natl. Acad. Sci.*, 2012, **109**, E111–E118.
168. C. J. Paddon, P. J. Westfall, D. J. Pitera, K. Benjamin, K. Fisher, D. McPhee, M. D. Leavell, A. Tai, A. Main, D. Eng, D. R. Polichuk, K. H. Teoh, D. W. Reed, T. Treynor, J. Lenihan, H. Jiang, M. Fleck, S. Bajad, G. Dang, D. Dengrove, D. Diola, G. Dorin, K. W. Ellens, S. Fickes, J. Galazzo, S. P. Gaucher, T. Geistlinger, R. Henry, M. Hepp, T. Horning, T. Iqbal, L. Kizer, B. Lieu, D. Melis, N. Moss, R. Regentin, S. Secrest, H. Tsuruta, R. Vazquez, L. F. Westblade, L. Xu, M. Yu, Y. Zhang, L. Zhao, J. Lievense, P. S. Covello, J. D. Keasling, K. K. Reiling, N. S. Renninger and J. D. Newman, *Nature*, 2013, **496**, 528–532.
169. J. Kong, Y. Yang, W. Wang, K. Cheng and P. Zhu, *RSC Adv.*, 2013, **3**, 7622–7641.
170. J. Turconi, F. Griolet, R. Guevel, G. Oddon, R. Villa, A. Geatti, M. Hvala, K. Rossen, R. Göller and A. Burgard, *Org. Process Res. Dev.*, 2014, **18**, 417–422.
171. M. Peplow, *Nature*, 2016, **530**, 389–390.
172. M. Demiray, X. Tang, T. Wirth, J. A. Faraldos and R. K. Allemann, *Angew. Chem., Int. Ed.*, 2017, **56**, 4347–4350.
173. X. Tang, M. Demiray, T. Wirth and R. K. Allemann, *Bioorg. Med. Chem.*, 2017, **56**, 4347–4350.
174. O. Cascon, S. Touchet, D. J. Miller, V. Gonzalez, J. A. Faraldos and R. K. Allemann, *Chem. Commun.*, 2012, **48**, 9702–9704.
175. J. A. Faraldos, D. J. Miller, V. González, Z. Yoosuf-Aly, O. Cascón, A. Li and R. K. Allemann, *J. Am. Chem. Soc.*, 2012, **134**, 5900–5908.
176. D. E. Cane and Y. S. Tsantrizos, *J. Am. Chem. Soc.*, 1996, **118**, 10037–10040.
177. Q. Jin, D. C. Williams, M. Hezari, R. Croteau and R. M. Coates, *J. Org. Chem.*, 2005, **70**, 4667–4675.
178. Y. Jin, D. C. Williams, R. Croteau and R. M. Coates, *J. Am. Chem. Soc.*, 2005, **127**, 7834–7842.
179. A. Vattekkatte, N. Gatto, E. Schulze, W. Brandt and W. Boland, *Org. Biomol. Chem.*, 2015, **13**, 4776–4784.

180. D. J. Miller, F. Yu, D. W. Knight and R. K. Allemann, *Org. Biomol. Chem.*, 2009, 7, 962–975.
181. R. Lauchli, K. S. Rabe, K. Z. Kalbarczyk, A. Tata, T. Heel, R. Z. Kitto and F. H. Arnold, *Angew. Chem., Int. Ed. Engl.*, 2013, **52**, 5571–5574.
182. W. Bowers, C. Nishino, M. Montgomery, L. Nault and M. Nielson, *Science*, 1977, **196**, 680–681.
183. D. B. Little and R. B. Croteau, *Arch. Biochem. Biophys.*, 2002, **402**, 120–135.
184. Y. Yoshikuni, T. E. Ferrin and J. D. Keasling, *Nature*, 2006, **440**, 1078–1082.
185. F. Alberti, K. Khairudin, E. R. Venegas, J. A. Davies, P. M. Hayes, C. L. Willis, A. M. Bailey and G. D. Foster, *Nat. Commun.*, 2017, **8**, 1831.
186. T. Kuzuyama, J. P. Noel and S. B. Richard, *Nature*, 2005, **435**, 983–987.
187. T. Itoh, K. Tokunaga, Y. Matsuda, I. Fujii, I. Abe, Y. Ebizuka and T. Kushiro, *Nat. Chem.*, 2010, 2, 858–864.
188. F. Leipoldt, P. Zeyhle, A. Kulik, J. Kalinowski, L. Heide and L. Kaysser, *PLoS One*, 2015, **10**, e0143237.
189. B. Botta, G. D. Monache, P. Menendez and A. Boffi, *Trends Pharmacol. Sci.*, 2005, **26**, 606–608.
190. N. C. Perellino, J. Malyszko, M. Ballabio, B. Gioia and A. Minghetti, *J. Nat. Prod.*, 1992, **55**, 424–427.
191. N. Flieger, P. Sedmera, J. Vokoun, Z. Řeháček, J. Stuchlik, Z. Malinka, L. Cvak and P. Harazim, *J. Nat. Prod.*, 1984, **47**, 970–976.
192. A. Baumert, D. Erge and D. Gröger, *Planta Med.*, 1982, **44**, 122–123.
193. H.-Y. Lee, N. Yerkes and S. E. O'Connor, *Chem. Biol.*, 2009, **16**, 1225–1229.
194. C. Portmann, C. Prestinari, T. Myers, J. Scharte and K. Gademann, *ChemBioChem*, 2009, **10**, 889–895.
195. E. McCoy and S. E. O'Connor, *J. Am. Chem. Soc.*, 2006, **128**, 14276–14277.
196. L. J. Wigley, P. G. Mantle and D. A. Perry, *Phytochemistry*, 2006, **67**, 561–569.
197. S. Chen, M. C. Galan, C. Coltharp and S. E. O'Connor, *Chem. Biol.*, 2006, **13**, 1137–1141.
198. A. Lang, S. Polnick, T. Nicke, P. William, E. P. Patallo, J. H. Naismith and K. H. van Pee, *Angew. Chem., Int. Ed. Engl.*, 2011, **50**, 2951–2953.
199. W. Runguphan, X. Qu and S. E. O'Connor, *Nature*, 2010, **468**, 461–464.
200. P. Bernhardt, E. McCoy and S. E. O'Connor, *Chem. Biol.*, 2007, **14**, 888–897.
201. E. McCoy and S. E. O'Connor, *J. Am. Chem. Soc.*, 2006, **128**, 14276–14277.
202. Z. Časar and T. Mesar, *Org. Process Res. Dev.*, 2015, **19**, 378–385.
203. C. F. C. Lam, A. C. Giddens, N. Chand, V. L. Webb and B. R. Copp, *Tetrahedron*, 2012, **68**, 3187–3194.

204. Z. Qing, H. Cao, P. Cheng, W. Wang, J. Zeng and H. Xie, *Org. Chem. Front.*, 2018, **5**, 353–357.
205. M. C. Baird, S. G. Pyne, A. T. Ung, W. Lie, T. Sastraruji, A. Jatisatienr, C. Jatisatienr, S. Dheeranupattana, J. Lowlam and S. Boonchalermkit, *J. Nat. Prod.*, 2009, **72**, 679–684.
206. H. Liu and Y. Jia, *Nat. Prod. Rep.*, 2017, **34**, 411–432.
207. S. Galanie, K. Thodey, I. J. Trenchard, M. F. Interrante and C. D. Smolke, *Science*, 2015, **349**, 1095–1100.
208. A. Nakagawa, E. Matsumura, T. Koyanagi, T. Katayama, N. Kawano, K. Yoshimatsu, K. Yamamoto, H. Kumagai, F. Sato and H. Minami, *Nat. Commun.*, 2016, **7**, 10390.
209. K. Thodey, S. Galanie and C. D. Smolke, *Nat. Chem. Biol.*, 2014, **10**, 837.
210. J. W. Reed and T. Hudlicky, *Acc. Chem. Res.*, 2015, **48**, 674–687.
211. G. J. Williams, C. Zhang and J. S. Thorson, *Adv. Enzymol. Relat. Areas Mol. Biol.*, 2008, **76**, 55–119.
212. R. W. Gantt, P. Peltier-Pain and J. S. Thorson, *Nat. Prod. Rep.*, 2011, **28**, 1811–1853.
213. G. J. Williams, R. D. Goff, C. Zhang and J. S. Thorson, *Chem. Biol.*, 2008, **15**, 393–401.
214. G. J. Williams, R. W. Gantt and J. S. Thorson, *Curr. Opin. Chem. Biol.*, 2008, **12**, 556–564.
215. G. J. Williams, C. Zhang and J. S. Thorson, *Nat. Chem. Biol.*, 2007, **3**, 657–662.
216. J. M. Langenhan, B. R. Griffith and J. S. Thorson, *J. Nat. Prod.*, 2005, **68**, 1696–1711.
217. R. D. Goff and J. S. Thorson, *MedChemComm*, 2014, **5**, 1036–1047.
218. J. M. Langenhan, N. R. Peters, I. A. Guzei, F. M. Hoffmann and J. S. Thorson, *Proc. Natl. Acad. Sci. U. S. A.*, 2005, **102**, 12305–12310.
219. C. Kasey, M. Zerrad, Y. Li, T. A. Cropp and G. J. Williams, *ACS Synth. Biol.*, 2017, **7**, 227–239.
220. M. H. Medema, R. Breitling, R. Bovenberg and E. Takano, *Nat. Rev. Microbiol.*, 2011, **9**, 131–137.

CHAPTER 12

Site-specific Protein Modification and Bio-orthogonal Chemistry

MICHAEL E. WEBB,* ROBIN S. BON AND MEGAN H. WRIGHT

Astbury Centre for Structural Molecular Biology, University of Leeds, UK
*Email: m.e.webb@leeds.ac.uk

12.1 Modification of Native Amino Acids

The 20 commonly-occurring natural amino-acids provide a range of functional groups for specific chemical modification including: the β-thiol(ate) of cysteine, the ε-amino group of lysine, the β-hydroxy groups of serine and threonine, the ortho-positions of the phenol ring of tyrosine and the γ-thioether of methionine.[1] The field has been comprehensively reviewed in recent years, by Baran and co-workers[2] (from the point of view of peptide rather than protein modification and therefore including reactions incompatible with most proteins), by Boutureira and Bernardes[3] (an excellent comprehensive review), by Basle *et al.*[1] (a good summary of the key work up to 2010) and most recently by Xia and co-workers.[4] In this chapter, we will describe the key principles for protein labelling and some of the most recent highlights.

The selectivity of protein modification reactions is limited by both the abundance of these naturally-occurring amino-acid residues in proteins (Figure 12.1) and the innate chemoselectivity of the conjugation chemistry. Fortunately, the amino-acids with the most "chemically-interesting" functional groups are also the least abundant which facilitates selective

Chemical Biology No. 10
Chemical and Biological Synthesis: Enabling Approaches for Understanding Biology
Edited by Nick J. Westwood and Adam Nelson

Published by the Royal Society of Chemistry, www rsc.org

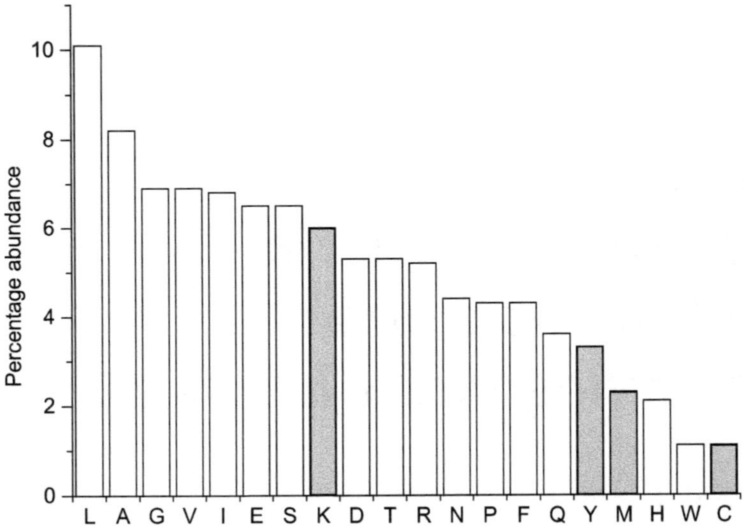

Figure 12.1 Frequency of amino-acid usage across the proteome—the most readily chemically modified amino acids are highlighted in grey. The relatively low abundance and high reactivity of cysteine mean that is the most commonly modified amino acid for selective modification. Although more common, the high reactivity of lysine makes it a popular choice for non-selective and multiple labelling reactions.

labelling.[5] Serine and threonine are largely too abundant (and unreactive) for selective modification whereas cysteine is sufficiently rarely incorporated that specific labelling can be achieved, often with little or no genetic modification of the protein to be labelled. Lysine is highly abundant and although modification of this amino-acid residue frequently competes with modification of the protein N-terminus, its high reactivity means that it is commonly used for non-selective modification. Chemoselective methods which can be used to label other amino acids with low abundance, such as tyrosine, histidine, tryptophan and methionine, are much more limited.

12.1.1 Modification of Lysine, Serine and Threonine

The ε-amino group of lysine can be conventionally modified with a wide range of reagents, including activated esters, such as *N*-hydroxysuccinimide (**1**) and pentafluorophenyl (**2**) esters, activated sulfonate esters and isothiocyanates.[6] This method is most frequently used for non-specific labelling and for covalent immobilization of proteins onto surfaces for biophysical techniques such as surface plasmon resonance (SPR) and a wide range of suitable labelling molecules are commercially available (See Figure 12.2). Selective labelling of the N-terminus over lysine residues can be achieved with careful control of pH but labelling is typically non-specific. This can be an advantage for applications where high label : protein ratios

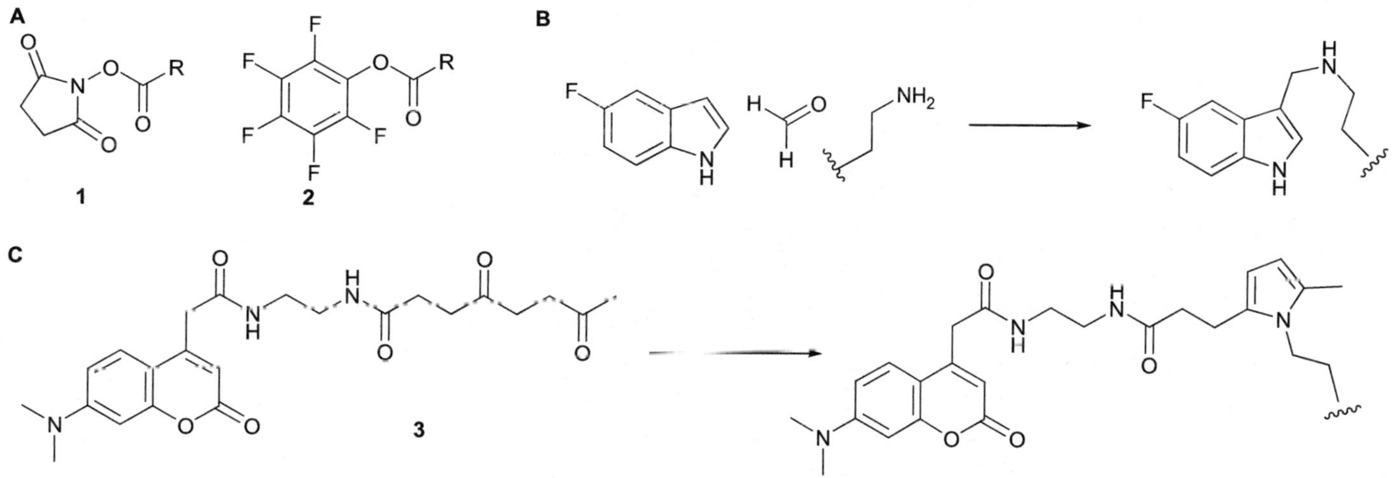

Figure 12.2 A. Examples of commercially-available activated labelling reagents available for labelling of lysine and the protein N-terminus. B. Mannich-type condensation for labelling of lysine and N-terminal amino residues. C. Use of Paal–Knorr cyclisation to fluorescently label lysine.

are required, such as during dye-labelling of secondary antibodies and where a range of labelled positions are an advantage (*e.g.* display of multiple surface-attached poses for phage display selection). In the case of lysine labelling, it has been possible to use kinetic approaches with limited reagents to selectively label single lysines in model proteins, however it is challenging to identify which residue this will be *a priori*, and it is quite likely this residue will be in the active site of an enzyme if such a protein is the labelling target.[7] Traditional synthetic reactions have also been adapted to bioconjugation chemistry of lysine, including recent approaches such as the Mannich-type condensation and Paal–Knorr cyclisation. In the former approach, lysine is labelled using formaldehyde and indoles at low concentration. This shows some kinetic site-selectivity but unless carefully controlled will probably lead to labelling of all lysines (Figure 12.2B).[8] In the latter approach, 1,4-diketones (*e.g.* **3**) condense with the lysine residues to generate 2,5-pyrrole systems.[9] In contrast, Che and co-workers have shown that ketene-derived reagents appear to demonstrate preferential labelling of the N-terminus over lysine side-chains.[10]

Selective labelling of the primary and secondary alcohols of serine and threonine is not typically feasible due to their high abundance in proteins and low reactivity. An exception is labelling of N-terminal serine and threonine residues where the Cα–Cβ bond between the α-amino group and β-hydroxyl group can be oxidatively cleaved with periodate to form an N-terminal oxalyl group (Figure 12.3A).[11] This aldehyde can then be selectively condensed with suitable nucleophiles such as hydrazines, hydrazides or amino-oxyethers to form the corresponding hydrazone, acylhydrazone or oxime.[12] This strategy has been used numerous times, for example Turnbull and co-workers recently used it to generate picomolar protein-based inhibitors of cholera toxin ligand binding.[13] An important

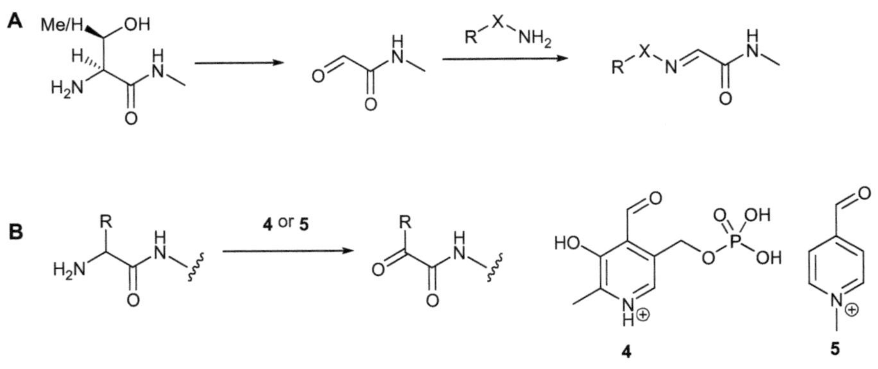

Figure 12.3 A. Oxidative cleavage of N-terminal serine and threonine residues can be used to generate electrophilic oxalyl groups, which can be functionalized with hydrazine and aminooxyethers to form hydrazones and oximes. B. Alternatively direct transamination of N-terminal amino-acids with pyridoxal phosphate **4** or methylpyridinium carboxaldehydes (Rapport's salt, **5**) can be used to generate N-terminal ketones.

experimental consideration is the preparation of suitable nucleophile coupling partners—frequently these are susceptible to sequestration as the acetone adduct and preparation of these with a last stage Boc-deprotection (without extensive purification) can be critical to the success of the labelling reaction. As an alternative, the N-terminus of many other proteins can be stoichiometrically transaminated using reagents based on pyridoxal phosphate (**4**)[14] (Figure 12.3B). The conditions required for the former reagent are typically too harsh for most proteins despite optimization, but alternatives such as Rapoport's salt (**5**)[15,16] work under milder conditions (Figure 12.3B).

12.1.2 Modification of Cysteine

Cysteine is the most readily chemoselectively-modified amino-acid—and the most frequently employed in selective protein labelling. A very large number of distinct methods to label this amino-acid residue have been reported. The relatively low abundance of free cysteine in proteins means that production of a protein with only a single accessible cysteine is frequently feasible. In many cases, the remaining cysteine residues are naturally protected from reaction by incorporation into disulphide cross-links, these can themselves be selectively reduced and cross-linked as described later in this chapter. In cases where more than one free cysteine residue is present in a protein construct it can still be possible to partially label the protein by careful manipulation of the reaction conditions and purification of the differentially labelled protein. Two different principal approaches to modification of isolated cysteine residues can be taken: either the natural nucleophilicity of the thiol/thiolate can be used with suitable electrophiles, such as maleimides, α-halocarbonyls or Michael acceptors, or a range of reagents can be used to convert the cysteine (or cysteines) into an electrophilic dehydroalanine residue which can then be treated with a suitable nucleophile to generate a wide range of unnatural amino-acids.

12.1.2.1 Modification of Cysteine Using Electrophilic Reagents

In the same manner as for lysine-modifying reagents, a wide range of cysteine-modifying reagents (Figure 12.4), particularly maleimides (**6**) are commercially available. Reaction with a maleimide is typically selective for thiols, though side-reactions with amine nucleophiles can occur upon prolonged incubation). One particular problem with maleimide-based labelling is that the product β-thioether is susceptible to cleavage *via* a retro-Michael elimination to regenerate the labelling reagent and the free maleimide reagent. This can be a particular problem for cellular delivery of *e.g.* fluorescent dyes or other reagents where it is possible that such deconjugation of the reagent may lead to artefactual observations. This problem can be resolved by the use of bromomaleimides[17] in which irreversible loss of the bromide regenerates the maleimide functional group, which is then not susceptible to the retro-Michael reaction.

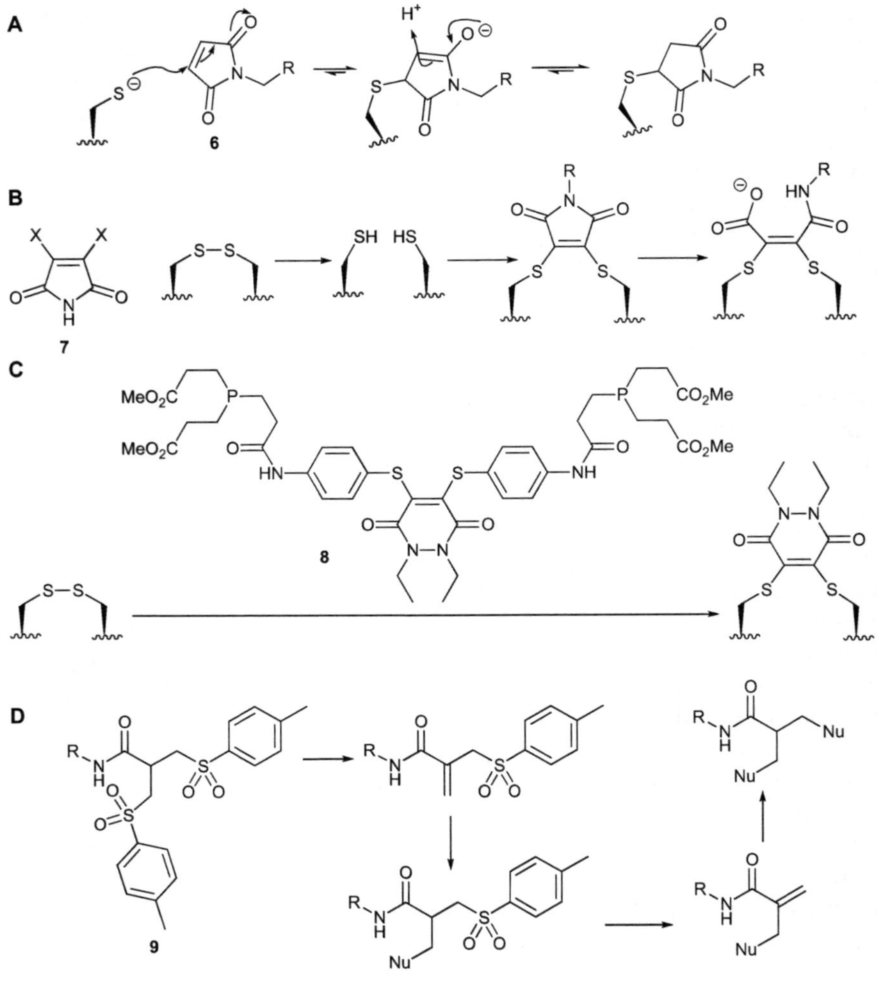

Figure 12.4 Chemistry of cysteine and cystine disulphide modification. A. The most common modification, alkylation *via* a maleimide **6** is subject to retro-Michael addition and cleavage. B. Bis-functionalised maleimides *e.g.* dibromomaleimide (**7**, X = Br) can be used for both non-reversible cysteine modification and for bridging of disulphide bonds. C. Inclusion of a phosphine reducing agent into the scaffold of the bridging reagent leads to efficient bridging of disulphides without an independent reduction step. D. Alternative masked bridging reagents based on a bis-sulfone architecture **9**, sequential elimination generates two distinct α,β-unsaturated Michael acceptors.

The bromomaleimide approach has been applied to good effect with the use of **7** as disulphide bridging reagents.[18,19] In this case, careful reduction of the cysteines in a disulphide bond reveals two cysteine residues, which react sequentially with a dibromomaleimide to generate a bridge between the two cysteines replacing the disulphide, or in the case of peptides generating a peptide 'staple'.[20] A range of alternative maleimides have been employed, for example bis(thiophenyl)-substituted maleimides have also been reported.[19] In all cases, the bridged disulphide still contains the maleimide functional group and is therefore susceptible to attack by cellular thiols and subsequent removal from the target protein. To alleviate this problem, mild alkaline conditions can be used to hydrolyse the maleimide ring to generate a β-carboxyacrylamide which is not susceptible to attack.[21] Molecules analogous to dibromomaleimides, such as dibromopyridazinediones,[22] can also be used and in this case combined reduction/cross-linking using the dibromopyridazinedione reagent **8** has been shown to effectively carry out both steps in a single pot.[23] An alternative approach for this kind of conjugation to disulphides is the use of masked Michael acceptors.[24] In this chemistry, a disulfone molecule of general structure **9** can undergo elimination to generate a Michael acceptor *in situ*. Following this first addition, a second elimination rapidly generates a new Michael acceptor which can undergo reaction to irreversibly replace the disulphide bond with the bridging, functionalized molecule. Alternatively, instead of trying to avoid the retro-Michael addition, other researchers have sought to make this reaction easier, 5-methylene pyrrolones can be easily generated from amino-functionalised compounds and undergo rapid attachment to and release from cysteine residues.[25]

In addition to these simple electrophiles, a range of more complex structures with enhanced reactivity and selectivity have been reported (Figure 12.5). The first of these, developed by the late Carlos Barbas III, was based on optimization of methylsulfonylbenzothiazole (MSBT,[26] **10**) a previously reported thiol-blocking reagent. They screened a range of alternative heterocycles and identified that phenyloxadiazolylsulfones **11** reacted much more rapidly and were capable of quantitatively labelling proteins and, unlike the corresponding maleimide reagents, did so largely irreversibly.[27] The other major class of cysteine-modifying reagents are the aminocyanobenzothiazoles **12**.[28,29] These reagents are most commonly used to generate luciferin, the substrate for luciferases but have also been used in protein conjugation reactions—either with the 1,2-aminothiol of N-terminal cysteines[30] or with genetically-encoded 1,2-aminothiols.[31]

The final class of direct addition reaction is the use of palladium-mediated arylation as reported by Buchwald and co-workers.[32] Conditions for catalytic palladium chemistry are generally not compatible with proteins, however, it is possible to use preformed Pd–aryl complexes **13** as direct arylating reagents, which can be used to selectively modify cysteine with fluorophores, bio-orthogonal functional groups, drug molecules and biotin. Optimisation of these reagents to generate fully water-compatible reagents **14** means that this approach can now be used on proteins which cannot withstand the organic co-solvents required for the initial study.[33]

Cysteine can also be modified by a variety of reagents to generate allyl cysteine.[34,35] For example molecules such as **15** generate a disulphide intermediate which undergoes sigmatropic rearrangement to form a reactive *S*-thiothioether intermediate which is quenched by a suitable phosphine

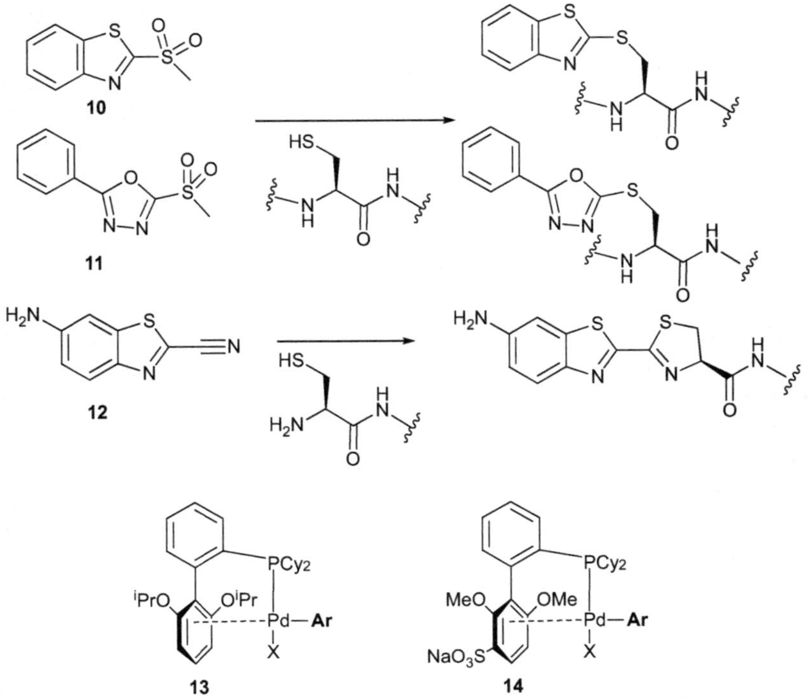

Figure 12.5 Examples of alternative reagents for direct modification of cysteine. Methylsulfonylbenzathiazole **10** and the related oxadiazolylsulfone **11** react rapidly and irreversibly with cysteine residues. Cyanobenzothiazole **12** selectively reacts with N-terminal cysteine residues. Structure of direct cysteine arylation reagents **13** and **14** reported by Pentelute, Buchwald and co-workers. (X = Cl, Br, I, OTf).

(Figure 12.6). These allyl ethers then present a suitable bio-orthogonal tag for reaction by metathesis chemistry. More recently the allyl cysteine has been used as a reaction partner for tetrazine in inverse electron demand Diels–Alder cross-linking.[36] Alternatively, the innate reactivity of the cysteine to alkenes can be harnessed in the thiol–ene reaction.[37] This chemistry, which is more commonly exploited in the polymer chemistry field, uses either UV light or a radical initiator to generate a thiol radical, which can attack alkenes to generate a stable thioether linkage. While not extensively employed for protein modification, it has been effectively used in other cases; for example for the generation of mimics of N-acetyl lysine,[38] for the generation of lipidated proteins,[39] for the insertion of paramagnetic tags,[40] for 'stapling' of cysteines in peptides and proteins[41] and for immobilization of farnesylated proteins to surfaces.[42]

12.1.2.2 *Modification of Cysteine* via *Dehydroalanine*

An alternative strategy for modification of a cysteine residue is chemical conversion to form a dehydroalanine residue. This can then act as a Michael

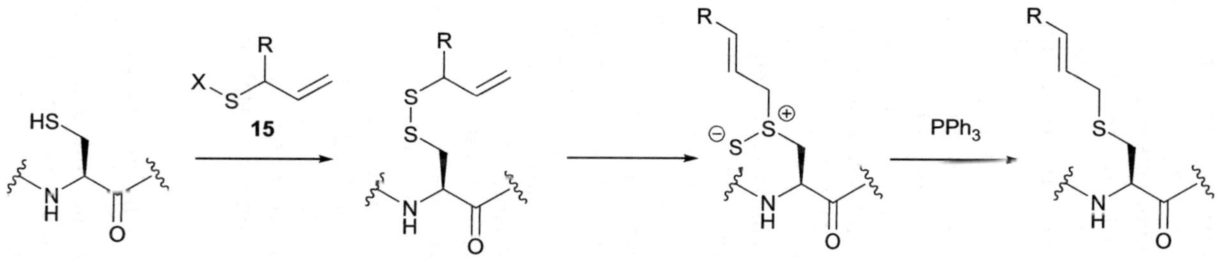

Figure 12.6 Formation of allyl cysteine species *via* use of activated vinyl donors **15**. (X = SO_3H,Br) Once formed the allyl cysteine can be used for metathesis chemistry.

acceptor with a range of thiol-containing nucleophiles to generate phosphorylated and glycosylated proteins, to generate non-natural amino acids in enzymes and to generate cyclised peptides and proteins. A wide range of chemical methods to generate dehydroalanine residues have been reported[43] including the use of aminooxy-sulfonates[44] and 1,4-dibromofunctionalised compounds.[43,45] The latter class of compounds, chiefly 2,5-dibromoadipamide **16**[43] but also methyl 2,5-dibromovalerate **17**,[45] have been used extensively for the generation of dehydroalanine residues in proteins and peptides (Figure 12.7). This class of electrophiles react twice with the cysteine residue to generate a tetrahydrothiazolium ion **18** which is an effective leaving group for E2 or E1cb elimination between the α and β positions of cysteine.

In this context, dehydroalanine residues have been used in tandem with a variety of thiol nucleophiles to generate mimics of post-translational modifications, including farnesylation, phosphorylation, glycosylation and methylation.[46] They have also been used to generate unnatural catalytic amino-acids, for example in aldolases, where they generate new or altered enzyme activity.[47,48] While this chemistry is effective at creating diversity, it does have some limitations. For example, for protein modification it is often necessary to unfold the protein and the subsequent Michael addition is not generally stereoselective. In the case of an unfolded protein, the protein containing the wrong enantiomer will frequently not refold and will therefore be lost at this stage. Alternatively (and if refolding is not required), the product will contain an unknown ratio of the two isomers. In some cases, incomplete formation of the tetrahydrothiazolium intermediate or elimination means that an adduct containing the adipamide backbone can also be observed.[49] Davis and co-workers have now extended this chemistry to allow the use of carbon nucleophiles using Zn-mediated radical formation to form C–C bonds to the dehydroalanine motif.[50] This chemistry requires careful exclusion of oxygen, which otherwise leads to chain cleavage and pyruvoyl-group formation at the modified cysteine residue. This chemistry has been shown to be compatible with a wide range of functional groups but still suffers from the key limitations of the dehydroalanine chemistry, including epimerisation at the peptide backbone and the need to refold the modified protein. Most recently, the chemistry has been extended to imidazole and 1,2,4-triazole nucleophiles, leading to installation of histidine mimics in proteins.[51]

12.1.3 Modification of Tyrosine, Tryptophan and Methionine

The remaining naturally occurring amino-acids are either rarely present in proteins (*e.g.* selenocysteine), are very abundant in proteins and therefore challenging to selectively label (*e.g.* arginine, where labelling with α-dicarbonyl systems in the presence of borate can lead to residue-specific labelling[52]), have limited selective chemistry (*e.g.* histidine, where histidine-specific labelling with epoxides is only possible in a ligand-dependent manner[53]) or require reaction conditions incompatible with most proteins

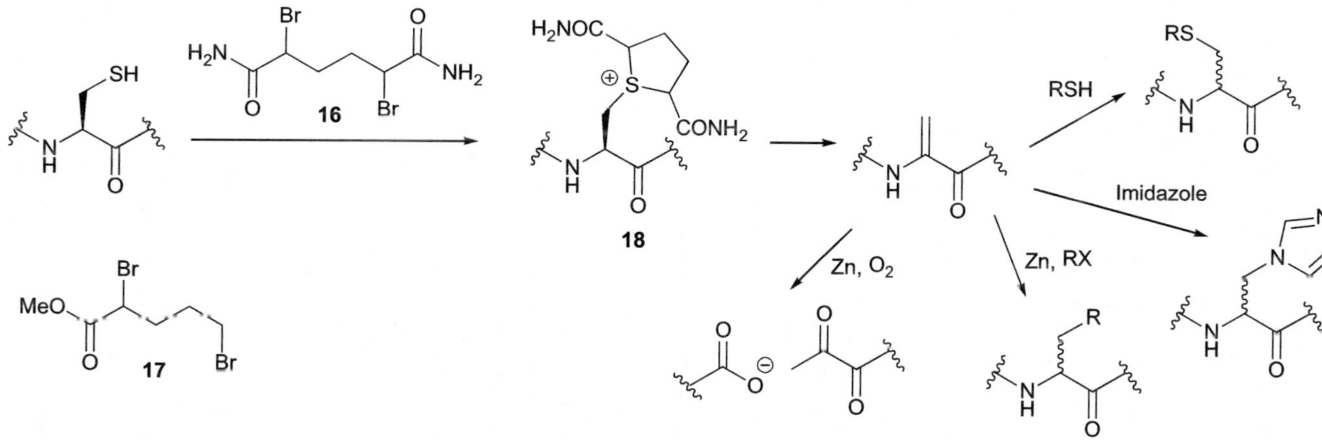

Figure 12.7 Example of reagents used for generation of dehydroalanine from cysteine residues. Dibromoadipamide **16** and methyl dibromovalerate **17** can both be used to generate dehydroalanine from cysteine residues. Addition of a suitable nucleophile then leads to thia- and aza-containing amino acids. In the presence of Zn, it is possible to use carbon nucleophiles to generate C–C bonds, but contamination with O_2 leads to chain cleavage, generating a pyruvoyl group.

(*e.g.* tryptophan[54,55]). For tyrosine, it is possible to use the phenolic hydroxyl to direct both diazonium crosslinking (*e.g.* Figure 12.8A)[56,57] and Mannich-type condensation[58] reactions as well as ceric ammonium nitrate-mediated crosslinking to dimethylanilines,[59] but these reactions have not been extensively exploited and cross-linking to tryptophan can also be observed. For tryptophan, Kanai and co-workers have shown that it is possible to insert strained ketone-containing N-oxides **19** into tryptophan residues using model protein systems which can be subsequently derivatised *via* oxime formation. The conditions required are still however relatively harsh (*e.g.* dilute acetic acid) for the majority of proteins.[60] Using milder conditions, diazonium compounds have been exploited for cross-linking to genetically-encoded 5-hydroxytryptophan[61] and to install tetrazine groups for subsequent bioorthogonal cross-linking.[62] This latter method is, however, dependent upon replacement of the tryptophan residue with the non-natural amino-acid. For methionine, which has a relatively low abundance, a new class of selective reagent, oxaziridines **20** have recently been reported by Chang and co-workers that selectively oxidise methionine residues to generate sulfinilimine linkages even in the presence of cysteine residues[63] (Figure 12.8). This class of reagents can be conveniently generated from aryl hydrazones with *m*CPBA and promise to provide an exciting new direction for controlled modification of native proteins.

12.2 Bio-orthogonal Chemistry

While the range of functional groups found in native proteins is fundamentally limited, chemical biologists have found numerous ways to incorporate additional bio-orthogonal functional groups into proteins for subsequent derivatization. These reactions exploit pairs of functional groups not usually found in nature and each of which do not react with those functional groups usually found in proteins. Their reaction is therefore orthogonal to the biological system. Several major classes of such bio-orthogonal reactions are now in common use, including oxime/hydrazone formation as described above, copper-catalysed 1,3-dipolar cycloaddition of azide and alkynes, inverse-electron-demand Diels–Alder reaction of tetrazines and, to a lesser extent, photoclick reaction of tetrazoles and alkenes.[64–66]

12.2.1 Staudinger Ligation and CuAAC

The first bio-orthogonal reactions to be applied were the Staudinger ligation[67] and copper-catalysed azide–alkyne cycloaddition (CuAAC).[68,69] In both cases, the chemistry could be applied using azides incorporated using labelling of native proteins, or by metabolic or Amber suppression (Figure 12.9) but each approach had limitations. The Staudinger ligation of ortho-azidobenzoates has a slow reaction rate ($<0.1\ M^{-1}\,s^{-1}$) and has therefore been superseded by more recently developed chemistries. CuAAC on the other hand is still one of the fastest bio-orthogonal

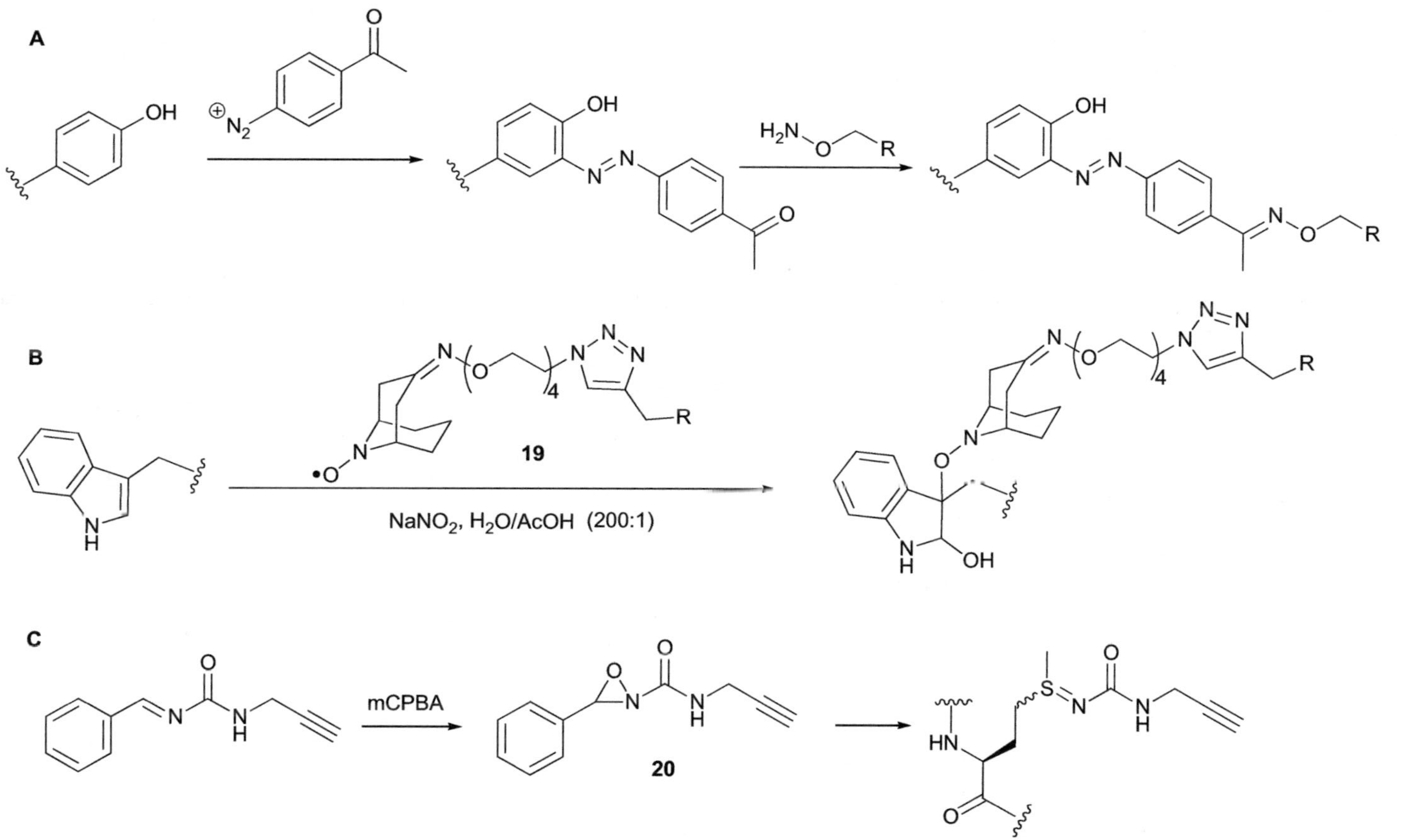

Figure 12.8 Selective chemistry for modification of tyrosine, tryptophan and methionine. A. In tyrosine, the phenol ring can be used for selective diazo coupling to a variety of reagents. This can be used to install either aldehyde or tetrazine functionality for subsequent bio-orthogonal functionalization. B. Radical-crosslinking to the 3-position of tryptophan using $NaNO_2$ and strained N-oxides *e.g.* **19**. C. Methionine can be selectively functionalized with acyl oxaziridines such as the propargyl example **20**.

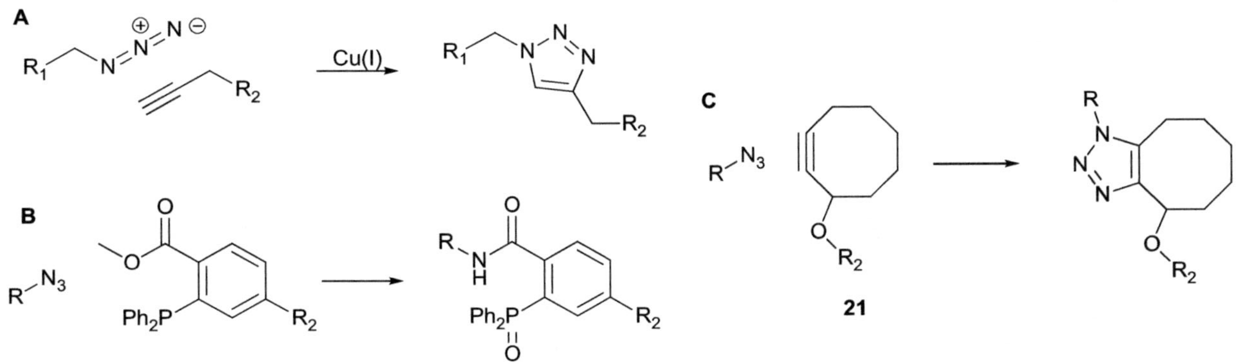

Figure 12.9 The earliest examples of bio-orthogonal reactions A. the copper-catalysed azide–alkyne cycloaddition (CuAAC), B. The Staudinger ligation and C. the first example of a strain-promoted azide–alkyne cycloaddition.

reactions(10^3–10^4 M^{-1} s^{-1}), but the requirement for Cu(I) species to form the copper acetylide required for reaction means that it is not compatible with labelling in the cell due to its toxicity. However, the use of suitable ligands such as tris-(benzyltriazolylmethyl)amine[70] enables effective protein labelling *in vitro* and on cell surfaces.[71]

12.2.2 SPAAC and SPANC

The use of strain-promoted azide–alkyne cycloaddition (SPAAC) was originally reported as a method to label sugars as discussed below, but has now been extensively adapted for use in protein modification *in vitro* and *in vivo*. In the original format, a 2-hydroxycyclooctyne derivative **21** (formed *via* oxidative ring opening of a dibromobicyclooctane) was used as a coupling partner for an azide.[72] A wide range of alternative and improved strained alkynes (Figure 12.10) with increased reaction rates to azides have now been reported, including fluorinated versions **22**[73] in which the electronegative fluorine inductively destabilizes the alkyne, more highly-substituted systems with increased ring-strain such as the dibenzocyclooctynes biarylazacyclooctyne (BARAC) (**24**)[74] and dibenzoazocyclooctyne (DIBAC) (**23**)[75] and the highly strained bicyclononynes **25**.[76] The reaction rates for these reagents (~1–100 M^{-1} s^{-1}) are compatible with convenient labelling of proteins in the laboratory, in cell imaging and in proteomics applications. Many of these reagents are now available commercially, pre-conjugated to amino- and thiol-reactive groups, or conjugated to fluorophores, biotin or other commonly used modifying groups. The choice of which 1,3-dipolarophile is used is largely dependent upon the application; while the dibenzocyclooctynes give the fastest reaction rates, their lipophilicity means that they are not always convenient and a hydrophilic dipolarophile such as bicyclononyne (BCN) or azacyclooctyne[77] might be preferred.

12.2.3 Inverse Electron-demand Diels–Alder Reactions (IEDDA) and Photoclick

The second major class of contemporary bio-orthogonal chemistry is that of the inverse-electron demand Diels–Alder reaction of 1,2,4,5-tetrazines which

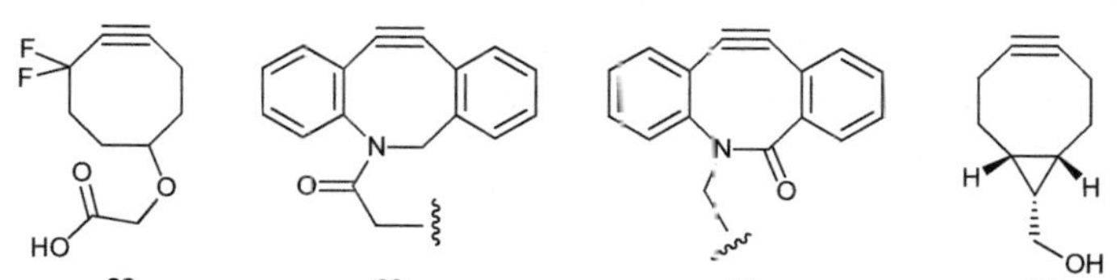

22 **23** **24** **25**

Figure 12.10 Range of available strained alkynes for protein labelling purposes include difluorocyclooctyne (DIFO, **22**), Dibenzoazocyclooctyne (DIBAC, **23**) Biarylazacyclooctyne (BARAC, **24**) and bicyclononyne (BCN, **25**).

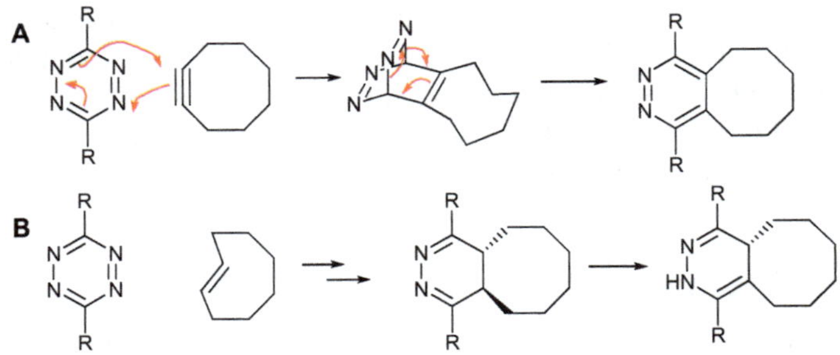

Figure 12.11 Inverse electron-demand Diels–Alder reaction of tetrazines with strained cycloalkynes A and cycloalkenes B. Both reactions are rapid and irreversible due to the loss of dinitrogen *via* a retro-Diels–Alder reaction after the initial cycloaddition.

provides an alternative, and faster, conjugation method.[78–80] Tetrazines react very rapidly (rates >2000 M^{-1} s^{-1}) with a wide range of suitable strained dienophiles *e.g.* cyclopropenes,[81] *trans*-cyclooctenes,[79,82,83] bicyclononynes[76] and even norbonene.[84] In all cases, the initial cycloaddition reaction is followed immediately by a cycloreversion to generate dinitrogen and either a dihydropyridazine (alkene dienophile) or pyridazine (alkyne) (Figure 12.11). This reaction has a number of additional advantages over the azide–alkyne cycloaddition in addition to speed. For example, the tetrazine can quench some red fluorophores and cycloaddition therefore leads to a fluorescent product which can be used for imaging purposes.[85] The formation of a dihydropyridazine with an alkene dienophile can be used to convert the cycloaddition into an 'uncaging' reaction *via* elimination of a conjugated molecule driven by rearomatization.[86] The corresponding 1,2,4-triazines (which can be synthetically more accessible) also react with the same set of reaction partners but with greatly reduced reaction rates.[87,88]

12.3 Strategies to Incorporate Amino-acids Containing Reactive Groups

While bio-orthogonal chemistry provides a wide-range of potential conjugation chemistries, it requires that the initial bio-orthogonal functional group be installed in the protein. There are several approaches to this, including direct incorporation into the protein by metabolite or Amber codon suppression, chemoenzymatic incorporation of the modification and incorporation into post-translational modifications such as glycosylation and lipidation.

12.3.1 Metabolite Suppression

The simplest method by which to incorporate bio-orthogonal probes into proteins is to use metabolite suppression approaches.[89–91] While cells have

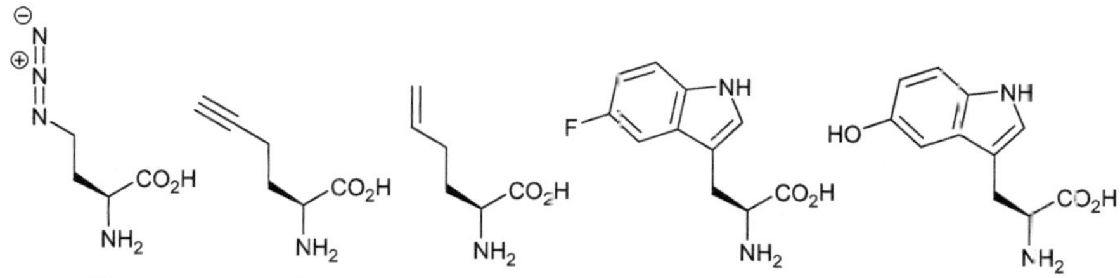

26 27 28 29 30

Figure 12.12 Examples of amino-acids which can be incorporated into proteins *via* metabolite suppression using WT *Escherichia coli* or auxotrophic strains include azide-containing amino-acids, such as azidohomoalanine **26**, but-3-ynylglycine **27** and but-3-enylglycine **28**, in place of methionine, and fluoro- and hydroxytryptophan derivatives **29** and **30**, in place of tryptophan.

naturally evolved to selectively interpret the genetic code to install only the naturally occurring amino-acids, this selection has only occurred against the pool of amino acids in the cell and it is therefore possible to incorporate structurally similar surrogates of amino acids into proteins. The interpretation of the genetic code in a particular cell is dependent upon the selectivity of the cell's pool of amino-acyl tRNA synthetases (AAtRS), which selectively activate the correct amino-acids and edit mis-acylated tRNAs. In many cases, they are able to activate and load tRNAs with similar amino-acids. Such methods for amino acid incorporation have been in development since the 1950s. In a small number of cases, such as selenomethionine, the amino-acid can be incorporated in place of the naturally-occurring amino-acid with no major effects on cell growth (though growth is reduced for SeMet).

For the majority of surrogates, widespread incorporation of an amino acid throughout the proteome is toxic. In this case, however, cells can be grown to a threshold level in minimal media supplemented with the natural amino acid before protein production is induced in the presence of the surrogate amino-acid. This allows incorporation of 'toxic' amino-acids. In cases where the surrogate is a poor substrate for the aminoacyl-tRNA synthetase, over-expression of the corresponding wild-type (or mutant) AAtRS can further improve incorporation at a sense codon. This approach has been used to globally incorporate a vast range of surrogate amino-acids. For example the azide-containing amino-acids azidohomoalanine **26**[71] and azidonorleucine.[92,93] These amino-acids can then be used in azide–alkyne cycloaddition reactions (Figure 12.12).

12.3.2 Amber Suppression

Only a limited pool of bio-orthogonally reactive amino-acids can be incorporated by direct metabolite suppression in this manner. Incorporation of more complex structures, such as tetrazines and strained cycloaddition substrates requires the use of engineered AAtRS. This approach is now

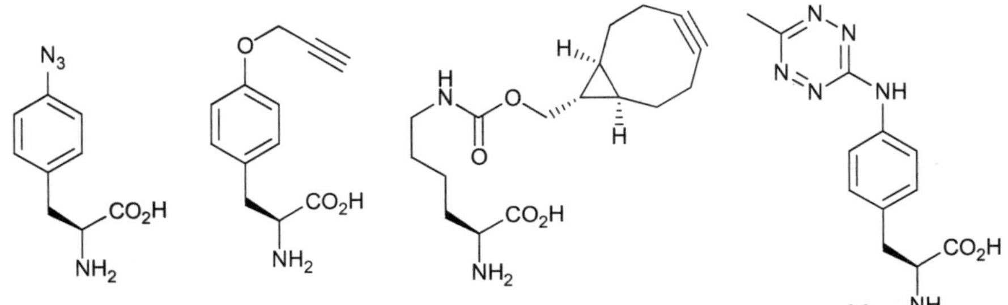

31 **32** **33** **34**

Figure 12.13 Examples of amino-acids containing bio-orthogonal reactive groups incorporated *via* amber suppression include molecules designed for CuAAC, SPAAC and IEDDA such as *p*-azidophenylalanine **31**,[102] propargyltyrosine **32**,[98] bicyclononylmethylcarboxylysine **33**[103] and 4-(4-methyltetrazinyl)aminophenylalanine **34**.[104]

becoming widespread, and by the use of an orthogonal pair of tRNA and AAtRS, in which the tRNA is not recognized by the cognate AAtRSs of the cell, it is possible to express proteins containing single unnatural amino-acids at a defined position using either a quadruplet codon or an Amber codon to encode the amino-acid. A range of such pairs have been developed, though the most common are those based on the pyrrolysyl-tRNA synthetase[94–97] from *Methanosarcina barkeri* and to a lesser extent the tyrosyl-tRNA synthetase from *Methanococcus janaschii*.[98] Some of the amino-acids that can be incorporated in this fashion are illustrated in Figure 12.13.[99–101]

12.4 Enzymatic Protein Modification

In addition to residue-specific strategies for protein modification there are a number of sequence- and structure-dependent strategies. These are principally dependent upon the use of particular enzymes which recognize the peptide sequence either *in vivo* or *in vitro* to generate a modified protein. This product can then sometimes be further elaborated with the use of bio-orthogonal approaches. These approaches have been adapted by exploitation and engineering of a wide variety of naturally occurring enzymes, including biotin ligases, lipoyl transferases and transpeptidases.

12.4.1 Modification Using Post-Translational Modification Enzymes

A wide range of naturally-occurring post-translational modifications are made to proteins in the cell. These include the addition of phosphate, glycosides and lipids (discussed in Section 12.4.2) as well as the addition of essential catalytic prosthetic groups, such as biotin. In most cases, the enzymes required for installation of the modification are either dependent upon recognition motifs presented as a result of the protein's tertiary structure (rather than being encoded by specific primary sequences) or are dependent upon the presence or absence of other defined post-translational modifications. In a few cases, however, it has been possible to identify defined peptide sequences. The simplest example of this is biotin ligase. This protein naturally adds biotin to the ε-amino group of a lysine residue in biotin carboxyl carrier proteins[105] and will add biotin, or analogues of it, to this sequence in expressed proteins. First used for *in vivo* labelling,[106] it was adapted by Ting and co-workers for *in vitro* and cell-surface labelling.[107,108] They optimized a minimum 15 amino-acid sequence to generate a sequence with a much higher propensity to be biotinylated. Subsequently they have explored the use of biotin ligases from other organisms and have, for example, demonstrated that it is possibly to selectively label proteins using either the yeast or *E. coli* biotin ligase in an orthogonal fashion (Figure 12.14).[109,110] In general expressed proteins are only partially biotinylated *in vivo* and subsequent ATP-dependent labelling using the purified enzyme *in vitro* is required for quantitative labelling.[111] The ability to use the enzyme *in vitro* and *in vivo* also opens the potential for *in vitro* (and *in vivo*) labelling with biotin analogues, *e.g.* the cyclopentanone derivative **35**. After labelling with this molecule, the ketone can be further derivatised with a range of amino-oxy compounds to generate site-specifically labelled proteins.

12.4.2 Post-translational Modification

Beyond the incorporation of unnatural amino acids, functionality can be introduced into proteins by exploiting the cellular post-translational modification (PTM) machinery. Enzyme-mediated PTMs enable the rapid tuning of protein function, and with over 200 chemical classes of PTM known the ensuing potential chemical diversity is very high. Approaches to label proteins *via* their PTMs can be categorized into two classes. Either the intrinsic chemistry of a PTM provides a handle for specific modification, or unnatural PTM analogues can be incorporated metabolically *in vitro* or in live cells. The latter approach has been particularly widely applied in recent years to label proteins that are modified by specific lipids or sugars, often with the aim of profiling and identifying these proteins.[112,113] In metabolic tagging approaches, the cell is fed a structurally similar PTM precursor or metabolite analogue, which is tolerated by cellular uptake pathways and the biosynthetic machinery, ideally becoming incorporated into proteins that are natively modified (Figure 12.15). In most cases the precursor is bioorthogonally tagged, typically using an azide or alkyne since these are small enough to be tolerated by the enzymes involved in processing the metabolite upon uptake and incorporating it into proteins. Here we focus on PTMs that are introduced by enzymatic processes in the cell and have been widely exploited to label proteins (namely, fatty acid acylation and glycosylation) and also discuss examples where the intrinsic functionality of the PTM has been exploited to introduce labels.

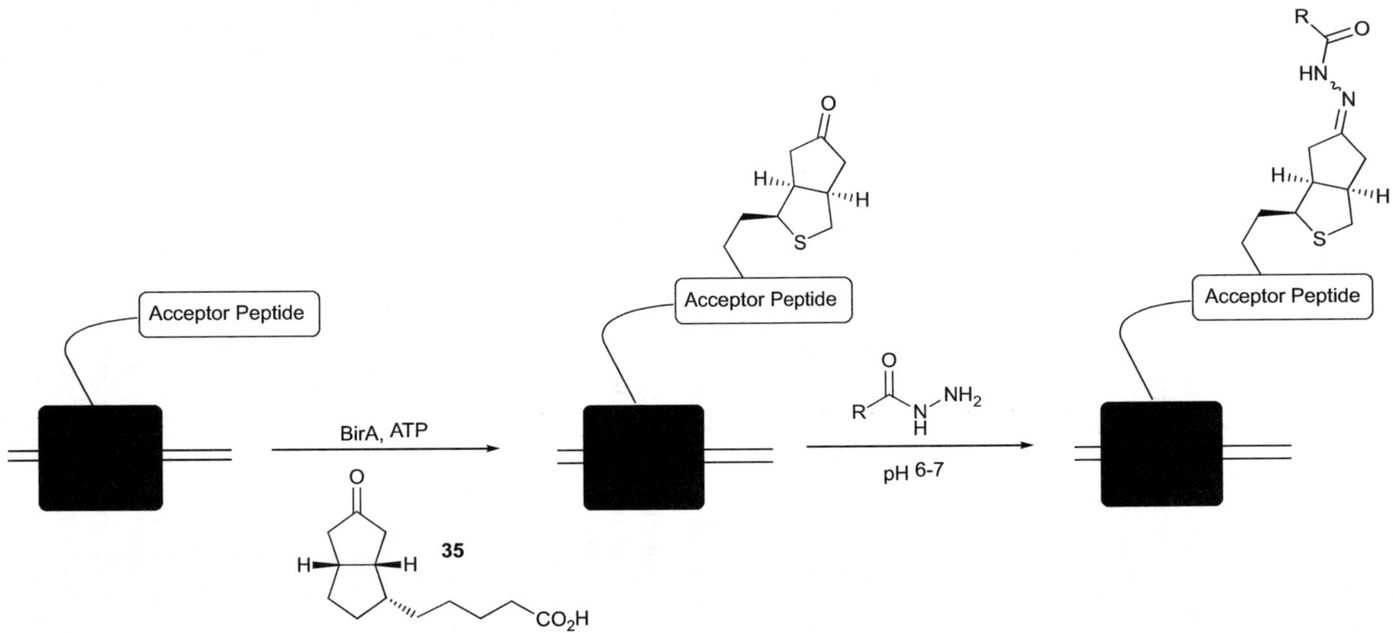

Figure 12.14 Example of using biotin ligase (BirA) for the bio-orthogonal labelling of cell-surface proteins.

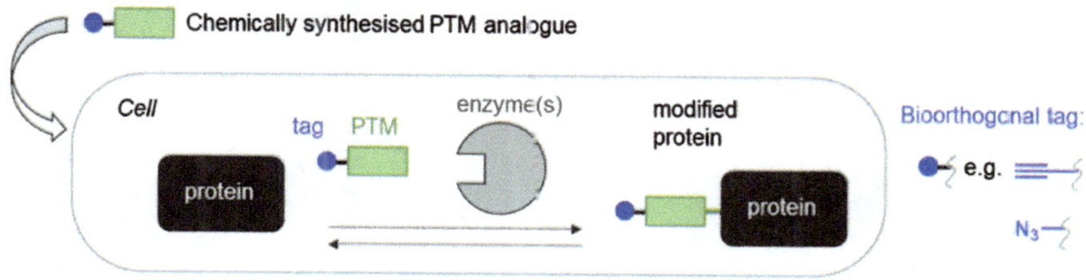

Figure 12.15 Metabolic tagging of proteins *via* post-translational modification (PTM) analogues.

12.4.2.1 *Lipidation*

The development of bioorthogonally tagged PTMs for the study of lipidation was motivated by a lack of robust biochemical approaches for determining whether a protein is lipidated in the cell. Radiolabelling with ^{3}H- or ^{125}I-analogues is insensitive and provides no means of identifying the lipidated protein unless this is expressed with a genetic tag, whilst antibodies against lipidated proteins are unavailable or exhibit poor specificity. Common protein lipidations include N-terminal *N*-myristoylation (modification with C14 : 0 fatty acid), cysteine *S*-acylation (often referred to as palmitoylation, since C16 : 0, palmitate, is the most common fatty acid incorporated), lysine *N*-acylation (which may be non-enzymatic), *O*-acylation of Wnt proteins, *S*-farnesylation and *S*-geranylgeranylation (Figure 12.16).

Bio-orthogonally tagged fatty acid analogues have been widely applied to identify and image acylated proteins,[114] particularly in protozoan parasites[115–118] (reviewed in ref. 119) and mammalian cells.[120–124] Bacterial lipoproteins have also been labelled *via* these methods in a handful of different organisms.[125,126] In eukaryotes, chain length determines which proteins are labelled in a manner consistent with direct attachment of the analogue to proteins in most cases. Alkynyl-fatty acids or azido-fatty acids may in some conditions enter fatty acid elongation or degradation pathways in cells, but this possibility is yet to be explored in any depth. Thus C14 analogues tetradec-13-ynoic acid (YnMyr) and 12-azidododecanoic acid (AzMyr) label *N*-myristoylated proteins, whilst longer chain octadec-17-ynoic acid (17-ODYA) or heptadec-16-ynoic acid (YnPal) are commonly used to profile *S*-acylation. Whilst both azides and alkynes are used for lipidation-mediated labelling, a preference for alkyne tags has emerged, mostly due to the observation that the orientation of the alkyne on the protein and an azido-fluorophore/biotin labelling partner results in lower background labelling.[121,127,128] There are also a few examples where azides are less well tolerated by the biosynthetic machinery than alkynes.[117] The requirement for small tags in such analogues is clearly illustrated in the case of *N*-myristoylation. The enzyme *N*-myristoyltransferase (NMT) modifies select N-terminal glycine residues using the coenzyme A thioesters of C14 : 0 and only close structural analogues of this fatty acid are tolerated.[129] Crystal structures reveal that the acyl chain of myristoyl-CoA is bound in a long narrow pocket of NMT,

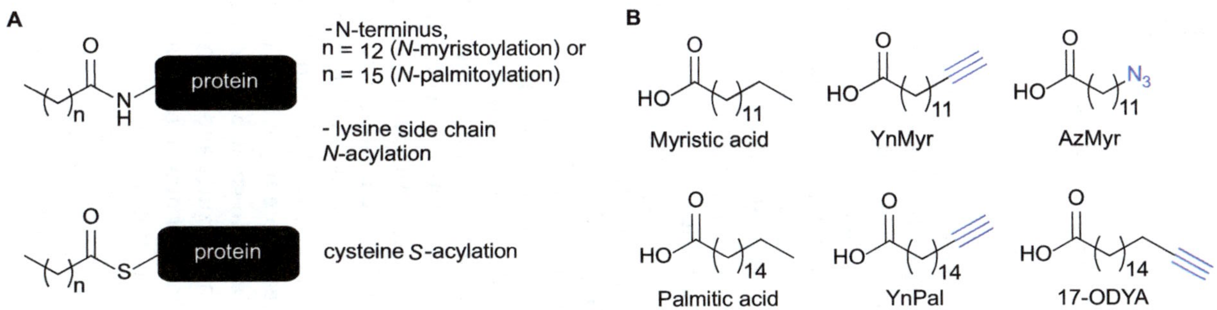

Figure 12.16 A. *N*- and *S*-acylation of proteins. B. Native fatty acids and commonly used probes for *N*-myristoylation (top) and *S*-acylation (bottom).

which can accommodate the alkyne tag of YnMyr-CoA but cannot tolerate modifications that are much larger.[115]

Due to the selectivity of *N*-myristoylation, its irreversibility and the absence of NMT activity in bacteria, myristate analogue tagging can be applied to site-specifically label a protein of interest in *E. coli*.[130] The recombinant protein, including a short N-terminal peptide sequence corresponding to a known NMT substrate, is co-expressed with NMT and bacteria are grown in the presence of tagged-myristate. Fatty acyl-CoA synthetases convert these analogues into coenzyme A thioesters, which can be used as substrates by NMT. This approach has recently been adapted for imaging of select proteins in bacterial cells, apparently without altering protein localisation despite the attachment of the lipid.[131]

The biology of other types of fatty acid acylation is less well understood and consequently less frequently applied in site-specific protein labelling. *S*-Acylation is mediated by multiple (23 in humans) acyltransferases, plus a smaller number of deacylases that render this PTM reversible.[132] *S*-Acyl sites also lack clear consensus sequences and fatty acid chain length is less clearly defined.[133] Metabolic labelling studies have revealed the broad scope of this PTM,[118,134,135] but which proteins are substrates for which enzymes *in vivo* is not known and further developments, such as acyltransferase subtype-specific inhibitors, will be required to unpick and exploit this biology.[132] Similarly, although specific examples of ε-*N*-lysine fatty acid acylation have long been known, this PTM is only now receiving renewed attention following the discovery that a sirtuin-family deacylase (previously associated with deacetylation activity) is able to remove long-chain acyl chains from lysine resides.[136] The Lin group recently reported functional deacylation of an oncoprotein,[137] indicating that many roles for deacylation remain to be discovered. However, the extent to which lysine acylation is enzyme-mediated and site-specific is not yet known.

A particularly widespread PTM is *N*-acetylation, known to control gene expression *via* modification of histones and other DNA regulatory proteins and mediated by lysine acetyltransferases (KATs) (Figure 12.17).[138] Bio-orthogonal probes of acetylation include 4-pentynoate **36** and its coenzyme A analogue, applied by the Hang group in cells and lysates respectively to acetylate and profile KAT substrates.[139] Longer chain analogues **37** and **38** have also been used.[139,140] One limitation of these tools is that several KAT family members are less tolerant of the longer chain length of tagged derivatives. To overcome this, the Zheng group engineered mutant KATs with enlarged acyl-CoA binding sites.[140,141] 3-azido-propionyl-CoA **39** was successfully applied to capture KAT substrates in cell lysates in combination with the exogenously added KATs: wild-type p300 and engineered GCN5.[142] Interestingly, 4-pentenoate **40** containing a terminal alkene as a bio-orthogonal tag has also been shown to be incorporated into proteins in cells, with subsequent detection *via* a Heck reaction.[143]

Prenylation is the incorporation of farnesyl (C15) or geranylgeranyl (C20) isoprenoid groups onto cysteine-containing sequences. Farnesyltransferase

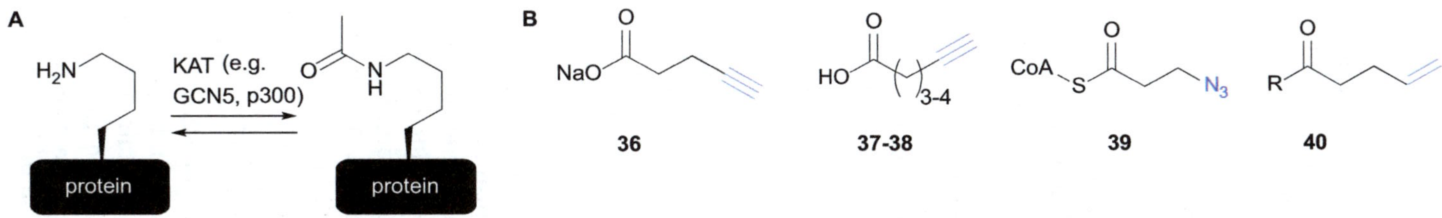

Figure 12.17 A. Acetylation of proteins catalysed by KAT enzymes. B. 4-pentynoate **36** for cell labelling; longer chain analogues **37–38**; Coenzyme A thioester of azido-probe **39**; alkenyl-probe **40**.

(FTase) and geranylgeranyl transferase type 1 (GGTase-1) modify C-terminal CAAX motifs, and have been applied to mediate attachment of bio-orthogonally tagged prenyl analogues to proteins in lysates and live cells (Figure 12.18).[144–149] Chemoenzymatic prenylation has also been utilised *in vitro* to site-specifically modify a non-antibody scaffold with geranyl ketone pyrophosphate **41**; the ketone subsequently enabled attachment of a drug *via* oxime ligation.[150]

Finally, some proteins contain rather uncommon PTMs and these can be exploited for selective cellular labelling. An example is cholesterylation, thus far described uniquely for hedgehog family proteins, although this PTM may be more widespread. Building on work with an azido-cholesterol analogue,[151] Tate and co-workers developed alkyne-tagged probes **42** and showed that these were incorporated into Sonic Hedgehog protein (Shh) in mammalian cells (Figure 12.19).[152] The unnatural modification did not affect Shh signalling activity and thus provided highly sensitive detection of Shh cholesterylation. This enabled in-depth characterisation of cholesterylated Shh in cells, including detection of a multimeric state, quantification of abundance in different cancer cell lines, single-cell imaging and visualisation in a whole organism.

12.4.2.2 Protein Glycosylation

The field of bio-orthogonal ligation has arguably been driven in large part by the challenges of characterising the non-genome-encoded portion of the cell, including PTMs. Glycosylation is the most prevalent and complex PTM, being highly cell-type-specific and heterogeneous. In a landmark paper in 2000, Saxon and Bertozzi reported the application of the modified Staudinger ligation for detection of cell-surface glycans displaying azide groups from metabolic incorporation of *N*-azidoacetylmannosamine **43**.[153] The Bertozzi group and others have subsequently reported numerous distinct sugar analogues and their wide application to understanding glycobiology across bacteria, plants and mammals (reviewed in ref. 113 and 154). This approach is termed metabolic oligosaccharide engineering (MOE) (Figure 12.20). Thus, there is an extensive toolbox of analogues to address glycosylation, including sugars tagged with ketones, azides, alkynes, cyclopropenes, terminal alkenes, isonitriles, diazo groups, nitrones and even norbornene.[154] However, the most common bio-orthogonal ligations in this field are still CuAAC and SPAAC, due to their speed, the relative availability and synthetic accessibility of reagents, and their use of small tags (namely alkynes and azides). MOE has dramatically improved our ability to interrogate and label sugar-modified proteins in biological systems.

As with lipidation, the success of MOE depends on the presence of scavenging pathways, to take up analogues from the media, and on the cellular machinery tolerating modifications to metabolite structure. To enable uptake, tagged sugars are fed to cells in fully acetyl-protected form, relying on cellular esterases to generate unprotected analogues inside the

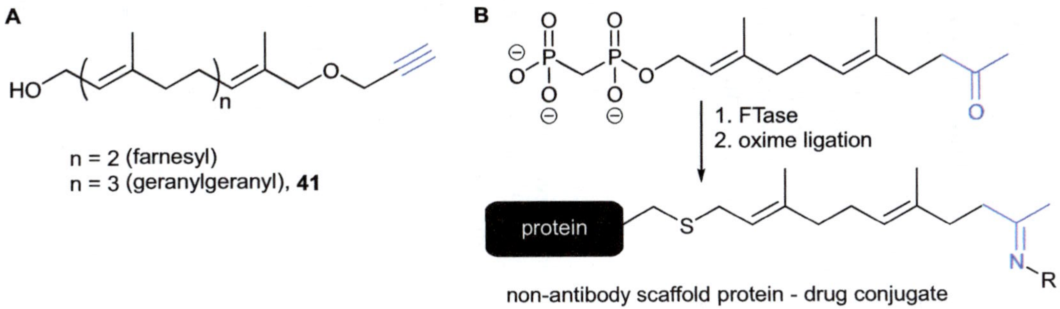

Figure 12.18 A. Bio-orthogonal probes for prenylation labelling in cells. B. Chemoenzymatic prenylation of proteins with an unnatural ketone farnesyl mimic, followed by oxime ligation to attach a drug.

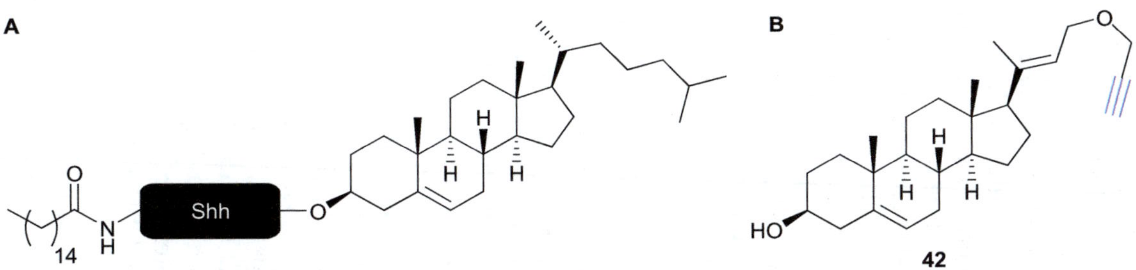

Figure 12.19 A. Sonic hedgehog protein is *N*-palmitoylated and *O*-cholersterylated. B. Alkynyl-cholesterol probe **42** reported by Ciepla *et al.*[152]

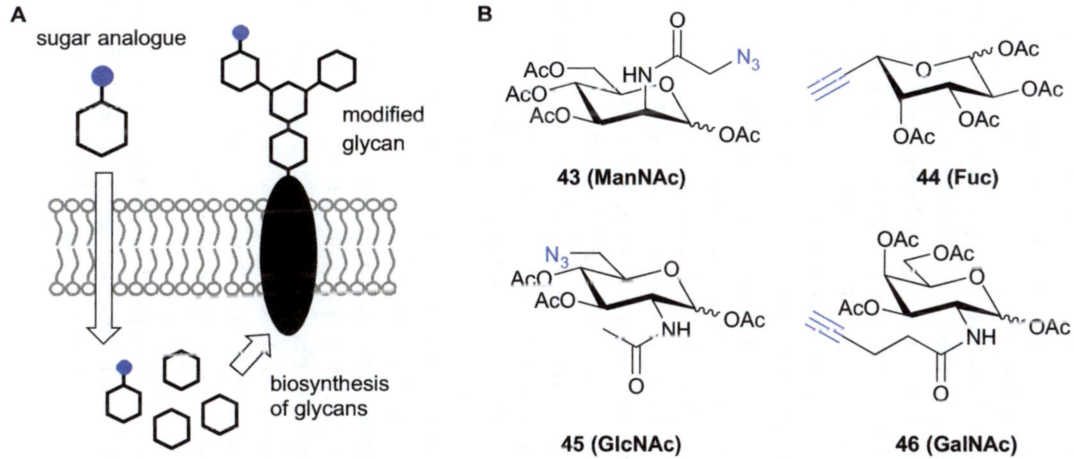

Figure 12.20 A. Metabolic oligosaccharide engineering (MOE). B. Examples of sugar analogues that have been applied in MOE include derivatives of N-acetyl mannosamine **43**, fucose **44**, N-acetyl glucosamine **45** and N-acetyl galactosamine **46**.

cell. However, *in situ* incorporation of analogues is further complicated by the presence of epimerases and other enzymes that can interconvert sugars in the cell and so an understanding of the biosynthetic machinery is important in designing experiments and interpreting data. The introduction of functionality onto native glycans *via* chemical or chemoenzymatic methods can overcome some of these limitations and thus provides a suite of orthogonal methods for tagging proteins *via* sugar PTMs (Figure 12.21). Due to the requirement for accessing the modification site directly, these approaches are typically applied on cell surface glycoconjugates or in lysates.

Chemically, the *cis*-diol feature of sugars can be exploited for selective labelling or enrichment. *Cis*-diols can be directly captured by boronic acids or oxidised by mild oxidants, such as sodium periodate, to generate aldehydes that can then be ligated to hydrazides or aminoxy groups (the hydrazone and oxime ligations, respectively).[113] Several groups have built on this approach to selectively oxidise the *cis*-diol of terminal sialic acid residues by optimising the conditions of periodate cleavage.[155,156] Another method introduces selectivity enzymatically, by using galactose oxidase to generate an aldehyde on terminal galactose and GalNAc residues.[156,157] This approach has been extended by directed evolution of the enzyme to generate a mutant that is able to address mannose and GlcNAc.[158]

Multiple chemoenzymatic approaches have been reported for glycan labelling (reviewed by Lopez Aguilar *et al.*[159]). These methods exploit specific glycosyltransferases to transfer unnatural sugar analogues onto native glycans. The specificity possible with enzymes enables selective labelling of higher order (*i.e.* more than one monosaccharide) glycan structures or of specific linkages. The first example of a chemoenzymatic method that transferred a bio-orthogonally tagged sugar was reported by Hsieh-Wilson and co-workers. They labelled *O*-GlcNAc modified proteins with a ketone-incorporating UDP-galactose donor sugar, which was subsequently detected by oxime ligation, transferred by a mutant β1–4 galactosyltransferase.[160] Wu and co-workers first applied chemoenzymatic methods to label surface glycans, using an engineered α(1,3)-fucosyltransferase to transfer 6-azidofucose to the disaccharide *N*-acetyllactosamine.[161] Subsequent work by many groups has since applied this method to other glycans[162–164] and for select disaccharides.[165–167]

In summary, the complex structures and biology of cellular glycans has led to the development of diverse methods for their selective labelling, ranging from direct incorporation of bio-orthogonally tagged monosaccharides to highly selective chemoenzymatic methods.

12.4.2.3 Specific Labelling of Other Endogenous PTMs

The unique chemical properties of a PTM can provide a handle for selective modification. Here we illustrate this with *S*-sulfenylation, a transient and unstable redox-related cysteine modification (Figure 12.22). As with the other PTM chemical technologies, the development of methods to label

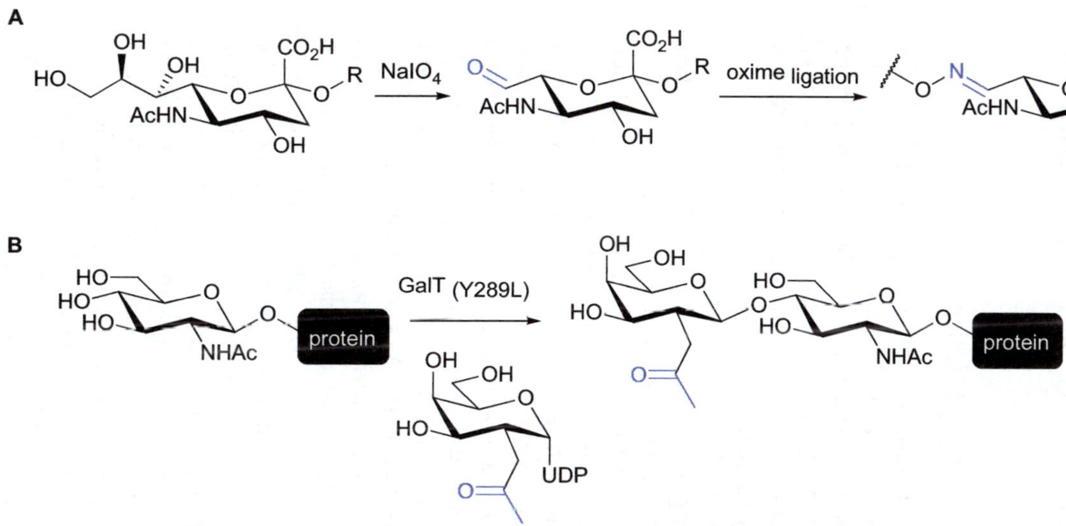

Figure 12.21 Examples of chemical and chemoenzymatic approaches to labelling glycosylated proteins. A. Selective periodate cleavage of terminal sialic acid residues to reveal an aldehyde, which can undergo an oxime ligation. B. Chemoenzymatic labelling of *O*-GlcNAc modified proteins with UDP-galactose–ketone sugar using a GalT mutant.

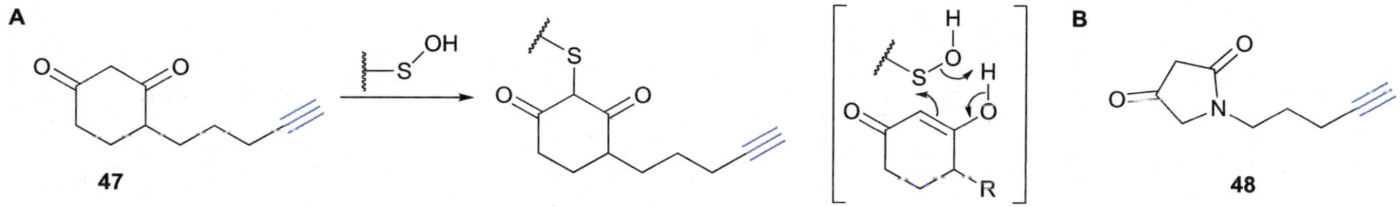

Figure 12.22 Site-specific labelling of *S*-sulfenylation sites with chemical probes. A. Dimedone probe **47**. B. pyrrolidinedione probe **48**.

S-sulfenic acids has been motivated by the need to profile this PTM in live cells and understand the functional consequences of modification at each site.[168] *S*-Sulfenic acid has interesting chemistry and can react with both electrophilic and nucleophilic reagents.[169] Probes based on the 1,3-diketone dimedone **47** have been widely applied to detect this modification, including an alkyne-tagged version developed by the Carroll group to profile *S*-sulfenylated proteins in intact cells.[170,171] Recent advances in this area include the development of diverse dimedone-inspired nucleophiles leading to the discovery of pyrrolidinedione **48**, which preferentially reacted with protein tyrosine phosphatases.[172]

12.4.3 Modification Using Transpeptidases

Catalytic transpeptidases can be effectively used for labelling both the N- and C-terminus of proteins. The strategy is to adapt one of the growing class of transpeptidase enzymes, such as sortase and butelase, or engineered proteases, such as subtiligase or trypsiligase. In all cases, a defined recognition sequence is required in the substrate protein or labelling reagent and the final product is a hydrolytically inert peptide bond.

The principal enzyme class exploited for this purpose are the sortases (described as a 'gift' to protein engineering in this regard[173]). This class of bacterial transpeptidases are responsible for anchoring of cell surface proteins to the cell walls of Gram-positive bacteria and are exemplified by Sortase A from *Staphylococcus aureus*[174,175] (saSrtA) (Figure 12.23). A C-terminal anchoring motif in the excreted protein (LPXT/G for saSrtA) is recognized by the enzyme, a cysteine residue then reacts with the peptide chain to form a thioacyl intermediate, and a pentaglycine chain attached to the peptidoglycan intercepts this intermediate to form a peptide linkage from the cell wall to the protein. This protein has now been extensively applied for modification of proteins both at the N- and C-terminus.[173,176,177] The enzyme has been particularly used for anchoring of proteins to surfaces.[178–181] For C-terminal labelling, the expressed protein must be modified with a suitable LPXTGX motif **49** (the additional residue is essential—LPXTG with a free carboxy-terminus is not a substrate for the enzyme). For *N*-terminal labelling, a suitable *N*-terminal glycine residue **50** is required. While in many cases a polyglycine sequence is used, this is not strictly required for labelling; a single, sterically-accessible, glycine residue is sufficient.[182,183] This can be formed either by action of a suitable protease to cleave the peptide chain, by use of excreted proteins where a glycine is left after removal of the signal peptide or by use of methionine aminopeptidase *in vivo*, which will remove the N-terminal methionine in bacterially-expressed proteins if it is followed by a small amino-acid residue.

The key challenge in using sortase is that the enzyme-catalysed reaction is intrinsically reversible. The products of the ligation reaction are also substrates for sortase and the reaction therefore continues to an equilibrium position. In the context of the cell, this is not a problem since the product

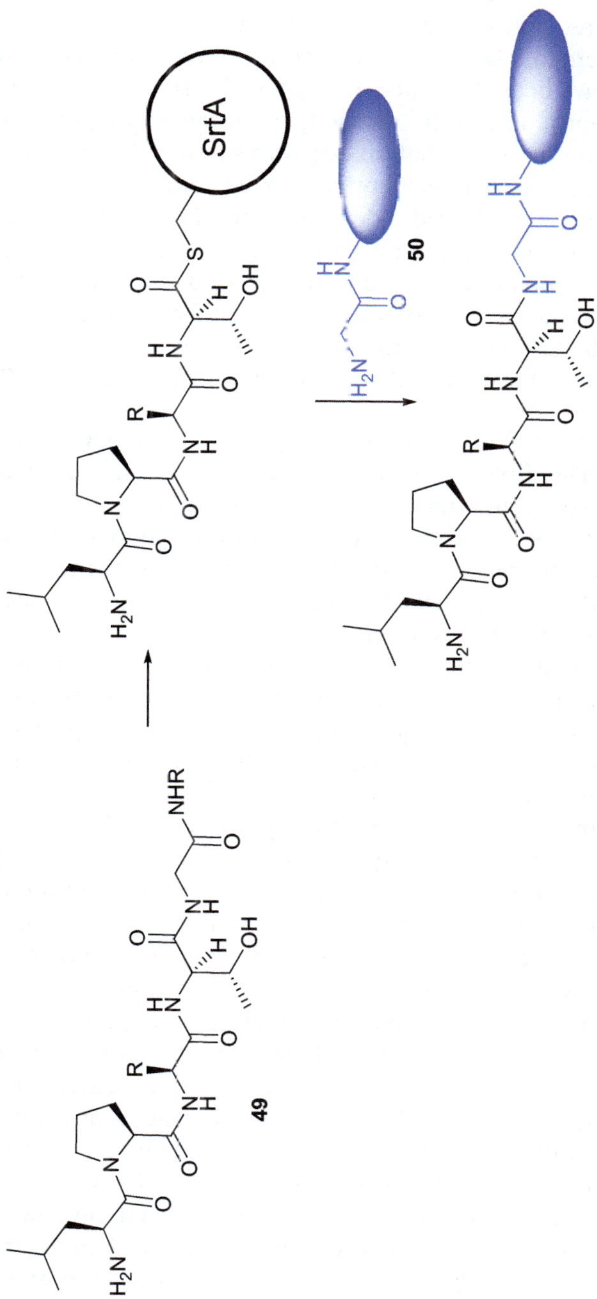

Figure 12.23 Transpeptidation reaction catalysed by Sortase A. Transpeptidation transfers a LPXT motif from one glycine residue (*e.g.* on a labelling substrate) to another (*e.g.* the N-terminus of a target protein).

freely diffuses away from the cell surface. However, in the context of a laboratory reaction, an excess of one coupling partner is required to drive the reaction to completion. In the case of C-terminal ligation an excess of the enzyme is also frequently required to avoid the effect of product inhibition as the enzyme remains bound to the product. Several strategies to mitigate this challenge have been described. One possible strategy is to mechanically remove the side-products of the reaction, either by carrying out the reaction in a spin concentrator or in a dialysis bag with a low-molecular-weight cut-off such that only the by-product peptide can be removed from the system. Alternatively, use of microfluidic devices is possible. In these systems the catalytic enzyme is immobilized on a surface,[184] saturated with the labelling substrate and the protein to be labelled is passed through, picking up the label. In all cases, the removal of affinity tags can be used to allow judicious removal of unlabelled protein from the reaction mixture. This approach has been used for the production of antibody–drug conjugates by C-terminal labelling.[185]

For N-terminal labelling, two other major strategies have been reported (Figure 12.24). The first is the use of metal-chelating amino-acids. For example, a released amino-terminal glycine-histidine motif can bind to Ni salts. This strategy can be used to drive C-terminal labelling but does require very high concentrations of Ni salts.[186] Quantitative N-terminal labelling of proteins is possible using depsipeptide substrates. The hydroxyacetamide by-product of the ligation reaction is not a substrate for the reverse reaction and so this makes the labelling reaction irreversible. It is therefore possible to label a chosen protein using only a small excess of the labelling reagent.[183,187] A related strategy is the use of isodepsipeptides in which cyclisation of the product (with loss of *N*-acetyl serine) generates a diketopiperazine driving the reaction to completion.[188]

The applications of sortase were initially limited by three other factors: Ca^{2+}-dependence, speed and sequence-specificity. The Ca^{2+}-dependence of sortase initially prevented application to *in cellulo* labelling as well as limiting the range of buffers that could be used *in vitro*. Simple point mutation to remove the Ca^{2+}-binding residues is sufficient to generate a Ca^{2+}-independent version which can act *in vitro* and *in vivo*.[189,190] The activity of the native enzyme is poor and a large quantity of enzyme (often super-stoichiometric) is required for labelling applications. Enzyme evolution has yielded variants with over 1000-fold improved activity. Finally, the sortase recognition motif (LPxTG) limits the range of proteins which can be labelled, and also initially limited the range of proteins that could be labelled orthogonally. This was initially mitigated by the use of sortases from other organisms with an altered substrate specificity, *e.g.* Sortase A from *Streptococcus pyogenes*, which recognizes an LPxTA motif in addition to LPxTG, but more recently by engineered sortases with altered specificity, including isoforms which recognize LAxTG, LPxSG,[191] FPxTG and APxTG.[192]

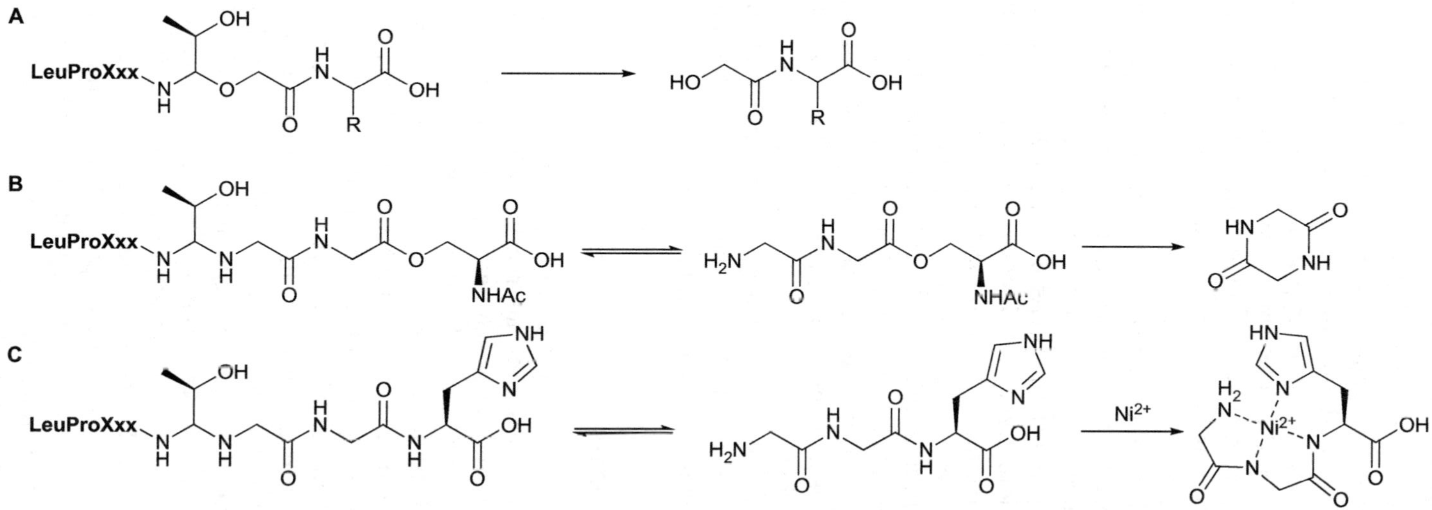

Figure 12.24 Strategies to enhance turnover by sortases for N-terminal labelling. In all cases, the by-product peptide is either unreactive hydroxyacetamide A, or can be sequestered either by further reaction to form the diketopiperazine B or as the Ni^{2+} complex C.

12.4.4 Other Ligating Enzymes

Several other ligating enzymes have been reported. Subtiligase[193–195] and trypsiligase[196,197] were both generated by engineering of the corresponding protease and can be used for effective peptide ligation. Trypsiligase, a quadruple-site-directed mutant of trypsin with enhanced affinity for a YRH motif, is able to cleave the substrate protein between the tyrosine and arginine before adding an activated peptide to the N-terminus of the RH motif after loss of the N-terminal fragment.[196] This surprising activity is dependent upon substrate-activated catalysis by the Arg–His terminated protein. The enzymes butelase and OaAEP1 have recently appeared as potential alternatives to sortase.[198–200] Butelase has the advantage that the majority of the peptide sequence (NHV) required for recognition by the enzyme is not retained in the product and therefore the product need contain only an Asn–Gly linker region. Using the butelase enzyme in combination with thiodepsipeptide substrates[201] led to quantitative labelling as observed for sortases. It has now been used in numerous applications similar to those in which sortase has been applied.[202,203] Unfortunately, the enzyme must be purified from source plant material and efforts to generate a heterologous expression system (and further evolve the enzyme) have not been successful. In contrast, the asparaginyl endopeptidase from *Oldenlandia affinis* (OaAEP1) can be cloned and overexpressed in *E. coli*[204] and a single Cys247Ala mutation[200] is sufficient to generate a fast protein ligase with a short NGL recognition motif. This enzyme therefore provides a potential alternative to sortase for single-labelling of proteins.

12.5 Conclusions

There are numerous other approaches to protein modification beyond the scope of this chapter, including approaches which use a small-molecule ligand to direct site-specific attachment of a label onto a protein. These approaches include activity-based protein profiling, which makes use of enzymatic mechanisms to label the active sites of proteins with reactive small molecules, often resulting in inhibition of activity,[205–208] and photoaffinity labelling, where probe attachment is mediated by a photoreactive group.[205,209] The Hamachi group have pioneered ligand-directed labelling approaches, which use a ligand to direct reactive functionality in close proximity to a nucleophilic residue on a protein surface. Both tosyl- and acyl-imidazole reactive functionalities can enable transfer of a label to the protein target.[210,211] Although not yet widely used, this approach is one of many alternative approaches that hold great promise for labelling proteins in live cells without perturbing the activity of the protein of interest. In this chapter, we have focused on how the intrinsic properties of amino-acids and PTMs can be exploited to label proteins site-specifically and on how functionality can be introduced *via* chemoenzymatic methods, unnatural amino acids and PTM precursor analogues. This account is inevitably a snapshot in time as the repertoire of available methods is being continually expanded as new chemistry is applied and discovered, as new classes of amino-acid and metabolite precursors are incorporated into proteins and as new enzymes are discovered and exploited. The contemporary challenge is now one both of complexity and successful exploitation—to find ways to apply multiple, orthogonal labelling methods in the same system as well as to find ways to exploit these diverse synthetic methods to answer contemporary biological questions.

References

1. E. Basle, N. Joubert and M. Pucheault, *Chem. Biol.*, 2010, **17**, 213–227.
2. J. N. deGruyter, L. R. Malins and P. S. Baran, *Biochemistry*, 2017, **56**, 3863–3873.
3. O. Boutureira and G. J. L. Bernardes, *Chem. Rev.*, 2015, **115**, 2174–2195.
4. W. H. So, Y. Zhang, W. Kang, C. T. T. Wong, H. Sun and J. Xia, *Curr. Opin. Biotechnol.*, 2017, **48**, 220–227.
5. S. Itzkovitz and U. Alon, *Genome Res.*, 2007, **17**, 405–412.
6. O. Koniev and A. Wagner, *Chem. Soc. Rev.*, 2015, **44**, 5495–5551.
7. X. Chen, K. Muthoosamy, A. Pfisterer, B. Neumann and T. Weil, *Bioconjugate Chem.*, 2012, **23**, 500–508.
8. S. T. Larda, D. Pichugin and R. S. Prosser, *Bioconjugate Chem.*, 2015, **26**, 2376–2383.
9. M. Chilamari, L. Purushottam and V. Rai, *Chem. – Eur. J.*, 2017, **23**, 3819–3823.
10. A. O.-Y. Chan, C.-M. Ho, H.-C. Chong, Y.-C. Leung, J.-S. Huang, M.-K. Wong and C.-M. Che, *J. Am. Chem. Soc.*, 2012, **134**, 2589–2598.
11. K. F. Geoghegan and J. G. Stroh, *Bioconjugate Chem.*, 1992, **3**, 138–146.
12. R. J. Spears and M. A. Fascione, *Org. Biomol. Chem.*, 2016, **14**, 7622–7638.
13. T. R. Branson, T. E. McAllister, J. Garcia-Hartjes, M. A. Fascione, J. F. Ross, S. L. Warriner, T. Wennekes, H. Zuilhof and W. B. Turnbull, *Angew. Chem., Int. Ed. Engl.*, 2014, **53**, 8323–8327.
14. J. M. Gilmore, R. A. Scheck, A. P. Esser-Kahn, N. S. Joshi and M. B. Francis, *Angew. Chem., Int. Ed. Engl.*, 2006, **45**, 5307–5311.
15. L. S. Witus, C. Netirojjanakul, K. S. Palla, E. M. Muehl, C. H. Weng, A. T. Iavarone and M. B. Francis, *J. Am. Chem. Soc.*, 2013, **135**, 17223–17229.
16. K. S. Palla, L. S. Witus, K. J. Mackenzie, C. Netirojjanakul and M. B. Francis, *J. Am. Chem. Soc.*, 2015, **137**, 1123–1129.
17. L. M. Tedaldi, M. E. Smith, R. I. Nathani and J. R. Baker, *Chem. Commun.*, 2009, 6583–6585.
18. M. E. B. Smith, F. F. Schumacher, C. P. Ryan, L. M. Tedaldi, D. Papaioannou, G. Waksman, S. Caddick and J. R. Baker, *J. Am. Chem. Soc.*, 2010, **132**, 1960–1965.

19. F. F. Schumacher, M. Nobles, C. P. Ryan, M. E. B. Smith, A. Tinker, S. Caddick and J. R. Baker, *Bioconjugate Chem.*, 2011, **22**, 132–136.
20. C. M. Grison, G. M. Burslem, J. A. Miles, L. K. A. Pilsl, D. J. Yeo, Z. Imani, S. L. Warriner, M. E. Webb and A. J. Wilson, *Chem. Sci.*, 2017, **8**, 5166–5171.
21. C. P. Ryan, M. E. B. Smith, F. F. Schumacher, D. Grohmann, D. Papaioannou, G. Waksman, F. Werner, J. R. Baker and S. Caddick, *Chem. Commun.*, 2011, **47**, 5452–5454.
22. V. Chudasama, M. E. B. Smith, F. F. Schumacher, D. Papaioannou, G. Waksman, J. R. Baker and S. Caddick, *Chem. Commun.*, 2011, **47**, 8781–8783.
23. M. T. W. Lee, A. Maruani, J. R. Baker, S. Caddick and V. Chudasama, *Chem. Sci.*, 2016, **7**, 799–802.
24. S. Shaunak, A. Godwin, J. W. Choi, S. Balan, E. Pedone, D. Vijayarangam, S. Heidelberger, I. Teo, M. Zloh and S. Brocchini, *Nat. Chem. Biol.*, 2006, **2**, 312–313.
25. Y. Zhang, X. Zhou, Y. Xie, M. M. Greenberg, Z. Xi and C. Zhou, *J. Am. Chem. Soc.*, 2017, **139**, 6146–6151.
26. D. Zhang, N. O. Devarie-Baez, Q. Li, J. R. Lancaster and M. Xian, *Org. Lett.*, 2012, **14**, 3396–3399.
27. N. Toda, S. Asano and C. F. Barbas, *Angew. Chem., Int. Ed.*, 2013, **52**, 12592–12596.
28. D. C. McCutcheon, W. B. Porterfield and J. A. Prescher, *Org. Biomol. Chem.*, 2015, **13**, 2117–2121.
29. J. R. Hauser, H. A. Beard, M. E. Bayana, K. E. Jolley, S. L. Warriner and R. S. Bon, *Beilstein J. Org. Chem.*, 2016, **12**, 2019–2025.
30. H. Ren, F. Xiao, K. Zhan, Y.-P. Kim, H. Xie, Z. Xia and J. Rao, *Angew. Chem., Int. Ed.*, 2009, **48**, 9658–9662.
31. D. P. Nguyen, T. Elliott, M. Holt, T. W. Muir and J. W. Chin, *J. Am. Chem. Soc.*, 2011, **133**, 11418–11421.
32. E. V. Vinogradova, C. Zhang, A. M. Spokoyny, B. L. Pentelute and S. L. Buchwald, *Nature*, 2015, **526**, 687.
33. A. J. Rojas, B. L. Pentelute and S. L. Buchwald, *Org. Lett.*, 2017, **19**, 4263–4266.
34. Y. A. Lin, J. M. Chalker, N. Floyd, G. J. Bernardes and B. G. Davis, *J. Am. Chem. Soc.*, 2008, **130**, 9642–9643.
35. J. M. Chalker, Y. A. Lin, O. Boutureira and B. G. Davis, *Chem. Commun.*, 2009, 3714–3716.
36. B. L. Oliveira, Z. Guo, O. Boutureira, A. Guerreiro, G. Jimenez-Oses and G. J. Bernardes, *Angew. Chem., Int. Ed. Engl.*, 2016, **55**, 14683–14687.
37. A. Dondoni, A. Massi, P. Nanni and A. Roda, *Chem. – Eur. J.*, 2009, **15**, 11444–11449.
38. F. Li, A. Allahverdi, R. Yang, G. B. J. Lua, X. Zhang, Y. Cao, N. Korolev, L. Nordenskiöld and C.-F. Liu, *Angew. Chem., Int. Ed.*, 2011, **50**, 9611–9614.

39. T. H. Wright, A. E. S. Brooks, A. J. Didsbury, G. M. Williams, P. W. R. Harris, P. R. Dunbar and M. A. Brimble, *Angew. Chem., Int. Ed.*, 2013, **52**, 10616–10619.
40. Q. F. Li, Y. Yang, A. Maleckis, G. Otting and X. C. Su, *Chem. Commun.*, 2012, **48**, 2704–2706.
41. Y. Wang, B. J. Bruno, S. Cornillie, J. M. Nogieira, D. Chen, T. E. Cheatham, C. S. Lim and D. H.-C. Chou, *Chem. – Eur. J.*, 2017, **23**, 7087–7092.
42. D. Weinrich, P.-C. Lin, P. Jonkheijm, U. T. T. Nguyen, H. Schröder, C. M. Niemeyer, K. Alexandrov, R. Goody and H. Waldmann, *Angew. Chem., Int. Ed.*, 2010, **49**, 1252–1257.
43. J. M. Chalker, S. B. Gunnoo, O. Boutureira, S. C. Gerstberger, M. Fernandez-Gonzalez, G. J. L. Bernardes, L. Griffin, H. Hailu, C. J. Schofield and B. G. Davis, *Chem. Sci.*, 2011, **2**, 1666–1676.
44. G. J. Bernardes, J. M. Chalker, J. C. Errey and B. G. Davis, *J. Am. Chem. Soc.*, 2008, **130**, 5052–5053.
45. P. M. Morrison, P. J. Foley, S. L. Warriner and M. E. Webb, *Chem. Commun.*, 2015, **51**, 13470–13473.
46. J. M. Chalker, L. Lercher, N. R. Rose, C. J. Schofield and B. G. Davis, *Angew. Chem., Int. Ed.*, 2012, **51**, 1835–1839.
47. N. Timms, C. L. Windle, A. Polyakova, J. R. Ault, C. H. Trinh, A. R. Pearson, A. Nelson and A. Berry, *ChemBioChem*, 2013, **14**, 474–481.
48. C. L. Windle, K. J. Simmons, J. R. Ault, C. H. Trinh, A. Nelson, A. R. Pearson and A. Berry, *Proc. Natl. Acad. Sci. U. S. A.*, 2017, **114**, 2610–2615.
49. R. Nathani, P. Moody, M. E. B. Smith, R. J. Fitzmaurice and S. Caddick, *ChemBioChem*, 2012, **13**, 1283–1285.
50. T. H. Wright, B. J. Bower, J. M. Chalker, G. J. Bernardes, R. Wiewiora, W. L. Ng, R. Raj, S. Faulkner, M. R. Vallee, A. Phanumartwiwath, O. D. Coleman, M. L. Thezenas, M. Khan, S. R. Galan, L. Lercher, M. W. Schombs, S. Gerstberger, M. E. Palm-Espling, A. J. Baldwin, B. M. Kessler, T. D. Claridge, S. Mohammed and B. G. Davis, *Science*, 2016, **354**, 597.
51. J. Dadová, K.-J. Wu, P. G. Isenegger, J. C. Errey, G. J. L. Bernardes, J. M. Chalker, L. Raich, C. Rovira and B. G. Davis, *ACS Cent. Sci.*, 2017, **3**, 1168–1173.
52. M. A. Gauthier and H.-A. Klok, *Biomacromolecules*, 2011, **12**, 482–493.
53. Y. Takaoka, H. Tsutsumi, N. Kasagi, E. Nakata and I. Hamachi, *J. Am. Chem. Soc.*, 2006, **128**, 3273–3280.
54. J. M. Antos and M. B. Francis, *J. Am. Chem. Soc.*, 2004, **126**, 10256–10257.
55. A. Foettinger, M. Melmer, A. Leitner and W. Lindner, *Bioconjugate Chem.*, 2007, **18**, 1678–1683.
56. T. L. Schlick, Z. Ding, E. W. Kovacs and M. B. Francis, *J. Am. Chem. Soc.*, 2005, **127**, 3718–3723.

57. J. Gavrilyuk, H. Ban, M. Nagano, W. Hakamata and C. F. Barbas, 3rd, *Bioconjugate Chem.*, 2012, **23**, 2321–2328.
58. N. S. Joshi, L. R. Whitaker and M. B. Francis, *J. Am. Chem. Soc.*, 2004, **126**, 15942–15943.
59. K. L. Seim, A. C. Obermeyer and M. B. Francis, *J. Am. Chem. Soc.*, 2011, **133**, 16970–16976.
60. Y. Seki, T. Ishiyama, D. Sasaki, J. Abe, Y. Sohma, K. Oisaki and M. Kanai, *J. Am. Chem. Soc.*, 2016, **138**, 10798–10801.
61. P. S. Addy, S. B. Erickson, J. S. Italia and A. Chatterjee, *J. Am. Chem. Soc.*, 2017, **139**, 11670–11673.
62. J. Zhang, Y. Men, S. Lv, L. Yi and J.-F. Chen, *Org. Biomol. Chem.*, 2015, **13**, 11422–11425.
63. S. Lin, X. Yang, S. Jia, A. M. Weeks, M. Hornsby, P. S. Lee, R. V. Nichiporuk, A. T. Iavarone, J. A. Wells, F. D. Toste and C. J. Chang, *Science*, 2017, **355**, 597–602.
64. W. Song, Y. Wang, J. Qu and Q. Lin, *J. Am. Chem. Soc.*, 2008, **130**, 9654–9655.
65. J. Wang, W. Zhang, W. Song, Y. Wang, Z. Yu, J. Li, M. Wu, L. Wang, J. Zang and Q. Lin, *J. Am. Chem. Soc.*, 2010, **132**, 14812–14818.
66. Z. Yu, L. Y. Ho and Q. Lin, *J. Am. Chem. Soc.*, 2011, **133**, 11912–11915.
67. M. Kohn and R. Breinbauer, *Angew. Chem., Int. Ed. Engl.*, 2004, **43**, 3106–3116.
68. R. Huisgen, *Angew. Chem., Int. Ed. Engl.*, 1963, **2**, 565–598.
69. H. C. Kolb, M. G. Finn and K. B. Sharpless, *Angew. Chem., Int. Ed. Engl.*, 2001, **40**, 2004–2021.
70. T. R. Chan, R. Hilgraf, K. B. Sharpless and V. V. Fokin, *Org. Lett.*, 2004, **6**, 2853–2855.
71. A. J. Link and D. A. Tirrell, *J. Am. Chem. Soc.*, 2003, **125**, 11164–11165.
72. N. J. Agard, J. A. Prescher and C. R. Bertozzi, *J. Am. Chem. Soc.*, 2004, **126**, 15046–15047.
73. J. A. Codelli, J. M. Baskin, N. J. Agard and C. R. Bertozzi, *J. Am. Chem. Soc.*, 2008, **130**, 11486–11493.
74. J. C. Jewett, E. M. Sletten and C. R. Bertozzi, *J. Am. Chem. Soc.*, 2010, **132**, 3688–3690.
75. X. Ning, J. Guo, M. A. Wolfert and G. J. Boons, *Angew. Chem., Int. Ed. Engl.*, 2008, **47**, 2253–2255.
76. J. Dommerholt, S. Schmidt, R. Temming, L. J. A. Hendriks, F. P. J. T. Rutjes, J. C. M. van Hest, D. J. Lefeber, P. Friedl and F. L. van Delft, *Angew. Chem., Int. Ed.*, 2010, **49**, 9422–9425.
77. E. M. Sletten and C. R. Bertozzi, *Org. Lett.*, 2008, **10**, 3097–3099.
78. B. L. Oliveira, Z. Guo and G. J. L. Bernardes, *Chem. Soc. Rev.*, 2017, **46**, 4895–4950.
79. M. L. Blackman, M. Royzen and J. M. Fox, *J. Am. Chem. Soc.*, 2008, **130**, 13518–13519.
80. F. Thalhammer, U. Wallfahrer and J. Sauer, *Tetrahedron Lett.*, 1990, **31**, 6851–6854.

81. J. M. J. M. Ravasco, C. M. Monteiro and A. F. Trindade, *Org. Chem. Front.*, 2017, **4**, 1167–1198.
82. M. T. Taylor, M. L. Blackman, O. Dmitrenko and J. M. Fox, *J. Am. Chem. Soc.*, 2011, **133**, 9646–9649.
83. R. Selvaraj and J. M. Fox, *Curr. Opin. Chem. Biol.*, 2013, **17**, 753–760.
84. K. Lang, L. Davis, J. Torres-Kolbus, C. Chou, A. Deiters and J. W. Chin, *Nat. Chem.*, 2012, **4**, 298.
85. N. K. Devaraj, S. Hilderbrand, R. Upadhyay, R. Mazitschek and R. Weissleder, *Angew. Chem., Int. Ed. Engl.*, 2010, **49**, 2869–2872.
86. R. M. Versteegen, R. Rossin, W. ten Hoeve, H. M. Janssen and M. S. Robillard, *Angew. Chem., Int. Ed.*, 2013, **52**, 14112–14116.
87. K. A. Horner, N. M. Valette and M. E. Webb, *Chem. – Eur. J.*, 2015, **21**, 14376–14381.
88. D. N. Kamber, Y. Liang, R. J. Blizzard, F. Liu, R. A. Mehl, K. N. Houk and J. A. Prescher, *J. Am. Chem. Soc.*, 2015, **137**, 8388–8391.
89. N. Budisa, *Angew. Chem., Int. Ed. Engl.*, 2004, **43**, 6426–6463.
90. J. T. Ngo, J. A. Champion, A. Mahdavi, I. C. Tanrikulu, K. E. Beatty, R. E. Connor, T. H. Yoo, D. C. Dieterich, E. M. Schuman and D. A. Tirrell, *Nat. Chem. Biol.*, 2009, **5**, 715–717.
91. J. T. Ngo and D. A. Tirrell, *Acc. Chem. Res.*, 2011, **44**, 677–685.
92. A. J. Link, M. K. Vink, N. J. Agard, J. A. Prescher, C. R. Bertozzi and D. A. Tirrell, *Proc. Natl. Acad. Sci. U. S. A.*, 2006, **103**, 10180–10185.
93. J. T. Ngo, E. M. Schuman and D. A. Tirrell, *Proc. Natl. Acad. Sci. U. S. A.*, 2013, **110**, 4992–4997.
94. T. Mukai, T. Kobayashi, N. Hino, T. Yanagisawa, K. Sakamoto and S. Yokoyama, *Biochem. Biophys. Res. Commun.*, 2008, **371**, 818–822.
95. T. Yanagisawa, R. Ishii, R. Fukunaga, T. Kobayashi, K. Sakamoto and S. Yokoyama, *Chem. Biol.*, 2008, **15**, 1187–1197.
96. P. R. Chen, D. Groff, J. Guo, W. Ou, S. Cellitti, B. H. Geierstanger and P. G. Schultz, *Angew. Chem., Int. Ed. Engl.*, 2009, **48**, 4052–4055.
97. D. P. Nguyen, H. Lusic, H. Neumann, P. B. Kapadnis, A. Deiters and J. W. Chin, *J. Am. Chem. Soc.*, 2009, **131**, 8720–8721.
98. A. Deiters and P. G. Schultz, *Bioorg. Med. Chem. Lett.*, 2005, **15**, 1521–1524.
99. K. Lang and J. W. Chin, *Chem. Rev.*, 2014, **114**, 4764–4806.
100. K. Wals and H. Ovaa, *Front. Chem.*, 2014, **2**, 15.
101. R. Brabham and M. A. Fascione, *ChemBioChem*, 2017, **18**, 1973–1983.
102. J. W. Chin, S. W. Santoro, A. B. Martin, D. S. King, L. Wang and P. G. Schultz, *J. Am. Chem. Soc.*, 2002, **124**, 9026–9027.
103. T. Machida, K. Lang, L. Xue, J. W. Chin and N. Winssinger, *Bioconjugate Chem.*, 2015, **26**, 802–806.
104. J. L. Seitchik, J. C. Peeler, M. T. Taylor, M. L. Blackman, T. W. Rhoads, R. B. Cooley, C. Refakis, J. M. Fox and R. A. Mehl, *J. Am. Chem. Soc.*, 2012, **134**, 2898–2901.
105. J. E. Cronan, Jr., *J. Biol. Chem.*, 1990, **265**, 10327–10333.

106. E. de Boer, P. Rodriguez, E. Bonte, J. Krijgsveld, E. Katsantoni, A. Heck, F. Grosveld and J. Strouboulis, *Proc. Natl. Acad. Sci. U. S. A.*, 2003, **100**, 7480–7485.
107. I. Chen, M. Howarth, W. Lin and A. Y. Ting, *Nat. Methods*, 2005, **2**, 99–104.
108. M. Howarth, K. Takao, Y. Hayashi and A. Y. Ting, *Proc. Natl. Acad. Sci. U. S. A.*, 2005, **102**, 7583–7588.
109. I. Chen, Y. A. Choi and A. Y. Ting, *J. Am. Chem. Soc.*, 2007, **129**, 6619–6625.
110. S. A. Slavoff, I. Chen, Y. A. Choi and A. Y. Ting, *J. Am. Chem. Soc.*, 2008, **130**, 1160–1162.
111. M. Fairhead and M. Howarth, *Methods Mol. Biol.*, 2015, **1266**, 171–184.
112. E. W. Tate, K. A. Kalesh, T. Lanyon-Hogg, E. M. Storck and E. Thinon, *Curr. Opin. Chem. Biol.*, 2015, **24**, 48–57.
113. K. K. Palaniappan and C. R. Bertozzi, *Chem. Rev.*, 2016, **116**, 14277–14306.
114. E. Thinon and H. C. Hang, *Biochem. Soc. Trans.*, 2015, **43**, 253–261.
115. M. H. Wright, B. Clough, M. D. Rackham, K. Rangachari, J. A. Brannigan, M. Grainger, D. K. Moss, A. R. Bottrill, W. P. Heal, M. Broncel, R. A. Serwa, D. Brady, D. J. Mann, R. J. Leatherbarrow, R. Tewari, A. J. Wilkinson, A. A. Holder and E. W. Tate, *Nat. Chem.*, 2014, **6**, 112–121.
116. M. H. Wright, D. Paape, H. P. Price, D. F. Smith and E. W. Tate, *ACS Infect. Dis.*, 2016, **2**, 427–441.
117. M. H. Wright, D. Paape, E. M. Storck, R. A. Serwa, D. F. Smith and E. W. Tate, *Chem. Biol.*, 2015, **22**, 342–354.
118. M. L. Jones, M. O. Collins, D. Goulding, J. S. Choudhary and J. C. Rayner, *Cell Host Microbe*, 2012, **12**, 246–258.
119. M. Ritzefeld, M. H. Wright and E. W. Tate, *Parasitology*, 2017, 1–18.
120. H. C. Hang, E. J. Geutjes, G. Grotenbreg, A. M. Pollington, M. J. Bijlmakers and H. L. Ploegh, *J. Am. Chem. Soc.*, 2007, **129**, 2744–2745.
121. G. Charron, M. M. Zhang, J. S. Yount, J. Wilson, A. S. Raghavan, E. Shamir and H. C. Hang, *J. Am. Chem. Soc.*, 2009, **131**, 4967–4975.
122. R. N. Hannoush and N. Arenas-Ramirez, *ACS Chem. Biol.*, 2009, **4**, 581–587.
123. M. C. Yap, M. A. Kostiuk, D. D. Martin, M. A. Perinpanayagam, P. G. Hak, A. Siddam, J. R. Majjigapu, G. Rajaiah, B. O. Keller, J. A. Prescher, P. Wu, C. R. Bertozzi, J. R. Falck and L. G. Berthiaume, *J. Lipid Res.*, 2010, **51**, 1566–1580.
124. E. Thinon, R. A. Serwa, M. Broncel, J. A. Brannigan, U. Brassat, M. H. Wright, W. P. Heal, A. J. Wilkinson, D. J. Mann and E. W. Tate, *Nat. Commun.*, 2014, **5**, 4919.
125. T. M. Charlton, A. Kovacs-Simon, S. L. Michell, N. F. Fairweather and E. W. Tate, *Chem. Biol.*, 2015, **22**, 1562–1573.

126. K. J. Rangan, Y. Y. Yang, G. Charron and H. C. Hang, *J. Am. Chem. Soc.*, 2010, **132**, 10628–10629.
127. N. J. Agard, J. M. Baskin, J. A. Prescher, A. Lo and C. R. Bertozzi, *ACS Chem. Biol.*, 2006, **1**, 644–648.
128. A. E. Speers and B. F. Cravatt, *Chem. Biol.*, 2004, **11**, 535–546.
129. W. P. Heal, S. R. Wickramasinghe, R. J. Leatherbarrow and E. W. Tate, *Org. Biomol. Chem.*, 2008, **6**, 2308–2315.
130. W. P. Heal, M. H. Wright, E. Thinon and E. W. Tate, *Nat. Protoc.*, 2011, 7, 105–117.
131. S. H. Ho and D. A. Tirrell, *J. Am. Chem. Soc.*, 2016, **138**, 15098–15101.
132. T. Lanyon-Hogg, M. Faronato, R. A. Serwa and E. W. Tate, *Trends Biochem. Sci.*, 2017, **42**, 566–581.
133. K. Lemonidis, C. Salaun, M. Kouskou, C. Diez-Ardanuy, L. H. Chamberlain and J. Greaves, *Biochem. Soc. Trans.*, 2017, **45**, 751–758.
134. B. R. Martin, C. Wang, A. Adibekian, S. E. Tully and B. F. Cravatt, *Nat. Methods*, 2011, **9**, 84–89.
135. J. P. Wilson, A. S. Raghavan, Y. Y. Yang, G. Charron and H. C. Hang, *Mol. Cell. Proteomics*, 2011, **10**, M110 001198.
136. J. L. Feldman, J. Baeza and J. M. Denu, *J. Biol. Chem.*, 2013, **288**, 31350–31356.
137. X. Zhang, N. A. Spiegelman, O. D. Nelson, H. Jing and H. Lin, *eLife*, 2017, **6**, e25158.
138. E. Verdin and M. Ott, *Nat. Rev. Mol. Cell Biol.*, 2015, **16**, 258–264.
139. Y. Y. Yang, M. Grammel and H. C. Hang, *Bioorg. Med. Chem. Lett.*, 2011, **21**, 4976–4979.
140. C. Yang, J. Mi, Y. Feng, L. Ngo, T. Gao, L. Yan and Y. G. Zheng, *J. Am. Chem. Soc.*, 2013, **135**, 7791–7794.
141. Z. Han, Y. Luan and Y. G. Zheng, *ChemBioChem*, 2015, **16**, 2605–2609.
142. Z. Han, C. W. Chou, X. Yang, M. G. Bartlett and Y. G. Zheng, *ACS Chem. Biol.*, 2017, **12**, 1547–1555.
143. M. E. Ourailidou, P. Dockerty, M. Witte, G. J. Poelarends and F. J. Dekker, *Org. Biomol. Chem.*, 2015, **13**, 3648–3653.
144. A. F. Berry, W. P. Heal, A. K. Tarafder, T. Tolmachova, R. A. Baron, M. C. Seabra and E. W. Tate, *ChemBioChem*, 2010, **11**, 771–773.
145. G. Charron, L. K. Tsou, W. Maguire, J. S. Yount and H. C. Hang, *Mol. BioSyst.*, 2011, 7, 67–73.
146. M. Brioschi, A. Martinez Fernandez and C. Banfi, *Expert Rev. Proteomics*, 2017, **14**, 515–528.
147. C. C. Palsuledesai, J. D. Ochocki, M. M. Kuhns, Y. C. Wang, J. K. Warmka, D. S. Chernick, E. V. Wattenberg, L. Li, E. A. Arriaga and M. D. Distefano, *ACS Chem. Biol.*, 2016, **11**, 2820–2828.
148. G. Charron, M. M. Li, M. R. MacDonald and H. C. Hang, *Proc. Natl. Acad. Sci. U. S. A.*, 2013, **110**, 11085–11090.
149. J. E. Gisselberg, L. Zhang, J. E. Elias and E. Yeh, *Mol. Cell. Proteomics*, 2017, **16**, S54–S64.

150. J. J. Lee, H. J. Choi, M. Yun, Y. Kang, J. E. Jung, Y. Ryu, T. Y. Kim, Y. J. Cha, H. S. Cho, J. J. Min, C. W. Chung and H. S. Kim, *Angew. Chem., Int. Ed. Engl.*, 2015, **54**, 12020–12024.
151. W. P. Heal, B. Jovanovic, S. Bessin, M. H. Wright, A. I. Magee and E. W. Tate, *Chem. Commun.*, 2011, **47**, 4081–4083.
152. P. Ciepla, A. D. Konitsiotis, R. A. Serwa, N. Masumoto, W. P. Leong, M. J. Dallman, A. I. Magee and E. W. Tate, *Chem. Sci.*, 2014, **5**, 4249–4259.
153. E. Saxon and C. R. Bertozzi, *Science*, 2000, **287**, 2007–2010.
154. T. J. Sminia, H. Zuilhof and T. Wennekes, *Carbohydr. Res.*, 2016, **435**, 121–141.
155. J. Nilsson, U. Ruetschi, A. Halim, C. Hesse, E. Carlsohn, G. Brinkmalm and G. Larson, *Nat. Methods*, 2009, **6**, 809–811.
156. T. N. Ramya, E. Weerapana, B. F. Cravatt and J. C. Paulson, *Glycobiology*, 2013, **23**, 211–221.
157. Y. Taga, M. Kusubata, K. Ogawa-Goto and S. Hattori, *Mol. Cell. Proteomics*, 2012, **11**, M111 010397.
158. J. B. Rannes, A. Ioannou, S. C. Willies, G. Grogan, C. Behrens, S. L. Flitsch and N. J. Turner, *J. Am. Chem. Soc.*, 2011, **133**, 8436–8439.
159. A. Lopez Aguilar, J. G. Briard, L. Yang, B. Ovryn, M. S. Macauley and P. Wu, *ACS Chem. Biol.*, 2017, **12**, 611–621.
160. N. Khidekel, S. Arndt, N. Lamarre-Vincent, A. Lippert, K. G. Poulin-Kerstien, B. Ramakrishnan, P. K. Qasba and L. C. Hsieh-Wilson, *J. Am. Chem. Soc.*, 2003, **125**, 16162–16163.
161. T. Zheng, H. Jiang, M. Gros, D. S. del Amo, S. Sundaram, G. Lauvau, F. Marlow, Y. Liu, P. Stanley and P. Wu, *Angew. Chem., Int. Ed. Engl.*, 2011, **50**, 4113–4118.
162. S. H. Yu, P. Zhao, T. Sun, Z. Gao, K. W. Moremen, G. J. Boons, L. Wells and R. Steet, *J. Biol. Chem.*, 2016, **291**, 3982–3989.
163. T. Sun, S. H. Yu, P. Zhao, L. Meng, K. W. Moremen, L. Wells, R. Steet and G. J. Boons, *J. Am. Chem. Soc.*, 2016, **138**, 11575–11582.
164. N. E. Mbua, X. Li, H. R. Flanagan-Steet, L. Meng, K. Aoki, K. W. Moremen, M. A. Wolfert, R. Steet and G. J. Boons, *Angew. Chem., Int. Ed. Engl.*, 2013, **52**, 13012–13015.
165. J. L. Chaubard, C. Krishnamurthy, W. Yi, D. F. Smith and L. C. Hsieh-Wilson, *J. Am. Chem. Soc.*, 2012, **134**, 4489–4492.
166. L. Wen, Y. Zheng, K. Jiang, M. Zhang, S. M. Kondengaden, S. Li, K. Huang, J. Li, J. Song and P. G. Wang, *J. Am. Chem. Soc.*, 2016, **138**, 11473–11476.
167. Q. Li, Z. Li, X. Duan and W. Yi, *J. Am. Chem. Soc.*, 2014, **136**, 12536–12539.
168. J. Yang, K. S. Carroll and D. C. Liebler, *Mol. Cell. Proteomics*, 2016, **15**, 1–11.
169. V. Gupta, H. Paritala and K. S. Carroll, *B*ioconjugate Chem., 2016, **27**, 1411–1418.

170. C. E. Paulsen, T. H. Truong, F. J. Garcia, A. Homann, V. Gupta, S. E. Leonard and K. S. Carroll, *Nat. Chem. Biol.*, 2011, **8**, 57–64.
171. J. Yang, V. Gupta, K. S. Carroll and D. C. Liebler, *Nat. Commun.*, 2014, **5**, 4776.
172. V. Gupta, J. Yang, D. C. Liebler and K. S. Carroll, *J. Am. Chem. Soc.*, 2017, **139**, 5588–5595.
173. S. Tsukiji and T. Nagamune, *ChemBioChem*, 2009, **10**, 787–798.
174. S. K. Mazmanian, G. Liu, T. T. Hung and O. Schneewind, *Science*, 1999, **285**, 760–763.
175. H. Ton-That, G. Liu, S. K. Mazmanian, K. F. Faull and O. Schneewind, *Proc. Natl. Acad. Sci. U. S. A.*, 1999, **96**, 12424–12429.
176. H. Y. Mao, S. A. Hart, A. Schink and B. A. Pollok, *J. Am. Chem. Soc.*, 2004, **126**, 2670–2671.
177. R. Parthasarathy, S. Subramanian and E. T. Boder, *Bioconjugate Chem.*, 2007, **18**, 469–476.
178. L. Y. Chan, H. F. Cross, J. K. She, G. Cavalli, H. F. P. Martins and C. Neylon, *PLoS One*, 2007, **2**, e1164.
179. F. Clow, J. D. Fraser and T. Proft, *Biotechnol. Lett.*, 2008, **30**, 1603–1607.
180. T. Tanaka, T. Yamamoto, S. Tsukiji and T. Nagamune, *ChemBioChem*, 2008, **9**, 802–807.
181. S. Wu and T. Proft, *Biotechnol. Lett.*, 2010, **32**, 1713–1718.
182. R. G. Kruger, B. Otvos, B. A. Frankel, M. Bentley, P. Dostal and D. G. McCafferty, *Biochemistry*, 2004, **43**, 1541–1551.
183. D. J. Williamson, M. A. Fascione, M. E. Webb and W. B. Turnbull, *Angew. Chem., Int. Ed.*, 2012, **51**, 9377–9380.
184. R. L. Policarpo, H. Kang, X. L. Liao, A. E. Rabideau, M. D. Simon and B. L. Pentelute, *Angew. Chem., Int. Ed.*, 2014, **53**, 9203–9208.
185. N. Stefan, R. Gebleux, L. Waldmeier, T. Hell, M. Escher, F. I. Wolter, U. Grawunder and R. R. Beerli, *Mol. Cancer Ther.*, 2017, **16**, 879–892.
186. R. D. Row, T. J. Roark, M. C. Philip, L. L. Perkins and J. M. Antos, *Chem. Commun.*, 2015, **51**, 12548–12551.
187. P. M. Morrison, M. R. Balmforth, S. W. Ness, D. J. Williamson, M. D. Rugen, W. B. Turnbull and M. E. Webb, *ChemBioChem*, 2016, **17**, 753–758.
188. F. Liu, E. Y. Luo, D. B. Flora and A. R. Mezo, *J. Org. Chem.*, 2014, **79**, 487–492.
189. H. Hirakawa, S. Ishikawa and T. Nagamune, *Biotechnol. J.*, 2015, **10**, 1487–1492.
190. K. Strijbis, E. Spooner and H. L. Ploegh, *Traffic*, 2012, **13**, 780–789.
191. B. M. Dorr, H. O. Ham, C. H. An, E. L. Chaikof and D. R. Liu, *Proc. Natl. Acad. Sci. U. S. A.*, 2014, **111**, 13343–13348.
192. K. Piotukh, B. Geltinger, N. Heinrich, F. Gerth, M. Beyermann, C. Freund and D. Schwarzer, *J. Am. Chem. Soc.*, 2011, **133**, 17536–17539.
193. J. A. Wells, T. Chang and L. Abrahmsen, *Biochemistry*, 1992, **31**, 2191.

194. T. K. Chang, D. Y. Jackson, J. P. Burnier and J. A. Wells, *Proc. Natl. Acad. Sci. U. S. A.*, 1994, **91**, 12544–12548.
195. H. A. I. Yoshihara, S. Mahrus and J. A. Wells, *Bioorg. Med. Chem. Lett.*, 2008, **18**, 6000–6003.
196. S. Liebscher, M. Schopfel, T. Aumuller, A. Sharkhuukhen, A. Pech, E. Hoss, C. Parthier, G. Jahreis, M. T. Stubbs and F. Bordusa, *Angew. Chem., Int. Ed. Engl.*, 2014, **53**, 3024–3028.
197. C. Meyer, S. Liebscher and F. Bordusa, *Bioconjugate Chem.*, 2016, **27**, 47–53.
198. G. K. T. Nguyen, S. J. Wang, Y. B. Qiu, X. Hemu, Y. L. Lian and J. P. Tam, *Nat. Chem. Biol.*, 2014, **10**, 732–738.
199. G. K. T. Nguyen, A. Kam, S. Loo, A. E. Jansson, L. X. Pan and J. P. Tam, *J. Am. Chem. Soc.*, 2015, **137**, 15398–15401.
200. R. Yang, Y. H. Wong, G. K. T. Nguyen, J. P. Tam, J. Lescar and B. Wu, *J. Am. Chem. Soc.*, 2017, **139**, 5351–5358.
201. G. K. T. Nguyen, Y. Cao, W. Wang, C. F. Liu and J. P. Tam, *Angew. Chem., Int. Ed.*, 2015, **54**, 15694–15698.
202. X. Hemu, Y. B. Qiu, G. K. T. Nguyen and J. P. Tam, *J. Am. Chem. Soc.*, 2016, **138**, 6968–6971.
203. X. B. Bi, J. Yin, G. K. T. Nguyen, C. Rao, N. B. A. Halim, X. Hemu, J. P. Tam and C. F. Liu, *Angew. Chem.Int. Ed.*, 2017, **56**, 7822–7825.
204. K. S. Harris, T. Durek, Q. Kaas, A. G. Poth, E. K. Gilding, B. F. Conlan, I. Saska, N. L. Daly, N. L. van der Weerden, D. J. Craik and M. A. Anderson, *Nat. Commun.*, 2015, **6**, 10199.
205. M. H. Wright and S. A. Sieber, *Nat. Prod. Rep.*, 2016, **33**, 681–708.
206. L. I. Willems, H. S. Overkleeft and S. I. van Kasteren, *Bioconjugate Chem.*, 2014, **25**, 1181–1191.
207. C. M. Pichler, J. Krysiak and R. Breinbauer, *Bioorg. Med. Chem.*, 2016, **24**, 3291–3303.
208. A. M. Roberts, C. C. Ward and D. K. Nomura, *Curr. Opin. Biotechnol*, 2017, **43**, 25–33.
209. G. W. Preston and A. J. Wilson, *Chem. Soc. Rev.*, 2013, **42**, 3289–3301.
210. S. Tsukiji, M. Miyagawa, Y. Takaoka, T. Tamura and I. Hamachi, *Nat. Chem. Biol.*, 2009, **5**, 341–343.
211. S. H. Fujishima, R. Yasui, T. Miki, A. Ojida and I. Hamachi, *J. Am. Chem. Soc.*, 2012, **134**, 3961–3964.

CHAPTER 13

Chemical Protein Synthesis: Strategies and Biological Applications

MANUEL M. MÜLLER

King's College London, Department of Chemistry, 7 Trinity Street, London SE1 1DB, UK
Email: manuel.muller@kcl.ac.uk

13.1 Introduction

Proteins are amino acid polymers that carry out an enormous breadth of biological tasks. For example they serve as mechanical scaffolds, molecular power generators and precise copying machines for genetic material. Not surprisingly, chemists and biologists have long sought to synthesise proteins to better understand and manipulate their diverse properties. This challenging task has been made possible through a continuum of developments in synthetic peptide and protein chemistry, which will be the focus of this chapter. Equally important, although beyond the scope of this chapter, are analytical methods that generate information on the composition and structure of natural proteins to provide targets for synthetic undertakings and to validate the authenticity of products. Foremost, innovations in chromatography and mass spectrometry have facilitated the isolation and identification of natural and synthetic proteins.

This chapter will provide an overview of solid-phase peptide synthesis and convergent approaches that dramatically expedite protein total- and semi-synthesis. General applications, including the site-specific installation

Chemical Biology No. 10
Chemical and Biological Synthesis: Enabling Approaches for Understanding Biology
Edited by Nick J. Westwood and Adam Nelson

Published by the Royal Society of Chemistry, www.rsc.org

of non-natural building blocks and biochemical signals, as well as the generation of unusual topologies, will be discussed. I will conclude with a series of case studies on areas of biochemistry that were profoundly affected by systematic explorations fuelled by protein synthesis.

13.2 Overview of Chemical Protein Synthesis

Nature synthesises proteins through sequence-directed polymerisation of amino acids using a complex machine, the ribosome. Amino acids are pre-activated as transfer RNA (tRNA) esters using energy provided by ATP, iteratively recruited to the ribosome according to a sequence specified by a messenger RNA (mRNA) template. In the active site of the ribosome, ribosomal RNA (rRNA) catalyses amide bond formation, transferring the nascent polypeptide chain onto the incoming aminoacyl-tRNA. Ribosomes can easily synthesise proteins of hundreds of amino acids in length, yet biological protein synthesis works best with the standard 20 proteinogenic amino acids. The development of ingenious technologies to infiltrate cellular protein synthesis with non-standard building blocks has mitigated this limitation,[1] but access to polypeptides and proteins that deviate strongly from the archetypal ribosomal repertoire still require the deployment of chemical methods. The following sections will discuss in broad strokes how synthetic chemists carry out the functions of the ribosome to access chemically defined proteins.

13.2.1 Bottom Up: Solid-phase Peptide Synthesis

The formation of amide bonds is chemically simple, yet the sequence-specific synthesis of tens to hundreds of amide bonds is far from trivial. To address this issue, Merrifield developed a strategy where the carboxy (*C*)-terminal amino acid is immobilised on a macroscopic solid support, allowing the assembly of the peptide chain through the iterative coupling of residues towards the amino (*N*)-terminus (Figure 13.1).[2] By using solid-phase peptide synthesis (SPPS), excess reagents can be effectively removed *via* washing and filtration, which facilitates the formation of amide bonds in near quantitative yields and obviates the need for tedious purification steps after each amino acid coupling. This innovative and highly practical method enabled the total synthesis of entire proteins, culminating in the chemical synthesis of 124 amino acid RNase A.[3]

Amide bond formation is typically achieved upon activation of an amino acid with a carbodiimide in the presence of a nucleophilic catalyst such as hydroxybenzotriazole (HOBt) or Oxyma (Figure 13.1, inset).[4,5] A rich palette of alternative activating agents (*e.g.* uronium reagents such as HBTU, HATU and HCTU; phosphonium reagents such as PyBOP) is available for routine couplings as well as for particularly challenging amide-bond-forming reactions.[6] The use of a protecting group at *N*-α of the incoming N-terminal amino acid provides selectivity. The *tert*-butyloxycarbonyl (Boc)[7] and especially the fluorenylmethyloxycarbonyl (Fmoc)[8] groups have prevailed

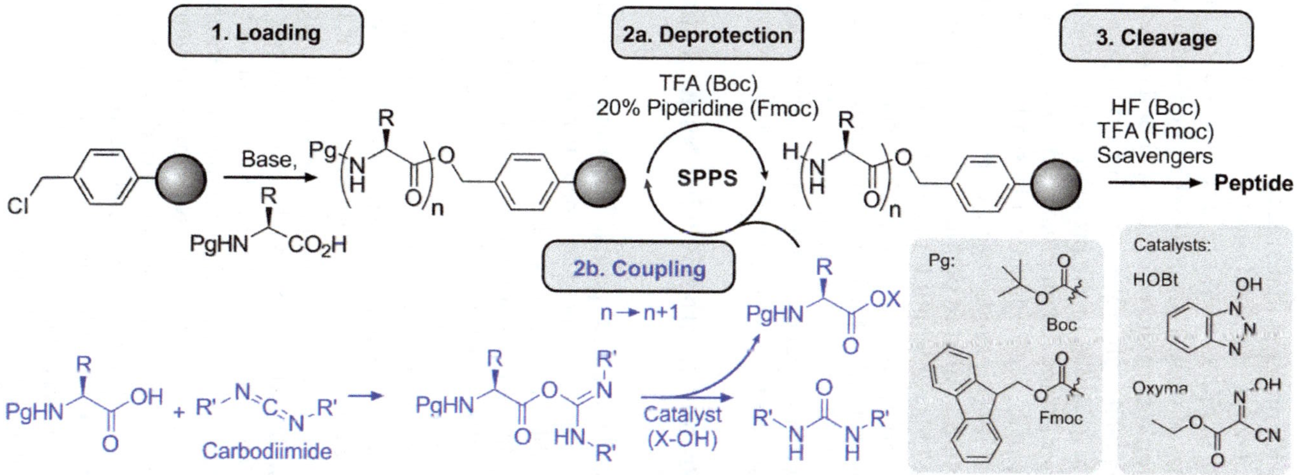

Figure 13.1 Solid-phase peptide synthesis. A functionalised resin (typically polystyrene or polyethylene glycol, grey sphere) is loaded with the C-terminal amino acid (1). Iterative cycles of *N*-α-deprotection (2a) and amide bond formation with an activated amino acid (2b) are performed. The most commonly used *N*-α-protecting groups, tert-butyloxycarbonyl (Boc) and fluorenylmethyloxycarbonyl (Fmoc), and nucleophilic catalysts, hydroxybenzotriazole (HOBt) and Oxyma, are depicted in grey boxes. Upon completion of the elongation cycles, the peptide is cleaved from the resin with concomitant side chain deprotections (3). Pg = Protecting group.

as the most popular choices. These groups are stable to the coupling conditions employed, but are easily removed in the presence of trifluoroacetic acid (TFA) or 20% piperidine, respectively. Functional groups on amino acid side chains are protected with moieties that are resilient to these treatments. Upon coupling and deprotecting the final amino acid, the peptide chain is cleaved from the solid support. Cleavage and removal of side-chain-protecting groups are achieved with HF in Boc-SPPS (the C-terminal residue is typically attached through a benzylic ester linkage) or TFA in Fmoc-SPPS (attachment of peptides as *p*-alkoxybenzylic esters or through a trityl linker renders the linkage more acid labile) in the presence of scavengers.[9,10]

SPPS is readily automated,[11] and heating *via* microwave irradiation[12] or other sources is often used to accelerate peptide synthesis. Nevertheless, despite the presence of highly optimized reagents, resins and protocols, routine peptide synthesis is often limited to peptides and small proteins of less than 50 residues in length. This constraint is due to the fact that even if every coupling and deprotection proceeds with near-perfect yields of 99%, after 50 residues (100 steps), the overall yield falls below 40%, and purification of the desired material from side products is often challenging. Besides this general limitation, particular amino acid sequences have a tendency to aggregate on resin, thereby precluding access of incoming reagents to the termini.

13.2.2 Convergent Protein Synthesis with Native Chemical Ligation

To facilitate the synthesis of full-length proteins, convergent methods to assemble peptidic fragments needed to be developed. Peptides with protected side chains are poorly soluble. Therefore, such methods must be chemoselective and proceed readily in aqueous buffers, compatible with unprotected, water-soluble peptides. Native chemical ligation (NCL; Scheme 13.1), developed by Kent and co-workers, provides an ingenious solution to this problem.[13] The N-terminal fragment is synthesised as a C-terminal α-thioester, providing an electrophile with excellent reactivity while mitigating epimerisation of the C-terminal amino acid. The C-terminal fragment is furnished with an N-terminal cysteine residue. Upon mixing of the fragments, the nucleophilic cysteine-sulphhydryl group attacks the thioester and *trans*-thioesterification ensues. The linear thioester rearranges *via* an energetically favourable *S*-to-*N* acyl shift, providing a native peptide bond. Importantly, the unique 1,2-amino thiol arrangement of an N-terminal cysteine provides exquisite selectivity: internal cysteine residues will undergo reversible *trans*-thioesterification but lack the adjacent amino-functionality to drive the reaction, and amino groups at lysine side chains do not benefit from the initial *trans*-thioesterification capture step. Owing to the simplicity of the approach, NCL had a tremendous effect on the field of protein synthesis, and is still regarded as the method of choice for many protein synthesis undertakings.

The preparation of peptides harbouring an N-terminal cysteine residue is straightforward, but how are peptide thioesters synthesised? Conveniently,

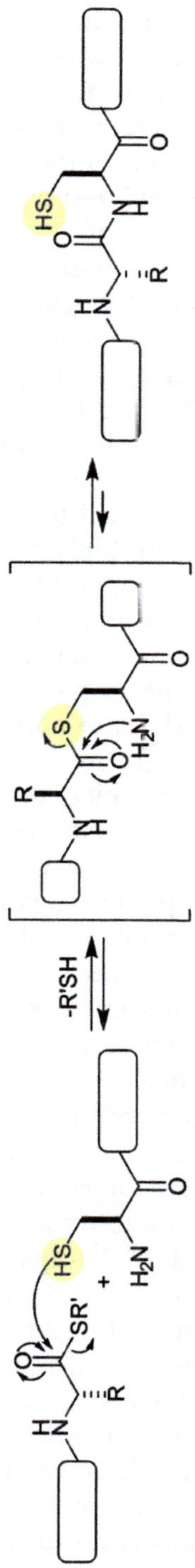

Scheme 13.1 Native chemical ligation (NCL).[13]

the protocols for Boc-SPPS are compatible with α-thioesters, allowing the direct access of ligation-ready fragments.[10] Nevertheless, the practical difficulties associated with HF cleavages and complications for the synthesis of phosphopeptides and glycopeptides fostered the development of routes to access peptide α-thioesters *via* Fmoc-SPPS. Here, repeated treatment of α-thioesters with piperidine deteriorates the quality of the peptides due to aminolysis of the α-thioester and epimerisation of the C-terminal residue. Specialised Fmoc deprotection protocols[14,15] and diverse auxiliaries[16–22]—usually operating *via* an *O*-to-*S* or *N*-to-*S* acyl shift—provided versatile solutions to this problem. In these strategies, peptides are assembled *via* a dedicated linker, which can be converted into a leaving group, which is, in turn, displaced by a small-molecule thiol to generate an α-thioester. Popular choices are Dawson's diaminobenzoic acid linker[18] (activated with p-nitrophenylchloroformate to yield an acyl-urea) and Liu's acyl hydrazide methodology[21] (activated with $NaNO_2$ to yield a reactive acyl azide, Scheme 13.2).

Native chemical ligation is not limited to the coupling of two fragments. Highly convergent assemblies of multiple fragments have been reported. Orthogonal protection of the N-terminal cysteine residue, usually as a thiazolidine moiety, enables directed ligation of multiple sequences. Moreover, kinetic control (aryl thioesters react considerably faster than alkyl thioesters) provides a means to program the sequence of multi-fragment condensations. These methods were used, for example, to assemble the 46-residue model protein Crambin from 6 fragments.[23]

Crambin contains six native cysteines, greatly facilitating such a convergent assembly strategy. In general, however, cysteine is a relatively rare amino acid, curtailing the applicability of NCL. Consequently, diverse strategies have been developed to perform ligations at non-cysteine residues. Alkylation of cysteines with 2-bromoethylamine or bromoacetic acid yields mimics of lysine and glutamate, respectively (Figure 13.2A).[24,25] Cysteine residues can also be chemically desulphurised (Figure 13.2B),[26,27] yielding alanine residues, which are far more frequent in proteins. Furnishing other amino acids with a sulphur atom provides new options for ligation junctions. Of particular interest is β-mercapto aspartate (Figure 13.2C), which can be selectively desulphurised even in the presence of cysteine residues.[28] Selenocysteine can act as a surrogate for cysteine in NCL reactions,[29] and can be deselenized into alanine or serine under reductive and oxidative conditions, respectively (Figure 13.2D),[30,31] thus further diversifying available ligation junctions. Furnishing the N-terminal residue with a sulphhydryl-containing ligation handle such as the 2-mercapto-2-phenethyl auxiliary (Figure 13.2E)[32] can mitigate the considerable challenges of synthesising thiolated amino acids. Such ligations are best performed with at least one of the flanking residues being glycine, due to the steric demand of the auxiliary. To bypass the need for a sulphhydryl group altogether, the reactivity of thioesters towards direct aminolysis can be specifically enhanced by Ag^+ ions, albeit at a loss of chemoselectivity, thus requiring partial protection of the fragments (Figure 13.2F).[33]

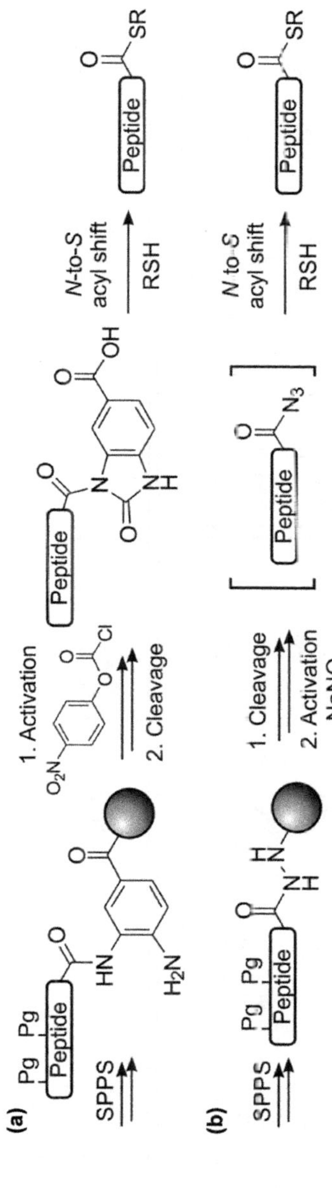

Scheme 13.2 Strategies to synthesise peptide α-thioesters *via* Fmoc SPPS. A. Dawson's acyl-urea method.[18] B. Liu's acyl hydrazide method.[21]

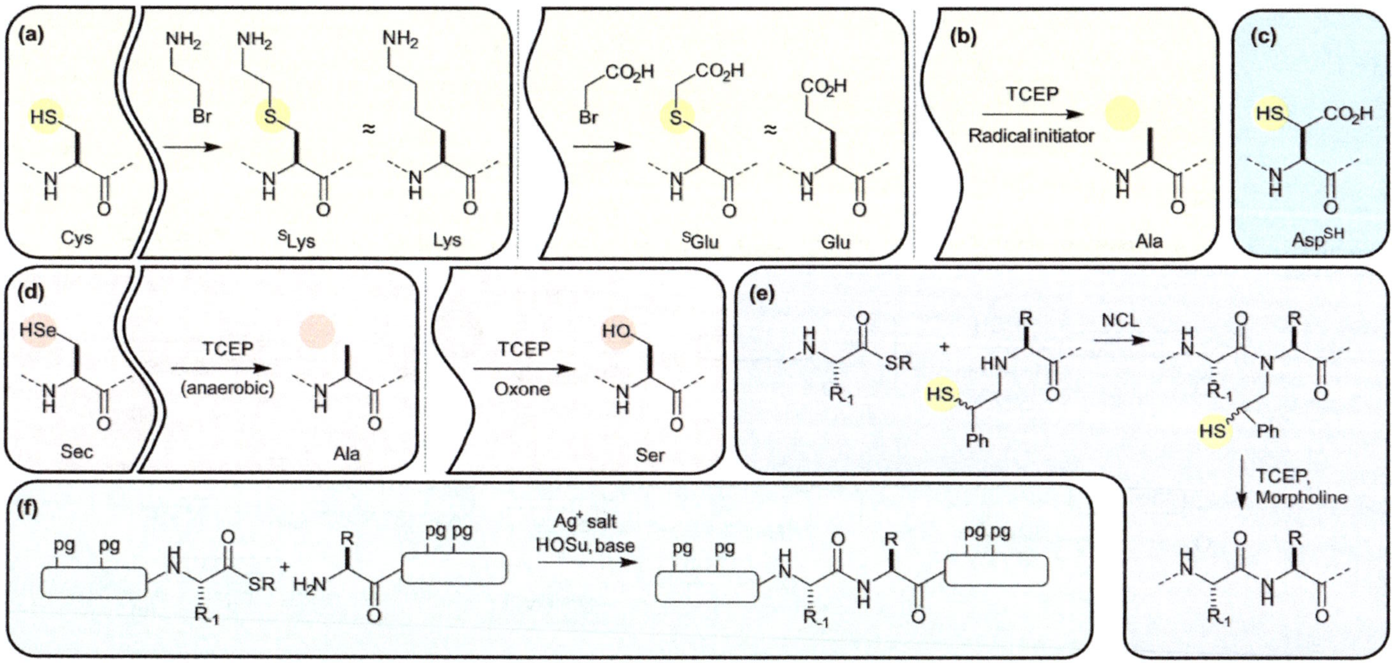

Figure 13.2 Native chemical ligation at non-cysteine residues. A. Cysteine alkylation strategies to generate analogues of Lys and Glu.[24,25] B. Desulphurisation of Cys into Ala.[26,27] C. β-Mercapto aspartate enables ligations at Asp residues. Note that this residue can be selectively desulphurised in the presence of Cys.[28] D. Deselenisation of selenocysteine yields Ala and Ser residues under reductive and oxidative conditions, respectively.[30,31] E. Native chemical ligation (NCL) with a 2-mercapto-2-phenethyl auxiliary.[32] F. Silver-promoted aminolysis of thioesters for ligation of partially protected peptide fragments.[33] Pg = Protecting group; TCEP = Tris(2-carboxyethyl)phosphine; HOSu = *N*-hydroxysuccinimide.

Many distinct amide bond formation chemistries have been developed and contribute valuable options to protein synthesis approaches. For example, the Staudinger ligation consists of a chemoselective condensation of a C-terminal phosphinothioester with an N-terminal azide (Scheme 13.3A).[34] Capture of the fragments under release of N_2 is followed by an intramolecular *S*-to-*N* acyl shift to form the amide bond. The auxiliary is subsequently released by hydrolysis, rendering the reaction traceless. Bode and co-workers developed a unique method of forming amide bonds based on the decarboxylative condensation of an α-ketoacid with a hydroxylamine (KAHA ligation, Scheme 13.3B).[35] Of particular interest for protein synthesis is the use of a 1,2-oxazetidine residue (Scheme 13.3C) as the nucleophilic component because this building block is converted to serine over the course of the ligation.[36]

13.2.3 Semi-synthesis: Coupling Chemical Synthesis with Recombinant Technology

Total chemical synthesis of proteins is a powerful method to assemble small proteins in the range of 100 amino acids. However, many proteins of biological interest are significantly larger and, thus, are challenging to synthesise with purely chemical approaches. Given that nature has solved this problem—cells can routinely assemble proteins with hundreds to thousands of residues—is it possible to combine biological and chemical methods for protein semi-synthesis? Indeed, semi-synthesis can be achieved by attaching a synthetic, site-specifically modified peptide to the N-terminus of a recombinantly produced protein. Native chemical ligation is ideally suited for this union because fragments containing an N-terminal cysteine can easily be accessed recombinantly (Figure 13.3A). Endogenous cleavage of the initiator methionine or furnishing the fragment with a purification tag followed by a protease cleavage site for *in vitro* unmasking of an N-terminal cysteine have been exploited.[37] In this way, large proteins with tailored N-termini can be synthesised. To semi-synthesise proteins with custom-made C-termini, strategies for recombinant production of protein α-thioesters are required. Inteins—proteins that mediate their own excision from recombinant polypeptides in a process known as protein splicing—are ideal tools for this purpose (Figure 13.3B).[38] Intein excision is initiated by an *N*-to-*S* acyl shift at the N-terminal cysteine of the intein. Trans-thioesterification to the N-terminal cysteine of the C-terminal flanking protein is followed by peptide bond cleavage and a spontaneous *S*-to-*N* acyl shift, yielding the full-length protein without the intein. To generate a protein α-thioester, Muir and co-workers exploited this process by genetically fusing a crippled intein (which cannot perform the final cleavage step) to a protein of interest. The intein thioester intermediate can then be intercepted with an excess of a small-molecule thiol (Figure 13.3C).[38] Upon purification, the resulting α-thioester can be used in NCL. This process has been named expressed protein ligation. These semi-synthesis methods tremendously facilitated access to chemically defined proteins and, accordingly, have been

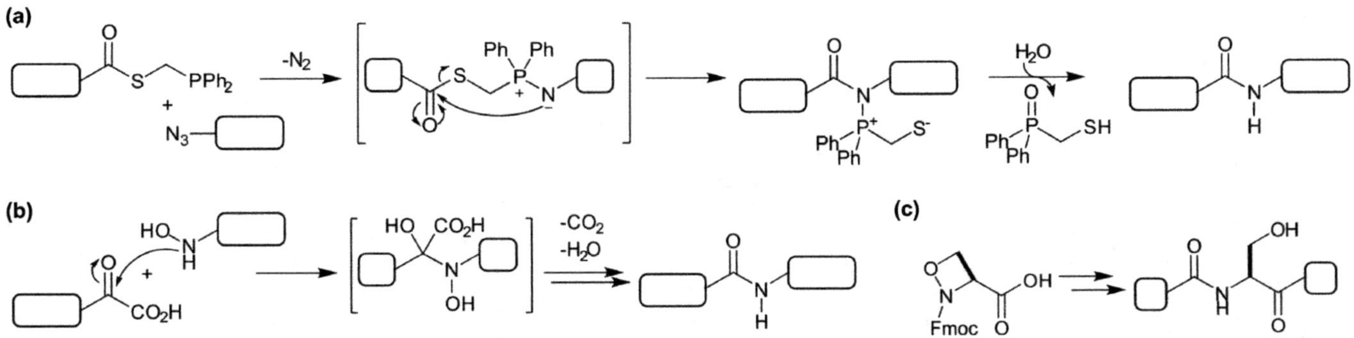

Scheme 13.3 Bioorthogonal peptide ligation strategies. A. Traceless Staudinger ligation.[34] B. KAHA-ligation.[35] C. An oxazetidine residue enables KAHA-ligation at serine residues.[36]

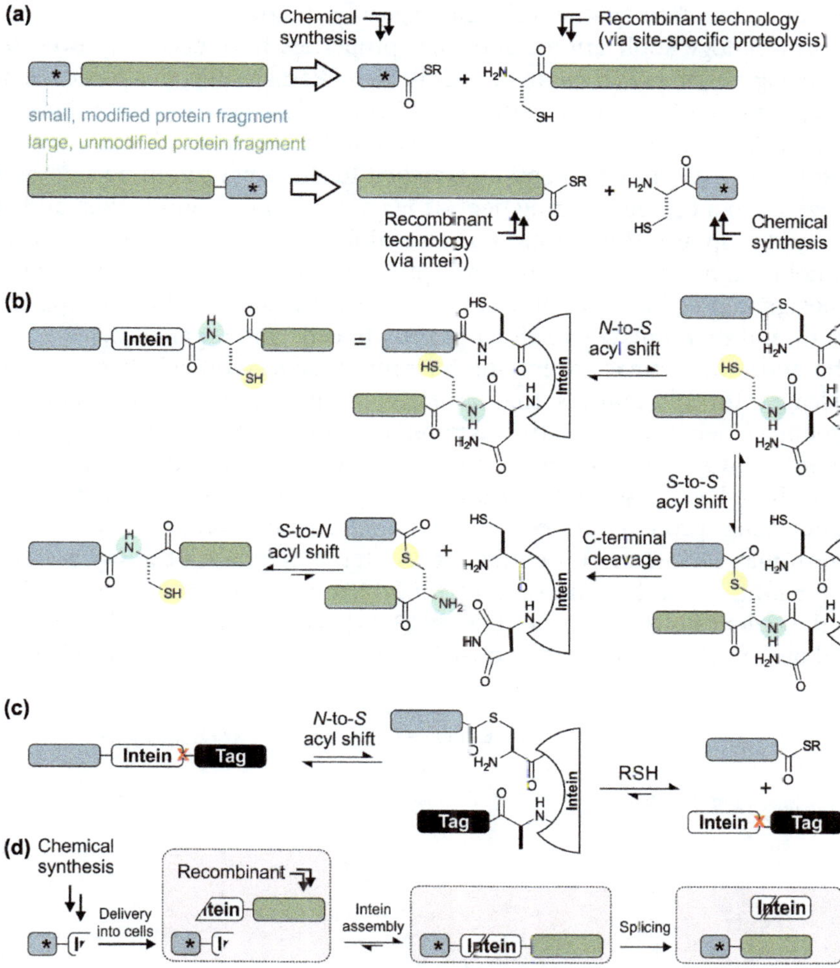

Figure 13.3 Protein semi-synthesis and intein mechanism of action. A. Native chemical ligation can be conducted using a small synthetic peptide (blue)-containing a site-specific modification (asterisk) and a large recombinant protein fragment (green). In the protein component, N-terminal cysteine residues are typically exposed *via* site-specific proteolysis and C-terminal α-thioesters are accessed *via* intein fusions. B. Protein splicing *via* a functional intein (white). C. Recombinant production of protein α-thioesters using an intein with an Asn-to-Ala mutation.[38] D. *Trans*-splicing of a split intein enables protein semi-synthesis in cells.[40,41]

applied to gain key insights into diverse biological signalling processes (Sections 13.4.3–13.4.5).

Can inteins also directly be harnessed to ligate separate polypeptide fragments? Indeed, some inteins are naturally split; they are produced as two separate polypeptide chains and self-assemble to form a functional intein to ligate their flanking peptides in *trans* (Figure 13.3D). Split inteins featuring a short, and therefore synthetically manageable, fragment are attractive for chemical biology applications, and their properties have been improved by protein engineering to facilitate *trans*-splicing.[39] Remarkably, this process can be adapted to perform semi-synthesis in living cells.[40,41] An intein fragment is genetically fused to a protein of interest and the resulting chimera is produced in cells. A synthetic peptide, attached to the other intein portion and conjugated to a cell-penetrating peptide *via* a disulphide bond is then added to the cells. Upon cellular uptake, the disulphide bond is cleaved, allowing the intein fragments to assemble and splice to ligate the synthetic peptide to the designated cellular protein. *In vivo* semi-synthesis can thus be applied to study and manipulate signalling directly in cells and organelles.[40,41]

The enormous impact of inteins for protein semi-synthesis has sparked the search for biocatalysts that accelerate the coupling of two peptide fragments—ideally without the requirement for pre-activation or specific flanking residues. Several promising natural and engineered peptide ligase systems have been described. Sortase A is a bacterial transpeptidase involved in anchoring proteins to cell walls.[42] To do so, Sortase A processes a C-terminal LPXTG motif of substrate proteins, liberating the Gly residue and forming an acyl-enzyme intermediate between its active site cysteine and the substrate Thr residue (Figure 13.4A). The substrate is then transferred onto the N-terminus of a poly-Gly chain of the peptido-glycan cell wall.

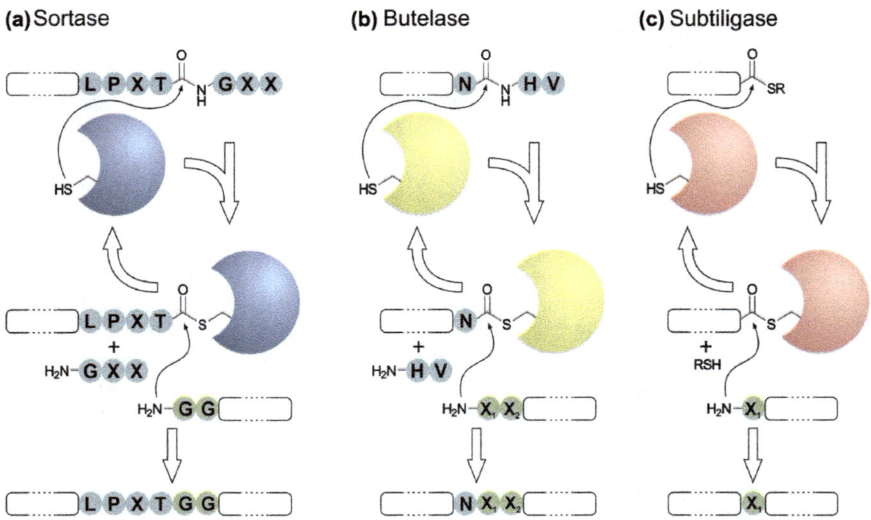

Figure 13.4 Biocatalytic peptide ligation reactions. A. Sortase (blue), an enzyme involved in anchoring proteins to cell walls, catalyses the transpeptidation of glycine-rich peptides.[42] B. Transpeptidase activity of Butelase, an Asn-specific plant peptide cyclase.[45] C. Subtiligase, a reengineered version of the protease Subtilisin, catalyses the aminolysis of peptide α-thioesters.[47] G = Gly; H = His; L = Leu; N = Asn; P = Pro; T = Thr; V = Val; X_1 = Any amino acid except Asp, Glu and Pro; X_2 = hydrophobic amino acid, preferentially Ile, Leu, Val or Cys.

This reaction can be exploited for protein semi-synthesis and to generate proteins of unusual topologies.[43] While natural sortases suffer from low rates and considerable substrate sequence requirements, reengineered versions have partially addressed these problems.[44] Butelase, a transpeptidase involved in plant peptide cyclisations, represents another exciting candidate biocatalyst for protein semi-synthesis.[45] Compared with Sortase, Butelase imposes fewer restrictions on substrate sequence (Figure 13.4B) and displays increased reaction rates. A third promising peptide ligase is Subtiligase, an engineered variant of the protease Subtilisin (Figure 13.4C).[46] Substitution of the active site Ser for a Cys residue results in a preference for aminolysis over hydrolysis of acyl-enzyme intermediates, thus favouring peptide ligation over proteolysis. Notably, coupling an α-thioester and a peptide with Subtiligase does not hinge upon the presence of an N-terminal Cys residue in the nucleophilic peptide fragment.[47] Collectively, intein-based and biocatalytic systems harbour great potential to facilitate peptide fragment couplings, which, along with further optimised variants, will undoubtedly enrich the field of synthetic protein chemistry.

13.3 General Applications of Synthetic Protein Chemistry

Synthetic approaches provide unique control over the chemical makeup of proteins beyond the standard proteinogenic amino acids. This exacting control over complex molecules enables a host of studies, ranging from site-specific incorporation of small probes for mechanistic investigations to the construction of novel functionalities and topologies. This section will cover some of the most common applications of protein synthesis technologies in the areas of biochemistry, biophysics and chemistry.

13.3.1 Precise Installation of Probes

The ability to precisely manipulate proteins enables detailed structure–function studies, thus providing an unprecedented view of how proteins work. A key application of chemical protein synthesis is the site-specific introduction of probes. Examples include isotope labels for NMR- and IR-spectroscopy,[34,48,49] spin labels for EPR spectroscopy,[50] fluorescent probes,[51] photo-caged amino acids[52] and photo-crosslinkers.[53] In the same manner, specific probes that reveal the mechanism of action of proteins in unprecedented detail can be incorporated. For example, SH2 domains—important for mediating protein–protein interactions in signal transduction—were synthesised *via* a three-piece ligation to address how the geometry of a salt bridge determines the selectivity of ligand binding.[54] Semisynthetic enzymes with non-charged arginine isosteres in their active site served to ascertain the importance of electrostatic catalysis.[55,56] Semi-synthesis also enables metalloprotein engineering, providing great flexibility to introduce unusual ligands to metal centres. The redox potential of the electron transporter azurin was tuned by substitution of the copper ligands from a thioether (Met) to a thiolate (homocysteine) and from a sulphur (Cys) to a selenium atom (selenocysteine).[57,58] Similarly, the importance of the nickel ligands in the detoxifying enzyme superoxide dismutase was demonstrated by the loss of activity in a modified protein featuring a secondary amine in place of the native backbone amide ligand.[59]

Without genetic constraints, chemical protein synthesis enables the modification of the protein backbone in further ways. This ability contributed greatly to the understanding of ion channels. The introduction of a D-amino acid in place of Gly77 into the K^+ channel KcsA revealed that a closed channel conformation contributes to preventing inadvertent Na^+ transport.[60–62] By sterically preventing this closure, a Gly77DAla mutation caused Na^+ to leak through the channel in the absence of the preferred K^+ ions. Structural and functional analysis of another semi-synthetic KcsA variant, bearing an ester in place of an amide at a K^+ coordination site, revealed that the backbone dipole is important for arranging K^+ ions in the selectivity filter.[63] Amide-to-ester mutations have also been applied to measure the energetic contributions of backbone hydrogen bonding in orienting functional groups,[64] promoting protein folding[65] and stabilising protein–protein interactions.[66]

The trajectory of the protein backbone can be steered by proline residues. Peptide bonds involving proline can adopt both the *cis*- and the *trans*-configuration which can induce profoundly different conformations of the peptide chain (Scheme 13.4). Synthetic proline analogues with different *cis–trans* equilibria and interconversion rates have been incorporated into proteins *via* (semi-)synthesis. Studies using synthetic β2-microglobulin revealed that the rate of proline isomerisation directly influences the misfolding and aggregation behaviour of this protein.[67,68] Proline analogues with strong *cis*-preferences induce β-turn formation, and can therefore be used to accelerate protein folding as demonstrated by semisynthetic

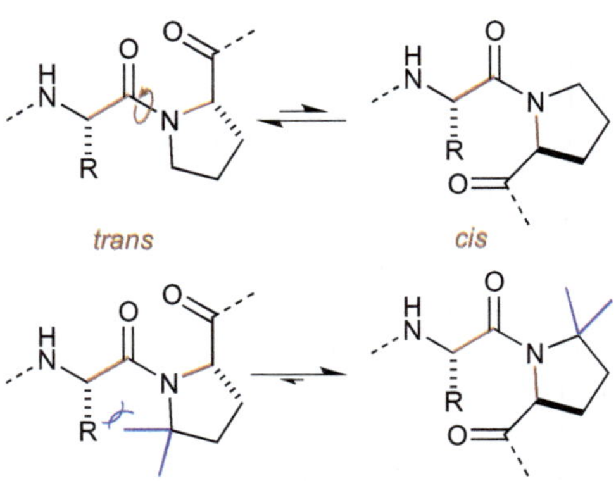

Scheme 13.4 Synthetic proline analogues control the *cis–trans* equilibrium. For example, 5,5-dimethyl-L-Proline (bottom) favours the *cis* conformation due to steric clashes in the *trans* conformation.[69]

ribonuclease A variants.[69,70] Entirely artificial turn mimics stabilize Ribonuclease A in an analogous manner.[71]

Collectively, these reports demonstrate the precision with which proteins can be manipulated using synthetic chemistry. By adding numerous spectroscopic and mechanistic probes to the palette of protein building blocks, chemical protein engineering provides unprecedented insights into the inner workings of diverse proteins. The development of novel synthetic strategies and creative implementation of physicochemical approaches will certainly continue to fuel exciting discoveries in protein science.

13.3.2 Access to Post-translationally Modified Proteins

Proteins are regulated by diverse covalent modifications. These post-translational modifications (PTMs) modulate the physicochemical properties of proteins or serve as docking platforms in protein–protein interactions, allowing dynamic switching of protein function. Enzymes that install or remove PTMs make up approximately 5% of the human proteome,[72] demonstrating the importance of protein modifications in biology. Access to site-specifically modified proteins using synthetic methods has revolutionised the ability to directly measure the functional consequences of PTMs. Among PTMs, protein phosphorylation has been researched most intensively, and will be discussed in more detail in Section 13.4.4. Synthesis of peptides carrying phosphorylated amino acids is usually achieved *via* the Fmoc strategy, and protected building blocks for the most common phosphoamino acids (phosphoserine, phosphothreonine and phosphotyrosine) are commercially available. Strategies to prepare peptides featuring unusual phosphoamino acids (*e.g.* phosphoarginine[73] and phospholysine[74]) and stable analogues thereof (*e.g.* phosphohistidine[75] and pyrophosphoserine[76]) have also been developed (Scheme 13.5). Importantly, the mild conditions employed in NCL are compatible with these building blocks.

Many proteins require constitutive post-translational modifications for their function. The anticoagulant microprotein S contains 11 γ-carboxyglutamic acid residues, crucial for Ca^{2+} binding and its blood coagulation activity. A synthetic version of microprotein S containing all naturally occurring PTMs, prepared *via* two consecutive NCL reactions, exhibited anticoagulant activity, demonstrating that functional proteins with multiple modifications can be accessed chemically.[77] In other instances, it is desirable to access proteins devoid of any modifications. To obtain a crystal structure of the acetyltransferase p300/CBP, it was necessary to use protein semi-synthesis to avoid auto-acetylation of fully functional p300/CBP in *Escherichia coli*.[78] Over their lifetimes, many proteins also accumulate spontaneous non-desirable modifications. To probe the effects of such lesions, several groups have developed synthetic approaches to access site-specifically "damaged" proteins. Lashuel and co-workers probed the effect of tyrosine nitration on the folding and aggregation of α-synuclein, a protein known to form oligomers implicated in neurodegenerative

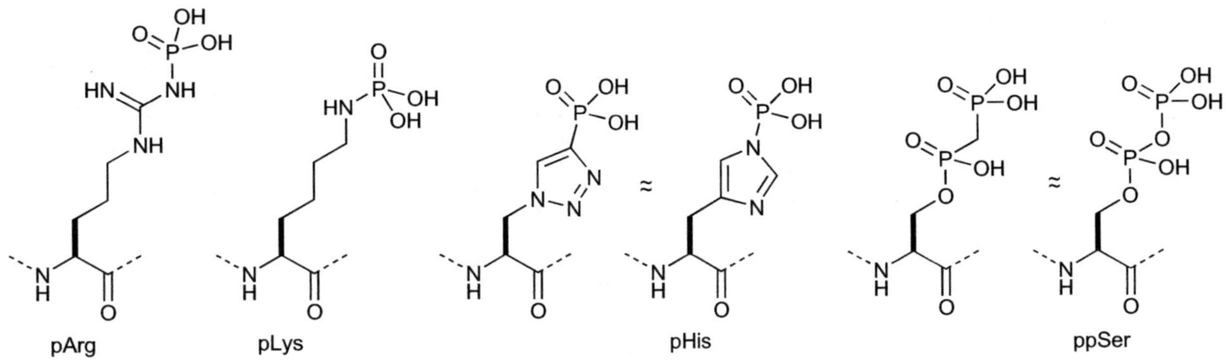

Scheme 13.5 Selected phosphoamino acids and their analogues.[73–76]

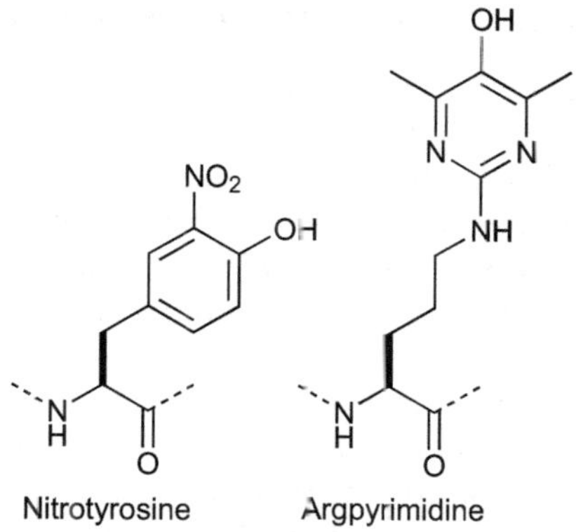

Scheme 13.6 'Damaged' residues.[79,80]

diseases.[79] Semi-synthetic α-synuclein containing nitrotyrosine residues (Scheme 13.6) formed different aggregates than their unmodified counterparts, demonstrating a direct effect of nitrative stress on protein misfolding. Electrophilic metabolites readily react with nucleophilic proteins. Becker and co-workers applied protein semi-synthesis to reveal that the modification of a specific arginine residue in the heat-shock protein Hsp27 to argpyrimidine (Scheme 13.6) decreases this chaperone's ability to prevent misfolding of substrate proteins.[80]

Secreted proteins are often stabilised by disulphide bonds between two cysteine residues. The importance of disulphide connections is easily demonstrated by site-directed mutagenesis to change one or both of the cysteine residues to alanine or serine, typically strongly destabilising the protein. Synthetic methods provide a pathway to strengthen such linkages through strategic replacement of cysteine residues with selenocysteine; diselenides are more stable and form more rapidly than disulphides.[81] Metanis and Hilvert harnessed diselenides to steer the folding pathway of a protein with four disulphide bridges through a non-native connectivity, ultimately accelerating oxidative protein folding.[82] Selenocysteine is found in several mammalian enzymes, including thioredoxin reductase, which contributes to the reduction of intracellular disulphide bonds. To explore the role of selenium in this enzyme, Hondal and co-workers synthesised selenocysteine- and cysteine-containing versions of Thioredoxin reductase, and found that the former is much more resilient toward oxidative inactivation than the latter.[83]

Post-translational modifications play diverse roles in biology. They represent constitutive additions to enhance the properties of proteins and can serve as dynamic on/off switches, but also act to cause protein damage. By providing direct access to site-specifically modified proteins, chemical total and semi-synthesis technologies have enabled direct comparison between proteins in distinct states. Detailed examples of how protein chemistry enhances our understanding of the diverse functions of PTMs are showcased in Sections 13.4.2–13.4.5.

13.3.3 Synthesis of Unusual and Artificial Structures

Protein synthesis has been used extensively to probe the function of existing proteins. Excitingly, synthetic protein chemistry has also driven the creation of proteins with novel properties. These features can be achieved through linking existing building blocks in unusual topologies, incorporating individual building blocks with unique functionalities, substituting secondary structure elements and even constructing entire proteins from synthetic building blocks.

Most natural proteins are linear polymers. Nevertheless, branched proteins have been observed, most notably in the post-translational ubiquitination of proteins. Ubiquitin, a 76-amino-acid protein, can serve as a biochemical signal upon covalent attachment of its C-terminus (Gly76) to the *N*-ε-amino group of a lysine side chain *via* an isopeptide bond.[84] Further derivatisation with additional ubiquitin moieties *via* lysines on ubiquitin generates complex poly-ubiquitin chains that can dramatically affect the properties of the substrate. A prominent example is the recruitment of polyubiquitinated proteins to the proteasome for degradation. The synthesis of site-specifically ubiquitinated proteins has shed light on the functions of diverse types of ubiquitin modifications. Typically, a ubiquitin thioester (generated *via* intein technology) is ligated to a synthetic peptide with a lysine analogue featuring a sulphhydryl group (mercaptolysine),[85] or *via* a thiolated auxiliary[86,87] (Figure 13.5). Alternative routes resulting in non-native linkages have been developed to facilitate the synthesis of these highly complex molecules and provide resilience against isopeptide bond cleavage.[88–90] Semi-synthetic monoubiquitinated histones revealed that this modification directly controls the activity of a multitude of proteins involved in engaging and modifying chromatin[91,92] and acts as a molecular crowbar to prevent the formation of tight chromatin fibres (see also Section 13.4.5).[93,94] Synthetic, well-defined

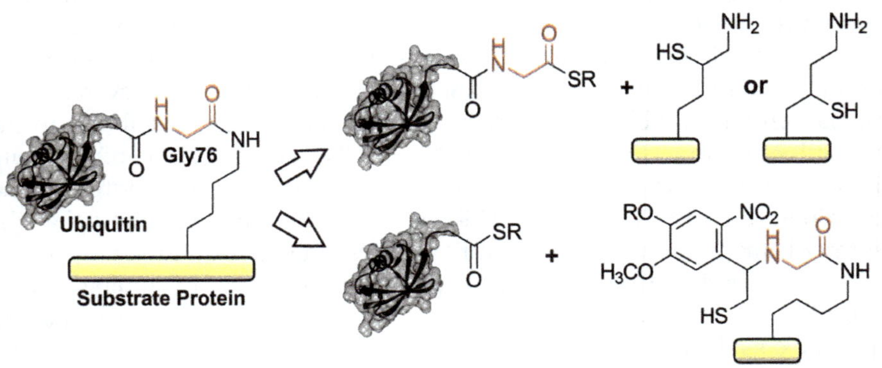

Figure 13.5 Synthesis of site-specifically ubiquitinated proteins. Mercaptolysine[85] (top) or thiol-containing cleavable auxiliaries[86] (bottom) enable ligation of an ubiquitin thioester to a substrate protein. Desulphurisation and (photolytic) auxiliary removal, respectively, result in native isopeptide connections.

polyubiquitinated proteins illuminated how chain length and topology control deubiquitination, protein degradation and molecular recognition by ubiquitin binding proteins.[95–97]

Branched topologies also occur as intermediates in protein splicing (Figure 13.3B). Capturing such intermediates for detailed structural and functional studies is difficult, but protein semi-synthesis can be harnessed to access sufficient quantities of branched inteins carrying an inactivating mutation, and thus provide a glimpse into the inner workings of inteins.[64] Similarly, (semi-)synthetic dimeric,[98,99] circular[100] and looped[101] proteins have shed light on the interplay between topology and stability.

Synthetic building blocks can serve to endow proteins with unique functionalities. Diazobenzene-containing amino acids represent a means to control proteins with light, complementing natural and synthetic methods for generating photoswitches.[102–104] Building blocks with a meta-substituted diazobenzene (Scheme 13.7) are ideally suited to control hairpin conformations due to the spacing of the amino and carboxy groups in the *cis-* and *trans*-conformation.[105] This photoswitch was introduced into Aβ to study how the preorganisation of a turn affects amyloid formation.[106] Synthetic protein chemistry also enables the production of functional proteins with secondary structure elements (10–20 residues) replaced by synthetic helix analogues (foldamers). Examples include functionally important helices of the hormone Interleukin 8 (ref. 107) and the transcription activation domain of the transcription factor HIF-1α (ref. 108) and even a part of the active site of an enzyme, chorismate mutase.[109]

Total chemical synthesis is entirely compatible with D-amino acid building blocks, thus granting access to mirror image proteins. A landmark proof of principle study from 1992 demonstrated that the enantiomer of HIV protease can fold and, as expected, cleave D-peptide substrates.[110] Similarly, a mirror-image polymerase efficiently catalyses the assembly of mirror-image nucleic acids, representing an exciting step toward complex artificial self-replicating systems.[111] Chaperones, protein machines that help their substrate proteins fold are "ambidextrous", *i.e.* they can fold both natural and synthetic mirror image proteins.[112] Because mirror-image peptides and proteins are resilient to protease digestion, they are attractive alternatives to biopharmaceuticals. Synthetic D-proteins have aided the discovery of new peptide ligands through mirror-image phage display.[113] In this process, vast L-peptide libraries are screened for high-affinity binders to a synthetic D-protein. Mirror image D-peptides of selected ligands are then synthesized as potential inhibitors of the naturally occurring L-protein. Mirror image

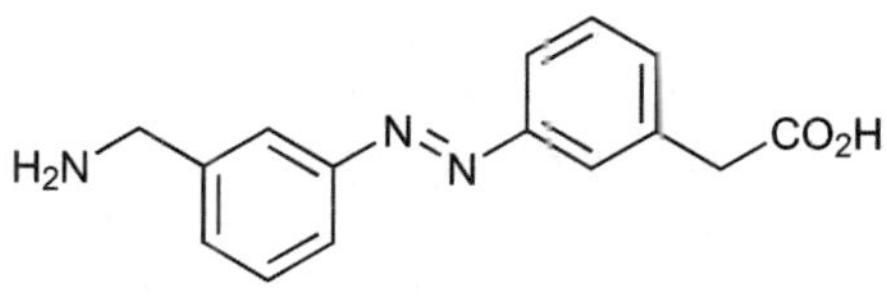

Scheme 13.7 Photoswitchable amino acid.[105]

proteins have also benefitted structural biology. Racemic mixtures crystallise in more space groups than the pure enantiomers, and the structures of numerous difficult-to-crystallise proteins have been solved using racemic protein crystallography.[114] This technology can be extended to quasi-racemic molecule pairs where the individual components differ slightly.[114] An example is the co-crystallisation of monomeric D-ubiquitin with L-ubiquitin polymers for structural characterisation of ubiquitin chains.[115]

Synthetic protein chemistry adds exciting building blocks to the toolkit of protein engineers. The design of novel protein topologies and functions, limited only by our imagination, will certainly continue to be a fruitful area of research with immense technological and pharmaceutical potential.

13.4 Case Studies: Biological Insights Through Synthetic Protein Chemistry

Chemical methods have transformed our ability to understand and manipulate proteins. Diverse protein families and types of functions have been probed by synthetic approaches, revealing their biochemical and biophysical underpinnings in unprecedented detail. A selection of topics where protein chemistry has and continues to have an especially profound effect–ranging from mechanistic enzymology to biological signalling and epigenetics–will be discussed in detail in the following sections.

13.4.1 HIV Protease—Probing Structure, Mechanism and Asymmetry

HIV protease catalyses the cleavage of a viral polyprotein precursor into mature viral proteins. It is a homodimer comprised of two 99-amino acid chains where each polypeptide contributes one aspartate residue, Asp25, to form an aspartic protease dyad (Figure 13.6A and B). Given its central role in viral maturation, it has been the target of many inhibitor development projects which yielded an array of successful antiviral drugs. Chemical protein synthesis efforts have facilitated structural studies that served as a basis for rational inhibitor engineering and provided unprecedented insight into the mechanism and dynamics of HIV protease.

In 1988, when recombinant technology was still in its infancy, Schneider and Kent reported the first total chemical synthesis of HIV protease.[116] The peptide chain was assembled on a polystyrene solid support using Boc-chemistry. Upon cleavage and side-chain deprotection with HF, the protein was purified by size exclusion chromatography. The synthetic protein was found to be active and was used to determine the first structure of inhibitor-bound HIV protease *via* X-ray crystallography.[117,118] The synthetic protocol served as a template to access site-specifically modified HIV protease variants, including a mirror image version built from D-amino acids[110] (Section 13.3.3) and a stabilized variant featuring a constrained beta-turn

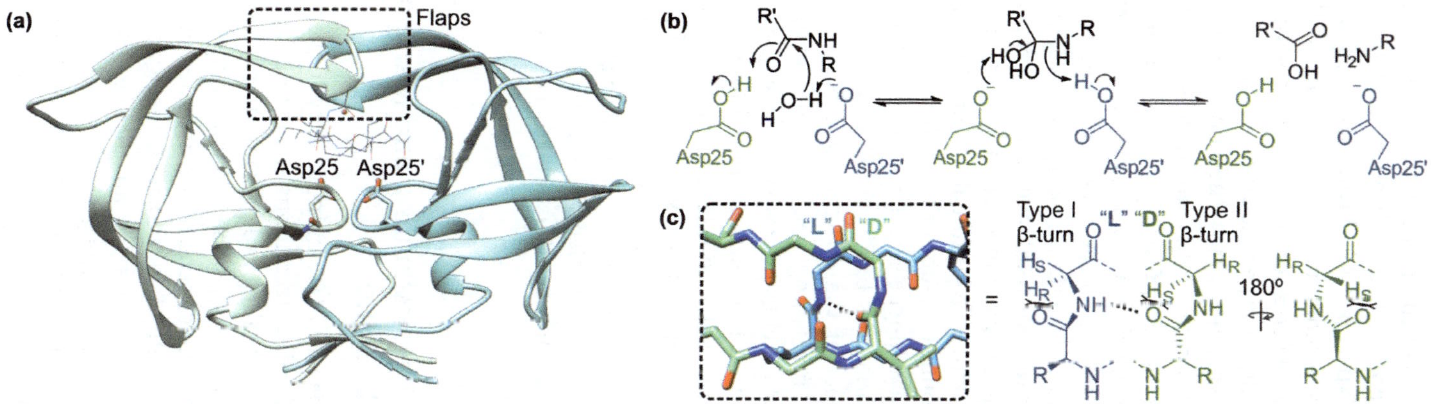

Figure 13.6 Structure and mechanism of HIV protease. A. X-ray structure of the HIV protease dimer in complex with the reduced peptide inhibitor MVT-101 (black wire; pdb code 3HAU).[121] The catalytic aspartate residues are shown as sticks for each subunit. The mobile flap regions are indicated by a black box. B. Mechanism of peptide bond cleavage by aspartic proteases. C. Conformational asymmetry in the flap region. One flap adopts a type I β-turn (blue, compatible with an L-amino acid), the other adopts a type II β-turn (green, compatible with a D-amino acid). The flaps are connected through an inter-chain hydrogen bond (dashed line).

mimic.[119] The site-specific introduction of ^{13}C-labels into Asp25, the catalytic acid/base of each subunit, enabled detailed mechanistic studies.[120] The presence of two distinct chemical shifts observed by ^{13}C-NMR as well as chemical shift perturbation experiments indicated that Asp25 of one subunit is ionized while the other is protonated when a mimic of the tetrahedral intermediate is bound. This observation is consistent with a mechanism where the two otherwise identical aspartate residues assume distinct roles as catalytic base and acid, respectively. How then does a symmetric enzyme mediate this functional asymmetry? A hint came from structural studies: the flaps covering the substrate (Figure 13.6C) adopt different conformations.[117] The geometry of a conserved glycine (Gly51) is typical for L-amino acids in one subunit and for D-amino acids in the other subunit. To establish how the flap conformation relates to function, a series of covalent HIV protease dimers was prepared *via* convergent total synthesis.[121] Each monomer was assembled from two fragments by native chemical ligation, and joined through a five-residue linker with an N-terminal cysteine (originally protected as a thiazolidine). Kinetic control ensured the desired sequence of peptide ligations (Section 13.2.2). The wild-type covalent dimer displayed full activity. Mutating Gly51 in both monomers to either L-Ala or D-Ala rigidified the flaps and reduced the activity of the enzyme by a factor of 10 and 100, respectively.[121] Remarkably, furnishing one monomer with L-Ala and the other with D-Ala at position 51 provided a protease with wild-type-like activity. These results demonstrated that flap asymmetry is important for catalysis. Moreover, synthetic variants with amide-to-ester[121,122] and amide-to-thioester[123] substitutions in the peptide backbone revealed that a functionally important hydrogen bonding network connects the flaps *via* a water molecule with the substrate and the active site aspartate residues. These detailed mechanistic studies have revealed the role of several components of the function of HIV protease and have thus provided a valuable framework for rational inhibitor design.

13.4.2 Erythropoietin: Understanding and Manipulating Protein Serum Lifetime

Erythropoietin, also called EPO, is a 166-amino-acid cytokine that stimulates the production of red blood cells. Recombinant EPO has proven useful in treating anaemic patients, thus fuelling programmes to engineer more stable and potent derivatives. This process has been challenging because the biological activity of EPO is critically dependent on its extensive glycosylation, amounting to approximately half of its molecular weight. Four branched polysaccharides are attached to asparagine and serine residues to protect EPO in serum, ensuring safe passage through the endocrine system. Indeed, removal of three glycosylation sites through mutation of the target asparagine residues reduces biological activity by an order of magnitude.[124] Following these discoveries, chemists have tackled the synthesis of homogeneous EPO and various derivatives to shed light on the role of glycosylation and pave the way for engineering simplified EPO variants.

Bioconjugation strategies provided early access to EPO analogues. Macmillan *et al.* recombinantly produced mutant EPO with cysteines in place of the asparagine glycosylation sites.[125] Upon oxidative protein folding (native EPO contains two disulphide bonds), iodoacetamide-modified glycan units could be conjugated selectively to the free cysteines at the native glycosylation sites. Hirano *et al.* subsequently combined this process with native chemical ligation to facilitate the production of more densely glycosylated EPO variants.[126,127] A peptide α-thioester corresponding to the N-terminal region of EPO was alkylated with branched undecasaccharide units and subsequently condensed with a recombinant EPO fragment bearing an N-terminal cysteine residue. A variant containing two polysaccharides stimulated the growth of cultured erythrocyte cells, but the addition of an extra polysaccharide at a non-native position was detrimental for activity, demonstrating that the positioning of glycan units is important for function.

Kochendoerfer *et al.* set out to synthesise an EPO variant where the glycan units were replaced by branched anionic polymers.[128] Four peptide fragments were prepared by solid-phase synthesis. Where appropriate, hydrophilic polymers were attached to the peptides *via* an oxime bond between an aminooxy moiety on the polymer and the ketone group of a Lys(N^{ε}-levulinyl) residue. Three iterative NCL steps yielded homogeneous, full-length EPO that displayed superior biological activity to recombinant EPO due to enhanced stability of the synthetic variant in serum. Attachment of unbranched chains or polymers with an insufficient or excessive number of anionic units reduced the activity of synthetic EPO.[129]

In 2013, the Danishefsky group reported a tour de force total synthesis of EPO with a native-like glycosylation pattern.[130] Key to their success was a combination of glycosylamino acid synthesis for incorporation of the *O*-linked tetrasaccharide,[131] aspartylation reactions on solid phase to generate the *N*-linked glycopeptides,[132,133] and a series of strategic convergent peptide ligation reactions (Figure 13.7).[134] Upon completion of the ligations, Wang *et al.* performed radical desulphurisation reactions to convert non-native cysteine residues, exploited as temporary ligation handles, into alanine and subsequently unmasked the native cysteine residues by Ag^+-mediated removal of Acm protecting groups.[130] Once refolded, synthetic EPO displayed the expected erythropoietic activity. Using a modified strategy, Murakami *et al.* achieved the synthesis of EPO with varying numbers of the N-linked oligosaccharides.[135] A total of six (glyco-)peptides, synthesised using Boc and Fmoc chemistries, were condensed by NCL. Characterisation by mass spectrometry and circular dichroism spectroscopy confirmed the identity and overall fold of the synthetic glycoproteins. The authors were also able to confirm that reducing the number of glycans diminished the erythropoietic effect of the glycoprotein. These remarkable achievements demonstrate that even highly complex biomacromolecules can be synthesised and thus probed *via* detailed structure–activity relationships.

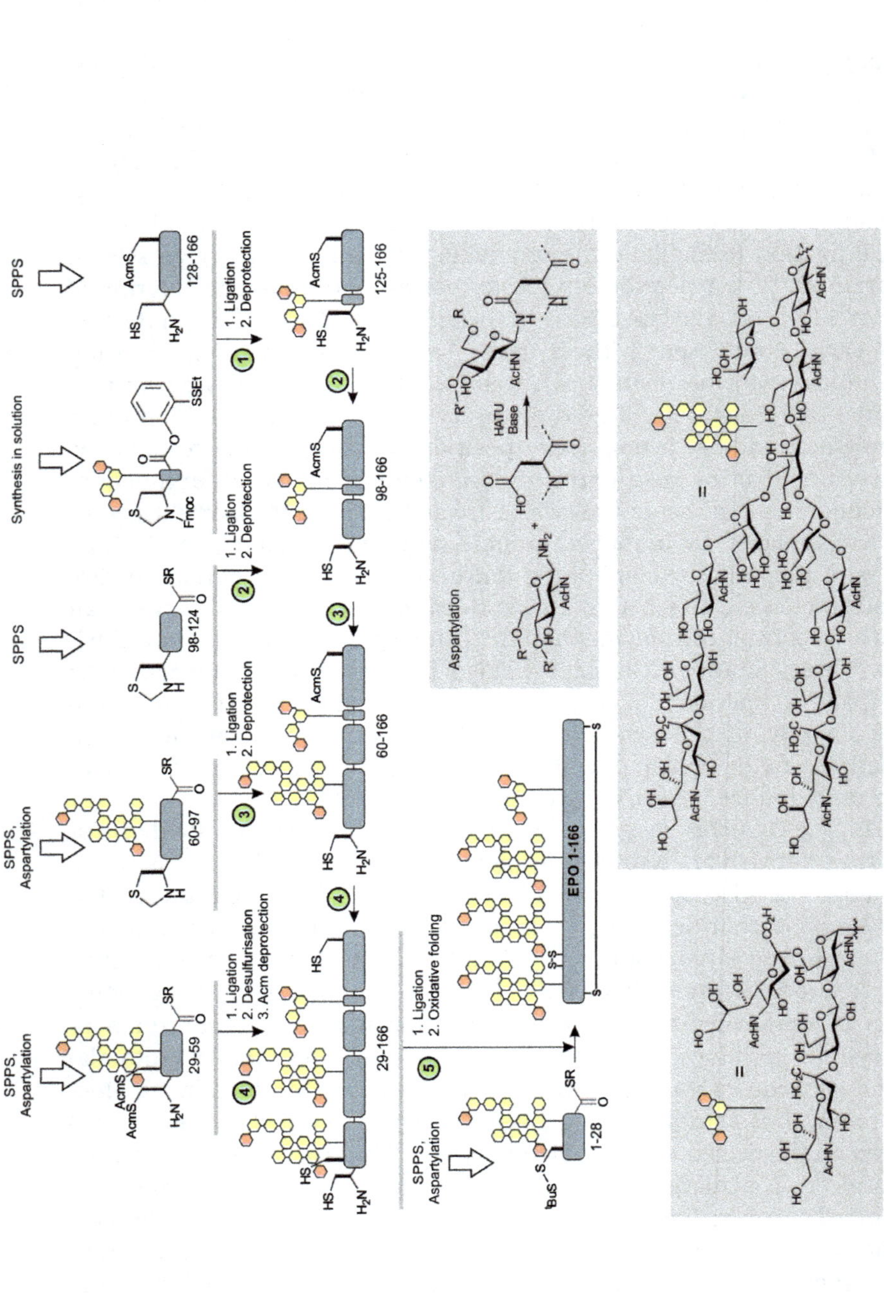

Figure 13.7 Total synthesis of the glycoprotein erythropoietin (EPO). Iterative ligation and deprotection steps are performed to join six (glyco-)peptides into full length EPO.[130] Glycans are depicted as yellow (neutral monosaccharides) and orange (anionic monosaccharides) hexagons. The chemical structures are detailed in the grey boxes (bottom). Circled numbers refer to the sequence of ligation steps.

13.4.3 GTPases: Mechanism of Membrane Attachment of Molecular Address Tags

Many biochemical processes in eukaryotic cells are compartmentalised in membranous organelles. As the functions and hence the contents of organelles are unique, so are their membranes. To ensure transport of cargo to the correct destination, cells harness membrane-associated proteins as address tags, specifying the identity of associated organelles. Key constituents of this system are lipidated GTPases of the Rab family.[136] Rab proteins act as molecular switches because they display distinct conformations, and consequently distinct functions, when binding guanosine triphosphate (GTP) or diphosphate (GDP). Their properties have been inherently difficult to probe due to the lipidated residues at their C-termini, which mediate membrane anchoring. Several groups at the Max Planck Institute for Molecular Physiology joined forces to overcome this challenge. They developed approaches to synthesise site-specifically modified GTPases by furnishing C-terminally truncated GTPase variants with inteins, and condensing the resulting thioesters with tailored peptides carrying N-terminal cysteine residues (Figure 13.8A).[137] Semi-synthetic, fluorescently labelled Rab7 variants served to aid in the analysis of the mechanism of membrane anchoring through prenylation by Geranyl-Geranyl-transferase (GGTase).[138] A change in dansyl fluorescence enabled quantification of the interaction of Rab7 with GGTase to reveal that the two proteins interact relatively weakly but that the presence of a Rab escort protein (REP-1) drives the formation of a ternary complex. Lipidated Rab variants can be constructed using the same approach, although the native chemical ligation step has to be performed in a suitable detergent.[139] Refolding can be achieved in the presence of GDP, GTP or an analogue thereof to provide direct access to Rab variants in different functional states. By using site-specifically prenylated variants as substrates, Durek *et al.* found that GGTase catalyses the prenylation of multiple cysteine residues at the C-terminus of Rab by a random sequential mechanism.[140]

Once lipidated, Rab GTPase isoforms can be distributed to their cognate membranes. This delivery is mediated by escort proteins, such as REP-1 or the guanine nucleotide inhibitor RabGDI. Semi-synthetic lipidated GTPases were invaluable in characterising how RabGDI binds prenylated GTPases and is able to mediate their distribution. A crystal structure of RabGDI with a semi-synthetic monoprenylated GTPase revealed that the prenyl chain is inserted into a hydrophobic pocket distal from the core protein–protein interaction surface (Figure 13.8B).[141] This arrangement facilitates promiscuous binding of diverse Rab family members with distinct C-termini. The binding energies of prenylated Rab7 to the escort proteins REP-1 and RabGDI further provide a mechanistic basis for how GTPases are targeted to membranes.[142] In its GDP-bound form, Rab7 engages its escorts with nanomolar affinities, allowing extraction of Rab7 from membranes (Figure 13.8C). In contrast, when Rab7 harbours a stable GTP analogue, these interactions weaken by approximately three orders of magnitude.

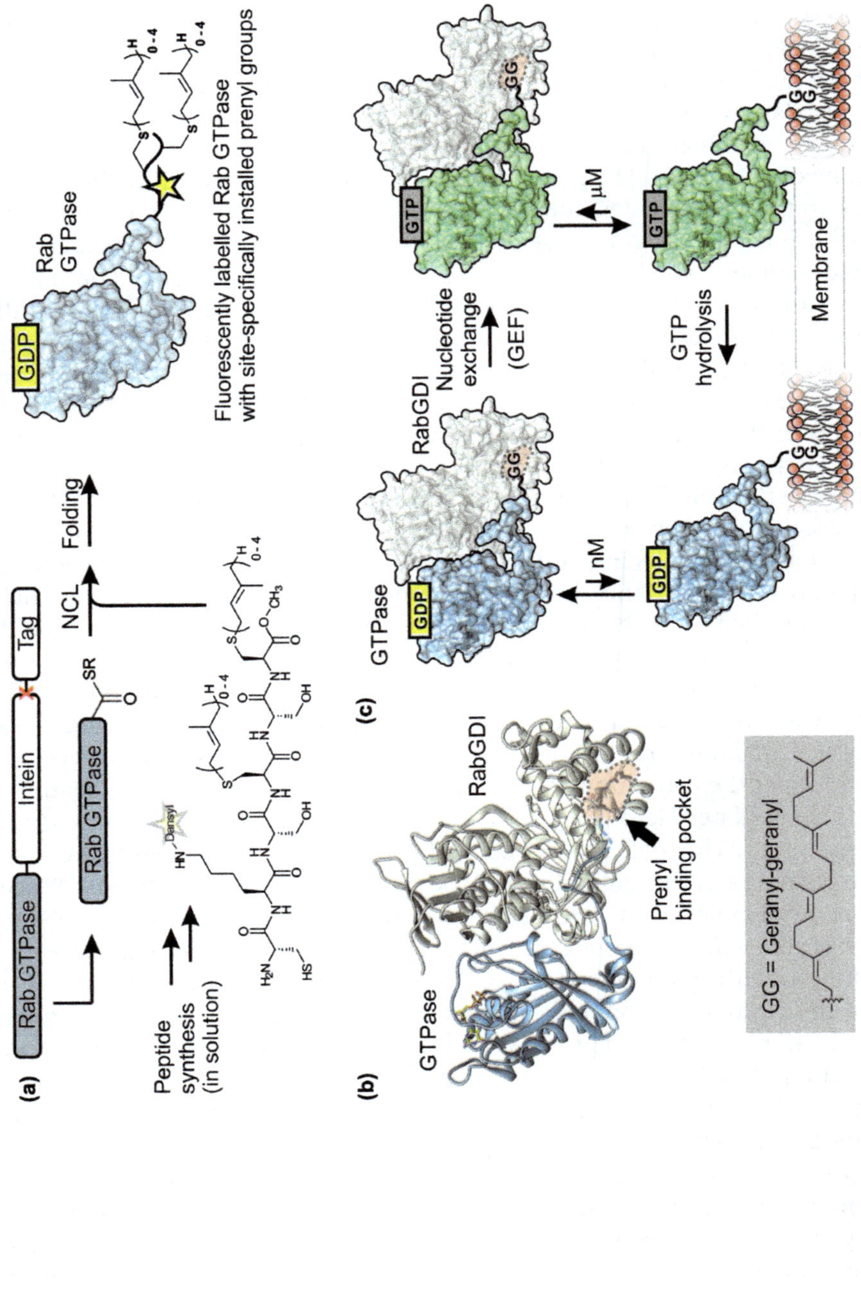
(a)
Rab GTPase
Intein
Tag
Rab GTPase
SR
Peptide synthesis (in solution)
Dansyl
NCL
Folding
GDP
Rab GTPase
Fluorescently labelled Rab GTPase with site-specifically installed prenyl groups
(b)
GTPase
RabGDI
Prenyl binding pocket
GG = Geranyl-geranyl
(c)
GTPase
RabGDI
GDP
GG
nM
Nucleotide exchange
(GEF)
GTP
μM
GTP hydrolysis
Membrane

Thus, upon nucleotide exchange from GDP to GTP—mediated by a guanine nucleotide exchange factor—GTPases can be released from the tight grip of escort proteins and deposited in a cognate membrane. Collectively, these quantitative biochemical and biophysical experiments revealed in detail the mechanisms and driving forces for GTPase shuttling to and from cognate membranes.

13.4.4 Regulation of Regulators: Synthesis of Phosphorylated Kinases and Phosphatases

Many biological signalling networks rely on protein phosphorylation to transmit extracellular inputs into intracellular responses, such as changes in gene expression. Phosphorylation is dynamic, and the enzymes that catalyse the addition (kinases) and the removal of phosphate groups (phosphatases) are tightly regulated, often through phosphorylation of the kinases and phosphatases themselves. Mutations in enzymes that cause aberrant regulation of protein phosphorylation are a frequent cause of cancer and other pathologies. Therefore, there is great interest in elucidating how phosphorylation alters the structure and activity of signalling enzymes.

Besides their catalytic domains, many kinases and phosphatases harbour peptide-binding modules such as SH2 (recognition of phosphotyrosine-containing peptides) and SH3 (recognition of proline-rich peptides).[143] These domains not only recruit auxiliary proteins but also act intramolecularly to control enzymatic activity. For example, Src family kinases (tyrosine-specific kinases involved in a multitude of cellular signalling pathways) are autoinhibited when their SH2 domains engage a C-terminal phosphotyrosine residue (Figure 13.9A).[144] This regulatory feature can be extended to other enzymes with the same domain architecture: Appending an artificial phosphotyrosine C-terminal tail to CSK (*C*-terminal Src kinase) in the first application of expressed protein ligation led to a change in kinase activity, although in this case an increase (Figure 13.9A).[38] Following the same principle, the protein tyrosine phosphatases SHP-1 and SHP-2 are activated by intramolecular SH2-phosphotyrosine interactions. Due to the inherent phosphatase activity of these enzymes, Cole and co-workers assembled semi-synthetic versions with the non-hydrolysable phosphotyrosine analogue phosphonomethylene phenylalanine (Pmp, Figure 13.9B).[145,146] The

Figure 13.8 Synthesis and applications of site-specifically modified Rab GTPase. A. Semi-synthesis of lipidated Rab GTPases using expressed protein ligation.[137] The yellow star indicates a fluorescent dansyl label. B. Structure of a semi-synthetic prenylated GTPase with the escort protein RabGDI (pdb code 1UVK).[141] The prenyl binding pocket on RabGDI (shaded patch) is distal from the main protein–protein interaction surface. C. Model for membrane deposition and extraction of Rab GTPase based on quantitative measurement of RabGDI binding as a function of GTPase state.[142] GEF = Guanine nucleotide exchange factor.

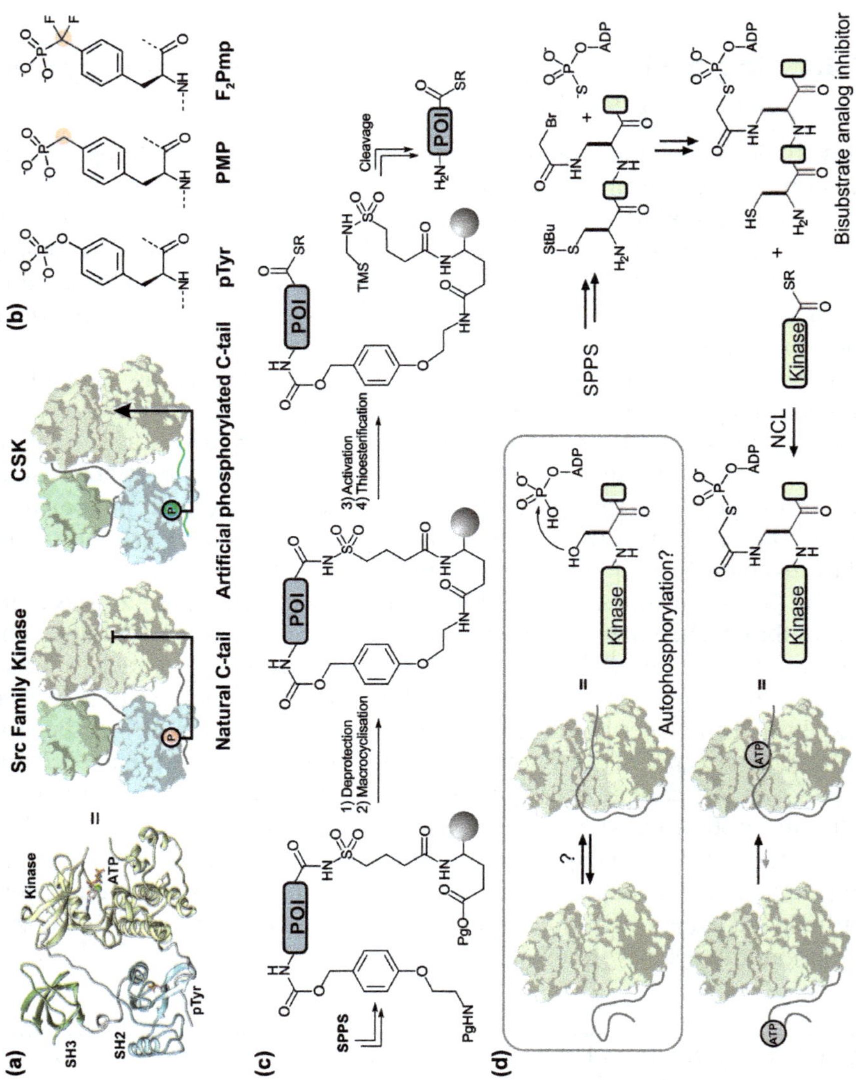
(a)
SH3
SH2
Kinase
ATP
pTyr
Src Family Kinase
CSK
Natural C-tail
Artificial phosphorylated C-tail
(b)
pTyr
PMP
F2Pmp
(c)
SPPS
POI
1) Deprotection
2) Macrocyclisation
3) Activation
4) Thioesterification
Cleavage
TMS
(d)
Autophosphorylation?
Kinase
SPPS
NCL
Bisubstrate analog inhibitor
ADP
ATP

difluoromethylene derivative F_2Pmp is a closer mimic of phosphotyrosine, and the authors found that its incorporation into SHP-1 and SHP-2 conferred a larger increase in phosphatase activity as compared with Pmp. Mechanistically distinct modes of kinase/phosphatase regulation by phosphorylation have been elucidated *via* protein semi-synthesis; examples include phosphorylation near the substrate binding site,[147] to regulate membrane association[148] and stability[149] and even substrate specificity.[149]

Phosphorylation of peptide binding domains themselves can also modulate kinase activity. To probe the function of three phosphotyrosine residues in the SH3 domains of Abl and Arg kinases (regulators of cell shape and motility), Zitterbart and Seitz synthesised a combinatorial array of 20 SH3 domains.[150] To bypass time-consuming purification steps, the authors synthesised peptides as C-terminal sulphonamides and performed an on-resin cyclisation to connect the N-terminus to a glutamate side-chain positioned at the C-terminus (Figure 13.9C). Activation of the sulphonamide and thiolysis liberate the C-terminus as an α-thioester; full-length peptides remain attached to the solid support *via* their N-terminus. Upon global deprotection and cleavage from the resin, the 'self-purified' (phospho)peptide α-thioesters were coupled *via* native chemical ligation directly on the array surface. Individual pTyr marks and their combinations differentially affected the affinity of the SH3 domains to various proline-rich peptide ligands, indicating that these phosphorylation events can direct protein–protein interaction pathways.

Autophosphorylation can occur intramolecularly or in *trans*, *i.e.* one kinase molecule phosphorylates the regulatory region of another. The former mechanism requires that the kinase tail can access its own active site. Support for the existence of such a conformation in protein kinase A (PKA) came from a semi-synthetic variant where the tail was converted into a bisubstrate inhibitor.[151] An ATP analogue was covalently tethered to the side chain of a regulatory serine residue within the tail (Figure 13.9D). The resulting PKA variant displayed reduced kinase activity and a distinct conformation from a control variant, consistent with the kinase active site

Figure 13.9 Protein synthesis approaches to study kinase and phosphatase phosphorylation. A. Domain architecture and regulation of Src family kinases. An SH3 domain (green, binds proline-rich peptides) and an SH2 domain (cyan, binds phosphotyrosine) regulate protein–protein interactions and control the activity of the kinase domain (yellow; pdb code 1AD5). Phosphorylation of the C-terminal tail diminishes kinase activity of archetypal Src family members (middle). Appending an artificial phosphorylated tail to CSK by expressed protein ligation activates this kinase.[38] B. Stable analogues of phosphotyrosine. C. Synthesis of self-purifying thioesters for convergent SH3 assembly.[150] Deprotection, 1% TFA; Macrocyclisation, PyBOP; Activation, Trimethylsilyldiazomethane; Thioesterification, NaSPh, BzlSH; Cleavage, TFA, scavengers. D. Capturing the autophosphorylation conformation in a semisynthetic kinase *via* covalent ATP tethering.[151]

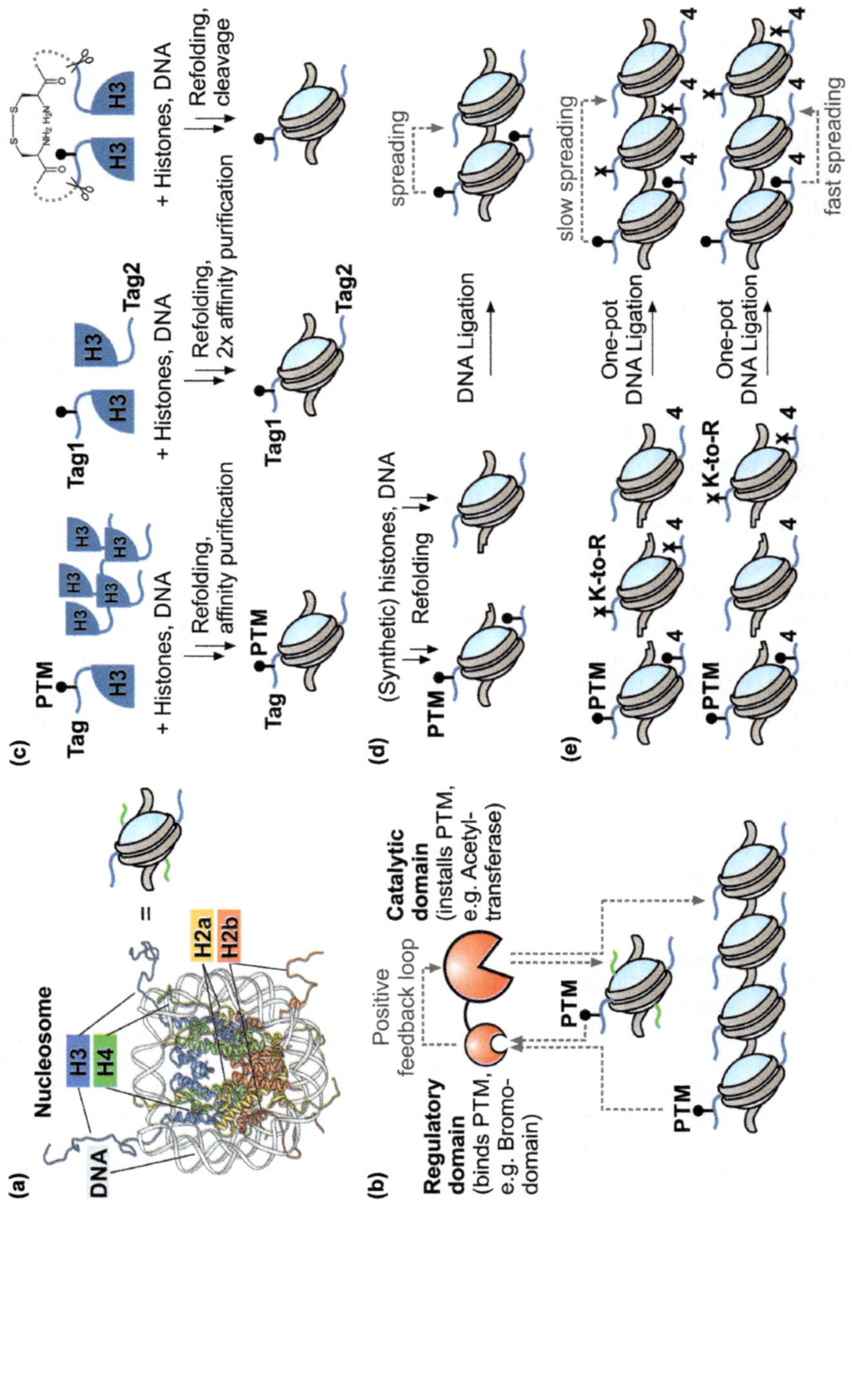
(a)
Nucleosome
DNA
H3
H4
H2a
H2b
(b)
Regulatory domain (binds PTM, e.g. Bromo-domain)
Positive feedback loop
Catalytic domain (installs PTM, e.g. Acetyl-transferase)
PTM
(c)
Tag
PTM
H3
+ Histones, DNA
Refolding, affinity purification
Tag1
Tag2
Refolding, 2x affinity purification
Refolding, cleavage
(d)
(Synthetic) histones, DNA
Refolding
DNA Ligation
spreading
(e)
K-to-R
One-pot DNA Ligation
slow spreading
fast spreading

'biting its tail'. Taken together, these semi-synthesis projects provided the mechanistic foundation of distinct molecular logic operations that control kinase and phosphatase activity in cellular signalling.

13.4.5 Histones: The Geometry of Positive-feedback Loops

In eukaryotic cells, the genome is packaged in a DNA–protein complex referred to as chromatin.[152] DNA is wrapped around protein spools composed of octamers of cationic proteins (two copies each of histones H2A, H2B, H3 and H4) to yield so-called nucleosomes (Figure 13.10A). Post-translational modifications of histones regulate access to the surrounding DNA sequence, culminating in the formation of heritable chromatin states that facilitate or restrict gene expression (epigenetics).[153] Many of the functions of histone PTMs have been scrutinized using synthetic chromatin templates, comprehensively reviewed elsewhere.[154] The preparation of 'designer chromatin' benefits from the small size of the core histones (less than 150 residues) and robust protocols for the reconstitution of nucleosomes from purified recombinant histones and DNA, either individually as mononucleosomes or in regular nucleosome arrays. However, an additional layer of complexity is represented by the spatial dimension: how does a PTM of one histone affect the properties of chromatin within the same *versus* neighbouring nucleosomes? Addressing this type of question requires designer chromatin with controlled asymmetries. The following section will cover the unique opportunities and challenges for the synthesis and application of defined non-covalent topologies, exemplified by the enzymology of histone modifications involving positive-feedback loops.

Many histone-modifying enzymes contain domains to install and bind the same PTM and thus possess the hardware for positive-feedback loops (Figure 13.10B). This feature indicates that enzymes can 'spread' chromatin states along the fibre through iterative modification, engaging of the product through the binding domain and further modifying adjacent histones. For example, the lysine acetyltransferases GCN5 and p300 contain a bromodomain, dedicated to recognition of acetyllysine

Figure 13.10 Synthesis of non-covalent assemblies to study PTM feedback loops on chromatin. A. Structure of a nucleosome, the repeating unit of chromatin (pdb code 1KX5). DNA is wrapped around an octamer of two copies each of histones H2a (orange), H2b (red), H3 (blue) and H4 (green). B. Positive feedback loops operate on chromatin intranucleosomally and internucleosomally. Histone-modifying enzymes (red) bind PTMs *via* regulatory domains, thus activating the catalytic domains to modify residues within the same, adjacent or even distant nucleosomes. C. Strategies to assemble asymmetrically modified nucleosomes.[155,158,159] D. Nucleosome ligation strategy to assemble site-specifically modified dinucleosomes.[160,161] E. Synthesis of tripartite designer chromatin. Spreading is inhibited by the presence of intervening K-to-R mutant nucleosomes that cannot be methylated.[162]

residues. Consistent with a positive-feedback loop, histone H3 of designer chromatin templates is acetylated more readily by GCN5 in the case of pre-installed acetyllysine residues on H4 as compared with unmodified chromatin.[155] To test whether this feedback loop even operates between the two copies of H3 within one nucleosome, Li and Shogren-Knaak assembled asymmetric mononucleosomes.[155] These reagents were prepared by mixing tagged, acetylated histone H3 and a tenfold excess of untagged, unmodified H3 with recombinant H2A, H2B, H4 and DNA (Figure 13.10C). Asymmetrically modified mononucleosomes could then be enriched through affinity chromatography. Acetyltransferase assays with asymmetric nucleosomes as substrates demonstrated that pre-acetylation of one copy of H3 indeed promoted GCN5 activity in a bromodomain-dependent fashion.[156]

Different acetyllysine inputs, *i.e.* which residue on which histone is acetylated, affect acetyltransferase outputs differently. To dissect this complex feedback loop, Nguyen *et al.* synthesised a library of over 50 site-specifically modified nucleosomes and furnished each variant with a unique DNA barcode.[157] Following incubation with the acetyltransferase p300 and isolation of products using antibodies that recognize H3 acetyllysine residues, preferred substrates were identified by DNA sequencing. The results revealed that acetyllysine marks on H4 stimulate p300-dependent acetylation of H3, and that this stimulation was enhanced when several acetyllysine marks were combined.

Several histone lysine methyltransferases are stimulated by methyllysine binding through a regulatory domain. Elucidating these feedback mechanisms benefited from streamlined syntheses of defined chromatin templates. For instance, asymmetric mononucleosomes can be prepared efficiently by harnessing tandem affinity tags,[158] where each copy of H3 is furnished with a unique purification tag, or *via* a covalent heterodimeric intermediate (Figure 13.10C).[159] In the latter strategy, uniquely modified histones are fused head-to-head with a cleavable linker, which is removed by proteolysis upon refolding of asymmetric histone octamers. Synthetic asymmetrically methylated nucleosomes were instrumental in quantifying the existence of natural asymmetric nucleosomes[158] and investigating the signalling propensities of 'bivalent' substrates containing both activating and repressive PTMs.[158,159]

Positive feedback loops also operate across nucleosomes, thereby facilitating the propagation of chromatin states along the chromatin fibre. Investigations of 'spreading' mechanism require nucleosome array substrates that contain chemically methylated 'input' nucleosomes and unmodified nucleosomes to measure the 'output'. Such topologies can be accessed from homogeneous mononucleosomes furnished with short, single-stranded DNA overhangs that allow directional ligation using DNA ligase enzymes (Figure 13.10D).[160] Dinucleosomes prepared in this way featuring a pre-methylated and an unmodified nucleosome are methylated more readily than variants with a Lys-to-Arg mutated and an unmodified nucleosome.[161]

The difference was manifested in an increase in catalytic constant (k_{cat}), indicating that product binding by the regulatory domain increases the activity of the methyltransferase domain. This concept can be extended to tripartite chromatin where three blocks of four (i) methylated 'input' nucleosomes (ii) unmodified 'readout' nucleosomes and (iii) Lys-to-Arg barrier nucleosomes are ligated in defined topologies (Figure 13.10E).[162] Orthogonal DNA overhangs facilitate a one-pot ligation to generate designer chromatin featuring site-specific modifications of specific domains. These types of arrays served to establish that methyllysine read–write feedback is strongest when input and substrate nucleosomes are adjacent and weaker when these blocks are separated by unmethylatable barrier nucleosomes.

Innovations in sample preparation and protein mass spectrometry continue to fuel discoveries of new protein-modification signatures.[163] Improvements in sensitivity and mass accuracy are pivotal for the identification of novel types of PTMs. Fragmentation methods facilitate pinpointing the locations of PTMs within a given histone as well as in the context of the chromatin landscape. The latter also hinges upon sophisticated means to enrich for specific chromatin domains according to DNA sequence, function and biochemical makeup. Collectively, these methods define key principles underlying the complex architecture of chromatin and the components involved in its regulation. The resulting blueprint sets the stage for chemists and biologists to develop synthetic 'designer' chromatin that mimics the complexity of natural chromatin as closely as possible. Milestone achievements have been realised through judicious combination of convergent chemical and biochemical syntheses, yielding unique platforms for the reconstitution of diverse chromatin-dependent processes *in vitro*. Such experiments have provided unprecedented insight into how histone PTMs fine-tune the structure and dynamics of chromatin and orchestrate the activities of molecular machines that operate on chromatin.

13.5 Concluding Remarks

Protein total- and semi-synthesis is enabled by highly efficient amide bond formation on solid phase and facilitated through chemoselective ligation reactions that join unprotected peptide fragments under mild conditions. Nowadays, small and relatively simple proteins can be routinely synthesised with site-specific modifications, especially if these modifications are located close to the protein termini. For example, over 100 chemically defined histones have been assembled *via* total- and semi-synthesis. More complex substrates, including those with unusual topologies, are nevertheless still challenging to synthesise at high yields. Thus, there is a continuing demand for innovation in the field of synthetic protein chemistry. In particular the high-throughput synthesis of proteins represents a grand challenge. Methods that bypass time-consuming purification steps based, for example, on self-purifying fragments[150] and solid-phase protein ligations are encouraging candidates for this development.[164] Increasing the speed of peptide synthesis *per se* is also highly desirable, and flow-chemistry technology promises to meet this challenge.[165]

Beyond total synthesis, exciting discoveries in the areas of bioconjugation and synthetic biology[1] increase the diversity of building blocks that can selectively be incorporated into proteins, and the speed with which site-specifically modified proteins can be accessed. Importantly, chemical bottom-up and top-down, as well as synthetic biological methods, are fully orthogonal, and their combination will probably continue to improve chemical control over proteins for diverse applications in basic research, materials science and biomedicine.

Acknowledgements

M.M.M. is supported by a Sir Henry Dale Fellowship, generously provided by the Wellcome Trust and the Royal Society. I thank Dr Daynea Richards, Karola Gerecht and Sofia Margiola for helpful comments.

References

1. L. Wang, J. Xie and P. G. Schultz, *Annu. Rev. Biophys. Biomol. Struct.*, 2006, **35**, 225–249.
2. R. B. Merrifield, *J. Am. Chem. Soc.*, 1963, **85**, 2149–2154.
3. B. Gutte and R. B. Merrifield, *J. Biol. Chem.*, 1971, **246**, 1922–1941.
4. W. Konig and R. Geiger, *Chem. Ber.*, 1970, **103**, 2034–2040.
5. R. Subiros-Funosas, R. Prohens, R. Barbas, A. El-Faham and F. Albericio, *Chemistry*, 2009, **15**, 9394–9403.
6. A. El-Faham and F. Albericio, *Chem. Rev.*, 2011, **111**, 6557–6602.
7. L. A. Carpino, *J. Am. Chem. Soc.*, 1957, **79**, 98–101.
8. L. A. Carpino and G. Y. Han, *J. Am. Chem. Soc.*, 1970, **92**, 5748–5749.
9. D. S. King, C. G. Fields and G. B. Fields, *Int. J. Pept. Protein Res.*, 1990, **36**, 255–266.
10. M. Muttenthaler, F. Albericio and P. E. Dawson, *Nat. Protoc.*, 2015, **10**, 1067–1083.
11. R. B. Merrifield, *Science*, 1965, **150**, 178–185.
12. H. M. Yu, S. T. Chen and K. T. Wang, *J. Org. Chem.*, 1992, **57**, 4781–4784.
13. P. E. Dawson, T. W. Muir, I. Clark-Lewis and S. B. Kent, *Science*, 1994, **266**, 776–779.
14. X. Q. Li, T. Kawakami and S. Aimoto, *Tetrahedron Lett.*, 1998, **39**, 8669–8672.
15. A. B. Clippingdale, C. J. Barrow and J. D. Wade, *J. Pept. Sci.*, 2000, **6**, 225–234.
16. P. Botti, M. Villain, S. Manganiello and H. Gaertner, *Org. Lett.*, 2004, **6**, 4861–4864.
17. T. Kawakami and S. Aimoto, *Chem. Lett.*, 2007, **36**, 76–77.
18. J. B. Blanco-Canosa and P. E. Dawson, *Angew. Chem., Int. Ed.*, 2008, **47**, 6851–6855.

19. N. Ollivier, J. Dheur, R. Mhidia, A. Blanpain and O. Melnyk, *Org. Lett.*, 2010, **12**, 5238–5241.
20. J. Masania, J. Li, S. J. Smerdon and D. Macmillan, *Org. Biomol. Chem.*, 2010, **8**, 5113–5119.
21. G. M. Fang, Y. M. Li, F. Shen, Y. C. Huang, J. B. Li, Y. Lin, H. K. Cui and L. Liu, *Angew. Chem., Int. Ed. Engl.*, 2011, **50**, 7645–7649.
22. F. Burlina, G. Papageorgiou, C. Morris, P. D. White and J. Offer, *Chem. Sci.*, 2014, **5**, 766–770.
23. D. Bang, B. L. Pentelute and S. B. Kent, *Angew. Chem., Int. Ed. Engl.*, 2006, **45**, 3985–3988.
24. A. Allahverdi, R. Yang, N. Korolev, Y. Fan, C. A. Davey, C. F. Liu and L. Nordenskiöld, *Nucleic Acids Res.*, 2011, **39**, 1680–1691.
25. L. Zhou, M. T. Holt, N. Ohashi, A. Zhao, M. M. Müller, B. Wang and T. W. Muir, *Nat. Commun.*, 2016, 7, 10589.
26. L. Z. Yan and P. E. Dawson, *J. Am. Chem. Soc.*, 2001, **123**, 526–533.
27. Q. Wan and S. J. Danishefsky, *Angew. Chem., Int. Ed. Engl.*, 2007, **46**, 9248–9252.
28. R. E. Thompson, B. Chan, L. Radom, K. A. Jolliffe and R. J. Payne, *Angew. Chem., Int. Ed. Engl.*, 2013, **52**, 9723–9727.
29. R. J. Hondal, *Protein Pept. Lett.*, 2005, **12**, 757–764.
30. L. R. Malins, N. J. Mitchell, S. McGowan and R. J. Payne, *Angew. Chem., Int. Ed. Engl.*, 2015, **54**, 12716–12721.
31. P. S. Reddy, S. Dery and N. Metanis, *Angew. Chem., Int. Ed. Engl.*, 2016, **55**, 992–995.
32. S. F. Loibl, Z. Harpaz and O. Seitz, *Angew. Chem., Int. Ed. Engl.*, 2015, **54**, 15055–15059.
33. S. Aimoto, *Biopolymers*, 1999, **51**, 247–265.
34. B. L. Nilsson, R. J. Hondal, M. B. Soellner and R. T. Raines, *J. Am. Chem. Soc.*, 2003, **125**, 5268–5269.
35. J. W. Bode, R. M. Fox and K. D. Baucom, *Angew. Chem., Int. Ed. Engl.*, 2006, **45**, 1248–1252.
36. I. Pusterla and J. W. Bode, *Nat. Chem.*, 2015, 7, 668–672.
37. D. S. Waugh, *Protein Expression Purif.*, 2011, **80**, 283–293.
38. T. W. Muir, D. Sondhi and P. A. Cole, *Proc. Natl. Acad. Sci. U. S. A.*, 1998, **95**, 6705–6710.
39. A. J. Stevens, Z. Z. Brown, N. H. Shah, G. Sekar, D. Cowburn and T. W. Muir, *J. Am. Chem. Soc.*, 2016, **138**, 2162–2165.
40. I. Giriat and T. W. Muir, *J. Am. Chem. Soc.*, 2003, **125**, 7180–7181.
41. Y. David, M. Vila-Perelló, S. Verma and T. W. Muir, *Nat. Chem.*, 2015, 7, 394–402.
42. S. K. Mazmanian, G. Liu, T. T. Hung and O. Schneewind, *Science*, 1999, **285**, 760–763.
43. H. Mao, S. A. Hart, A. Schink and B. A. Pollok, *J. Am. Chem. Soc.*, 2004, **126**, 2670–2671.
44. J. M. Antos, M. C. Truttmann and H. L. Ploegh, *Curr. Opin. Struct. Biol.*, 2016, **38**, 111–118.

45. G. K. Nguyen, S. Wang, Y. Qiu, X. Hemu, Y. Lian and J. P. Tam, *Nat. Chem. Biol.*, 2014, **10**, 732–738.
46. D. Y. Jackson, J. Burnier, C. Quan, M. Stanley, J. Tom and J. A. Wells, *Science*, 1994, **266**, 243–247.
47. S. H. Henager, N. Chu, Z. Chen, D. Bolduc, D. R. Dempsey, Y. Hwang, J. Wells and P. A. Cole, *Nat. Methods*, 2016, **13**, 925–927.
48. A. Romanelli, A. Shekhtman, D. Cowburn and T. W. Muir, *Proc. Natl. Acad. Sci. U. S. A.*, 2004, **101**, 6397–6402.
49. H. T. Kratochvil, J. K. Carr, K. Matulef, A. W. Annen, H. Li, M. Maj, J. Ostmeyer, A. L. Serrano, H. Raghuraman, S. D. Moran, J. L. Skinner, E. Perozo, B. Roux, F. I. Valiyaveetil and M. T. Zanni, *Science*, 2016, **353**, 1040–1044.
50. C. F. Becker, K. Lausecker, M. Balog, T. Kalai, K. Hideg, H. J. Steinhoff and M. Engelhard, *Magn. Reson. Chem.*, 2005, **43**, S34–S39.
51. C. F. Becker, C. L. Hunter, R. Seidel, S. B. Kent, R. S. Goody and M. Engelhard, *Proc. Natl. Acad. Sci. U. S. A.*, 2003, **100**, 5075–5080.
52. E. M. Vogel and B. Imperiali, *Protein Sci.*, 2007, **16**, 550–556.
53. M. Vila-Perello, M. R. Pratt, F. Tulin and T. W. Muir, *J. Am. Chem. Soc.*, 2007, **129**, 8068–8069.
54. S. Virdee, D. Macmillan and G. Waksman, *Chem. Biol.*, 2010, **17**, 274–284.
55. A. Kienhofer, P. Kast and D. Hilvert, *J. Am. Chem. Soc.*, 2003, **125**, 3206–3207.
56. N. Metanis, A. Brik, P. E. Dawson and E. Keinan, *J. Am. Chem. Soc.*, 2004, **126**, 12726–12727.
57. S. M. Berry, M. D. Gieselman, M. J. Nilges, W. A. van Der, Donk and Y. Lu, *J. Am. Chem. Soc.*, 2002, **124**, 2084–2085.
58. K. M. Clark, Y. Yu, N. M. Marshall, N. A. Sieracki, M. J. Nilges, N. J. Blackburn, W. A. van der Donk and Y. Lu, *J. Am. Chem. Soc.*, 2010, **132**, 10093–10101.
59. J. O. Campecino, L. W. Dudycz, D. Tumelty, V. Berg, D. E. Cabelli and M. J. Maroney, *J. Am. Chem. Soc.*, 2015, **137**, 9044–9052.
60. F. I. Valiyaveetil, M. Sekedat, R. Mackinnon and T. W. Muir, *Proc. Natl. Acad. Sci. U. S. A.*, 2004, **101**, 17045–17049.
61. F. I. Valiyaveetil, M. Leonetti, T. W. Muir and R. Mackinnon, *Science*, 2006, **314**, 1004–1007.
62. P. K. Devaraneni, A. G. Komarov, C. A. Costantino, J. J. Devereaux, K. Matulef and F. I. Valiyaveetil, *Proc. Natl. Acad. Sci. U. S. A.*, 2013, **110**, 15698–15703.
63. F. I. Valiyaveetil, M. Sekedat, R. MacKinnon and T. W. Muir, *J. Am. Chem. Soc.*, 2006, **128**, 11591–11599.
64. Z. Liu, S. Frutos, M. J. Bick, M. Vila-Perello, G. T. Debelouchina, S. A. Darst and T. W. Muir, *Proc. Natl. Acad. Sci. U. S. A.*, 2014, **111**, 8422–8427.
65. S. W. Pedersen, G. Hultqvist, K. Stromgaard and P. Jemth, *PLoS One*, 2014, **9**, e95619.

66. S. W. Pedersen, S. B. Pedersen, L. Anker, G. Hultqvist, A. S. Kristensen, P. Jemth and K. Stromgaard, *Nat. Commun.*, 2014, **5**, 3215.
67. V. Y. Torbeev, M. O. Ebert, J. Dolenc and D. Hilvert, *J. Am. Chem. Soc.*, 2015, **137**, 2524–2535.
68. V. Y. Torbeev and D. Hilvert, *Proc. Natl. Acad. Sci. U. S. A.*, 2013, **110**, 20051–20056.
69. U. Arnold, M. P. Hinderaker, J. Koditz, R. Golbik, R. Ulbrich-Hofmann and R. T. Raines, *J. Am. Chem. Soc.*, 2003, **125**, 7500–7501.
70. U. Arnold and R. T. Raines, *Org. Biomol. Chem.*, 2016, **14**, 6780–6785.
71. U. Arnold, M. P. Hinderaker, B. L. Nilsson, B. R. Huck, S. H. Gellman and R. T. Raines, *J. Am. Chem. Soc.*, 2002, **124**, 8522–8523.
72. C. Walsh, *Posttranslational Modification of Proteins: Expanding Nature's Inventory*, W. H. Freeman, 2006.
73. F. T. Hofmann, C. Lindemann, H. Salia, P. Adamitzki, J. Karanicolas and F. P. Seebeck, *Chem. Commun.*, 2011, **47**, 10335–10337.
74. J. Bertran-Vicente, M. Schümann, P. Schmieder, E. Krause and C. P. Hackenberger, *Org. Biomol. Chem.*, 2015, **13**, 6839–6843.
75. J.-M. Kee, B. Villani, L. R. Carpenter and T. W. Muir, *J. Am. Chem. Soc.*, 2010, **132**, 14327–14329.
76. L. M. Yates and D. Fiedler, *ACS Chem. Biol.*, 2016, **11**, 1066–1073.
77. T. M. Hackeng, J. A. Fernández, P. E. Dawson, S. B. Kent and J. H. Griffin, *Proc. Natl. Acad. Sci. U. S. A.*, 2000, **97**, 14074–14078.
78. X. Liu, L. Wang, K. Zhao, P. R. Thompson, Y. Hwang, R. Marmorstein and P. A. Cole, *Nature*, 2008, **451**, 846–850.
79. R. Burai, N. Ait-Bouziad, A. Chiki and H. A. Lashuel, *J. Am. Chem. Soc.*, 2015, **137**, 5041–5052.
80. M. Matveenko, E. Cichero, P. Fossa and C. F. Becker, *Angew. Chem., Int. Ed.*, 2016, **55**, 11397–11402.
81. H. J. Reich and R. J. Hondal, *ACS Chem. Biol.*, 2016, **11**, 821–841.
82. N. Metanis and D. Hilvert, *Angew. Chem., Int. Ed.*, 2012, **51**, 5585–5588.
83. G. W. Snider, E. Ruggles, N. Khan and R. J. Hondal, *Biochemistry*, 2013, **52**, 5472–5481.
84. D. Komander and M. Rape, *Annu. Rev. Biochem.*, 2012, **81**, 203–229.
85. K. S. Ajish Kumar, M. Haj-Yahya, D. Olschewski, H. A. Lashuel and A. Brik, *Angew. Chem., Int. Ed. Engl.*, 2009, **48**, 8090–8094.
86. C. Chatterjee, R. K. McGinty, J. P. Pellois and T. W. Muir, *Angew. Chem., Int. Ed.*, 2007, **46**, 2814–2818.
87. C. E. Weller, A. Dhall, F. Ding, E. Linares, S. D. Whedon, N. A. Senger, E. L. Tyson, J. D. Bagert, X. Li, O. Augusto and C. Chatterjee, *Nat. Commun.*, 2016, 7, 12979.
88. R. K. McGinty, M. Kohn, C. Chatterjee, K. P. Chiang, M. R. Pratt and T. W. Muir, *ACS Chem. Biol.*, 2009, **4**, 958–968.
89. C. Chatterjee, R. K. McGinty, B. Fierz and T. W. Muir, *Nat. Chem. Biol.*, 2010, **6**, 267–269.
90. S. K. Singh, I. Sahu, S. M. Mali, H. P. Hemantha, O. Kleifeld, M. H. Glickman and A. Brik, *J. Am. Chem. Soc.*, 2016, **138**, 16004–16015.

91. R. K. McGinty, J. Kim, C. Chatterjee, R. G. Roeder and T. W. Muir, *Nature*, 2008, **453**, 812–816.
92. M. T. Morgan, M. Haj-Yahya, A. E. Ringel, P. Bandi, A. Brik and C. Wolberger, *Science*, 2016, **351**, 725–728.
93. G. T. Debelouchina, K. Gerecht and T. W. Muir, *Nat. Chem. Biol.*, 2017, **13**, 105–110.
94. B. Fierz, C. Chatterjee, R. K. McGinty, M. Bar-Dagan, D. P. Raleigh and T. W. Muir, *Nat. Chem. Biol.*, 2011, **7**, 113–119.
95. M. Haj-Yahya, B. Fauvet, Y. Herman-Bachinsky, M. Hejjaoui, S. N. Bavikar, S. V. Karthikeyan, A. Ciechanover, H. A. Lashuel and A. Brik, *Proc. Natl. Acad. Sci. U. S. A.*, 2013, **110**, 17726–17731.
96. H. P. Hemantha, S. N. Bavikar, Y. Herman-Bachinsky, N. Haj-Yahya, S. Bondalapati, A. Ciechanover and A. Brik, *J. Am. Chem. Soc.*, 2014, **136**, 2665–2673.
97. J. Liang, L. Zhang, X.-L. L. Tan, Y.-K. K. Qi, S. Feng, H. Deng, Y. Yan, J.-S. S. Zheng, L. Liu and C.-L. L. Tian, *Angew. Chem., Int. Ed.*, 2017, **56**, 2744–2748.
98. V. Y. Torbeev and S. B. Kent, *Angew. Chem., Int. Ed. Engl.*, 2007, **46**, 1667–1670.
99. L. E. Canne, A. R. Ferredamare, S. K. Burley and S. B. H. Kent, *J. Am. Chem. Soc.*, 1995, **117**, 2998–3007.
100. J. P. Tam and C. T. Wong, *J. Biol. Chem.*, 2012, **287**, 27020–27025.
101. K. Mandal, B. L. Pentelute, D. Bang, Z. P. Gates, V. Y. Torbeev and S. B. Kent, *Angew. Chem., Int. Ed. Engl.*, 2012, **51**, 1481–1486.
102. A. Gautier, C. Gauron, M. Volovitch, D. Bensimon, L. Jullien and S. Vriz, *Nat. Chem. Biol.*, 2014, **10**, 533–541.
103. K. E. Brechun, K. M. Arndt and G. A. Woolley, *Curr. Opin. Struct. Biol.*, 2016, **45**, 53–58.
104. R. J. Mart and R. K. Allemann, *Chem. Commun.*, 2016, **52**, 12262–12277.
105. V. Krautler, A. Aemissegger, P. H. Hunenberger, D. Hilvert, T. Hansson and W. F. van Gunsteren, *J. Am. Chem. Soc.*, 2005, **127**, 4935–4942.
106. T. M. Doran, E. A. Anderson, S. E. Latchney, L. A. Opanashuk and B. L. Nilsson, *ACS Chem. Neurosci.*, 2012, **3**, 211–220.
107. R. David, R. Gunther, L. Baumann, T. Luhmann, D. Seebach, H. J. Hofmann and A. G. Beck-Sickinger, *J. Am. Chem. Soc.*, 2008, **130**, 15311–15317.
108. G. M. Burslem, H. F. Kyle, A. L. Breeze, T. A. Edwards, A. Nelson, S. L. Warriner and A. J. Wilson, *Chem. Commun.*, 2016, **52**, 5421–5424.
109. C. Mayer, M. M. Müller, S. H. Gellman and D. Hilvert, *Angew. Chem., Int. Ed.*, 2014, **53**, 6978–6981.
110. R. C. Milton, S. C. Milton and S. B. Kent, *Science*, 1992, **256**, 1445–1448.
111. Z. Wang, W. Xu, L. Liu and T. F. Zhu, *Nat. Chem.*, 2016, **8**, 698–704.
112. M. T. Weinstock, M. T. Jacobsen and M. S. Kay, *Proc. Natl. Acad. Sci. U. S. A.*, 2014, **111**, 11679–11684.
113. T. N. M. Schumacher, L. M. Mayr, D. L. Minor, M. A. Milhollen, M. W. Burgess and P. S. Kim, *Science*, 1996, **271**, 1854–1857.

114. T. O. Yeates and S. B. Kent, *Annu. Rev. Biophys.*, 2012, **41**, 41–61.
115. S. Gao, M. Pan, Y. Zheng, Y. Huang, Q. Zheng, D. Sun, L. Lu, X. Tan, X. Tan, H. Lan, J. Wang, T. Wang, J. Wang and L. Liu, *J. Am. Chem. Soc.*, 2016, **138**, 14497–14502.
116. J. Schneider and S. B. H. Kent, *Cell*, 1988, **54**, 363–368.
117. M. Miller, J. Schneider, B. Sathyanarayana, M. Toth, G. Marshall, L. Clawson, L. Selk, S. Kent and A. Wlodawer, *Science*, 1989, **246**, 1149–1152.
118. A. Wlodawer, M. Miller, M. Jaskolski, B. Sathyanarayana, E. Baldwin, I. Weber, L. Selk, L. Clawson, J. Schneider and S. Kent, *Science*, 1989, **245**, 616–621.
119. M. Baca, P. F. Alewood and S. B. H. Kent, *Protein Sci.*, 1993, **2**, 1085–1091.
120. R. Smith, I. M. Brereton, R. Y. Chai and S. B. H. Kent, *Nat. Struct. Mol. Biol.*, 1996, **3**, 946–950.
121. V. Y. Torbeev, H. Raghuraman, D. Hamelberg, M. Tonelli, W. M. Westler, E. Perozo and S. B. H. Kent, *Proc. Natl. Acad. Sci. U. S. A.*, 2011, **108**, 20982–20987.
122. M. Baca and S. B. Kent, *Proc. Natl. Acad. Sci. U. S. A.*, 1993, **90**, 11638–11642.
123. M. Miller, M. Baca, J. K. Mohana Rao and S. B. H. Kent, *J. Mol. Struct. THEOCHEM*, 1998, **423**, 137–152.
124. K. Yamaguchi, K. Akai, G. Kawanishi, M. Ueda, S. Masuda and R. Sasaki, *J. Biol. Chem.*, 1991, **266**, 20434–20439.
125. D. Macmillan, R. M. Bill, K. A. Sage, D. Fern and S. L. Flitsch, *Chem. Biol.*, 2001, **8**, 133–145.
126. K. Hirano, D. Macmillan, K. Tezuka, T. Tsuji and Y. Kajihara, *Angew. Chem., Int. Ed.*, 2009, **48**, 9557–9560.
127. K. Hirano, M. Izumi, D. Macmillan, K. Tezuka, T. Tsuji and Y. Kajihara, *J. Carbohydr. Chem.*, 2011, **30**, 306–319.
128. G. G. Kochendoerfer, S.-Y. Y. Chen, F. Mao, S. Cressman, S. Traviglia, H. Shao, C. L. Hunter, D. W. Low, E. N. Cagle, M. Carnevali, V. Gueriguian, P. J. Keogh, H. Porter, S. M. Stratton, M. C. Wiedeke, J. Wilken, J. Tang, J. J. Levy, L. P. Miranda, M. M. Crnogorac, S. Kalbag, P. Botti, J. Schindler-Horvat, L. Savatski, J. W. Adamson, A. Kung, S. B. Kent and J. A. Bradburne, *Science*, 2003, **299**, 884–887.
129. S.-Y. Y. Chen, S. Cressman, F. Mao, H. Shao, D. W. Low, H. S. Beilan, E. N. Cagle, M. Carnevali, V. Gueriguian, P. J. Keogh, H. Porter, S. M. Stratton, M. C. Wiedeke, L. Savatski, J. W. Adamson, C. E. Bozzini, A. Kung, S. B. Kent, J. A. Bradburne and G. G. Kochendoerfer, *Chem. Biol.*, 2005, **12**, 371–383.
130. P. Wang, S. Dong, J.-H. H. Shieh, E. Peguero, R. Hendrickson, M. A. Moore and S. J. Danishefsky, *Science*, 2013, **342**, 1357–1360.
131. J. B. Schwarz, S. D. Kuduk, X.-T. Chen, D. Sames, P. W. Glunz and S. J. Danishefsky, *J. Am. Chem. Soc.*, 1999, **121**, 2662–2673.
132. P. Wang, B. Aussedat, Y. Vohra and S. J. Danishefsky, *Angew. Chem., Int. Ed.*, 2012, **51**, 11571–11575.

133. V. Ullmann, M. Rädisch, I. Boos, J. Freund, C. Pöhner, S. Schwarzinger and C. Unverzagt, *Angew. Chem., Int. Ed.*, 2012, **51**, 11566–11570.
134. P. Wang, S. Dong, J. A. Brailsford, K. Iyer, S. D. Townsend, Q. Zhang, R. C. Hendrickson, J. Shieh, M. A. Moore and S. J. Danishefsky, *Angew. Chem., Int. Ed.*, 2012, **51**, 11576–11584.
135. M. Murakami, T. Kiuchi, M. Nishihara, K. Tezuka, R. Okamoto, M. Izumi and Y. Kajihara, *Sci. Adv.*, 2016, **2**, e1500678.
136. S. R. Pfeffer, *Trends Cell Biol.*, 2001, **11**, 487–491.
137. L. Brunsveld, J. Kuhlmann, K. Alexandrov, A. Wittinghofer, R. S. Goody and H. Waldmann, *Angew. Chem., Int. Ed.*, 2006, **45**, 6622–6646.
138. A. Iakovenko, E. Rostkova, E. Merzlyak, A. M. Hillebrand, N. H. Thomä, R. S. Goody and K. Alexandrov, *FEBS Lett.*, 2000, **468**, 155–158.
139. K. Alexandrov, I. Heinemann, T. Durek, V. Sidorovitch, R. S. Goody and H. Waldmann, *J. Am. Chem. Soc.*, 2002, **124**, 5648–5649.
140. T. Durek, K. Alexandrov, R. S. Goody, A. Hildebrand, I. Heinemann and H. Waldmann, *J. Am. Chem. Soc.*, 2004, **126**, 16368–16378.
141. A. Rak, O. Pylypenko, T. Durek, A. Watzke, S. Kushnir, L. Brunsveld, H. Waldmann, R. S. Goody and K. Alexandrov, *Science*, 2003, **302**, 646–650.
142. Y.-W. W. Wu, L. K. Oesterlin, K.-T. T. Tan, H. Waldmann, K. Alexandrov and R. S. Goody, *Nat. Chem. Biol.*, 2010, **6**, 534–540.
143. T. Pawson, *Nature*, 1995, **373**, 573–580.
144. F. Sicheri, I. Moarefi and J. Kuriyan, *Nature*, 1997, **385**, 602–609.
145. Z. Zhang, K. Shen, W. Lu and P. A. Cole, *J. Biol. Chem.*, 2003, **278**, 4668–4674.
146. W. Lu, K. Shen and P. A. Cole, *Biochemistry*, 2003, **42**, 5461–5468.
147. D. Schwarzer, Z. Zhang, W. Zheng and P. A. Cole, *J. Am. Chem. Soc.*, 2006, **128**, 4192–4193.
148. D. Bolduc, M. Rahdar, B. Tu-Sekine, S. C. Sivakumaren, D. Raben, L. M. Amzel, P. Devreotes, S. B. Gabelli and P. Cole, *eLife*, 2013, **2**, e00691.
149. M. K. Tarrant, H.-S. S. Rho, Z. Xie, Y. L. Jiang, C. Gross, J. C. Culhane, G. Yan, J. Qian, Y. Ichikawa, T. Matsuoka, N. Zachara, F. A. Etzkorn, G. W. Hart, J. S. Jeong, S. Blackshaw, H. Zhu and P. A. Cole, *Nat. Chem. Biol.*, 2012, **8**, 262–269.
150. R. Zitterbart and O. Seitz, *Angew. Chem., Int. Ed.*, 2016, **55**, 7252–7256.
151. K. A. Pickin, S. Chaudhury, B. C. Dancy, J. J. Gray and P. A. Cole, *J. Am. Chem. Soc.*, 2008, **130**, 5667–5669.
152. R. D. Kornberg, *Annu. Rev. Biochem.*, 1977, **46**, 931–954.
153. C. D. Allis, T. Jenuwein and D. Reinberg, *Epigenetics*, Cold Spring Harbor Laboratory Press, 2007.
154. M. M. Müller and T. W. Muir, *Chem. Rev.*, 2015, **115**, 2296–2349.
155. S. Li and M. A. Shogren-Knaak, *Proc. Natl. Acad. Sci. U. S. A.*, 2008, **105**, 18243–18248.

156. S. Li and M. A. Shogren-Knaak, *J. Biol. Chem.*, 2009, **284**, 9411–9417.
157. U. T. T. Nguyen, L. Bittova, M. M. Müller, B. Fierz, Y. David, B. Houck-Loomis, V. Feng, G. P. Dann and T. W. Muir, *Nat. Methods*, 2014, **11**, 834–840.
158. P. Voigt, G. LeRoy, W. J. Drury, B. M. Zee, J. Son, D. B. Beck, N. L. Young, B. A. Garcia and D. Reinberg, *Cell*, 2012, **151**, 181–193.
159. C. C. Lechner, N. D. Agashe and B. Fierz, *Angew. Chem., Int. Ed. Engl.*, 2016, **55**, 2903–2906.
160. C. Zheng and J. J. Hayes, *J. Biol. Chem.*, 2003, **278**, 24217–24224.
161. B. Al-Sady, H. D. Madhani and G. J. Narlikar, *Mol. Cell*, 2013, **51**, 80–91.
162. M. M. Müller, B. Fierz, L. Bittova, G. Liszczak and T. W. Muir, *Nat. Chem. Biol.*, 2016, **12**, 188–193.
163. K. A. Janssen, S. Sidoli and B. A. Garcia, *Methods Enzymol.*, 2017, **586**, 359–378.
164. L. Raibaut, O. Mahdi and O. Melnyk, *Top. Curr. Chem.*, 2015, **363**, 103–154.
165. A. J. Mijalis, D. A. Thomas, M. D. Simon, A. Adamo, R. Beaumont, K. F. Jensen and B. L. Pentelute, *Nat. Chem. Biol.*, 2017, **13**, 464–466.

Subject Index